Perspectives on European Development Co-operation

Events of the past twenty years, including the Cold War and the War on Terror, have meant that the environments of international development co-operation have changed extensively, with dramatic consequences for development policies and North–South relations in general.

Perspectives on European Development Co-operation takes stock of such changes in the millennium, describing and analysing the new European development agenda, including the role of the European Union. Essays by prominent authorities in the field examine the development policies of individual donor countries focusing on the principles and objectives governing aid strategies, and the performances of these policies.

This book will be of interest to students of development studies and those involved in determining development policy.

Paul Hoebink is Associate Professor at the Centre for International Development Issues Nijmegen (CIDIN), the Catholic University of Nijmegen. His research has focused on development co-operation policies, aid effectiveness and evaluation. He has published widely on the aid programmes of the Netherlands, the EU and other donors, and is a member of several advisory boards.

Olav Stokke is the Senior Researcher at the Norwegian Institute of International Affairs, where he has also served as Acting Director, Deputy Director and the Research Director. He has written extensively on development issues including *Food Aid and Human Security* (2000) and *Policy Coherence in Development Co-operation* (1999).

Routledge Research EADI Studies in Development

1. Perspectives on European Development Co-operation
Policy and performance of individual donor countries and the EU
Edited by Paul Hoebink and Olav Stokke

Previous titles to appear in the Routledge Research EADI Studies in Development include:

Food Aid and Human Security
Edited by Edward Clay and Olav Stokke

Regions and Development
Politics, security and economics
Edited by Sheila Page

Policy Coherence in Development Co-operation
Edited by Jacques Forster and Olav Stokke

Latin American and East European Economics in Transition
A comparative view
Edited by Claude Auroi

Enterprise Clusters and Networking in Developing Countries
Edited by Meine Pieter van Dijk and Roberta Rabellotti

Multilateralism versus Regionalism
Trade issues after the Uruguay Round
Edited by Meine Pieter van Dijk and Sandro Sideri

Foreign Aid Toward the Year 2000
Experiences and challenges
Edited by Olav Stokke

Perspectives on European Development Co-operation

Policy and performance of individual donor countries and the EU

Edited by Paul Hoebink and Olav Stokke

LONDON AND NEW YORK

First published 2005
by Routledge
2 Park Square, Milton Park, Abingdon, Oxon OX14 4RN

Simultaneously published in the USA and Canada
by Routledge
270 Madison Ave, New York, NY 10016

Routledge is an imprint of the Taylor & Francis Group

© 2005 Paul Hoebink and Olav Stokke editorial matter and selection; the contributors their contributions

Printed and bound in Great Britain by
TJ International Ltd, Padstow, Cornwall

All rights reserved. No part of this book may be reprinted or reproduced or utilised in any form or by any electronic, mechanical, or other means, now known or hereafter invented, including photocopying and recording, or in any information storage or retrieval system, without permission in writing from the publishers.

British Library Cataloguing in Publication Data
A catalogue record for this book is available from the British Library

Library of Congress Cataloging in Publication Data
A catalog record for this title has been requested

ISBN 0-415-36854-5

Contents

	List of Tables, Figures and Boxes	vii
	Introduction: European Development Co-operation at the Beginning of a New Millennium *Paul Hoebink and Olav Stokke*	1
1.	The Changing International and Conceptual Environments of Development Co-operation *Olav Stokke*	32
2.	Austrian Aid Policy *Michael Obrovsky*	113
3.	Belgian Aid Policies in the 1990s *Nathalie Holvoet and Robrecht Renard*	136
4.	British Aid Policy in the 'Short–Blair' Years *Oliver Morrissey*	161
5.	Danish Aid Policy in the Post-Cold War Period: Increasing Resources and Minor Adjustments *Gorm Rye Olsen*	184
6.	Finland: Aid and Identity *Juhani Koponen with Lauri Siitonen*	215
7.	French Development Co-operation Policy *Jean-Jacques Gabas*	242
8.	Germany's Development Co-operation Policy since the Early 1990s: Increased Conceptual Ambitions in Times of Severe Financial Constraint *Guido Ashoff*	267

9.	The Foreign Aid Policy of Ireland *Helen O'Neill*	303
10.	Trends in the Debate on Italian Aid *José Luis Rhi-Sausi and Marco Zupi*	336
11.	A New Member of the G-0.7: Luxembourg as the Smallest and Largest Donor *Paul Hoebink*	378
12.	All in the Name of Quality: Dutch Development Co-operation in the 1990s *Lau Schulpen*	406
13.	Norwegian Aid Policy: Continuity and Change in the 1990s and Beyond *Olav Stokke*	448
14.	Spanish Foreign Aid: Flaws of an Emerging Framework *José Antonio Alonso*	493
15.	Swedish Development Co-operation in Perspective *Anders Danielson and Lennart Wohlgemuth*	518
16.	Swiss Development Co-operation: Major Changes since the Early 1990s and Future Challenges *Catherine Schümperli Younossian*	546
17.	The European Union's Development Policy: Shifting Priorities in a Rapidly Changing World *Gorm Rye Olsen*	573
	Glossary	609
	Notes on Contributors	622
	Index	627

Tables, Figures and Boxes

TABLE IN.1	Official Development Assistance from DAC countries, 1970–2002 (US$m.)	23
TABLE IN.2	ODA from European DAC members and the EU, 1970–2003 (US$m. and %)	24
TABLE IN.3	Relative Official Development Assistance from European DAC members (1970–2003) (% of GNP/GNI)	25
TABLE 2.1	Austria's ODA, 1981–2001 (US$m. and % of GNP/GNI)	115
TABLE 2.2	Financial requirements (US$m.)	118
TABLE 2.3	DDC programme and project aid 1991–2002 (US$m.)	120
TABLE 2.4	DDC programme and project aid by region, 1992–2002 (%)	121
TABLE 3.1	Governments in power in the 1990s	140
TABLE 3.2	Breakdown of total Belgian ODA and the expenditures of the DGIC/BADC by type of action (1990–2000, US$m., current prices)	145
TABLE 3.3	Relative importance of Belgium in its ten major ODA recipients (1999–2000)	150
TABLE 4.1	Allocation of British aid, 1997/98 and 2000/1	168
TABLE 4.2	British bilateral aid by type, 1997/98 and 1999/2000	171
TABLE 5.1	Danish development assistance 1965–2003 (DKm. (current prices and %)	185
TABLE 5.2	Official Danish development aid. Selected budget lines (DKm. (current prices) and %)	189
TABLE 5.3	Danish bilateral aid 2000. Geographical distribution and types of aid (percentage distribution)	194

TABLE 5.4	Sectoral distribution of Danish bilateral aid 2001 and 2002	197
TABLE 6.1	Finland's ODA disbursements 1961–2002	224
TABLE 6.2	Finland's ODA composition (%)	226
TABLE 7.1	French Official Development Assistance (€m.)	243
TABLE 7.2	Bilateral debt relief for highly indebted poor countries (HIPC) (€m.)	245
TABLE 7.3	Net disbursements of ODA and OA to major regions (US$m.)	248
TABLE 7.4	Percentage of bilateral ODA for LDCs and LICs	248
TABLE 7.5	Trend of France's relative importance in sub-Saharan Africa	250
TABLE 7.6	The importance of major ministries in the administration of ODA (%)	253
ANNEX 7.1	List of reports on French development co-operation since 1960	263
ANNEX 7.2	Ex-post evaluations and studies financed by the Ministry of Foreign Affairs – DGCID – during the period 2000–2003	265
TABLE 8.1	Long-term quantitative trends in German development co-operation, 1960–2001 (€m.)	281
TABLE 8.2	Bilateral German ODA by main purposes in 1985/86, 1990/91, 1995/96 and 2000/2001 (% of total bilateral allocable ODA commitments)	287
TABLE 8.3	Geographical distribution of German ODA by main developing regions and income groups in 1995/96 and 2000/2001 (% of gross disbursements of bilateral allocable ODA)	288
TABLE 9.1	Ireland's ODA, selected years 1974–2002 (€m. and %)	308
TABLE 9.2	Ireland's bilateral aid, selected years 1992–2002 (€m.)	312
TABLE 9.3	Ireland's multilateral aid, selected years 1992–2002 (€m.)	319

TABLE 10.1	Net disbursements of ODA (US$m.) and annual variations (%), 1990–2003	343
TABLE 10.2	Net payments of ODA 1997–2000 (US$m.)	346
TABLE 10.3	Regions which benefited most from aid 1980–2001 (% of total gross disbursements)	347
TABLE 10.4	ODA disbursements in Europe, 1990–2001 (US$m.)	348
TABLE 10.5	ODA disbursements North of Sahara, 1999–2001 (US$m.)	348
TABLE 10.6	ODA disbursements in the Middle East, 1999–2001 (US$m.)	348
TABLE 10.7	ODA to least developed countries, 1986–2001	351
TABLE 10.8	Weight of ten main beneficiaries 1979–2001 (in US$m. and as % of bilateral ODA excluding unallocated component)	353
TABLE 10.9	HIPC countries' foreign debt cancelled (Oct. 2001–Oct. 2002)	357
TABLE 10.10	Fifteen main beneficiaries of bilateral aid during 2000 (US$m.)	358
TABLE 10.11	Share of debt relief compared with DAC total, 2000 (US$m.)	359
TABLE 10.12	Total ODA per destination sector 1999–2001 (US$m.)	363
TABLE 10.13	Bilateral ODA commitments, excluding technical assistance and administrative costs, 2000–2001 (US$m.)	365
TABLE 11.1	Luxembourg's development co-operation: Aid volumes	382
TABLE 11.2	Aid distribution by countries and sectors (in % of total ODA)	389
TABLE 12.1	Political parties and the public on the aid budget and the Minister	417
TABLE 12.2	Target countries of Dutch bilateral aid (2003)	425
TABLE 13.1	Norway's bilateral commitments as distributed on main purposes (%)	467
TABLE 13.2	Norway's ODA and its distribution	

	on bilateral and multilateral aid channels, 1980–2002	469
TABLE 13.3	The programme countries' share of total bilateral ODA, 1970–2001 (%)	472
TABLE 13.4	Bilateral ODA to ex-Yugoslavia and Palestine, 1990–2001 (%)	472
TABLE 14.1	The distribution of ODA, 1981–2002 (US$m. and %)	496
TABLE 14.2	Priority countries according to the *Strategic Guidelines*	501
TABLE 14.3	Regional distribution of net ODA (weight average)	502
TABLE 14.4	Distribution of net ODA according to level of income	503
TABLE 14.5	Bilateral aid: distribution according to sectors (%)	506
TABLE 15.1	Swedish aid flows, 1996–2001. US$m. (current prices) and %	532
TABLE 15.2	Bilateral aid by major purpose, 1988/89–99 (%)	537
TABLE 15.3	Geographical distribution of bilateral ODA (number of countries)	537
TABLE 15.4	Ten largest recipients of Swedish bilateral aid (% of total ODA)	542
TABLE 16.1	List of main programme appropriations currently in force (US$m.)	549
TABLE 16.2	ODA for developing countries (DAC, List 1), 1986–2002 (US$m. and %)	552
TABLE 16.3	ODA from Swiss cantons and municipalities for developing countries, 1991–2002 (US$m.)	552
TABLE 16.4	Economic and trade policy measures of Seco 1997–2001 (US$m. and %)	560
TABLE 16.5	Distribution of SDC bilateral co-operation activities by sector 1999–2001 (US$m. and %)	561
TABLE 16.6	Geographical distribution of humanitarian aid fund, multiannual fund 1996–2000 (US$m. and %)	565
TABLE 17.1	Major recipient geographical areas for European development assistance,	

TABLE 17.2	1980/81–2000/1 (% of total ODA) ODA from the European Community, selected years (US$m., current prices)	575 575
FIGURE 2.1	Austria's ODA 1980–2001	116
FIGURE 6.1	Finland's ODA disbursements 1961–2002	223
FIGURE 10.1	Evolution of bilateral ODA, 1997–2002	352
FIGURE 10.2	Bilateral ODA commitments, historical trend (1980–2001)	366
FIGURE 12.1	Total net ODA (in current US$ and % of GNP)	417
FIGURE 12.2	Total bilateral net ODA to recipient countries (US$m., in current and constant dollars)	419
FIGURE 12.3	Number of recipient countries (positive net ODA) – 1960–2000	425
FIGURE 12.4	Distribution of bilateral ODA per income category (current US$) – 1960–2000	427
FIGURE 12.5	Growth of the co-financing programme (€m.)	437
FIGURE 15.1	How Swedish aid reduces poverty	536
BOX 7.1	Reports submitted to the Prime Minister during the first HCCI mandate 1999–2002	254
BOX 7.2	French NGOs	256
BOX 8.1	Priority areas for action to be taken by the German Government in implementing its Programme of Action 2015 for Poverty Reduction	276
BOX 8.2	Earlier proposals for greater policy coherence in Germany	279
BOX 8.3	Changes in recent years to improve policy coherence in Germany	280
BOX 8.4	Main forms and implementing organisations of German bilateral ODA	285
BOX 8.5	'Priority' and 'partner' countries of German bilateral development	

	co-operation (as of 2002)	292
BOX 12.1	Technical assistance	412
BOX 12.2	History of the Dutch concentration country policy (mid-1960s to mid-1990s)	422

Introduction:
European Development Co-operation at the Beginning of a New Millennium

PAUL HOEBINK AND OLAV STOKKE

During recent years, the environments of international development co-operation have changed extensively. In the late 1970s and the early 1980s, development paradigms that had their heyday in the second part of the 1970s – related to the demands for a new international economic order and basic needs strategies – came under attack from the neo-liberal orthodoxy prescribing 'more market and less state'. Also the transformation of the international system in the late 1980s and early 1990s has affected development co-operation profoundly.

Three features have been particularly visible in the debate on development co-operation and the implementation of aid. First, in the mid-1990s, some commentators concluded that the end of development assistance had come. It was suggested, by some observers, that the end of the Cold War would remove an important rationale for aid; others suggested that private financial flows to developing countries would take over the role of financing development. In 1996, private flows made up almost five times the amount of official development assistance (ODA). Aid stagnated during the 1990s (Table IN.1). But from 1997 onwards, private capital flows began to stagnate. International bank lending to developing countries ran into negative figures. In recent years, also, private investments decreased sharply due to the economic recession. At the start of the twenty-first century, it thus appears that ODA is the only stable flow of financial resources to developing countries. What is more, since most bank lending and most private investments go to upper middle-income countries (UMCs) or large countries, ODA is nearly the only external source of finance for development for the least developed countries (LLDCs) and other low-income countries (LICs).[1]

---

1. Of the total flow of private investments since 1995, approximately 40 per cent has gone to China alone, while Africa has received 3 per cent or less, mainly going to oil exploration and gold mining. In 2002, gross foreign direct investments to developing countries amounted to US$162bn., of which 32.7 per cent went to China and only 6.8 per cent to Africa (main recipients were Angola, Nigeria, Algeria and Chad) [*UNCTAD, various years*]. In stark contrast, nearly two-thirds of ODA have gone to the low-income countries (LICs) and an additional 30 per cent to low middle-

In recent years, a second element has come to the fore: although public support for development co-operation has remained at a high level in most European countries, scepticism about the results of aid has also been growing. Influential publications, such as the World Bank's policy paper on assessing aid [*World Bank, 1998*], coming out with the conclusion that aid only works in so-called 'good governed countries', feed this scepticism. However, at the same time we have seen that the donor community is becoming more united than ever and is, in the declared policy, showing less self-interest and more eagerness in working together among themselves and with governments in developing countries. The Millennium Development Goals, adopted by the Millennium Summit of the United Nations on 8 September 2000, offer for the first time an ambitious common framework for donors and developing countries. Donor co-ordination and the harmonisation of donors' procedures are high on the international agenda and on those of most individual donor countries as well.

Third, development planning is back on the agenda. We might not call it so, but in the poorest countries, the Poverty Reduction Strategy Papers (PRSPs) could be seen as a national planned effort to eradicate poverty. And more importantly, donors, at least in words, adhere to these plans and promise to align their aid investments with them. More and more donors are willing to funnel their aid through the sector plans of local governments and via sector-wide approaches. Unfortunately, in most countries, such plans are, at best, only formulated for the social sectors.

A fourth feature might be added, commonly known as 'The War against Terrorism'. So far, the influence of 'September 11' is less visible in European aid, and is affecting the developing regions differently. There is a slight diversion of aid flows to the Middle East and Central Asia, but this is less the case for most European donors. In EU policies the worries about Islamic fundamentalism, most probably expressed more in words than in deeds, are older than September 11. It was expressed in the so-called 'Barcelona Process', launched at the Euro-Mediterranean Summit in November 1995 and was again polished in the new Neighbourhood Policy in March 2003. The fact that the Barcelona Process is considered to be slow, by some even to be a

---

income countries (LMICs). In 1989–90, 44.1 per cent of total ODA went to LLDCs and another 31.2 per cent to other LICs, while 17.6 per cent went to LMCs, adding up to 92.9 per cent. In 1999–2001, the distribution on these groups of countries was, respectively, 36.0, 27.5 and 29.7 per cent, adding up to 93.2 per cent of total ODA and leaving only 6.8 per cent for the UMCs. However, the share provided to the poorest groups (LLDCs and LICs) had declined from 72.3 per cent in 1989–90 to 63.5 per cent in 1999–2001 [*OECD, various years*].

failure, might demonstrate that, for Europe, September 11 seems to be less urgent, at least when it comes to the development co-operation programmes of most European countries.

It was against this background that we planned this book. The present volume results from a research and publishing project under the auspices of the Working Group (WG) on Aid Policy and Performance of the European Association of Development Research and Training Institutes (EADI). In the planning session of the WG during the Ninth General Conference of EADI in Paris in September 1999, it was decided that the time was ripe for giving the aid policy of the individual European donor countries – and that of the European Union (EU) – a new hard look. This effort represents a follow-up on the first major work of the group in the early 1980s.

I. BACKGROUND AND TRACK RECORD

In 1982, the Working Group initiated a project with the purpose of taking stock of the development co-operation policy of European donor countries. Two years later, it resulted in two volumes published in the EADI Book Series. The first volume focused on the policy of the European donor countries – the philosophy on which their policy was based, the overall objectives set for their policy, major guidelines established and the volume targets set. In the individual country studies, the simple overall research questions to be answered were the following: what was the stated policy? To what extent was the stated policy followed up (the performance)? [*Stokke (ed.), 1984a*]. The second volume focused on European development assistance as viewed from the outside, with particular reference to three commissioned contributions, from Bangladesh, Kenya and Tanzania, respectively. In addition, there were a contribution on trends in the European aid contributions and, interesting enough at that point in time, a contribution on aid conditionality [*Stokke (ed.), 1984b*]. The new effort was to be narrating much the same features as those focused on two decades earlier. The main focus this time was to be on policies and performance in the 1990s and beyond.

The mode of work followed when the first two volumes were produced set the standard for when the EADI WG on Aid Policy and Performance later on engaged in a series of more thematic studies. A project is initiated with a project proposal submitted by the Convenor(s) and modified after comments, *inter alia*, from those invited to write the component studies. Then the first drafts of the component studies included are presented and scrutinised in a workshop where, in addition to the authors, other researchers in the field are also invited along with top experts, which in most cases involves senior administrators within international and bilateral aid agencies (ministries) and often also major NGOs active and profiled in the particular field. Although difficult for budgetary reasons, efforts are always made to ensure high-quality participation from the South. The first project set a high standard in this regard, difficult to live up to later on because of the costs involved. After the workshop, the papers are revised and those selected for inclusion go through a tough editorial process before they are presented to the EADI Publishing Committee to be evaluated (external referees) for inclusion in the EADI Book Series.

Since the two first volumes in 1984, the WG has a solid track record. A volume on trade and development was published in 1988 [*Stokke (ed.), 1988*], followed by a programme on the evaluation of development assistance that produced three volumes in the EADI Book Series – one with a particular focus on food aid [*Clay and Stokke (eds), 1991, reprint 1995*], the second with a focus on the evaluation policies of European donors [*Stokke (ed.), 1991*], and the third on evaluation approaches and methods [*Berlage and Stokke (eds), 1992*].

Four more volumes were produced during the 1990s: one on aid and political conditionality [*Stokke (ed.), 1995*], another with an overall perspective on the evolving aid policy into the next millennium [*Stokke (ed.), 1996*], a third (a co-operation between Institut universitaire d'études du développement (IUED), Geneva, and the Norwegian Institute of International Affairs (NUPI), Oslo) on policy coherence in development co-operation [*Forster and Stokke (eds), 1999*], and the fourth (a co-operation between the Overseas Development Institute (ODI), London and NUPI) on food aid and human security [*Clay and Stokke (eds), 2000*].

The volumes produced by the EADI Working Group in 1984 represented the first effort to provide a broad, comprehensive overview of European policies within this policy area. Surprisingly, few other efforts have been made in providing this kind of comprehensive overview, the valuable evaluations of the OECD Development Assistance Committee aside.[2] However, the development co-operation policy of several individual European countries has been portrayed. And some efforts have also been made, on a comparative basis, to portray impor-

---

2. In 1985, two volumes appeared, written by German authors in German. However, only four of the largest DAC countries and Italy were included [*Holthus and Kebschull (eds)*]. The second volume was entirely devoted to the development policy of France.

INTRODUCTION

tant policy aspects within this policy area of European countries and the EU – including those referred to above by the EADI WG on Aid Policy and Performance.[3] The new undertaking by the group therefore fills a gap. When the policies of the 1990s and beyond are described and discussed, most of the component studies included here draw the lines back to past policies and performance.

The present project has followed the patterns established for the EADI WG. It has been organised under the umbrella of this WG as a co-operative effort between the Centre for International Development Issues Nijmegen (CIDIN) at the Catholic University of Nijmegen and the Norwegian Institute of International Affairs (NUPI), Oslo, and directed by the two editors. The first drafts of several of the chapters included in this volume were first presented and discussed in a workshop organised by CIDIN in Nijmegen 4–5 July 2002, and followed up through the editorial process described above, organised from NUPI.

II. ABOUT THE PRESENT VOLUME

In Chapter 1, *Olav Stokke* gives a comprehensive overview of the changing international and conceptual environments of development co-operation, guided by the questions: What motives have been driving international development assistance? What have been its main purposes? Although development assistance is part of a multitude of contexts, the discussion is, by and large, narrowed down to two frameworks; namely, the international setting in which aid is operating and the philosophical (ideological) environment with particular reference to predominant development paradigms. Both settings have been changing over time – and are interrelated. Although the long lines are drawn, the main focus is on the 1990s and beyond. Against this background, some prominent themes on the international development agenda and major actors on the stage during these years are identified and discussed. The main function of the chapter is to provide a general contextual setting for the other chapters in this volume.

The following chapters portray a picture of the development policies of individual European donor countries and the European Union. As already noted, the project was started in the first half of 2001 and

---

3. For an effort to outline and analyse the poverty reduction strategies of several European countries, involving ten European research institutes, see Cox *et al.* [2000]. The Overseas Development Institute (ODA), London, published the country studies.

the drafts of most country studies were presented and scrutinised at a workshop in early July the following year.

As was the case of the volumes on European aid policy appearing in 1984, the editors aimed at creating a basis for comparative analysis. The authors of the country studies, when invited to take part in the project, were asked to describe and analyse the major features of the development co-operation policy of their countries, as stated and implemented, particularly the motives (justifications) and overall objectives; the strategies, major principles and guidelines; targets set for the volume of official development assistance (ODA); the major structure of the development assistance programme (the geographical and sector distribution and the distribution on aid channels); and the administrative and institutional setting. Since the fight against poverty had appeared as a priority concern towards the end of the last century, special attention should be directed towards that issue. And since the EU was aiming at a common foreign and security policy, the authors were also asked to attend to this issue. Variations in the public support should also be identified, as should, to the extent possible, the major actors within this policy area. However, the authors were, at the same time, encouraged to adapt their analyses to the particulars of the countries they were dealing with.

Attention was to be given to continuity and change in the development co-operation policies of European countries in the 1990s and beyond. These policies vary, as is well known, from one country to another with regard to ambition level as well as actual performance. In some countries, development co-operation is a high-profiled issue, while in others this policy area attracts little attention from policy-makers and the general public alike. The historical context of this policy area also varies extensively from one country to another – and within countries policies vary over time, depending, *inter alia*, on the values underpinning the government.

The country study appearing first (Chapter 2) portrays a pessimistic picture of the prevailing situation with regard to development assistance and kindles little hope for a brighter future. In Austria, the development policy does not any longer attract great political or public interest, in stark contrast to what was the case in the 1970s. Since the 1980s, governments have not provided development assistance at a level that reflects the country's economic strength, *Michael Obrovsky* argues. He sees no prospect in the near future of an overall aid budget or a development policy framework that is binding on all activities reported as ODA. No sustainable quantitative or qualitative improvements can be expected, even if HIPC debt reduction efforts bring

Austrian ODA close to or above the average EU contribution for the next couple of years, he argues.

Nevertheless, the accession to the EU has led to an internationalisation of the policy and, as a consequence, to modifications in the general framework and conditions for implementing development co-operation activities. Programmes and projects of the Department of Development Co-operation (DDC) have been progressively focused on the goals adopted by the international community and, through the establishment of regional co-ordination offices in priority countries, have improved considerably in qualitative terms. However, the gradual qualitative and programmatic modifications in the DDC since 1992 cannot obscure the fact that there is a lack of coherence in overall Austrian aid – only a small share of the ODA is administered by the DDC.

In the 1990s, the aid policies of Belgium were being pursued in a climate of major upheavals, some of them externally caused, such as the dramatic events in Central Africa, *Nathalie Holvoet* and *Robrecht Renard* argue (Chapter 3). These events marked the end of the post-colonial period in which Belgium had been providing the major part of its bilateral aid to DR Congo, Rwanda and Burundi. In the past Belgium emphasised its 'special ties' with Central Africa, leading to unclear and unarticulated aid policies. All that has changed. The three countries went through dramatic crises, and the whole region remained highly unstable during most of the period. Belgian aid dwindled, and diplomacy became more important.

The aid that no longer went to the three traditional beneficiaries was spread very widely, without any clear sense of direction or a convincing effort at imposing some geographical concentration, it is argued. At the level of motives and overall objectives, aid became more complex, with all the major political parties involved in government giving emphasis to human rights and governance issues. In a significant break with past practice, aid tying was, if not abandoned, then at least greatly attenuated, except for the bilateral aid administered by the Ministry of Finance. Indirect actors, especially NGOs and the universities, became major channels for official aid, to the detriment of direct bilateral aid.

Administrative management remained the Achilles' heel despite several efforts at restructuring. A major reform at the end of the decade, splitting the administration into two entities, DGDC for planning and BTC for implementation, has not been in operation long enough for drawing a balanced judgment, the authors observe, although they find reasons for doubts about the good it may bring. The major cloud hanging over development policies at the turn of the new millennium, however, was the proposal to hand over important parts to the regional governments. This would, in practice, result in three governments in Belgium being in charge of development co-operation, one federal and two regional, and would call for yet another major rethinking of management structures.

In Chapter 4, *Oliver Morrissey* discusses the British aid policy under the 'New Labour' government. The seven years since Tony Blair won power in Britain have seen a dramatic change in British aid policy under the stewardship of Clare Short as Secretary of State for International Development, he observes. The transformation from the Overseas Development Administration to the Department for International Development (DFID) was more than a change of name for the aid agency. Its political status increased, with a seat at the Cabinet table, its budget has grown significantly and two White Papers have been published to announce directions in development policy. DFID is an aid agency with a purpose and a mission, a genuine attempt to increase the effectiveness of aid in tackling poverty and deprivation in developing countries. Britain has emerged as a donor that practises what it preaches. It is now one of the leading bilateral donors on aid policy thinking and it has set an example for other donors to follow, Morrissey argues.

During most of the 1990s and beyond, Denmark has been the leading country among aid providers – since the mid-1990s and until 2001, it was the only donor country that maintained aid at a level beyond the 1 per cent of GNI target. In Chapter 5, *Gorm Rye Olsen* argues that Danish aid policy in the post-Cold War era has been characterised by two main features: increasing budget allocations and a remarkable continuity as far as the goals and policy are concerned. This dual situation is explained by two circumstances. The existence of a small and closed policy community explains why it has been so difficult to make radical changes of policies, priorities and goals, it is argued. And the prominent position of moral arguments involved in aid has lifted the policy issue out of the general public debate, making it almost impossible to question the basic features of the policy including debating the remarkable automatic process involving the financial allocations.

With the new liberal-conservative government that came into power in November 2001, some of these features have been changed. On the other hand, the continuity is more striking than the changes. That is valid even for the level of Danish development assistance that has been reduced only slightly, Olsen points out.

INTRODUCTION

Finnish developmment co-operation looks very different today from its heyday in the late 1980s and early 1990s, *Juhani Koponen* with *Lauri Siitonen* argue (Chapter 6). The volume of ODA has been shrinking, and although an upward trend may be discerned recently, it is proceeding at a rather slow pace. At the same time, the whole approach to development co-operation has changed. Big projects maximising the use of Finnish resources have given way to smaller undertakings geared towards sustainability and backed by a new concern for aid effectiveness. Yet the transition from donor-driven aid modalities to genuine partnership represents more of an ideal than reality, and very little is known about the actual impacts of Finnish aid, it is argued.

The authors, employing a 'constructivist' approach, analyse Finnish aid policy by discussing the developmental roots and justifications and the instrumental use of aid, and their interplay. The main argument is that, until recently, a genuine development orientation has not been holding a strong position. It has been legitimate to justify aid by a multitude of motives and a belief in the compatibility of different interests. In addition to furthering economic development in poorer countries or contributing to poverty reduction, Finland has historically used its aid for a number of other purposes, such as, in the past, to build a profile of Finland as a Western country and supporting Finnish companies and, more recently, to portray Finland as a loyal member of the European Union. The shifting use of aid funds is explained by Finland's geopolitical position and the insecurity resulting from the country's ambivalent position between the East and the West. If recent trends are towards giving more emphasis to development justifications and purposes, this is due to the disillusionment with earlier instrumental approaches, the authors argue.

French aid policy cannot be separated from the country's colonial past, the close relations between public authorities and private interests and, more generally, the role France wants to play on the international stage. This is the point of departure for *Jean-Jacques Gabas* (Chapter 7), in search of the main characteristics and recent trends in the French policy within this area, starting with the 1998 reform, the strategy for resource allocations and with a focus on its main orientations.

Two factors contributed to the decrease of French aid in the 1990s, according to Gabas, namely the decrease in funding when the CFA franc was devalued in 1994 followed by a sharp reduction of funds allocated to structural adjustment programmes during the second half of the 1990s. In 2001, France ranked fifth among donors based on transfers in absolute terms.

In the 1998 reform, a priority solidarity zone was defined: bilateral aid should be selective and concentrated and include the least developed countries with no access to financial markets. In February 2002, a suspension regime for countries that hindered international co-operation was added, with particular reference to human rights violations, corruption and violations of the rule of law. However, the declared commitment to reduce poverty and inequalities has not really been followed up in the allocation of aid, Gabas argues; the LLDCs have received a decreasing share of total aid. Although a large share of the bilateral aid is still concentrated on African countries, the old relationship (*la France–Afrique*) – involving military support for some regimes in the region as well – is brought to an end, it is argued.

France has all through taken great interest in and exerted a strong influence on the development policy of the EU. To what extent are the old strategic positions *vis-à-vis* developing countries still viable? Gabas finds French development policies within the EU setting and *vis-à-vis* the multilateral system, the UN in particular, ambivalent and in need of greater clarity.

In Chapter 8, *Guido Ashoff* presents and analyses the aid policy of another major bilateral donor – Germany, ranking third after the US and Japan if measured in absolute transfers (2001). Since the early 1990s, its development co-operation policy has found itself in a paradoxical situation: while its conceptual approach has clearly become much more comprehensive and ambitious, net disbursements of ODA have declined both in real terms and as a share of GNI. However, in spite of the resource constraints, the past decade and especially the last few years have witnessed a number of reforms, Ashoff argues.

Development co-operation attracts far less attention than domestic and economic issues, particularly since the early 1990s when Germany first had to manage the process of reunification and now is facing the unprecedented challenge of reforming major pillars of its postwar growth and welfare model. The sharp decline in Germany's ODA/GNI ratio in the 1990s is seen as a clear sign of the reduced importance attached to development assistance. On the other hand, German aid policy, which originated from and was shaped by a mix of quite different motives throughout the previous decades, has increasingly developed a profile of its own. This is attributed to the growing developmental professionalism of the aid administration and a strong reform drive resulting particularly from the challenges of globalisation and the need for aid policy to redefine its role in Germany's response to these challenges. The conceptual shift towards a policy that, in addition to improving living conditions in partner countries, contributes to safeguarding our global future, has provided an opportunity for development co-operation to gain a more central political role. In contrast to most other ministries, the Ministry of Development Co-operation (BMZ) has not only a sizeable budget that may be used for global action but also relevant expertise (accumulated also in its executing agencies) that is frequently asked for by other government departments and now puts the BMZ in a better position to advocate more policy coherence, it is argued.

Yet, considerable challenges need to be faced if its new objectives are to be achieved, including raising the ODA/GNI ratio to 0.33 per cent by 2006, as committed by the Chancellor at the Monterrey Summit in 2002. Developing a detailed strategy for implementing the ambitious Programme of Action 2015 for poverty reduction will be another major challenge to which further reform of the institutional set-up has to be added, Ashoff argues. The size of its aid programme in absolute terms gives Germany a considerable weight among aid donors and in efforts to improve donor co-ordination. Germany has been slower than some other reform-minded countries in changing its development co-operation policy because of the influence of other interests, at times including institutional self-interests within the aid administration. But once a change of direction has occurred, Germany has followed this course fairly steadily and reliably, Ashoff points out.

Ireland established its bilateral ODA programme in 1974, a year after it entered the European Economic Community. In the period up to 2003, the programme has grown in volume terms from just under €2 million to over €450 million. As a percentage of gross national income (GNI), it has grown from 0.05 to 0.4, putting it in seventh place in the DAC donor list today. Although it has failed to meet all the targets set for its growth over the years, that of reaching the UN target of 0.7 per cent of GNI is supported at the highest levels of government and among the general public, *Helen O'Neill* observes (Chapter 9).

From its beginnings in the 1970s – and not surprisingly, given that missionaries had pioneered Ireland's involvement in developing countries – the Irish aid programme has been imbued with a strong humanitarian motivation and an overarching focus on poverty reduction. This focus is reflected in its choice of programme countries, all of which are among the poorest in sub-Saharan Africa, and in its sector assistance, which is directed towards primary education, primary health care, subsistence agriculture and rural infrastructure, it is argued.

In common with many other donors, Ireland has moved away from supporting 'project islands' and is now delivering its assistance mainly through targeted support to sub-national districts, to specific sectors and, in co-operation with other donors, to general support to central government budgets. During the past few years, it has been providing significant amounts to prevention and alleviation of HIV/AIDS both at bilateral and multilateral levels. Although Ireland provides all its ODA in grant form, it has begun to make significant contributions to debt relief, bilaterally to some of its programme countries and multilaterally to the World Bank/IMF HIPC scheme. Ireland Aid has always worked in close co-operation with Irish NGOs and now spends a higher proportion of its ODA budget through them than does any other donor. The 1999 DAC review of Irish aid noted that Ireland has had a 'very lively engagement' with international organisations. Its recommendation that contributions to UN agencies should become more focused is being followed, as is its recommendation that Ireland should become more involved in preparations for international conferences and in the councils of these agencies.

The present phase of Italian development co-operation follows a decade characterised by crisis. *Clean Hands*, the judicial process against generalised corruption initiated in 1992, also involved development co-operation, leaving it discredited in the public opinion. Since then, it has gradually recovered, although never returning to the volume levels of the 1980s. In the mid-1990s, Italy adopted the development priorities stated at the international level, first of all poverty reduction, and translated them into strategic guidelines for its aid policy, *José Luis Rhi-Sausi* and *Marco Zupi* argue (Chapter 10). However, its volume performance – in 2001, ranking tenth among DAC countries in absolute terms (US$1.6bn.) and second to last in relative terms (0.15 per cent of GNI) – has not been up to such standards, although increasing substantially in 2002.

The year 2000, with the Jubilee campaign, marked a turning-point: public opinion came out in favour of debt cancellation and showed renewed interest in development co-operation issues. The government responded with a law on bilateral debt cancellation of poor countries' external debt, placing Italy in the forefront. The debt cancellations are considered an indication of a cultural change. However, although they improve an otherwise poor bilateral performance, a reversal of the trend towards reduced ODA is not guaranteed, it is argued.

INTRODUCTION

In recent years, a trend towards multilateralisation of aid has been confirmed: Italian ODA, debt relief aside, continues to flow mainly through multilateral channels. A low level of aid combined with mandatory contributions to multilateral institutions (*inter alia*, to the EU) may be part of the explanation. Another recent trend is that Italian ODA has increasingly been switched towards neighbouring countries (the Balkans and the Mediterranean Basin). Led by national geopolitical priorities, this trend is in line with the EU 'Wider Europe approach' and has been translated into a reciprocity-based strategy that opens new space for decentralised co-operation as well as for Italian enterprises, SMEs in particular, it is argued.

Italian decentralised co-operation represents the emerging actor in Italian development policy. It provides an interesting opportunity to implement the objectives of participation and involvement of different stakeholders from local communities through international partnership. On the other hand, NGOs strive hard to gain a more prominent role in Italian international co-operation, trying to overcome the many limitations posed by the slow and complex bureaucratic aid management system.

The role of Italian entrepreneurship (which has followed the descending curve of Italian ODA in the last decade and is now small) in future aid policy depends on the outcome of the currently ongoing debate on aid reform and on the concrete implementation of the EU-sponsored neighbourhood policy.

Until the mid-1980s, Luxembourg scored low as an aid donor. In the 1990s, however, it managed to attain the target of 0.7 per cent of GNI in ODA and recently the government has promised to reach 1 per cent of GNI by 2005. *Paul Hoebink* gives the background of Luxembourg's evolving development co-operation programme, arguing that pressure from NGOs was decisive for its creation (Chapter 11). Business interests have been almost absent in forming the aid rationale. External pressure, from the DAC in particular, in combination with a sense of embarrassment and responsibility made politicians finally change the aid policy.

Luxembourg copied the structures and systems of the large donors without looking for its own niche and with an emphasis on project aid, it is argued. The aid to its largest recipient, the Cape Verde Islands, shows that project aid dominates, with an emphasis on hardware and visibility. Most aid goes to education and health. Although Luxembourg, when it comes to volume performance, choice of sectors and absence of aid tying may set an example for other donors, other features of the aid programme and its administration should not inspire other newcomers in this policy field, Hoebink argues. Problems are caused by stretching the programme to cover too much, paying little attention to policy development and overburdening the staff, it is concluded.

In the 1990s, focus was strongly on improving the quality of the development co-operation of the Netherlands. To a large extent, however, these efforts still belong to the declared policy – it remains to be seen if practice will also make a major leap forward, *Lau Schulpen* argues (Chapter 12). The Netherlands has, as always, followed mainstream international thinking on development, although with some differences in emphasis. This applies, for instance, to the sector-wide approach: in contrast to other donors, this approach was set as a norm for all bilateral aid.

Four other central features of the Dutch development policy in the 1990s, in addition to the sector-wide approach, are identified and discussed. The concentration of bilateral aid to a limited number of countries has been part and parcel of the policy since the 1960s. Until the late 1990s, however, the number of aid-receiving countries did not decrease at all. The change resulted from a more strict concentration policy in which good governance was used as the main criterion. Private-sector development in terms of involving the Dutch private sector has also always been part of the development co-operation policy. However, a more clearly designed policy was first presented in the second half of the 1990s. Although the role of Dutch enterprises remains important, the new policy concentrates on ways of developing the private sector within developing countries themselves. In the 1990s, major changes also took place with regard to the private aid channel. Two new organisations were added to the four major NGOs that traditionally had received the major part of their funding from the aid budget and a new budget line was opened for so-called thematic NGOs. Even more important, all NGOs had to make the quality of their work more visible than they had before.

In the 1990s, policy coherence was brought to the forefront and seems likely to remain a priority issue. This is illustrated by a recent policy paper that focuses on development co-operation and agriculture and calls, *inter alia*, for trade liberalisation and further reform of the agricultural sector in Europe. The 1990s have been a stimulating decade for Dutch aid, also with regard to the greater understanding of poverty and poverty reduction that has been attained, it is concluded. Nevertheless, it remains to be seen to what extent policy declarations will be transformed into implemented policy. The Netherlands has remained among the top four bilateral aid providers in relative terms – above 0.8 per cent of GNI.

INTRODUCTION

Although the aid policy of Norway changed in many respects during the last decade, the basic values underpinning it remained much the same, *Olav Stokke* concludes, based on an analysis of the *stated* policy as expressed in a series of White Papers (Chapter 13). The policy has increasingly been based on 'the Washington–New York consensus' (Chapter 1), adapting Norway's development policy to internationally agreed objectives, strategies, norms and mechanisms – with particular reference to poverty alleviation and promotion of liberal polititical ideas (human rights, democracy and good governance). At the same time, the aid policy became increasingly integrated into the foreign policy, administratively and politically.

The aid strategy of the 1990s has four core components, namely, the principle of recipient responsibility in combination with programme aid that was to be anchored in the recipient countries and co-ordinated with the assistance of other aid providers. Whereas the first three components are rooted in the past, the emphasis on international co-ordination represents the main novelty. It was to be combined with an active role in international bodies where policies are formulated, a tall order for a small nation. It became part of an active humanitarian foreign policy involving conflict resolution (mediation and peace-keeping) with aid as the primary instrument: throughout the 1990s, humanitarian aid stayed at an extraordinary high level.

In the course of two decades, Spain has radically changed its position from being an aid recipient (1979) to being the twelfth largest donor in absolute terms (2002). During this brief time span, the human resources, instruments and institutions necessary to develop a framework for development co-operation have been created, *José Antonio Alonso* observes (Chapter 14). Three creative stages are identified, with steps backwards in-between: in the first (1984–86), a specialised development assistance policy and the institutions to manage it were established; the second stage (1988–92) was one of consolidation in terms of increased resources and institution building – and Spain became a member of the DAC; the third stage (1997–99) brought Spain into line with international norms and the technical capacity was improved: a law on international development co-operation was approved, Spain adapted its policy to the international aid doctrine emphasising poverty alleviation, and mechanisms to improve the technical management of aid were established, including the adoption of strategic guidelines.

However, the author finds more recent developments disturbing: the aid system shows retrogradation in a number of areas. In terms of doctrine, the commercial and cultural interests of the donor are prevailing in the conceptualisation and orientation of aid. In the field of management, reforms needed to improve the integration of the aid policy and to strengthen some major institutions involved have been postponed. New management tools aimed at making aid more flexible – such as programme aid and the sector-wide approach – have been shunned in favour of project-oriented aid. And in social terms, recent years have been characterised by a weakening of dialogue and of trust between the government and other actors in the aid system, Alonso argues.

Summarising the Swedish aid philosophy (Chapter 15), *Anders Danielson* and *Lennart Wohlgemuth* conclude that although there have been significant changes over the years, many of these have been of a purely rhetorical nature. International solidarity is identified as the underlying motivation – in Sweden, commercial motives do not dominate the debate on aid. The membership of the EU is seen as the most significant change, providing Sweden with an opportunity to influence the development assistance of what is considered a very large, potentially important, and usually very old-fashioned aid apparatus. Most of the bilateral aid has been provided to countries with which Sweden has cultivated long-standing relations, particularly countries with a social-democratic vision. The idea of giving 'help to self-help' to countries whose visions were shared by Sweden has been important in guiding country allocations.

Not much evidence is found to suggest that an explicit development theory is determining the volume or allocation of modality: only scattered ideas about links between aid and aid objectives are found, an observation that finds support in several background papers produced for the 2001 review of Sweden's policy for global development. The guidelines that govern development co-operation, the selection of main partner countries in particular, suggest that history matters more than economic rationality, the authors argue.

During the last decade, Switzerland has adapted its bilateral and multilateral development co-operation activities to global changes and new requirements, and development co-operation has gradually become an integral part of the political, economic and social relations with the countries receiving aid. Through a more global approach, yet one that remains centred on technical co-operation and humanitarian aid, development co-operation is now better integrated within Switzerland's foreign policy, *Cathrine Schümperli Younossian* argues (Chapter 16).

The North–South Guidelines, adopted by the Swiss government in 1994, form the basis of a coherent policy towards the South, it is argued. They open new fields to development co-operation (more particularly the environment, and peace and security) and highlight the importance of looking at the totality of relations with a country 'in order to guarantee an integral and coherent development policy'. The institutional consequences are explored with particular reference to the way the federal administration has modified its practices and operations. Another feature that attracts particular interest is the background to and challenges resulting from Switzerland joining the International Monetary Fund and the World Bank in 1992 and the United Nations in 2002. The national debate ahead of these decisions, especially the role of NGOs, is highlighted.

Swiss NGOs play a particularly prominent role. The Swiss Confederation maintains close relations with the NGOs by providing financial support for their programmes, and some NGOs are also used by the aid administration (SDC) as implementing agents for its own programmes. Associations, which are well structured in Switzerland, play a particular role in this political context. The opportunity of launching popular initiatives or rejecting, by way of referendum, certain government decisions provides the NGOs with substantial power, it is pointed out.

Switzerland has not reached the target of 0.4 per cent of GNI to which the government has committed itself: its ODA has for several years stayed at about 0.34 per cent of GNI.

This volume ends with a chapter that puts the post-Cold War development policy of the European Union into perspective. At the start of the new century, the development assistance from the European Union (EU) and its member states, put together, represents more than half of the global ODA. High priority in the aid policy of the EU is given to fighting poverty and the promotion of sustainable economic and social development. It has development co-operation agreements with more than 100 countries. The aim of the common aid policy of the EU is to supplement the bilateral aid policies of the 15 (soon 25) member states.

Gorm Rye Olsen (Chapter 17) focuses on the development cooperation arrangement with the ACP countries. This represents the single most important arrangement of the EU in this policy field, covering no less than 77 developing countries. The co-operation has been directed under the Lomé conventions and, since 2000, it has been directed under the Cotonou agreement. He launches two hypotheses to explain the development policy of the Union in the post-Cold War era. The first argues that the dramatic changes that took place during the 1990s and the first years of the new century have to be understood against the background of the fact that European interests in the developing world changed after the end of the Cold War. The second hypothesis claims that it is impossible to understand the more specific changes and initiatives related to the European development policy unless it is accepted that bureaucratic and thus elite-based interests exert strong influence on policy formulations within this particular policy field. The influence of bureaucratic interests explains the increasing similarities between European aid policy and bilateral aid policies. The strong mutual influence among donors may also be explained with reference to the existence of a transnational aid policy community that has many of the same characteristics as national policy communities.

The empirical analysis shows that the European interests in Africa have decreased dramatically since the beginning of the 1990s, indicated by the conspicuous reduction in the European aid transfers to the region. Parallel to this development, the EU has to an increasing extent involved itself in issues related to security and stability questions in the region. Therefore, development aid is seen increasingly as an instrument that can contribute to promote stability and security. As a result, aid has gradually become one among a number of foreign policy instruments. This is so not only because the European Union is seriously concerned about security issues in Africa and other developing countries. It is also because promoting an independent EU security and foreign policy profile in Africa has become part of the overall strategy to establish a common foreign and security policy, Olsen argues.

He then turns to more specific changes in the policy of the EU. The appointment of the new Commission and a new commissioner for development assistance in mid-1999 initiated a reform of the whole aid administration. The reform came after years with increasing criticism of the common aid programmes for being inefficient. An evaluation report published shortly before the appointment of the new commissioner launched a serious critique of the common aid programmes for being without any effect, at best. The general overhauling of the aid programming process was initiated mainly with the aim of ensuring a better match between resources and political priorities. It also aimed at ensuring a stronger focus on poverty alleviation. A new implementing unit (AIDCO) was established to improve the quality of the project administration and to reduce the time from policy formulation to implementation. Along with the establishment of AIDCO, the Commission started to work on a general decentralisation of its development administration implying that staff based in

Brussels should be transferred to the Commission's delegations in the recipient countries. Two and a half years after the start of the reform process, the EU commissioner for aid has claimed that significant results have been accomplished.

The reform of the aid administration was an integrated element in a general reform of the European Union's external services, Olsen argues. The reform of the entire external services of the Union has been driven by strategic interests and by European interests in making the Union a significant international actor. The ambition manifested itself in the abolition of the separate development council and the transfer of the discussion on development issues to a new General Affairs and External Relations Council. Within this new framework it becomes a clear European interest to run a development aid programme that is efficient, Olsen concludes.

III. DOES EUROPE SPEAK WITH A COMMON VOICE OR SEVERAL WITHIN THIS POLICY AREA?

It appears from the brief presentations of these component studies that Europe does not yet speak with one voice – the aid policy differs from one country to the next and the importance accorded to the policy area by the government and general public also varies extensively from one government and country to another – few common trends can be identified.

Since member states of the European Union provide about half of total development assistance, as reported by the DAC, it might be expected that the EU – and the European Commission – would play an important role in forming and streamlining policies within this policy area. The commitments and efforts to arrive at a common foreign and security policy during recent years make such expectations natural. In the 'dialogue' with the US government, particularly after the Bush administration came to power, European governments (and the EU) have portrayed a different foreign policy (and security policy) approach from the one pursued by the US, giving emphasis to negotiations and dialogue, indirect pressure and 'positive measures' instead of coercive power in conflict resolution. This rhetoric contributes to raising such expectations further, since the aid policy has increasingly become integrated into the foreign policy of most donor countries. It would be natural to expect ideas and intellectual leadership to emerge from Brussels.

So far, however, little of the kind is appearing. The European Commission has not been capable of providing intellectual leadership to bring the member states to new ideas and positions. On the contrary, it has been struggling with the implementation of its own aid programme. Brussels has therefore been looked at with scepticism by the more 'dynamic' capitals – as conveyed in the component studies on Britain, Denmark, the Netherlands and (particularly interesting, since it is a 'newcomer', with ambitions and a determination to reform the system) Sweden – both in its capacity as aid provider and as a source of new ideas and inspiration within this particular policy area. As a new member of the EU, Finland became 'liberated' from the pressure emerging from the standards set by its Nordic neighbours – particularly involving the ODA volume: the DAC or the somewhat higher EU average might suffice. However, for some new member states (such as Austria and Spain), the membership of the EU inspired improvements of the aid policy: towards norms and explicit and tacit expectations set by the extended community. In addition, Spain was able to stimulate the Union to undertake steps in the direction of deepening the relations with Latin America. The probability is that the EU will exert similar influences involving the new member countries from Central and East Europe on the verge of joining the Union. As for other member states, contributions to the development activities of the EU will be mandatory.

A nearby conclusion, therefore, is that the EU as such has not added much to the international discourse on development and development co-operation, with the possible exception of the discussion on the coherence of European policies *vis-à-vis* developing countries involving several policy areas. The limited administrative and research resources that have been made available may be part of the explanation, although lacking resources cannot explain everything – small centres elsewhere have been able to generate and mobilise ideas and resources, making a difference. Some EU member governments (and their administrations) have, however, contributed directly and indirectly to this discourse within a variety of frameworks, including the EU setting. The importance of co-ordination of policy positions within the EU, particularly ahead of international conferences with a focus on issues that also affect development and developing countries, should not be underestimated: such co-ordinated positions affect both outcomes of international negotiations and policies of member states. Such influence has been manifest in the past; the potential for exerting such influence in the future seems even larger. The same is true for co-ordination within the aid-receiving countries. Co-ordination between EU member states is a given here with regard to political issues, and although some member states are reluctant to co-ordinate all aid issues in the same setting, thus leaving other like-minded donors out, other member states try to promote this type of co-ordination.

However, other arenas seem so far to have been been considered more attractive for most member governments, with particular reference to the DAC, the UN and its global conferences, and the international financial institutions, the World Bank in particular (Chapter 1). However, this situation cannot automatically be extrapolated into the future: with a common foreign and security policy in place, and a corresponding drive for a common European identity (where the aid policy may represent an important component), the EU has the potential to emerge in a more important role. Given the influx of a number of new member states into the community, however, this is not the only probable future scenario.

The turn of the century represented a new opportunity for many European governments that had been lagging behind: the UN millennium development goals (MDGs), particularly the commitment to fighting poverty, were generally acclaimed and, as will be seen in the country studies, led several heads of government to commit their governments to increase the volume of their ODA – although attaining the MDGs demands an integrated approach that goes beyond aid. In many ways, the MDGs represent a silent revolution because of this wider orientation and by identifying results (expressed in quantitative terms) to be obtained within an ambitious time frame instead of expressing general aspirations. It will be seen from the country studies that a number of governments have followed up on these commitments by also adopting integrated strategies for poverty alleviation. However, a varied picture emerges on this score: for most governments, it remains a challenge to transform rhetoric and general commitments into operative strategies.

Coherence between aid policy and other policy areas of relevance to developing countries, explored in a previous EADI project [*Forster and Stokke (eds), 1999*], has been high on the international development agenda for years. Improving this coherence, both of policies within the framework of the donor countries and between national and multilateral actors on the scene, will be a *sine qua non* for attaining the MDGs. As will be seen from the country studies, the profiles vary extensively. However, it bodes well for the future that this issue has been brought to the forefront as a regular part of the DAC peer review reports on individual donor countries.

During the 1990s, all donors were confronted with the need to enhance the effectiveness and efficiency of their delivery systems. In the rhetoric, there has been an emphasis on ownership, participation and recipient responsibility: the recipients should be in the driving seat. This affected the forms of aid towards more emphasis on programme aid (sector-wide support and ultimately budget support) and less emphasis on traditional project aid. It follows from such an approach that donor agencies are expected to pool their resources at the level of a recipient country or region, and in practice remove their flag from aid. Such an approach has already affected technical assistance: less use of traditional experts. Several donors have already shifted to a more extensive use of local experts and local consultants, although the main reason has been that they proved more cost-effective.[4] Delivery systems have also been reformed, towards a decentralisation of the aid administration from the donor capitals to the recipient countries, as happened recently in the EU, France and the Netherlands and is also under way in Norway.

Again, the country studies reveal a varied picture. At the conceptual level, there seems to be broad agreement among many European donors on the principles involved and even on some of the tools – such as the comprehensive development framework and poverty reduction strategy papers. However, for a range of different reasons, practice seems difficult to change. Nevertheless, some governments have taken decisive steps in the prescribed directions, particularly involving the rhetoric (declarations of intent) but also involving the practice, with even reforms of their aid administration. But these are exceptional cases.

IV. THE VOLUME OF ODA: PERFORMANCE AND COMMITMENTS

Also with regard to the volume of ODA, European countries display a varied picture, involving both intentions and practice, as described and analysed in the component studies. Tables IN. 1–3 give an overview of the ODA performance of European donors in absolute ($bn.) and relative terms (ODA as a percentage of GNP/GNI), with particular reference to the 1990s and beyond. Although most European donors compare well with the performance of the only remaining superpower, measured in relative terms, some of them are performing below the DAC average on this account. The top performers on this account are, however, all Europeans. The causes that may explain these varied performances are addressed in the component studies.

---

4. The HIV/AIDS pandemic may, however, create a new need for experts from developed countries in several African countries.

TABLE IN. 1
OFFICIAL DEVELOPMENT ASSISTANCE FROM DAC COUNTRIES, 1970–2003 (US$m.)

	1970	1980	1985*	1990	1995	1996	1997	1998	1999	2000	2001	2002	2003
1. USA	3,153	7,138	9,483	11,394	7,367	9,377	6,878	8,786	9,145	9,955	11,429	13,290	15,791
2. Japan	458	3,353	4,716	9,069	14,489	9,439	9,358	10,640	15,323	13,508	9,847	9,283	8,911
3. France	971	4,162	3,588	7,163	8,443	7,451	6,307	5,742	5,637	4,105	4,198	5,486	7,337
4. Germany	599	3,566	3,387	6,320	7,524	7,601	5,857	5,581	5,515	5,030	4,990	5,324	6,694
5. UK	482	1,854	1,633	2,638	3,202	3,199	3,433	3,864	3,401	4,501	4,579	4,924	6,166
6. Netherlands	196	1,630	1,438	2,538	3,226	3,246	2,947	3,042	3,134	3,135	3,172	3,338	4,059
7. Italy	147	683	1,751	3,395	1,623	2,416	1,266	2,278	1,806	1,376	1,627	2,332	2,393
8. Canada	337	1,075	1,663	2,470	2,067	1,795	2,045	1,707	1,699	1,744	1,533	2,006	2,209
9. Sweden	117	962	965	2,007	1,704	1,999	1,731	1,573	1,630	1,799	1,666	1,991	2,100
10. Norway	37	486	686	1,205	1,244	1,311	1,306	1,321	1,370	1,264	1,346	1,696	2,043
11. Spain	...	...	186	965	1,348	1,251	1,234	1,376	1,363	1,195	1,737	1,712	2,030
12. Belgium	119	595	493	889	1,034	913	764	883	760	820	908	1,072	1,887
13. Denmark	59	481	567	1,171	1,623	1,772	1,637	1,704	1,733	1,664	1,634	1,643	1,747
14. Switzerland	30	253	362	750	1,084	1,026	911	898	969	890	908	939	1,297
15. Australia	212	667	751	955	1,194	1,121	1,061	960	982	987	873	989	1,237
16. Finland	7	110	262	846	388	408	379	396	416	371	389	462	556
17. Ireland	...	...	51	57	153	179	187	199	245	235	287	398	510
18. Austria	10	178	223	394	767	557	527	456	527	423	533	520	503
19. Greece	...	...	-	-	-	184	173	179	194	226	202	276	356
20. Portugal	...	...	16	148	258	218	250	259	276	271	268	323	298
21. Luxembourg	...	...	10	25	65	82	95	112	119	123	141	147	189
22. New Zealand	...	...	65	95	123	122	154	130	134	113	112	122	169
Total DAC	6,949	27,266	32,296	52,961	58,926	55,485	48,324	52,084	56,378	53,734	52,336	58,274	68,483

Source: OECD [*several years*] and Development Committee [*2004: Table 1*]. *Notes*: *Average; ... not reported; - less than US$1 million.

TABLE IN. 2
ODA FROM EUROPEAN DAC MEMBERS AND THE EU, 1970–2003 (US$m. AND %)

	1970	1980	1990	1995	2000	2001	2002	2003
Total ODA (DAC)	6,949	27,267	52,960	58,894	53,734	52,336	58,274	68,483
Total European ODA	2,741	14,991	30,511	33,654	27,428	28,450	32,583	40,165
European (% of total DAC)	39.4	55.0	57.6	57.1	51.0	54.4	55.9	58.7

Source: OECD [*1977, 1987, 1996, 1997, 2004*] and Development Committee [*2004: Table 1*].

During recent decades, European donors have become more and more important as aid providers. As shown in Table IN. 2, their share of total ODA grew from about 40 per cent at the end of the 1960s (35 per cent at the beginning of the 1960s) to about 58 per cent by 1990 and 59 per cent by 2003. In 2000 and 2001, the EU and its member countries alone provided about half of total ODA (50.1 per cent in 2001 and 53.8 per cent in 2003).[5] In 2001, the European Commission became the third largest donor in absolute terms, passing Germany.

Europe's prominent position was partly due to diminishing aid flows from Japan and, in particular, the US. The picture changed slightly during the 1990s. The volume of US aid declined further in the first part of the 1990s, down to 0.1 per cent of GNP where it stayed into the new century, but increased from this low level, both in absolute and relative terms, in 2002 and 2003. Japan's aid has been volatile, decreasing sharply in 1996 but increasing towards 2000 and then decreasing again.

In addition, the performance of most European countries varied during the second part of the 1990s, as may be seen in more detail in Table IN.1. But aid from major European donors like Germany (down 26 per cent between 1995 and 1998), France (down 32 per cent in the same period) and Italy (down 48 per cent in 1997)[6] was decreasing even faster. For some smaller ones (Austria, Belgium and Switzerland), the trend went downwards while others (such as the UK and Luxembourg) increased their ODA. Among the latecomers, Spain slowly increased its aid during the 1990s. For most of the remaining European donor countries, ODA stagnated during the second part of the 1990s, although with variations from one year to the other. Part of

---

5. These figures are calculated on basis of OECD [*various years*] and Development Committee [*2004: Table 1*].
6. Italy's aid went up and down during the 1990s, never again reaching the peak of the end of the 1980s.

INTRODUCTION

this may be explained by the strong dollar and the relative performance (Table IN.3) of several countries may substantiate this.

Interesting trends can, however, be discerned from the performance during the two last years for which development assistance is recorded, namely 2002 and 2003. Most of the major donor countries, with the exception of Japan, increased their ODA (in absolute terms) – some of them quite substantially, including the US, France, Germany and the Netherlands – and so did some of the smaller countries, including Belgium and Norway.

TABLE IN.3
RELATIVE OFFICIAL DEVELOPMENT ASSISTANCE FROM EUROPEAN DAC MEMBERS (1970–2003)
(% OF GNP/GNI)

		1970	1980	1990	1995	2000	2001	2002	2003
1.	Norway	0.32	0.87	1.17	0.87	0.80	0.83	0.89	0.92
2.	Denmark	0.38	0.74	0.94	0.96	1.06	1.03	0.98	0.84
3.	Netherlands	0.61	0.97	0.92	0.81	0.84	0.82	0.81	0.81
4.	Luxembourg	...	...	0.21	0.36	0.72	0.82	0.77	0.80
5.	Sweden	0.38	0.78	0.91	0.77	0.80	0.81	0.83	0.70
6.	Belgium	0.46	0.50	0.46	0.38	0.36	0.37	0.43	0.61
7.	France	0.66	0.63	0.60	0.55	0.32	0.32	0.38	0.41
7.	Ireland	0.03	0.16	0.16	0.29	0.30	0.33	0.40	0.41
9.	Switzerland	0.15	0.24	0.32	0.34	0.34	0.34	0.32	0.38
10.	Finland	0.07	0.22	0.65	0.32	0.31	0.32	0.35	0.34
10.	UK	0.36	0.35	0.27	0.29	0.31	0.32	0.31	0.34
12.	Germany	0.32	0.44	0.42	0.31	0.27	0.27	0.27	0.28
13.	Spain	...	0.07	0.20	0.24	0.22	0.30	0.26	0.25
14.	Greece	...	...	...	...	0.20	0.17	0.21	0.21
14.	Portugal	0.45	...	0.25	0.25	0.26	0.25	0.27	0.21
16.	Austria	0.07	0.23	0.25	0.33	0.23	0.29	0.25	0.20
17.	Italy	0.16	0.17	0.31	0.15	0.13	0.15	0.20	0.16
For comparison:									
USA		0.31	0.27	0.21	0.10	0.10	0.11	0.13	0.14
Japan		0.23	0.32	0.31	0.28	0.28	0.23	0.23	0.20
Canada		0.45	0.42	0.44	0.38	0.25	0.22	0.28	0.26
Australia		0.61	0.48	0.34	0.36	0.27	0.25	0.26	0.25
New Zealand		...	0.32	0.22	0.23	0.25	0.25	0.22	0.23
Average DAC		0.34	0.38	0.33	0.27	0.22	0.22	0.23	0.25

Source: OECD [*1977, 1981, 1987, 1991, 1996, 1997, 2001, 2004*], Stokke (ed.) [*1984a*], UNCTAD [*1972*] and Development Committee [*2004: Table 1*].
Note: ... not reported.

Only four of these countries have, over time, lived up to the international target set by the UN already in 1970 – 0.7 per cent of GNP in official development assistance: Denmark, the Netherlands, Norway and Sweden (Table IN.3). Over a long period, they have been 'competing' among themselves over who was to be the most generous donor in terms of ODA as a share of GNP/GNI. During most of the 1980s, Norway was in the lead with the Netherlands a close second. In the latter part of the 1990s, Denmark took the lead. Sweden, at the top during the 1970s, has since then ranked fourth (third in 2002 and fifth in 2003) – since the mid-1970s, its performance has, most of the time, been above 0.8 per cent of GNP. Since 2000, Luxembourg has joined the 'club', with the intention of reaching the 1.0 per cent target by 2005.

In the 1980s, also France attained the 0.7 per cent target, if its aid to its 'overseas departments' was included.[7] In 1981, when this kind of aid was no longer accepted for inclusion in DAC statistics, France decided to attain the 0.7 per cent target anew (starting from the 0.5 per cent level). However, that aspiration was never met. On the contrary, its ODA has been gradually but drastically reduced to 0.32 per cent of GNI by 2000. It started to increase again in 2002, reaching 0.41 per cent of GNI the following year. Other former colonial powers, such as the UK (with peaks of 0.59 per cent in 1961 and 0.51 per cent in 1979) and Belgium (peaking at 0.59 per cent in several years in the 1980s) show the same pattern – Belgium reaching a new peak in 2003 (0.61 per cent of GNI).[8] Austria, Italy and Switzerland have never scored particularly high on this account, but Switzerland was close to 0.4 per cent of GNI in 2003. The performance of Germany has varied: above the 0.4 per cent mark in the 1960s (0.45 per cent in 1962) under the Christian Democrat governments, going down, surprisingly, under the Social Democrats in the 1970s, and up again in the 1980s (0.48 per cent in 1982 and 1983). As already noted, Germany's aid was reduced in the 1990s to its present low level (0.27 per cent of GNI in 2002 and 0.28 per cent in 2003).

Under way to the Monterrey conference in Mexico in 2002, the European Councils in Gothenburg (June 2001) and Laeken (December 2001) adopted resolutions to substantially increase European aid

---

7. Départements d'Outre Mer/Territoires d'Outre Mer, such as Réunion, Martinique and French Polynesia were receiving about 40 per cent of total French aid during the 1970s and 1980s [*OECD, 1986 and 1991*].
8. Non-colonial powers such as Canada and Australia show more or less the same pattern, going down from relatively high aid volumes (about 0.5 per cent in the mid-1970s and 0.6 per cent in the beginning of the 1970s, respectively) to low ones (0.28 per cent and 0.26 per cent, respectively, in 2002).

for 'the fight against poverty'. The European Commission was asked to produce a document to substantiate the plans for this increase. It turned out that France was most strongly opposed to a binding timetable for achieving the 0.7 per cent target. Of the member states that had not reached the 0.7 per cent target, only the governments of Belgium and Ireland committed themselves unequivocally to increase their aid regularly [*European Commission, 2002*].[9]

The European Union, whose member states in 2003 on average provided 0.35 per cent of their GNI in ODA (the DAC average was 0.25 per cent), has committed its member states to provide 0.39 per cent of their GNI as ODA by 2006. This commitment also applies to the ten new member states of the EU. Most of the new members are small economies; however, that does not apply to Poland. The EU commitment is probably the most important one of them all because it aims to bring all the countries of a region, particularly those which so far have been lagging behind, up to a higher level and because it will be part of the Union obligations of the member states, old as well as new ones.

Several of the European governments have made commitments to increase their ODA during the coming years. Among those countries that have already achieved the 0.7 per cent target, Luxembourg and Norway have committed themselves to reach the 1 per cent target by 2005 and Sweden has committed itself to reach that target the following year. The Netherlands has not made further commitments in this regard. Denmark (whose ODA in 2003 was reduced in relative but not in absolute terms) has committed itself to keep its ODA above the 0.7 per cent target.

Among the countries that so far have not reached the 0.7 per cent target, four countries have committed themselves to reach the target within a set date. Belgium has committed itself to reach this target by 2010; Finland to reach 0.4 per cent of GNI by 2007 and the 0.7 per cent target by 2010; France to reach 0.5 per cent by 2007 and the 0.7 per cent target by 2012; and Ireland to reach 0.45 per cent by 2002 (managed to reach 0.4 per cent) and the 0.7 per cent target by 2007. Other countries have also committed themselves to increase their ODA with set time frames. The United Kingdom has committed itself to reach 0.4 per cent of GNI in 2005 to 2006. The UK government has

---

9. These and the other commitments mentioned in the following paragraphs were made for the International Conference on Financing for Development, Monterrey, March 2002 [*European Commission, 2002*] and in the follow-up of that conference [*ODA Prospect After Monterrey: Update, 9 April 2003, DCD/DAC(2003)8; Development Committee, 2004: Table 1*], in particular the latter and most recent source.

a plan for stepped-up budgetary allocations – 0.02 per cent of GNI per year – in order to attain this target. Austria, Germany, Greece, Italy, Portugal and Spain – EU countries whose ODA has been lagging behind – have committed themselves to reach 0.33 per cent of GNI by 2006. Austria and Italy are among countries that once committed themselves to the 0.7 per cent target, but their performance has been far from convincing – during recent years the trends have pointed in the opposite direction. Switzerland has committed itself to bring its ODA up to 0.4 per cent of GNI by 2010, and made a great leap in that direction in 2003.[10]

The political position of a government on a left–right axis is assumed to influence a country's willingness to provide development assistance. As indicated above, this does not always apply. In more recent years, such shifts of government have occurred in Denmark, the Netherlands and Norway, making trends there particularly interesting. In Denmark, the new liberal–conservative government abolished the Minister for Development Co-operation and proposed cuts in ODA. Nevertheless, up until 2002, Denmark remained the frontrunner on this account, retaining ODA at around 1.0 per cent of GNI. However, as already indicated, the level aimed at was reduced and so was the relative performance in 2003, although Denmark remained among the top two aid providers in relative terms. In addition, the new conservative government in the Netherlands has retained aid at the 0.8 per cent level. And the Norwegian coalition government has increased its ODA quite substantially in absolute terms during recent years and has committed itself to again attain the 1 per cent target by 2005 and to keep ODA above that level once it has been reached.

Over the years, most European development assistance has been distributed to least developed countries and other low-income countries. In more recent years, however, there has been a trend, best illustrated by the aid of the European Union, to disburse more aid to low middle-income countries and upper middle-income countries. In the second half of the 1990s, the EU and some of its member states turned the focus more towards the East than towards the South. The United Kingdom has been an exception to this rule: it has followed up its new poverty reduction policy by a clear shift in its country alloca-

---

10. Among the non-European donor countries, the US has committed itself to 0.15 per cent of GNI by 2006; Canada to increase its ODA budget by 8 per cent annually so as to double its ODA by 2010 (from 0.28 per cent of GNI in 2002); Australia has already attained the level committed for 2006 (0.25 per cent of GNI), while New Zealand is reviewing its future level of ODA (0.23 in 2003). The prospect of increased ODA from Japan seems bleak.

tions towards economically poor countries. Spain represents an exception in the opposite direction: more than half of its aid has been distributed to middle-income countries in Latin America. China (being an important market for capital goods and construction works) also became a larger recipient of European development assistance. However, this is less the case with Afghanistan, Pakistan or Jordan, which are important recipients of US aid. It remains to be seen if these countries, together with Iraq, will also become prominent recipients of European development assistance.

V. SOME CONCLUDING OBSERVATIONS

The performance of European countries therefore varies extensively with regard to the volume provided in development assistance – in absolute and relative terms. As already noted, most are providing ODA well above the average DAC level and, in their stated policy, several governments aspire to increase their development assistance in both absolute and relative terms. The case studies in this volume will indicate that this generosity is true, for most European countries, also with regard to the conditions on which aid is provided, its distribution among countries and sectors, and in the ways they try to adhere to the goal of poverty reduction. But on these accounts too, European donors differ extensively.

Looking back at the last decade, however, some optimism for the future is due. Although conservative political parties have traditionally pressured for budget cuts also involving development assistance, the most generous European donors have been able to maintain a high ODA volume in both absolute and relative terms – even when conservatives have been part of the government, as illustrated recently in the cases of Denmark, the Netherlands and Norway. Several other countries are on their way back to higher levels and some newcomers are gradually increasing their ODA volumes. There is also a general tendency to shy away from putting business interests first, as illustrated in the trend towards untying of aid, and to give priority to the long-term foreign policy goals of a stable world and international social justice. During the 1990s and beyond, the commitment to a deeper co-operation with governments in developing countries as well as with other European donors has gained momentum. For donors and recipients alike, the Millennium Development Goals indicate the agreed direction for the years ahead.

VI. ACKNOWLEDGEMENTS

The research and publishing project has received much help and goodwill. The Netherlands' Ministry of Foreign Affairs has generously contributed to the project by covering major expenses involved in organising the workshop in Nijmegen and some of the costs related to the publication of this volume. This grant is acknowledged with pleasure and appreciation. The two co operating institutes – CIDIN and NUPI – have carried costs associated with this process, too. We are also indebted to the institutes with which the authors of the component studies included here are associated for their valuable contributions to this project: the salaries of the researchers involved and the costs of travel to attend the workshop in Nijmegen. From whatever perspective, their contributions represent, by far, the most valuable input in making the project possible, for which we are highly grateful. Special thanks are due to CIDIN for arranging the workshop and for hospitality rendered. The fertile interaction between some key administrators of development assistance and researchers at that workshop resulted in a significant strengthening of the papers presented in that workshop and now presented in this volume.

From the very beginning, the project had the privilege of being integrated into the research programmes of the two co-operating institutes – CIDIN and NUPI – and has benefited from their institutional support. The present volume has been edited from and produced at NUPI. The project has, in its final stage, benefited from the editorial support of Eilert Struksnes and the efficient secretarial support of Liv Høivik, both at NUPI. Liv Høivik has produced the camera-ready manuscript. Warm thanks also go to Margaret Cornell for assistance in editing the present volume, language revision and for providing it with an index.

REFERENCES

Berlage, Lodewijk and Olav Stokke (eds), 1992, *Evaluating Development Assistance: Approaches and Methods*, London: Frank Cass, EADI Book Series 14.

Clay, Edward and Olav Stokke (eds), 1991, 1995 (second edition), *Food Aid Reconsidered: Assessing the Impact on Third World Countries*, London: Frank Cass, EADI Book Series 11.

Clay, Edward and Olav Stokke (eds), 2000, *Food Aid and Human Security*, London: Frank Cass, EADI Book Series 24.

Cox, Aiden and John Healey, with Paul Hoebink and Timo Voipio, 2000, *European Development Cooperation and the Poor*, Houndmills, Basingstoke, Hampshire: Macmillan Press and Overseas Development Institute.

Development Committee, 2004, Statement by Mr. Richard Manning, Chairman, OECD Development Assistance Committee (DAC), (Joint Ministerial Committee of the Boards of Governors of the Bank and the Fund On the Transfer of Real Resources to Developing Countries), Sixty-ninth Meeting, Washington, DC, 25 April (DC/S/2004–0001).

European Commission, 2002, Commission Staff Working Paper, Preparation of the International Conference on Financing for Development (Monterrey, Mexico, 18–22 March), Brussels: COM (2002) 87, 13 February.

Forster, Jacques and Olav Stokke (eds), 1999, *Policy Coherence in Development Co-operation*, London: Frank Cass, EADI Book Series 22.

Holthus, M. and D. Kebschull (eds), 1985, *Die Entwicklungspolitik wichtiger OECD-länder* (The Development Policy of Important OECD Countries), 2 volumes, Hamburg: Verlag Weltarchiv.

OECD, various years, *Development Co-operation, Efforts and Policies of the Members of the Development Assistance Committee,* Paris: OECD DAC.

Stokke, Olav (ed.), 1984a, *European Development Assistance*, Volume I: *Policies and Performance,* Tilburg: EADI Secretariat, EADI Book Series 4.

Stokke, Olav (ed.), 1984b, *European Development Assistance*, Volume II: *Third World Perspectives on Policies and Performance*, Tilburg: EADI Secretariat, EADI Book Series 4.

Stokke, Olav (ed.), 1988, *Trade and Development: Experiences and Challenges*, Tilburg: EADI Secretariat, EADI Book Series 7.

Stokke, Olav (ed.), 1991, *Evaluating Development Assistance: Policies and Performance*, London: Frank Cass, EADI Book Series 12.

Stokke, Olav (ed.), 1995, *Aid and Political Conditionality*, London: Frank Cass, EADI Book Series 16.

Stokke, Olav (ed.), 1996, *Foreign Aid Towards the Year 2000: Experiences and Challenges*, London: Frank Cass, EADI Book Series 18.

UNCTAD, various years, *World Investment Report*, Geneva: UNCTAD.

UNCTAD, 1972, *Handbook of International Trade and Developopment*, New York: United Nations.

World Bank, 1998, *Assessing Aid, What Works, What Doesn't, and Why*, A World Bank Policy Research Report, New York: Oxford University Press and the World Bank.

1

The Changing International and Conceptual Environments of Development Co-operation

OLAV STOKKE

I. INTRODUCTION

What motives have been driving international development assistance?[1] What have been the main purposes? These are the principal questions to be addressed in this chapter, leaving aside other equally important questions, such as the most crucial one of all: Does aid work?

Development assistance is part of a multitude of contexts, and the discussion in this chapter will to a large extent be limited to two such environments, namely the international setting in which aid is operating, and the philosophical setting – the development paradigms or

---

Note of acknowledgement. A draft of this chapter has benefited from the scrutiny of a workshop organised by the EADI Working Group on Aid Policy and Performance in Nijmegen in early July 2002. I am grateful to participants for valuable comments. Warm thanks also go to Guido Ashoff, Paul Hoebink and Robrecht Renard for comments on a later draft. The usual disclaimers apply.

1. The term has shifted over time, and these shifts have had ideological connotations. The term 'development assistance' – 'aid' for short – reflects a donor perspective on the transfer of resources according to the DAC definition of 'official development assistance' (ODA). The term 'foreign aid' is part of this tradition, although it is broader and may also include assistance that is not normally associated with 'development'. 'Development co-operation', in contrast, alludes also to the recipient party, emphasising the co-operative aspects of aid: it is the term in common use during the last three decades. In spite of their different connotations, these terms are used as if they were synonymous in this chapter.

The DAC definition of ODA: 'GRANTS or LOANS to countries and territories on Part I of the DAC List of Aid Recipients (developing countries) which are: Undertaken by the official sector. With promotion of economic development and welfare as the main objective. At concessional financial terms (if a loan, having a GRANT ELEMENT of at least 25%). In addition to financial flows, TECHNICAL CO-OPERATION is included in aid. Grants, loans and credits for military purposes are excluded' [*OECD, 2003: 322*].

ideologies, with a focus on the role of aid – leaving more or less aside a lot of other interesting aspects associated with development co-operation. Both the international setting and the development paradigms are changing over time. In this chapter, the main focus will be on the 1990s and beyond, although long lines will be drawn. The purpose is to provide a general contextual setting for the other chapters in this volume.

The motives for providing development assistance vary from one actor to another and they change over time. Domestic actors representing a variety of values and interests influence the aid policy of a donor country. The aid policy is also part of the foreign policy of a donor country, thus adding actors and concerns influencing the policy outcome. These concerns may vary from one donor country to another.

The domestic arenas of donor countries are not the only arenas for policy formation within this policy area, nor are domestic actors the only actors: international organisations, both regional and multilateral ones and international non-governmental organisations (INGOs) as well, influence the aid policy of a donor country, and so does the aid policy of other donors. Public and private sector actors within these various arenas are not alone in setting the political agenda: in our time, the media may be equally important in this regard.

Tracing predominant motives, we shall start with the international setting. We shall first try to identify the motives of the main players when international development assistance was first conceived of in the late 1940s and early 1950s. Four decades later, a new transformation of the international system took place. The focus will be on how this transformation affected development assistance and aid policy (section II). Then the trends in development thinking will be traced, again with the main focus on the 1990s and beyond and the role assigned to development assistance (section III). This leads us to the international development agenda of the 1990s and beyond (section IV), before concluding with some general observations (section V).

II. THE INTERNATIONAL ENVIRONMENT: AID AND INTERNATIONAL POLITICS

1. The Beginning

Development assistance may offer itself as a response to challenges in the arena of foreign and international politics – a tool in states' foreign policy. In the late 1940s and early 1950s, when assistance to promote economic development in the South first emerged as an idea, the international system was in transition. Two major trends characterised the period: the decolonisation process and the process leading towards a bipolar international system. During the evolving Cold War, these processes became interlinked.

The decolonisation process was driven by ideas of freedom, democracy, equity and justice. These were old ideas in Europe, inspired by the French Revolution. They were central in the political, social and economic development within most countries, whether colonial powers or not, although developments were not always linear. However, these ideas seemed to be mainly for internal consumption, not for export; this applied, in particular, to the colonial powers.

In the late 1940s, a world war had recently been fought against regimes with totalitarian ideologies (in Europe, Nazi Germany and Fascist Italy) under the banner of these ideas. More mundane national interests had also been strongly involved, including the most vital one; namely, to survive as independent nations. In this struggle between ideologies, soldiers from the colonies also played an active part.

After World War II, it was therefore not unproblematic to close a chapter in history and return to pre-war 'normality'. The demands for freedom (self-rule, independence) met with strong resistance from most colonial powers. After more than five years of war, however, their strength had been weakened and they were confronted with the immense task of reconstructing and rebuilding their own economies.

The demands for independence were not met in the same way by all colonial powers. Their colonial policies and administrative systems differed, and such differences also influenced the line they took when confronted with the demand for sovereignty.[2] Some European metropoles followed a strategy of adaptation, agreeing to constitutional reforms that involved various degrees of self-rule, but not independence – although sovereignty twinkled in the future. This was the major strategy of the British. Gandhi turned out to be the politician who was most effective in playing on the domestic values of the colonial power in the fight for sovereignty. Others did not succeed in a similar way. In North African countries with extensive French settlements, the demands for independence were crushed by French mili-

---

2. A distinction can be made between the British policy of indirect rule (relying on the traditional systems and rulers that were subordinated to a colonial superstructure) and the French policy of direct rule (an administrative system modelled on the French system). The Portuguese government considered its colonies integral parts of Portugal.

tary power. In general terms, colonial systems may be differentiated into those applying indirect rule (most of the British Empire) and those exercising direct rule (most of the other colonial powers) on the one hand, and into those with large settler communities and those with small settler communities, on the other. In colonies with large settler communities, the transition turned out to be both difficult and violent. There should perhaps also be a differentiation between colonies rich in natural resources and those lacking such assets: wealth of natural resources did not always turn out to be a blessing, as developments in Congo (Belgium) and Angola (Portugal) can testify.

Given its ideological background, the Salazar regime in Portugal was not particularly impressed by appeals for sovereignty based on the values referred to above. Several of the Portuguese colonies, in Africa and elsewhere, were characterised by extensive Portuguese settlements, which complicated the decolonisation process. Portugal's economy (and its self-esteem) were dependent on incomes emerging from its colonies. The liberation struggle turned out to be violent with strong repercussions for the development process. In addition, with regard to the remaining European colonial powers, such as Belgium and the Netherlands, the decolonisation process proved difficult and involved conflicts, with particular reference to Congo and Indonesia. South Africa constituted a problem of its own, especially after the government of the independent republic based its policy on apartheid and white minority rule.

The United Nations gradually emerged as both the institution that was driving the decolonisation process and the prime arena for conflict resolution involving conflicts between the colonial powers and the new states that were emerging as a result of this process.

From the international setting of the late 1940s and early 1950s, it is possible to identify the following main sources of the idea of development assistance:

(i) The Marshall Plan involved huge financial transfers from the US government to European countries after World War II to enable them to rebuild their economies and recover from the damage caused by five years of war [*Lancaster, 1993; Clay, 1995*]. This generosity was coupled with enlightened self-interest: to create an external economic and political environment that coincided with the ideological aspirations and the economic and security interests of a great power aspiring to global hegemony [*Stokke, 1996*]. Typically, the USSR and countries under its influence declined the offer, with its associated conditions.

(ii) The Cold War. Development assistance became an instrument in the international power play between the emerging superpowers in the bipolar international system evolving after World War II. During the late 1940s and early 1950s, this justification for aid was stated openly: for the Western powers, promoting democracy and an open market economy and containing communism from penetrating the South were given as the prime arguments for aid. The US stood out as the most generous aid donor, in both absolute and relative terms. Aid was also used as a foreign policy instrument by the USSR and its allies in the Eastern bloc, and targeted on strategically important governments and social structures in the South.[3]

(iii) Maintaining relations with former colonies. For many of the colonial powers that gradually agreed to self-rule for their former colonies – under pressure from nationalist movements in the emerging countries concerned, the international society (the UN and the emerging superpowers – the pressure from the US had the strongest impact), and increasingly also from within the colonial powers themselves – aid became an instrument for reinforcing economic interests and cultural relations established under colonial rule. This applied even where the struggle for sovereignty had involved violence and even war. Thus, in the 1960s and early 1970s, the newly independent countries that had previously been their colonies turned out, almost without exception, to be the main recipients of the bilateral development assistance of the former metropoles (see Stokke [*1984a*] and several other chapters in Stokke (ed.) [*1984*]). For the colonial powers, bilateral aid became an instrument in a strategy to maintain their influence in former colonies and dependencies.

(iv) The crucial role of the UN. The UN came to play a crucial role in formulating the objectives and norms of development assistance and in promoting it during these early years, both in its role as an arena for international politics and as an actor in its own right.[4] The main

---

3. Examples are legion. As formulated by Crawford Young [*1991:15*], the 'competitive pursuit of strategic clients in the superpower cold war competition led both sides in the conflict to help sustain regimes supplying geopolitical services whose damage to polity and economy is now universally acknowledged (Mobuto Sese Seko for the Americans, Mengistu Haile Mariam for the Soviets). The possibility of relying on external protection to preserve power removed whatever shreds of accountability might exist in various patrimonial autocracies. ... With rare exceptions ... human rights abuses in Africa until the 1980s were ignored, and autocracy tacitly supported.'
4. In the autumn of 1948, the UN General Assembly adopted two resolutions with a bearing on aid to the so-called underdeveloped countries. One resolution expressed the hope that the World Bank would begin to provide development loans to these

actors already identified – the parties to the Cold War and the colonial powers – exerted strong influence in the UN bodies that shaped the policy within this area (ECOSOC and the UN General Assembly). This applied, in particular, to the US government, which played the most prominent role with regard to ideas, initiatives and funding. President Truman, announcing his Point Four Programme for development in 'underdeveloped' regions when initiating his second term in office in January 1949, exerted a strong impact on further developments, with implications beyond the emerging East–West divide. However, in the UN setting these actors were not alone: from the outset, some Asian (India, in particular) and Latin American governments put their stamp on the norms set to guide international development assistance. In the 1950s and increasingly in the 1960s, this 'group' of countries grew rapidly in numbers and gradually came to regard the UN system as an important arena for their foreign policy. For their governments, it was important to protect the sovereignty of recipient governments and to remove the 'flag' from multilateral aid. The development strategies adopted for the first and second development decades (the 1960s and 1970s) established norms that influenced both newcomers in the field of development assistance and the old ones as well.

*

Although, at an overall level, the emerging bipolar international framework affected the foreign policy of nations that were part of the two security alliances, the origin of aid differed from one country to another, varying also with other factors, such as historical relations with one or more developing countries involving trade, culture (in particular, religious bonds) and more integrated relations of the kind described above (colonialism of different brands). Even within industrialised countries, the sources varied extensively, from one social, cultural or economic group to another and from one political party

---

countries; the other made available the first allocation in the UN budget to a regular programme of technical assistance in the fields of economic development and public administration. In January the following year, Harry S. Truman opened his second term as President with an announcement that the United States would embark on a programme for the improvement and growth of 'underdeveloped areas', emphasising the transfer of knowledge, skills and technical know-how through the United Nations and its specialised agencies. Only two months later, the UN Economic and Social Council started the process leading to the creation of the Expanded Programme of Technical Assistance (EPTA), the forerunner of the United Nations Development Programme (UNDP). EPTA started operations in July 1950, financed by voluntary contributions from UN member states. In January 1950, the foreign ministers of the Commonwealth countries met in Colombo and launched the so-called Colombo Plan, arranging for development co-operation within that group of partner countries.

(with its primary basis in one or more of these groups and in different ideologies) to another.

Many industrial countries had no recent past as colonial powers, and some had themselves been under foreign rule and were relatively new as sovereign states. Their more recent relations with the South were based on trade and (Christian) missionary activities. In the case of the Scandinavian countries and a number of others, development assistance emerged primarily as a consequence of their commitments as members of the United Nations, although the new relationship was also rooted in commercial as well as humanitarian traditions, including relief operations [Forster, 1984; Stokke, 1984ab]. Such factors played an important part in motivating aid in other industrial countries of the North as well.

2. Foreign Politics and Development Assistance: Competing Paradigms

The Cold War – and the two superpowers' competition for global hegemony – left the governments of emerging countries in the South with two main options: they could align themselves with one or other of the superpowers (or their allies) and pick up the rewards in terms of foreign aid and (if needed) political support; or they could stay non-aligned, playing off one superpower against the other and hoping to pick up rewards from both, while at the same time maintaining a high degree of foreign policy independence. However, not all countries were equally well situated in this regard: the strategic geopolitical positions varied, influencing how attractive they were in the power play between the superpowers. Most countries of the South belonged to the group of non-aligned countries, although their commitment to real non-alignment varied.

(i) The realist paradigm. As observed earlier [Stokke, 1989a, 1996], the aid policy of the superpowers could, during the heyday of the Cold War, best be analysed within the paradigm of *international realism*, the school of thought that dominated the analysis of international relations during the post-World War II era, where aid is seen primarily as a means of pursuing the national interests of the donor.[5]

---

5. The roots of the realist school may be traced back to Machiavelli [1532]. The classic of this approach, however, is the seminal work of Hans Morgenthau [1948]. Important contributions to this school, developing the perspective, have also been made by, *inter alia*, Kenneth Waltz [1959, 1979] and Robert O. Keohane [1984].

In 1962, at a time when development economists held an almost hegemonic position in the development discourse (see section III), Hans Morgenthau developed his political theory of foreign aid [Morgenthau, 1962]. His main message was that the

Within this perspective, what really matters is the power to exert influence over the philosophy guiding actions, and over the actual actions and priorities of other governments. Although national interests may be defined differently by different regimes and the emphasis of what really matters in such a perspective may vary from one government to another, most governments, particularly those aspiring to a hegemonic position globally or regionally, will rank security interests high, along with predominant economic interests. As indicated, the predominant ideology and basic norms and values are part of the extended national interest concept, although they are not always easily captured and are not given the same weight by all governments seeking influence outside their primary domain. Ideology may, however, stand out as the prime national interest within this paradigm for instrumental reasons: its ability to direct the actions of those under its influence.

In its original classical form, even given this broad interpretation, the paradigm has obvious limitations for understanding both the foreign policy of nations in general and their aid policy in particular. As implicit in the title of Morgenthau's seminal work – *Politics Among Nations* – the focus is on interactions between *states*. Increasingly, other actors have entered the scene. *International organisations*, multilateral global organisations and regional organisations with wide responsibilities as well as organisations with more specialised responsibilities, have come to play an increasing role, both in their capacity as arenas for foreign policy and in their own right. As argued by Cohen *et al.*[*1972*], organisations tend to assume power beyond that of their individual members. A more complex reality than the one perceived by classical realism surfaced also within states. While the foreign policy of a country, in the classic interpretation, was the

---

various forms of foreign aid were instruments in the foreign policy of nation states and that '[m]uch of what goes by the name of foreign aid today is in the nature of bribes. The transfer of money and services from one government to another performs here the function of a price paid for political services rendered or to be rendered' [*ibid.: 302*]. The objective of providing aid for economic development, in order to adapt aid to the prevailing ideology, distorted it and made it less effective as an instrument in the foreign policy of the donor state, he argued, with particular reference to the foreign policy of the US. 'Bribery disguised as foreign aid for economic development makes of giver and recipient actors in a play which in the end they may no longer be able to distinguish from reality' [*ibid.: 303*]. Insofar as this kind of aid affected the political stability of recipient countries by undermining the political and economic status quo, aid for economic development might be counter-productive as a foreign policy instrument. And foreign aid to states (the ruling groups) might, on the other hand, be counter-productive *vis-à-vis* the objective of economic development: such types of foreign aid strengthened the very factors that stand in the way of economic development, he argued [*ibid.: 308*].

domain of a handful of major players (the minister and ministry of foreign affairs and the president (prime minister) and the president's (PM's) office in particular), a host of other players gradually entered the scene – including ministries with a more particular 'domestic' profile, institutions and organisations identified with civil society and the private sector in particular, but also 'nations', within but also transcending the borders of the nation state, associated with culture (ethnicity). Such actors were not easily handled by classical realism.[6]

Besides, political actors do not always act with the sole purpose of furthering their individual or collective self-interest. Interests may be defined in the short and long term, and these may be conflicting. Long-term interests are part of the national interest concept as defined by the realist school; however, how well they are integrated may depend on the time perspective and the kinds of interest. Interests may also be given a broader interpretation than the individual or collective interest of the community concerned. It is more difficult to see how such interests may be contained within the national interest concept of the classical realist school. It follows that it is unclear to what extent international common goods, in the 1980s and 1990s increasingly associated also with the concern for the global environment and poverty alleviation, belong to the core concept as defined by the realist school.

During the Cold War era, the foreign aid policy of the superpowers may best be understood and analysed under this paradigm. The aid policy of the United States has been driven by its security interest, observers have argued [*Gills et al., 1993; Griffin, 1991; Lancaster, 1993; Nelson and Eglinton, 1992; for an overview see also Ruttan, 1996*].[7] Even so, more altruistic (ideal) motives have also lain behind

---

6. For a critique of the perspectives of classical and neo-classical realism, with an emphasis on the role of values in international relations, see, *inter alia,* Lumsdaine [*1993*]. For a critical appraisal of the feasibility of this 'dominant body of theory ... with its core image of states in a system of anarchy' in the post-Cold War era, see Holm and Sørensen [*1995*].

7. According to Carol Lancaster [*1993: 13*], security concerns were the first and remained the principal pillar on which the US aid programme has rested. In the late 1940s, the US efforts had as primary objective 'to bolster the economies of Greece and Turkey in the face of communist-led insurgency in the former country and Soviet pressures for territorial concessions from the latter country. The Marshall Plan soon followed aid to Greece and Turkey and was also motivated by the concern that lagging economic recovery in Western Europe would lead to an expansion of communist influence there. From the very beginning, US foreign aid has rested very heavily on the need to respond to external threats to US and world security emanating from the Soviet Union. Indeed, it seems likely that without such threats, an insular American public and a sceptical Congress would have never acquiesced to a sizable foreign aid programme in the years following the Second World War.' Robert Wood [*1986*] argues in the same vein with reference to the US policy towards Europe immediately

the US aid package, and the mix of realist and altruistic motives has varied over the years with the administration in power and its interpretation of the international challenges with which the US was confronted.

(ii) The humane internationalism paradigm. The aid policies of other Western small and medium-sized powers, whether members of the Western security organisation (NATO) or not, were also characterised by a blend of altruism and self-interest: economic interests certainly applied when the aid policy was formed. However, as argued earlier [*Stokke, 1989bc, 1996*], for these countries self-interest, whether related to security or the economy, was not the main determinant of their aid policy, which could be explained primarily as a reflection of the basic domestic values prevailing, with particular reference to those connected with the welfare state. For them, traditions related to relief operations and (Christian) missionary activities that combined social activities (education, health) with the dissemination of the gospel set the agenda, along with the ideology of solidarity associated with socialist and social democratic traditions. Several of the new donor countries that became the most ardent providers of development assistance in the late 1960s and the 1970s had governments based on a Christian or social democratic tradition – or a combination of the two. This applied, in particular, to the Scandinavian countries and the Netherlands. The development assistance was driven mainly or partly by *values and norms* on which these traditions were founded. Such norms are not for domestic consumption alone and do not stop at the borders of the nation state.[8]

Paradigms of the realist tradition fail to capture these basically altruistic features of development co-operation – still an integral part of the donor countries' foreign policy, even increasingly so in several countries. This triggered the formulation of an alternative paradigm; namely, that of *humane internationalism* [*Stokke, 1989a, 1996; Pratt,*

---

after WWII: two major objectives are identified; namely, the containment of communist influence and the reintegration of the European economies into an open international economic system.

For analyses of the emerging development thinking in the US as related to aid, see Krueger *et al.* [*1989*], particularly chapters 2 and 3 (Krueger and Ruttan [*1989ab*]), and Ruttan [*1996*].

8. For an excellent overview and discussion of the ethical case for development assistance, see Roger C. Riddell [*1987*], a classic on foreign aid. Riddell presents the moral basis established for aid and confronts it with its main critics who represent other value positions or perspectives to the road towards 'development' and/or different interpretations of the 'reality'. This discussion addresses a core issue in the aid discourse, and this debate continues. See also Gasper [*1999, 2004*].

1989, 1990], the core of which is an acceptance of the principle that citizens of industrial nations have moral obligations towards peoples and events beyond their borders, which, in turn, has a bearing on the duties of governments. Cosmopolitan values constitute the foundation. Humane internationalism implies sensitivity to such universal rights as those included in various human rights conventions, and the obligation to refrain from the use of force in pursuit of national interests. Within a North–South context it implies, in particular, responsiveness to the needs of the South with regard to economic and social development, but does not stop at that point: civil and political rights are also considered to be universal. Such ethical obligations are combined with what is considered to be in the best long-term interests also of the industrialised countries concerned, related to cherished values associated with international common goods such as enhanced equity, social justice (welfare), internal and international stability and peace.[9] Although the emphasis is on values and long-term common interests (international common goods), the ethical thrust is combined with, and considered to be instrumental in, the promotion of the longer term, overall interest of the industrial countries, thus bridging the gap to the realist approach.

*

Both paradigms were developed as analytical frameworks for the study of states' foreign policy more generally (the realist school) or particular components of this policy, especially aid policy (humane

---

9. For a comprehensive description and discussion of the humane international concept, see Stokke [*1996: 22–5*]. In a North–South context, it implies 'an acceptance of an obligation to alleviate global poverty and to promote economic, social and political development in the South; a conviction that a more equitable world would be in the best long-term interest of Western, industrial nations; and the assumption that meeting these international responsibilities is compatible with the maintenance of a socially responsible national economic and social welfare policy. It is ... associated with a set of objectives, such as the promotion of economic, social and political development in the South, including the promotion of human rights over a broad spectrum (also civil and political rights) and the alleviation of human suffering' [*ibid.: 23, n. 7*]. It takes many forms: liberal internationalism, reform internationalism and radical internationalism. In contrast to liberal internationalism, which combines the core concept of humane internationalism with a strong commitment to an open, multilateral trade system, reform internationalism does not consider the market to be the most efficient instrument for determining production priorities or for solving the problem of income distribution. Although gradualist in approach and basically belonging to a liberal tradition, it favours state and inter-state intervention to correct inequities and alleviate poverty. Radical internationalism, in contrast, presupposes an obligation to show solidarity with the poor and oppressed in other countries, ultimately even at the expense of more narrow interests in the home country. It is associated with a set of objectives such as the attainment of full economic, social and political equity, coupled with ecological concerns, and favours state and inter-state intervention where such objectives are involved.

internationalism), within a bipolar international system. How well can the evolving policy of the post-Cold War era be analysed within one or the other of these paradigms – or both?

3. A New World Order: Implications for Aid

Towards the end of the 1980s, the international environment changed dramatically. The collapse of the Soviet bloc and the disintegration of its dominant power, the USSR, brought to an end the bipolar international security system that had obtained since the late 1940s. For more than 40 years, 'East' and 'West' had been competing for global hegemony, involving all fields – security, economy, ideology (including development paradigms) and culture in the widest sense of the concept. What did the new international framework mean for North–South relations, including development co-operation?

The new system was characterised by an almost hegemonic position of the 'West' with regard to economic and military power, and ideological power as well. A further feature was the trend towards increasing globalisation of the economy. The new framework conditions, in combination with the revolution in technology particularly with regard to communications, also facilitated globalisation within other fields, including culture.

The increasing globalisation under the new world order, although under way for decades, probably represents the change with the strongest impact on the situation for developing countries. Globalisation, particularly in respect of the economy, does not mean the same for all parties involved in the process – whether in the North or the South: within and between countries, it also opens up increased marginalisation. In the intense discourse on the blessings of globalisation, which will not be dealt with here, ideological positions are sharply divided. However, there seems to be broad agreement that less developed poor countries are among those benefiting the least and in need of protective measures, particularly in the shorter term.

Within the general trend of globalisation, where the emerging international trade regime institutionalised in the World Trade Organisation represents both a symbol and a driving force, an equally important trend towards regional integration is typical of this period, with particular reference to economic integration, resulting in regional centres of power and creating a framework for inter-regional competition. In Europe, important strides have been made to extend the scope of regional co-operation both geographically and in depth – towards a political union that also includes security and a joint foreign policy.

The US stands out as the only remaining superpower, particularly with regard to military power. However, in the new international setting, military capability has lost some of its previous value in terms of political power, although a spill-over from military to political and economic power will always be manifest. The events of 11 September 2001 reversed the situation in this regard.

Some of the more direct effects for aid policy and performance of this transformation of the international system are interlinked. They will be briefly identified under the following heading. Some of them will be explored and developed further in section IV, with particular reference to the policy dimensions.

(i) A new security situation. In the previous bipolar world order, violent conflict was, to a large extent, conceived of as conflicts between states. In West and East alike, the main security concern was related to the situation in Europe. However, the rest of the world was affected in several ways through systematic efforts to build formal and tacit security alliances and reciprocal efforts by both parties to contain each other's influence in terms of security arrangements and political alignments. Violent conflicts in the South between – and at times also within – states were sometimes inspired by the parties to the East–West conflict, in particular the two superpowers (war by proxy) or, in a few cases, incurred active involvement beyond supplies of finance and/or military equipment. However, occasionally the risk of a conflict escalation and a more direct superpower involvement also led them to moderate, and even contain, their clients. The vast majority of Third World governments were committed to the non-aligned movement, although some also cultivated close relations with one or other of the parties in the East–West conflict.

In the post-Cold War era, the pattern changed from predominantly inter-state to intra-state conflicts, at times involving regions. This security situation can only up to a point be attributed to the end of the Cold War and the disappearing 'discipline' exerted by the hegemonic powers. However, with the disintegration of the Soviet bloc, conflicts *within* states, which had been contained under the previous system, now came out into the open, involving both social conflicts and competition for power within the new entities, and conflicts rooted in a nationalism that for generations had survived Soviet rule. Multi-ethnic state formations – where each nationality had a history of its own and a long-standing record of conflict with others in the same region – were particularly vulnerable when politicians, in their struggle for power, exploited latent ethnic antagonisms. Tensions based on such rivalries add fuel to more 'normal' conflicts involving distribution of resources, political leadership and tensions created by misgovernment, abuse of power, discrimination against particular minority groups and a policy of social exclusion.[10] Developments in the former Yugoslavia illustrate the point. Africa also is all too rich in intra-state and regional conflicts that follow cultural dividing lines, which often transcend state borders. The conflicts in the Horn of Africa, in southern Sudan, in the Great Lakes region in Central Africa and in Congo are but a few examples.[11] However, as already noted, the main sources of many of these conflicts cannot be attributed to the new global security system: they must be sought in 'domestic' politics and conflicts. All the same, they had a strong impact on the post-Cold War aid policy – on the geographical distribution, the purposes and objectives, and the forms of aid.

Intra-state wars, like other forms of violent conflict, were increasingly seen as a major threat to development. As stated elsewhere, 'violent conflicts may, in the course of weeks, destroy material resources that have taken generations to build, cause immense human suffering, make millions of innocent people refugees in their own or neighbouring countries, cause states to collapse and wreck societies' [*Stokke, 1997: 196*]. In the 1990s, therefore, peace was increasingly seen as a precondition for development to take place: development depends on an enabling environment that includes the absence of violent conflict. Up to the late 1980s, the concept of security had been interpreted as security of territory from external aggression, as protection of national interests in foreign policy, or as global security, in the first place from the threat of nuclear holocaust. With the end of the Cold War, other aspects of security came to the fore. The emphasis has increasingly been on an extended security concept, that of human security, as defined in the *Human Development Report 1994* [*UNDP, 1994*].[12]

---

10. The new focus affected perceptions of the concept of 'peace'. As observed by Senghaas [*1995: 307*], the main point 'is to find and institutionalize lasting forms of constructive and nonviolent conflict management' – to design mechanisms for securing internal peace in the face of an always latent imminent civil war. He distinguishes between a traditional and a modern peace concept, both related to entirely different types of societies: 'the one non-politicized/elitist, the other broadly politicized on a mass basis' [*ibid.: 316*].

11. This argument is developed elsewhere [*Stokke, 1997: 203ff.*]. A distinction should be made between cultural patterns ('nationalism', religion) as a potential source of conflict and actors (including 'conflict entrepreneurs' [*Eide, 1997: 44*]) who exploit such conflict potential.

12. The concept of human security has two aspects: safety from chronic threats such as hunger, disease and repression, and protection from sudden and hurtful disruptions in the pattern of daily life. Such security allows people to exercise their expanded

The new security situation and the extended security concept posed new challenges to aid, to be explored further in sections IV and V.

(ii) A new ideological power balance, involving economic policy. Within the West, neo-liberal ideologies had been gaining strength during the 1980s, influencing the policy of most West European governments, with particular reference to the state–market dimension. This ideological development was only indirectly related to the old security situation. In the late 1980s, the pendulum was on its way back. The international system transformation of the early 1990s strengthened these trends globally. This will be discussed in the following section.

(iii) A lost rationale for aid. During the Cold War, development assistance became an instrument in the hegemonic competition for power. As argued above, the aid policy of the United States has been driven by security concerns. With the Cold War over, the main rationale for providing development assistance had therefore disappeared for the remaining superpower: it was no longer as necessary to buy allies and strategically important bases or to ensure that these did not fall into the hands of the adversary. However, security interests were never the only ones driving US aid. There has always been an ethical strand and US interests in the South have always included economic interests, and access to strategically important resources (oil and gas) in particular.[13]

---

choices and develop their capabilities. The absence of such security undermines the processes of development and may lead to social disintegration and humanitarian catastrophe. The threats to human security, as identified by the UNDP Report [*1994*], relate to the following categories: economic security (unemployment, job insecurity, inequality and poverty), food security (inadequate availability of food and access by all people), health security (communicable diseases and pollution-related illnesses, drug abuse), environmental security (degradation of air, water, soil, forests), personal security (conflict, terrorism, violence against women and children), community security (ethnic clashes) and political security (violation of human rights).

13. These arguments have been developed elsewhere [*Stokke, 1996: 48ff.*]. For an insightful analysis of the evolving US aid policy and the pillars on which it is/should be based towards the end of the 1990s, see Lancaster [*2000ab*]. The security rationale is reinvented, but is now adapted to a new security setting and also to an extended security concept. She argues that '[t]he most basic challenge facing the United States today is helping to preserve peace. The end of the Cold War eliminated a potential threat to American security, but it did not eliminate conflict.

Peacemaking will therefore be an important focus of U.S. diplomacy in the twenty-first century. As in the past, the United States is likely to take a lead in conflict prevention and peacemaking in regions that have high priority but lack effective alternatives to U.S. leadership: parts of Europe, the Middle East, countries near U.S. borders and the Pacific Rim, especially Korea and Taiwan. Foreign aid will become more important to peacemaking diplomacy than before – as a symbol of U.S. engagement,

It should be added that although substantial in absolute terms, the ODA of the US in the 1980s had stayed at a low level in terms of its share of GNP: slightly above 0.2 per cent. In the early 1990s, it declined even further – to 0.10 per cent of the gross national income (GNI) in 1995, and it remained at about that level during the second half of the 1990s.[14]

However, the so-called war against terrorism after 'September 11' represented a new turning point, visible already in the 2002 performance: the USA increased its ODA to 0.13 per cent of GNI and to 0.14 per cent of GNI in 2003 [*OECD, 2004: Table 14; Development Committee, 2004: 2*].

(iv) Shrinking ODA. In the years that followed the end of the Cold War, the total amount of ODA was reduced as was its share of the gross national product of the OECD countries. Because of its huge GNP, the decline in US aid also brought the DAC average down. In the 1980s, it had varied at around 0.35 per cent of GNP. In the early 1990s, it tended towards stagnation and decline. In 1995, the DAC average had declined to 0.27 per cent of GNI, and this average declined further in the following five years, varying between 0.25 and 0.22 per cent.[15]

---

an incentive for warring parties to negotiate peace, and a source of resources to help countries recover from the legacies of war' [*2000b: 75–6*]. However, since this was written, a new Republican Administration has taken over and perspectives have changed (Professor Lancaster had previously served in various positions in Democratic Administrations, including as deputy administrator of USAID (1993–96)); 11 September 2001 added 'new' security concerns. At an early stage, development assistance became a part of the US strategy to meet that challenge: the war against terrorism.

14. US development assistance declined from 0.2 per cent of GNP in 1991 to 0.15 per cent of GNP two years later. This decline represented US$2 billion (current prices). In 1992 prices and exchange rates, US aid in 1993 declined by US$2,325m. from the previous year [*OECD, 1995a: Table 8*]. In 1995, it amounted to $9.4bn. and varied between $9.4bn. and $6.9bn. the following four years (current prices) [*OECD, 2001: Table 4*]. In 2000, it was close to $10bn., and increased to $11.4bn. the following year, making up 0.10 and 0.11 per cent of GNI, respectively [*OECD, 2003: tables 12 and 13*].

15. In 1989, ODA made up 0.32 per cent of GNP (DAC countries), rising to 0.33 per cent during the two following years and then declining to 0.30 per cent in 1993. In 1992 prices and exchange rates, total ODA amounted to US$52bn. in 1985, $53bn. in 1986 and 1987, $57bn. in 1988, $56bn. in 1989, and $58, 60 and 61bn., respectively, in the following three years. In 1993, it declined to $56bn. [*OECD, 1995a*]. In 1995, their ODA amounted to $58.9bn., declining to $55.6bn. in 1996 and 48.5bn. in 1997, before increasing to $52.1bn. in 1998, and $56.4bn. in 1999. It then fell again: $53.7bn. in 2000 and $52.3bn. in 2001 (current prices and exchange rates), amounting to 0.22 per cent of GNI [*OECD, 2001: Table 4; 2003: tables 12 and 13*].

However, in the new century the tide turned. The new growth of international development assistance, in absolute as well as in relative terms (percentage of GNI), had many and different reasons, varying from one donor country and region to another. In 2001, the DAC countries contributed US$52bn. in ODA, 0.22 per cent of their total GNI. The following year, the amount increased to US$58bn., 0.23 per cent of their GNI, increasing further to US$68.5bn., 0.25 per cent of GNI, in 2003 [*OECD, 2004: tables 12 and 13; Development Committee, 2004: Table 1*]. The increase in US foreign aid, for reasons alluded to above, contributed to this new growth. For most other donor countries, the awareness of the need to fight global poverty and the solemn commitment by world leaders meeting in the UN Millennium Assembly in September 2000 to do so, followed up in Monterrey in March 2002, probably represented a stronger incitement for increasing the ODA than 'September 11'.

The performance of the individual DAC countries during the 1990s and the first two years of the new century varied extensively, as may be seen from tables IN.1 and IN.3 (Introduction, this volume). Some of them (Denmark in particular) increased their aid as a share of GNP, while the ODA of others declined for a variety of reasons. For most European governments, the main rationale for providing development assistance has not been security concerns related to the East–West conflict. This may explain why their ODA performance in the early 1990s, after the Cold War, became less affected when the 'security rationale' for aid associated with the East–West divide disappeared.

However, earlier hopes, even expectations, of a 'peace dividend' – whereby resources previously spent on security would be converted to development assistance after the end of the Cold War – did not materialise during the early 1990s.[16] On the contrary: in the 1990s, development assistance declined in both absolute and relative terms.

---

16. In the period 1989–94, world military budgets were reduced by about 23 per cent, representing around US$200bn. During this period, aggregate global aid increased by US$1bn. [*German and Randel, 1995*].

The idea of a peace dividend was that (part of) the saving that resulted from reduced military expenditures should be used for development. In the so-called Stockholm initiative [*PMO, 1991*], prominent leaders from all over the world proposed (i) a pledge by the governments of industrialised countries to allocate a specific part of the peace dividend for international co-operation, investments in the environment and development, for instance one-third of the military savings, which would release US$30–40bn. yearly; and (ii) a commitment by governments in the South to substantially reduce their armed forces, with the purpose of creating a peace dividend to be invested in human development. This idea was later picked up by the UNDP [*1994*].

(v) A lost source of development assistance. The collapse of the Soviet bloc had in itself a number of consequences that affected, directly or indirectly, the resources available for development cooperation globally. An almost immediate effect was that a source of development assistance disappeared. Up to the mid-1980s, aid from Central and Eastern Europe and the USSR was substantial, although it was not easily measured and was concentrated on a relatively small group of countries.[17]

The internal distribution of resources within the former Soviet Union aside, the main losers were a group of governments of 'countries in socialist development' (such as Cuba and Vietnam) and other political structures in the South that were aligned with the East. In the latter half of the 1980s, the financial support they received deteriorated and dwindled to nothing in the early 1990s. Although this support (including trade arrangements) was important, their loss of political 'capital' probably hit them even more strongly. A global political ('socialist') system – in which these governments and political movements were not only recipients but also partners – fell apart with the disintegration of its hegemonic power.

(vi) Aid providers turned into aid recipients. The disintegration of the Soviet bloc revealed, in addition to environmental degradation, a glaring poverty in some of the countries in Central and Eastern Europe, and in the former USSR as well. Several of the new states that emerged were equally poor: they revealed characteristics associated with developing countries. Previously, the USSR had been responsible for facilitating economic and social development, if necessary by the transfer of resources from the centre. In the new situation, the donor community saw poverty alleviation in these new countries and their development needs as an additional challenge.

(vii) Aid for political and economic transition in Central and Eastern Europe. In the early 1990s, the West regarded the transition system in Central and Eastern Europe as a political revolution that needed to be sustained by means of economic support. Normally, government budgets are not easily adapted to new tasks, almost regardless of their

---

17. Aid statistics from the Second World have been contested, particularly with regard to value assessments. According to OECD statistics [*OECD, 1995a: Table VI-1*], net disbursements from Central and Eastern Europe amounted to US$2,827m. in 1980 (of which $2,313m. came from the USSR), $3,610m. in 1985 (of which $3,065m. from the USSR), and $3,386m. in 1989 (of which $2,960m. from the USSR). In 1990, this aid declined to $2,178m. ($2,000m. from the USSR), and these amounts were halved the following year. For later years, no estimates were provided.

importance. This applies, in particular, to international tasks, where the ODA budget (for small and medium powers, in particular) easily becomes exposed as the major source to be relied on. Despite frequent statements to the contrary, such transfers were to compete for resources with the development assistance to traditional partners. For some donor countries, these financial transfers became an addition to traditional development assistance. However, both directly and indirectly, the new recipients were competing with the traditional recipients of development assistance for shrinking aid resources.[18]

(viii) ODA norms adapted to a new international setting. These developments, in turn, also affected the definition of development assistance, as set by the OECD Development Assistance Committee, with particular reference to countries eligible for ODA. Several DAC member governments that provided support for the 'countries in transition' wished to register their contributions as ODA in the OECD statistics. Already in 1990 it was suggested that Poland and Hungary should be included among countries eligible for development assistance. However, the majority of DAC members objected to this. Later,

---

18. Official aid disbursements from OECD countries to Central and East European countries (CEEC) and to the new independent states (NIS) during the early 1990s were as follows: in 1991, US$7,259.7m., declining to US$7,078.9m. in 1992 and US$6,900m. in 1993. These contributions amounted to 0.04 per cent of the gross national product of the DAC countries [*OECD, 1995a: Table IV–4*]. This aid came in addition to the contributions reported to and registered by DAC as official development assistance.

The burden sharing among the OECD countries provides additional information of importance. Germany was the greatest contributor in real terms. In the three years identified (1991–93) it provided about one-third of the total financial transfers (US$2,637.3m., US$3,344.4m. and US$2,416m., respectively). These contributions came on top of the far more resource-demanding task of transforming the economy of former East Germany (DDR). In absolute terms also the United States provided substantial contributions (US$1,832m., US$682m., and US$1,217m., respectively). As a share of GNP, Austria came out on top these years (between 0.18 and 0.22 per cent), followed by Germany (between 0.17 and 0.13 per cent), Sweden (0.15 per cent in 1992 and 1993), Denmark (0.14 per cent in 1993) and Switzerland (0.10 per cent in 1993). The US contribution varied between 0.01 and 0.03 per cent, the French between 0.03 and 0.05 per cent, the British stayed at 0.03 per cent and the Japanese at 0.01 per cent of GNP.

ODA disbursements by bilateral and multilateral organisations for the period 1990–93 to six Central and East European countries that had signed Association Agreements with the European Union have been estimated at US$19.1bn. by Ners and Buxell [*1995*]. Poland alone received US$8.5bn. However, ODA transfers constituted only part of the total financial transfers during these years. The cumulated financial commitments by the G-24 countries and international financial institutions amounted, for all 11 European ex-communist countries, to US$95.8bn. for the period 1990–94, of which US$14.5bn. in grants.

THE CHANGING INTERNATIONAL AND CONCEPTUAL ENVIRONMENTS

several of the Central Asian republics of the Commonwealth of Independent States (CIS) were included.[19]

In other ways also, ODA was redefined to allow for the inclusion of expenses that followed in the wake of the new security situation, with particular reference to conflicts and refugees, and which eroded the previous focus on economic and social development in developing countries.[20] The wave of new refugees accepted into European donor

---

19. A new two-part List of Aid Recipients (countries and territories) was introduced by DAC with effect from 1 January 1994, containing 48 least developed countries (LLDCs) and 17 other low-income countries (LICs) (GNP per capita below US$675 in 1992), including Tajikistan; 60 lower middle-income countries (LMICs) (GNP per capita between US$676 and US$2,695), including Armenia, Kyrgyzstan, Georgia, Uzbekistan, Azerbaijan, Turkmenistan, Kazakhstan, Albania and states of ex-Yugoslavia; 30 upper middle-income countries (UMICs) (GNP per capita between US$2,696 and US$8,355); and 16 high-income countries (HICs) (GNP per capita above US$8,355). The new list represented a modest revision of the previous one, beyond the newcomers: development assistance to the HICs was to be registered as ODA until 1996; at that time, all countries above the threshold were 'to progress towards more advanced status' and six of the most affluent of these countries were earmarked for that status (part II of the list). Part II included 'countries and territories in transition': 13 Central and East European countries (CEECs) and new independent states (NISs) [*OECD, 1995a: L4*]. In DAC statistics, there was a distinction between ODA and official aid (OA) – 'flows which meet the conditions of eligibility for inclusion in OFFICIAL DEVELOPMENT ASSISTANCE, except that the recipients are on Part II of the DAC List of Recipients' [*OECD, 2002: 294*]. The GNI per capita (upper and lower limits) for the various groups are adjusted at regular intervals.

20. DAC opened up for the inclusion of assistance to refugees (provided during the first year they stayed in the donor country) in its ODA statistics. This represented a 'masked' decline in 'real aid'. This issue became controversial and practices differed, as may be illustrated from a Nordic setting. In 1994, Denmark responded to NGO criticism by phasing out this kind of aid from its ODA budget [*Warburg, 1995*]. Sweden, in contrast, increased this form of aid in its 1995/96 budget to 10 per cent of total bilateral ODA (up from SEK665m. to SEK965m.) [*Sandberg, 1995*]. Finland, although reducing the amount budgeted for this purpose for 1995 by 40 per cent from the previous year, still allocated 11 per cent of its total ODA for refugees staying in Finland [*Sundman and Mattila, 1995*]. However, although some protests lingered on, these expenses eventually became a normal and growing item in the ODA budgets during the 1990s.

In addition to refugees, a variety of new purposes were also included in the list of purposes that qualified for inclusion as ODA, further eroding the previous focus on social and economic development in developing countries. These included aid for democratic development, contributions to combat narcotics, participation in UN peacekeeping operations (excluding purely military components), assistance to demobilisation efforts, and the lion's share of contributions to the Global Environmental Facility (GEF).

The DAC report for 2001 [*OECD, 2002: 297*] observes that, while the definition of ODA had not changed for over 25 years, 'some changes in interpretation have tended to broaden the scope of the concept'. The main one was the recording of administrative costs (from 1979), the imputation as ODA of the share of subsidies to educational systems representing the cost of educating students from aid-recipient countries (from 1984) and, as noted, the inclusion of assistance provided by donor countries in the first year after the arrival of a refugee from an aid-recipient country

countries resulted, to a large extent, from conflicts in Central Europe, and this affected the geographical distribution of aid, away from the traditional recipients. The real value of ODA as a means for development had also been depreciated because of an increasing share being used for debt rescheduling.[21]

A major effect of these several revisions was that the traditional recipients of development assistance, developing countries in the South, received less aid, and that the traditional focus on social and economic development was weakened in terms of resources allocated.

*

The focus of this subsection has not been on the dynamics involved when the dramatic transformation of the international system took place in the last years of the 1980s and the first of the 1990s, nor on the outcome of this revolution or its consequences for international politics in general. As promised above, we will return to some of these issues in sections IV and V, particularly in the concluding section, but also then with a more limited focus in mind: the effects of these dramatic changes on development policies and aid. The catastrophe of 11 September 2001, hitting the symbols of power of the new international hegemon, the International Trade Center in New York and the Pentagon in Washington – referred to only in passing in the above because our main concern here has been to deal with the 1990s, particularly the early 1990s – belongs to that discussion.

Our concern in this subsection has been confined to identifying some of the immediate consequences of the new world order on the willingness of donors to provide aid and how it has affected the actual distribution of development assistance and the definition of ODA. As we have seen, the responses to the new international setting have

---

(eligible to be reported from the early 1980s but widely used only since 1991). In the 1990s, this tended to increase the aid as reported (in relative and absolute terms). Other revisions, particularly the new System of National Accounts, broadened the coverage of GNP (renamed GNI), and tended to depress the ODA/GNI ratios of donors.

21. This use of ODA even involved debts incurred outside the guarantor's aid budget, thus converting losses on strictly commercial credits into ODA. It is generally accepted that debts constitute a major development problem, particularly for the poorest developing countries, the so-called severely indebted low-income countries (SILICs), with little if any possibility of extricating themselves from the debt trap. However, debt relief brings no new resources for development and does not meet recipient countries' need for investment in local capacity-building. In the early 1990s, an increasing share of total ODA was provided for this purpose; in 1992, it amounted to 14.7 per cent, including non-ODA debts (debts acquired for commercial or military purposes). However, this type of aid was not shared equally among DAC countries; the top four (above the average) were (percentages of total aid in parenthesis) Portugal (63.9), Austria (60.5), the United States (30) and Italy (16) [*OECD, 1995a: tables 26 and 6b*].

varied both between and within regions. In general terms, the willingness of most European governments to provide development assistance has been less affected by the change than has that of the US administration whose aid was halved, in relative terms, during the 1990s – from an already low level. For most European donor countries, other determinants affected their performance more strongly – in particular domestic policies, such as a change of government or a strained economy. Nevertheless, the actual distribution of aid, geographically and thematically, changed quite substantially – adapted, not least, to the new security situation of the post-Cold War era which has been briefly described above.

III. THE CONCEPTUAL ENVIRONMENT: DEVELOPMENT AND AID

In a foreign policy perspective, development assistance becomes an instrument in the pursuit of *foreign policy* objectives by a donor or group of donors. Such objectives may, as we have argued, go beyond narrow self-interests: also long-term interests related to international common goods – such as peace, stability and environmental concerns – fit into this perspective, along with the promotion of treasured values of the donor(s), including the predominant ideology and norms. In the search for motives driving development assistance, it becomes equally important to examine the philosophical setting: what kind of development is contemplated and how is it attained? The main focus then shifts to *development* objectives and strategies. In such a context, development assistance becomes one among several instruments. The focus in this section will be on the role assigned to aid; although long lines will be drawn, the emphasis will be on development paradigms predominant in the 1990s and beyond.

1. The Beginning

In the late 1940s and the 1950s, when development assistance was first conceived of and implemented, the United Nations contributed strongly in setting the agenda – by establishing the Expanded Programme of Technical Assistance (EPTA) in 1950 and the Special Fund nine years later, and by forming and adopting the development strategies for the first and second development decades, the 1960s and the 1970s. These efforts were crucial in shaping the ideological and normative foundation of this activity and in setting development objectives involving the South during the initial phase.

The development paradigms prevailing in these early years reflected, in the first place, European experiences of development and theories and models emerging from these experiences. In the post-World War II bipolar international system, these models varied considerably, with models associated with the main Western powers representing one pole and various (socialist) models associated with the Eastern powers, the USSR in particular, representing the other.[22] During the Cold War, they also became instruments in the power struggle outlined above, as competing ideologies involving the development policy.

The first efforts institutionalised by the UN reflected the development thinking that held a predominant position in the West at the time, continuing into the 1960s – the modernisation paradigm.[23] Economists indebted to Keynesian theory, with its emphasis on state intervention, dominated the development discourse. Their focus was almost exclusively on *economic growth.* As expressed by Björn Hettne [*1995*], development became the art of large-scale social engineering. Capital accumulation (saving) was seen as the prime factor in creating economic growth.[24] Development was conceived of

---

22. In Europe, the various development models had emerged in different historical settings. The liberal model, associated with the Industrial Revolution in Britain, relied on market forces, industrial growth with light industries, private investment resulting from high profits, to which low wages contributed, and a process of continuous technological innovation and improvement. Such a system would thrive with free international trade, also associated with the model. Latecomers, however, did not find free trade equally useful. In order to catch up, an alternative model was designed, that of state capitalism in combination with protective measures, particularly as regards infant industries. During the early years, this model was associated with Germany and Russia in particular. In actual practice, however, a model of mixed economies evolved in the West, with varying degrees of state intervention and participation – although the rhetoric projected a liberal model.

The Soviet model was qualitatively different, although some similarities existed with the state capitalism of the early stages of the Industrial Revolution. The authoritarian state used coercive powers to enforce accumulation. The main instrument was the five-year plan that set quantitative objectives to be attained during the plan period. The emphasis was on industrialisation, with priority given to heavy industry. Technologically advanced methods were used and large-scale industries developed. Resources were transferred from agriculture to industry; in this process, the agricultural sector was collectivised.

23. The modernisation paradigm includes broader approaches than those associated with development economics; other social sciences also contributed to the concept. The paradigm is summarised by Hettne [*1995: 50–1*] as follows: development is a spontaneous, irreversible process inherent in every society; it implies structural differentiation and functional specialisation; the process can be divided into distinct stages showing the level of development achieved by each society; and development can be stimulated by external competition or military threat and by internal measures that support modern sectors and modernise traditional sectors.

24. The so-called Harrod–Domar model played a prominent role: growth was seen as a continuous progress towards higher stages of development: each increase in output would provide the basis for further growth because part of the surplus would be reinvested. The ability to save (invest) would increase with the level of income: once the

as a national process. However, development economists closer to the neo-classical tradition stressed the importance of trade: foreign trade would, in addition to capital accumulation, also serve to promote the diffusion of new technology, skills and capital to the benefit of latecomers in the process.

Development was conceived of as a process whereby a society would move from one stage to a new, higher one, and these stages were identified and quantified in economic, political and cultural terms. Walt Rostow [1960] was probably most influential among the development economists in shaping the world-view of foreign policy actors concerned with development issues in those days and beyond. He saw development as primarily an endogenous process in which societies progressed through five stages, from traditional society to the mass consumption society.[25]

Capital formation in terms of saving and investment, particularly through industrialisation, constituted the key to development. During these early years, an active role was attributed to the state – in setting the rules of the game and following them up, in planning and following up social and economic development not only by providing the infrastructure but also by itself taking part in economic activities. Within this paradigm, the main function of aid was to fill the gaps in terms of capital and knowledge and to overcome the bottlenecks

---

process was started, economic growth would be self-sustaining. The several elaborations of the development economics tradition departed from this core model, along with parallel, neo-classical theories concerning the role of international trade.

25. The many characteristics of traditional society had to be removed during the second stage, termed the pre-take-off society: agricultural productivity increases rapidly (as subsistence farming is replaced by commercial agriculture), the infrastructure is improved and a new class of entrepreneurs emerges in order to establish the preconditions for the most vital phase, namely take-off. This third phase is characterised, in particular, by a high level of investment (national saving rising from 5 to over 10 per cent), resulting in an industrialisation process in which some sectors take on a leading role. This process, where advanced technology and skills are diffused from the leading sectors to society in general, in turn leads to the next stages: the road to maturity and the mass consumption society. Although development was conceived of as primarily an endogenous process, foreign capital and aid, particularly technical assistance, had a role to play in these stages.

The ethnocentrism of the model is obvious: the Western consumption society represented the ultimate objective to be strived for. Another function of this model was reflected in the subtitle of the volume: Rostow wanted to present an alternative to the Marxist development model.

Other development economists with a strong influence on development thinking during these formative years include Myrdal [1957], Hirschman [1958] and Prebisch [1962]. In the period 1955–65, the modernisation paradigm held a hegemonic position in the West, with a strong influence on development thinking and practice extending beyond that period. For an interesting retrospective on their major contributions, by the pioneers of development economics themselves, more than 20 years later, see Meier and Seers (eds) [1984]. See also Meier [1984].

inherent in a process of unbalanced growth in order to make the economy of the recipient country ready for take-off.

The development economists, although holding a dominant position at the time, were not unopposed; nor were the blessings attributed to aid.[26] Neo-classical traditionalists contested the wisdom of state intervention in markets and the instrumental (economic and political) as well as ethical arguments in favour of development assistance.[27] As will be seen, in the early 1980s the political environment was ripe for this school to surface to an almost hegemonic position.

The first efforts by the UN system to help 'underdeveloped countries' by way of development assistance focused precisely on the bottlenecks identified by the modernisation paradigm: both the EPTA and the Special Fund were established to assist governments in their economic development by providing technical assistance. The capital component was part of the initiative – a special UN fund for economic development (SUNFED). However, this fund never materialised within the UN system. Eventually, the international financial institutions, the World Bank in particular, were assigned the role of providers of development finance. The UN strategy for the first development decade (DD1) also reflected, to a large extent, the modernisation paradigm. The paradigm remained influential as the strategies of later development decades were formed.[28]

---

26. A competing perspective came, as already noted (*supra*, n.5), from outside the ranks of economists: Hans Morgenthau, analysing foreign aid for economic development from a realist foreign policy perspective, complained that the theoretical analyses within this field had been confined to economists. 'Economic thought, true to its prevailing academic tradition, tends to look at foreign aid as though it was a self-sufficient technical enterprise to be achieved with the instruments, and judged by the standards, of pure economics.' He finds their perspective, with a focus on capital formation and the accumulation of technical knowledge, all too simplistic when applied to the economic development of 'the underdeveloped nations of Asia, Africa and Latin America'.

A policy of foreign aid is no different from diplomatic or military policy or propaganda, he argued: 'They are all weapons in the political armory of the nation. As military policy is too important a matter to be left ultimately to the generals, so is foreign aid too important a matter to be left in the end to the economists. The expertise of the economists must analyze certain facts, device certain means, and perform certain functions of manipulation for foreign aid. Yet the formulation and over-all execution of foreign aid policy is a political function. It is the province of the political expert. It follows from the political nature of foreign aid that it is not a science but an art' [*Morgenthau, 1962: 304, 309 (quotations)*].

27. See, *inter alia*, Bauer [*1971, 1981, 1984*], particularly the essay entitled 'Foreign Aid: Issues and Implications' [*1984*], and Bauer and Yamey [*1972*]. For an excellent presentation and analysis of the main instrumental and ethical arguments levelled against aid (including those of the neo-classical traditionalists) as well as those in favour, see Riddell [*1987*].

28. Thus, the heritage is clear in the Pearson Report [*Pearson, 1969*], which expressed the view that expanded trade, investments and aid would give rise to self-

2. Predominant Development Paradigms in the 1990s and Beyond

(i) Old paradigms linger on. As observed earlier [*Stokke, 1996*], development paradigms tend to have long, overlapping lives. Old paradigms also adjust to more recent experiences and new realities, thus prolonging their lives. Any sequencing, therefore, runs the risk of portraying too simplistic a picture.[29]

The *modernisation paradigm* continued to exert influence in the 'donor community' long beyond its heyday. In the 1950s, development economists had, as craftsmen, set out to provide international agencies and Western governments with the answers to the new challenge: how to produce economic growth in the large number of new, independent countries – most of them considered poor and 'underdeveloped', although their resource bases varied extensively. Following Keynesian theory, the intervening (development) state was assigned a key role in these efforts: capital and technical assistance (aid) were to be channelled through governments. In the 1950s, the emphasis (in theory and practice) was on the transfer of technology. Transfer of capital (aid) as a supplement to domestic saving was part of the model from the beginning, but was not prominent in aid programmes. In the

---

generating social and economic development in the South before the turn of the century and that international agreements and institutions would be conducive to the creation of peace, equity and stability.

29. A condensed, sequenced exploration of the development discourse offered by Göran Hydén nevertheless deserves our attention (the warning remains valid!). He links the predominant approach of the period with the predominant theory, the discipline from which the theory emerged, the main strategy pursued, the institution assigned to implement this strategy, and the main instruments involved, with particular reference to development assistance. According to Hydén, in the period 1955–65, the hegemonic paradigm was the *modernisation approach*, with economists (Myrdal [*1957*], Hirschman [*1958*], Rostow [*1960*] and Prebisch [*1962*]) as the pioneers of development, stages of development as the main theory, the Big Push as the strategy, central government as the major actor, and the project as the main aid mechanism. The following ten years (1965–75) were dominated by the *dependency approach*, with sociology as the main discipline and the theory of underdevelopment influencing even the World Bank towards the basic needs strategy, with an emphasis on decentralised institutions, and programmes as the preferred aid mechanism. The 1975–85 period was dominated by the *popular participation approach*, involving a shift from the structuralist approach and derived from the theory of rational choice, with development anthropology as the leading discipline, 'small is beautiful' as the main strategy, grass-roots organisations emerging as the preferred development institutions, and policy as the main mechanism. Since 1985, the dominant approach has been to create an *enabling environment* for development, emphasising the role of institutions and bringing various schools of political science to the fore; governance has become the main strategy, with civic associations as the main development instrument – development is not likely to occur unless Third World governments are getting their policies right [*Hydén, 1994, 1999*].

1960s, however, the emphasis was on general (economic) growth, involving investments in the productive sectors and in infrastructure.

At the time, this approach was strongly contested by neo-classical traditionalists who saw no wisdom in state intervention in the market. It was also contested from the 'left' by neo-Marxists of the *dependencia* schools.[30] The structuralists, on the other side, argued for import substitution and protection of 'infant industries'.[31] This theme – the role of the state versus the market – has remained among the core issues in the development discourse ever since John Stuart Mill, favouring *laissez-faire*, indicated that state intervention might, at times, be required. As will be seen in the following subsection (and section IV), it remained a theme in the 1980s and 1990s, with the swing of the pendulum. Other components of the modernisation paradigm were also reflected in the aid discourse of the 1990s and beyond.

In the early 1970s, it became increasingly clear that the optimism of the 1950s and the 1960s, stimulated by the modernisation paradigm (and also by the enthusiasm of the new bilateral and multilateral institutions created to enhance development as well as the expectations aroused by the new-won independence in the Third World itself) was not matched by actual results. The development that had taken place in the South was skewed. The anticipated trickle-down effect of economic growth had not materialised. A large proportion of the population, the poorer sections in particular, had been left more or less unaffected by aid, and the gap between North and South was growing even wider in terms of indicators of aggregate economic growth. As a result, the one-sided focus of development economists on economic growth, anticipating that this would do the trick, came under fire. The social dimension of development came to the fore:

---

30. The critique of the *dependencia* schools was fundamental, involving the very structure of North–South relations. The centre–periphery model [*Galtung, 1971*] interpreted the core processes involved in a new way: integration of the South in the capitalist North would further deepen its dependence and increase its underdevelopment, except for the elites in the South that served as bridgeheads for Northern exploitation. The perspective was rooted in two main traditions, namely, Marxism, particularly neo-Marxism [*Frank, 1967*], and Latin American structuralism associated, in particular, with Raúl Prebisch [*Cardoso and Faletto, 1969*]. For an overview and analysis of this Latin American contribution to development theory, see Kay [*1989, 1991*]. See also Larrain [*1989*], where this tradition is related to evolving development theories. For an adaptation of the theory to African conditions, see Rodney [*1972*]. Samir Amin has also contributed extensively to this discourse [*inter alia, Amin, 1974, 1976, 1977, 1993ab*].

31. Given the present-day positions, it is interesting to note that, historically, the United States applied protection of 'infant industries' against competition from Europe as the main strategy of its industrial growth policy and remained a strongly protected economy until World War II. This illustrates a simple lesson: principles and policies are adapted to circumstances – and circumstances may change.

growth with redistribution became the answer of the mid-1970s [*Chenery* et al., *1974; McNamara, 1973*]. This had clear implications for aid discourse and practice.

Confronted with the harsh ideological and economic realities of the 1980s, the influence of the *dependencia* paradigm faded away. Nevertheless, at a very general level, some elements survived, reflected in the development ideology of the 1990s, with particular reference to the rhetoric: a broad consensus on avoiding aid dependency.

In addition, ideas that came to the fore in the mid-1970s – the demand for a new international economic order and the basic needs approach – were confronted, in various ways, with the new economic and ideological climate of the 1980s. The *NIEO approach* focused on improvements in North–South relations.[32] In the development discourse of the 1990s, this approach was hardly visible. At the level of practical politics, it had been buried for years.

Thus, the two major development paradigms that recommended radical structural reforms of the international economic and political system faded away in the 1980s and most of the 1990s. It is difficult, however, to see how the ambitious goals agreed on by the UN Millennium Assembly can be realised without radical reforms of the international system that would also affect the distribution of power. In addition, the paradigm that had its main focus on improvements in social services within individual developing countries, the *basic needs approach*,[33] experienced a similar fate, though with some variance for

---

32. The call for a new order emerged from governments of the South using the UN system as their primary arena. As it turned out, the call met a fertile ground within this system, which adopted the major claims, not least on normative grounds, and provided an institutional basis for their further development and promotion, with particular reference to the UN Conference on Trade and Development (UNCTAD). Although the perspective was general, involving most North–South relations, in practice the main focus was on improving trade relations, with particular reference to the primary products of the South. In the second half of the 1970s, it obtained support from the group of 'like-minded' Western small and medium-sized powers, dwindling in the 1980s [*Løvbræk, 1990*]. However, the major Western powers, the US in particular, consistently rejected the idea of this kind of intervention in the market. Besides, as argued by Fortin [*1988*], in the last quarter of the twentieth century commodities were not the key to prosperity. For an overview and analysis of the processes involved, see Adams [*1993*].

33. In the mid-1970s, the basic needs paradigm emerged as the main expression of the various approaches focusing on the social dimensions of development. In an aid context, the focus was on improved social services, largely along lines established in the North. Basic needs were interpreted in different ways. Some defined them to include what would be required for mere survival, such as access to food, shelter and health. Others broadened the concept to include facilities that allow the individual to take command of his/her fate, such as access to education and the exercise of civil and political human rights. For overviews and definitions, see Carter [*1977*] and Emmerij [*1977*]. The paradigm resulted from craftsmanship within the UN system, with

individual donors. The core of this approach, a concern for the social dimensions of development, was neglected by many of the largest donors during most of the 1980s. Nevertheless, the ideology underpinning it inspired the critique of the structural adjustment policy of the 1980s and beyond, and, towards the end of the 1980s, it came to the fore again and played a prominent part in the aid rhetoric of the 1990s, as we shall see (section IV).

The core of the modernisation paradigm continued to influence both developing thinking and aid practice into the 1990s and beyond. Nevertheless, it was not unaffected by the neo-liberal counter-revolution emerging in the 1990s and the effects of this policy on development in most developing countries, particularly through its main instrument, the structural adjustment policy. In the process, the modernisation paradigm has been 'modernised', and has adapted to the changes that have taken place during recent decades in the South. A crucial role is still attributed to capital formation and knowledge. But whereas, during the early years, the emphasis was on capital (savings and transfers) and technical assistance, with development assistance playing an important role in both, the emphasis has increasingly shifted from technological transfers to *self-generation* of innovation and knowledge. As we shall see in section IV, this had implications also for the role attributed to development assistance. In the 1990s, the emphasis was increasingly on human development, with particular reference to education and health, but also on institutional development. In these efforts, NGOs and civil society more generally in the donor countries were assumed to play a new and greater role. This approach was combined with the promotion of liberal political values, increasingly also a part of the development concept (see below, subsection iii). As in the past, the private sector, including the multinational corporations, was attributed an important role in the diffusion of technological innovation and knowledge – in the 1980s and 1990s with even fewer inhibitions than before.

---

particular reference to the International Labour Organisation, the World Health Organisation and the United Nations Children's Fund. For institutions adapting such ideas to their particular settings, see, in particular, the studies by ILO [*1976*], the joint study by WHO and UNICEF [*Djukanovic and Mach (eds), 1975*], and UNICEF [*1977*]. In the years immediately after 1975, it became a dominant paradigm within the donor community, particularly the rhetoric. Its basic weakness became increasingly clear towards the end of the 1970s when economies came under pressure: it presupposed an economy able to carry not only the investments in building and maintaining the appropriate structures but also the recurrent costs involved.

(ii) The neo-liberal counter-revolution. In the 1990s, the traditional brand of neo-classical economics that came to the fore in the early 1980s remained a predominant development paradigm, although weakened. In this paradigm, the role of the state as development agent and state and inter-state intervention in the economic sphere are de-emphasised. Basically, such intervention (including development assistance) is seen as a constraint, if not outright detrimental to development. However, the paradigm was by no means uncontested.[34]

The neo-liberal counter-revolution [*Toye, 1987*] had many sources. The ideological change involving major Western governments that had never been attracted by the ideas of international structural reform was important in this regard. But also developments in the developing countries themselves facilitated the advance of neo-liberalism, particularly the economic crisis in which most developing countries found themselves at this stage, resulting from events and policies in the second half of the 1970s.

For non-producers of oil and gas in the South, the rising cost of energy following the oil crises of the 1970s added to their burdens. Their economies had become dependent on the importation of oil and commodities that were affected by the rise in oil prices. When commercial financial institutions with petrodollars in abundance and a relaxed attitude towards traditional banking norms opened their purses, Third World governments grabbed the opportunity to meet rising expectations at home and the higher prices for imports in demand. As a result, they accumulated huge debts, the servicing of which became increasingly difficult as a consequence of a prolonged recession in industrial countries and deteriorating terms of trade for raw materials, and continuously deteriorating productivity. Rising interest rates added to the debt-servicing and balance-of-payments problems. As a result, commercial financial sources dried up and the banks became eager to have their loans repaid.[35]

The situation threatened the international financial system, involving, in particular, commercial financial institutions in North America and Western Europe. The International Monetary Fund came march-

---

34. The neo-liberal critique of structuralist development economics and the main elements of the 'new' paradigm are summarised by Ian Little [*1982*]. Prominent among the old warriors who were still going strong was Peter T. Bauer [*1971, 1981, 1984*]. For critical analyses, see, *inter alia,* John Toye [*1987, 1993*], Björn Hettne [*1995*] and Poul Engberg-Pedersen *et al.* (eds) [*1996*].
35. This is not the place for a detailed description of the deep-rooted economic crisis evolving in the late 1970s and the early 1980s, crippling the economy of many Third World countries for years, nor for analysing the causalities involved. The sources were many, with variations from one region to another.

ing in, as was its obligation in this type of financial crisis. However, the way it handled this crisis has been heavily criticised almost from day one.[36]

The declining legitimacy of many governments in the South in the wake of the rediscovery, in the mid-1980s, of the 'soft state' [*Myrdal, 1968, 1970, 1984*] also eased the advance of neo-liberalism to a dominant ideology. The ineffectiveness of several governments as channels for development assistance was no new discovery to the aid agencies: in the early days of development assistance and up to the mid-1980s, however, correcting this was considered part of the challenge to be met by way of development assistance. The emergence of the patrimonial – or neo-patrimonial – state[37] in many Third World countries shortly after independence had a fundamental impact on both development thinking in general and aid philosophy and practice in

---

36. The IMF, designed to help stabilise short-term balance-of-payments crises in Northern countries, followed much the same approach and prescriptions *vis-à-vis* the crisis-ridden Third World governments as those applied earlier to Northern governments in temporary crisis: they had to observe certain conditions before the IMF would agree to assist with short-term credits. According to the IMF analysis, the crisis in individual Third World countries was caused by excessive government spending (an imbalance between spending and revenue), resulting in budget deficits and inflation; overvalued currencies; disproportionate imports in relation to exports; and insufficient attention to factors on the supply side. The prescription, accordingly, was to correct these areas of neglect: the economic policy of individual countries had to be reformed, and economic reform was made a condition of assistance through structural adjustment loans.

The IMF (and the World Bank, particularly during the early structural adjustment period) was criticised for coming up with the same stereotyped recommendation, which failed to take account of the widely varying conditions applying in the countries concerned. For an early soft-spoken critic along these lines, see Fahy [*1984*]. The main problem, however, was that a structural economic crisis that developed into a fully fledged development crisis was conceived of as a temporary balance-of-payments crisis. The most fundamental criticism along these lines came recently from a Nobel laureate in economics, formerly the Chief Economist of the World Bank [*Stiglitz, 2002*]. For a critical analysis of the World Bank and policy-based lending, see Mosley *et al.*[*1991ab*]. For an analysis of what we know, do not know and need to know with regard to the completion rate of IMF programmes, departing from the recommendation by the International Financial Institution Advisory Commission that IMF conditionality be abandoned because it is ineffective, see Bird [*2002*].

37. Georg Sørensen [*1993a: 15–17*] defined neo-patrimonialism, with particular reference to post-colonial African states, as government based on personal loyalty especially to the leading figure of the regime, the 'strongman'. Important positions in the state – political, bureaucratic, military and police – are filled by loyal followers: relatives, friends, kinsmen and tribesmen. Their loyalty is reinforced by sharing the spoils of office. Political decisions are up for sale, and the capacity of the bureaucracy is destroyed by its participating in the system of spoils. The weakest point is the political elite itself. '[V]ery little in terms of *statecraft* seriously promoting economic development can be expected to emerge in this situation.' For other important contributions to this debate see, *inter alia*, Jackson and Rosberg [*1986*], Larry Diamond [*1988*] and Sørensen [*1995*].

particular. In previous paradigms, the state had been considered the main development agent, and aid was provided accordingly. In the 1980s, the state was increasingly considered a major barrier to development, and to be bypassed by aid.

This chain of events changed the main thrust of international development assistance. During the 1980s, aid increasingly became an instrument in the promotion of economic policy reform in developing countries, much in line with the neo-liberal orthodoxy. The linking of development finance to a commitment by the recipient government to structural adjustments in the general direction of a liberal economic regime (aid conditionality) became the most manifest expression of this policy. It was driven by, in particular, the Bretton Woods institutions, with strong support from major Western governments. The recipe prescribed was not restricted to the conventional wisdom of domestic and foreign housekeeping: the mechanisms prescribed reflected a neo-liberal perspective and were, therefore, highly political. Structural adjustment of *international* economic relations was not part of the perspective: as already noted, by the early 1980s, the NIEO programme already belonged to the distant past. The World Bank joined forces with the IMF and the major bilateral donors and, in the second part of the 1980s, 'like-minded' governments also fell into line. With few other sources of finance available (individual bilateral donors set agreement with the IMF as a condition for providing assistance), individual, debt-ridden Third World governments had little choice but to accept the conditions.[38] During this period, development assistance, particularly programme aid, came increasingly to be geared to realising the aims set for the structural adjustment policy – with an implicit if not explicit aim of improving the ability of developing countries to service their foreign debts.

In the 1980s, the focus centred on macroeconomics; concern for social development and the environment was clearly subordinated. As

---

38. The international political climate of the 1980s had changed dramatically as compared with the optimism reflected in the development paradigms of the mid-1970s. The predominant development philosophy of the Western industrial countries changed accordingly: former reform internationalists approached a liberal internationalism position and former liberal internationalists approached the realist fold. As summarised by one observer, realism ruled supreme in the inner councils of the foreign ministries of most Western states during the 1980s [*Pratt, 1990: 3*]: 'An operating premise of the world-view which predominates there is that states pursue their own, usually narrowly defined, self-interest in their international relations and need not, and probably should not, do otherwise. Any concern for cosmopolitan values, any wish, that is, that foreign policy should be responsive to the needs of those beyond their borders, has been marginalized – almost turned illegitimate – within these circles except when such concerns can be made to serve geopolitical, economic, or other traditional foreign policy interests.'

already described, aid policy became instrumental in persuading and forcing governments of the South into line – with particular reference to first-generation conditionality, using structural adjustment lending as the primary tool [*Stokke, 1995; Hewitt and Killick, 1996*]. There was little room for a strong, interventionist state in this paradigm. State intervention was considered detrimental to development: the market would do the job. However, towards the end of the decade, protests against the neo-liberal orthodoxy began to gain ground, since the prescribed recipe failed to deliver results; on the contrary, in many developing countries, the economic crisis developed into a development crisis. Events took a negative turn: progress achieved during the 1960s and 1970s within important areas such as education and health was retarded.

As the negative societal effects resulting from the structural adjustment policy became apparent, the strength of the counter-arguments increased. While the Bretton Woods institutions had driven the structural adjustment policy, the reorientation process was initiated by the UN system. A soft-spoken UNICEF study contributed strongly to turning the tide [*Cornia* et al. *(eds), 1987*]: remaining within the mainstream analysis that economic reform and structural adjustments were necessary, it argued strongly for a concern for the social dimension of development – adjustment with a human face. Fifteen years later, the critique of the policy pursued in the 1980s and 1990s was much more outspoken, emerging from a source with intimate insights and the authority of a Nobel laureate in economics [*Stiglitz, 2002*]: in the 'classic' conflict between Keynesian and neo-liberal economists, the pendulum had swung back again, although the debate continued. We shall return to one of the outcomes in the next section: a revival of the social dimension of development.

(iii) Political liberalism. In the 1990s, the neo-liberal orthodoxy, although challenged, remained in a strong position when it came to economic policy. With the dwindling of the power of the Second World, ideas associated with political liberalism came to the fore. There were also other reasons for this: the modest effects of the structural adjustment policy, the main prescription for solving the crises of the 1980s, were explained as the result of bad governance. Economic policy reforms, involving openness and market liberalism, had to be supplemented by reforms of the political system in order to work, the Bretton Woods institutions argued.

The new emphasis on political liberalism – democracy, human rights and 'good government' involving the rule of law, and economic and political transparency – may be considered a result of the changed global power balance. In the early 1990s, leading Western politicians, insisting on a positive correlation between development and liberal *democracy*, dominated the discourse.[39] In the theoretical discourse, this correlation was contested. The controversy involved both the notion of democracy and how it should be implemented by way of international development assistance. Should the promotion of democracy be limited to the formal, procedural aspects of government, such as free and fair elections in which two or more political parties were competing for power, or should the concept be defined more broadly, involving also the substance of the policy?

Most analysts found the narrow definition unsatisfactory. Elections may, in some cases, even mask systems of personal rule without changing its authoritarian features.[40] The broader concept of 'real democracy', involving openness and participation, with an emphasis on a thriving civil society – core components of the mainstream definition of the concept [*Diamond* et al., *1988–90*] – was advocated.[41]

---

39. For an overview, see Stokke [*1995: 21ff., especially notes 16 and 17*]. In 1989, the members of DAC claimed that 'there is a vital connection, now more widely appreciated, between open, democratic and accountable political systems, individual rights and the effective and equitable operation of economic systems', and that participatory development 'implies more democracy, a greater role for local organisations and self-government, respect for human rights including effective and accessible legal systems, competitive markets and dynamic private enterprise' [*OECD, 1992b, 1993*].

40. The role of the state in development re-entered the main stage of the development discourse in the second half of the 1980s, with particular reference to the neopatrimonial state (see n. 37). The lack of legitimacy of such states as development agents and trustees of development funds was brought forcefully to the fore by Jackson and Rosberg [*1986*], who argued that juridical statehood in sub-Saharan Africa tended to preserve ex-colonial jurisdictions regardless of their potential for development, with the frequent effect that state building was obstructed. 'In more than a few countries, sovereign rights have been purchased at the expense of human rights. The new international democracy is often disclosed, somewhat ironically, as democracy for the rulers only' [*ibid.: 27–8*]. Sørensen [*1995*] argued in the same vein: self-seeking political elites expected the fruits of sovereignty, but were not able or willing to pay the price by providing an effective development regime.

41. Diamond *et al.* [*1988, Vol. II: xvi*] argued for a definition that separates the political system from the economic and social system to which it is attached, insisting that economic and social democracy should be separated from the question of governing structure. On this basis, they suggest the following definition: 'a system of government that meets three essential conditions: meaningful and extensive *competition* among individuals and organized groups (especially political parties) for all effective positions of government power, at regular intervals and excluding the use of force; a highly inclusive level of *political participation* in the selection of leaders and policies, at least through regular and free elections, such that no major (adult) social group is excluded; and a level of *civil and political liberties* – freedom of expression, freedom of the press, freedom to form and join organizations – sufficient to ensure the integrity of political competition and participation.' The authors admit (in line with the concept created by Robert Dahl [*1971*] – *polyarchy*) that, confronted with the real

Others found even this definition too narrow, arguing that the concept of democracy would be meaningless if not associated with the policies pursued, bringing social justice and economic democracy at the local, national and even the international level on to the agenda [*Amin, 1993a; Gills* et al., *1993*].[42]

The argument that there was a positive correlation between 'democracy' and 'development', or more generally between the system of government and development, was contested. In an earlier period, the World Bank had indicated that there was a negative correlation between the two. In the discourse of the early 1990s, the controversy remained unsettled [*Sørensen, 1991, 1993ab, 1995; Healey* et al., *1993; Leftwich, 1995; White, 1995*]. However, the conclusion will not be unrelated to how 'development' is defined – narrowly, related to indicators of economic growth, or more broadly, including also indicators for human development. In the 1990s, several donor countries included democracy among the overall aims set for development co-operation, thus making democracy part of their development concept.

The concern for democracy, with aid as a tool, was not an invention of the early 1990s; as already noted, this concern was driving the development assistance of Western countries when it was initiated in the late 1940s and early 1950s. Nevertheless, during the following decades, it did not play an important role in aid policy, with a few exceptions. In the late 1980s and early 1990s, however, the promotion of democracy became an objective for development assistance in its own right. Democracy was seen as a precondition for development to take place. Most Western donors expected the recipients of development assistance to move in the direction of increased democracy, and such a commitment was increasingly set as a condition for providing this assistance. However, among the three main components included in second-generation aid conditionality, this was probably the one surrounded by the strongest ambiguity. Democracy was also promoted through so-called 'positive measures' [*Stokke, 1995*].[43]

---

world, the definition suggested presents a number of problems because systems that broadly satisfy the criteria established nevertheless do so to different degrees.

42. Gills *et al.* [*1993*] found a definition of democracy that excluded economic and social reality from the political missing the first and most important task of democratic regimes, namely, social reform: 'in the absence of progressive social reform, "democracy" is largely devoid of meaningful content' and may serve as a euphemism for sophisticated modern forms of neo-authoritarianism, they argued [*ibid.: 4*].

43. As implemented, a consistency problem appeared, particularly involving aid conditionality; this problem was most visible in the case of US policy. At an early stage, the US government considered democracy an objective in its own right: it was required by law that aid should be cut off to regimes coming to power by overthrowing a democratically elected government. However, aid was continued to non-

Human rights constituted the second pillar of the political liberalism coming to the fore in the early 1990s. For most Western donors, the concern for human rights was a separate foreign policy issue until the late 1980s, administered by units different from those concerned with development co-operation. For some countries, however, a link to aid was established in the mid-1970s, but then as a negation: respect for such rights was considered a condition for a government to receive aid or to be included as a main partner in development.[44] The events of the early 1990s therefore represented a continuation of what these frontrunners initiated 15 years earlier, but now in a more active, systematic and massive way. They also represented a new departure: the emphasis was much stronger, and the promotion of human rights became elevated to an objective set for development co-operation in its own right and seen increasingly as a precondition for 'development' in other areas, including social and economic development, to take place. The emphasis was increasingly on political and civil rights.[45] They were pursued by way of second-generation aid conditionality and, as for democracy, through so-called positive measures.

---

democratic regimes [*Lancaster, 1993*]. According to Gills *et al.*, the 'crusade for democracy' (Reagan's speech to the British Parliament in 1982) in Latin America was accompanied by 'state-sponsored campaigns of murder, torture and general barbarism. In most cases this war was waged from "within" by the military and paramilitary death squads. The main exception was Nicaragua. ... At best, the US tolerated, and at worst directly promoted the most grotesque abuses of human rights in recent history throughout Central America' [*1993:* 17]. Such inconsistencies were far from limited to the hegemonic superpower with worldwide security interests at stake. While Western small and medium-sized powers might agree to use aid conditionality against small and ugly African aid-dependent regimes, this instrument was not considered equally effective against stronger, more self-confident Asian authoritarian regimes. Nor were they particularly willing to make use of this mechanism when their own vested (commercial) interests might be at stake [*Stokke, 1995*].

44. In 1975, the US Congress passed legislation that established such a link [*Nelson and Eglinton, 1992: 26–7*]; the following year, the Norwegian Parliament set as a criterion for the selection as a new main partner country that the government contributed to economic, social and civil rights incorporated in the UN Declaration and Convention on Human Rights, and the Netherlands made compliance with human rights one criterion for the main partner countries ('concentration countries') for bilateral aid (see chapters on the two countries in Stokke (ed.) [*1984*].

45. The human rights perspective is based on the International Declaration of Human Rights (1948), followed by the International Covenant for Civil and Political Rights and the International Covenant for Economic, Social and Cultural Rights. In the post-World War II era, the major Western powers tended to give emphasis to the first one (civil and political rights) while the major Eastern powers gave emphasis to the second (economic and social rights) – and a few small and middle powers (including the Netherlands and the Nordic countries) gave equal importance to both. In addition, other treaties deal with special categories of people, such as children, women and the mentally ill, or with malpractice, such as torture, discrimination and disappearances. In 1986, the UN General Assembly adopted the Declaration on the Right to Development. Most countries, though not all, have ratified these treaties, although some

Most Western governments consider human rights as universal rights, to which all governments are committed and therefore obliged to follow up. This position is not uncontroversial.[46] At the national level, the responsibility for the follow-up rests with the government. Implementation becomes problematic, especially when the government itself is the main source of human rights violations and with a weak international regime around to enforce human rights.[47]

The notion of *good government*, broadly interpreted, includes both liberal democracy and human rights. However, the World Bank defined the concept somewhat narrowly as 'the manner in which power is exercised in the management of a country's economic and social resources for development' and 'good governance' as synonymous with 'sound development management' [*World Bank, 1992b: 1*]. The focus was on those aspects of government that had to do with public administration, such as effective financial accounting and

---

governments have expressed reservations; thus, the resolution on the right to development was adopted by 147 votes to one (the US) with eight governments abstaining.

An important distinction should be made between the various types of right. While there is a long tradition of codifying civil rights related to legal systems both domestically and internationally, other rights are not equally easily codified. Thus, economic, social and cultural rights – or the right to development – have been defined in less precise terms. Procedures for their implementation have not been specified and an international regime for their implementation is not in place.

46. In the early 1990s, when the discourse on these issues peaked, Peter Waller [*1992: 25–7; 1993: 54*] observed that, within the UN system, human rights and development issues had been administered by different bodies and kept apart; the UNDP had previously refused to make aid an instrument in pursuit of human rights. He argued strongly for a broader relationship between the two. So did the editors of the 1994 yearbook on human rights [*Baehr* et al., *1994: 3*], summing up the 1993 World Conference on Human Rights: 'there is ... hardly any need to further belabour the intimate relationship between development and human rights. The one cannot be without the other.' Leftwich offered a contrary view [*1994, 1995*]. He argued [*1995: 418*] that a 'grim feature shared by developmental states is the combination of their sometimes brutal suppression of civil rights, their apparently wide measure of legitimacy and their generally sustained performance in delivering developmental goods. I suggest these are intimately connected.'

47. As was the case with democracy, a consistency problem was also involved here: policy declarations have not always been followed up, particularly when competing national interests among donors were involved. Such inconsistencies have been particularly visible in the case of the US government, which took on a leading role in promoting (civil and political) human rights at an early stage [*Lancaster, 1993*]. For a survey on the follow-up on the human rights allocative conditionality, see Nelson and Eglinton [*1992: 26–32*], who argue that despite legislation mandating that economic and security aid be withheld from violators of human rights, the US record has at best been inconsistent because of, *inter alia*, human rights objectives often having been subordinated to other foreign policy concerns. See also Young [*1991*] and Gills *et al.* [*1993*]. Needless to say, such inconsistencies have not been restricted to US policies. They have been particularly visible when the commercial interests of donor governments have been involved – as generally illustrated in their trade and aid policies *vis-à-vis* China, emerging as one of the main recipients of development assistance.

auditing systems, an appropriate legal framework and open competition for contracts, emphasising accountability, transparency and predictability on the part of politicians and civil servants, and the rule of law. As defined by the World Bank, aspects that more explicitly belong to the political arena, such as political participation, open debate and political legitimacy, were excluded.[48] Others broadened the concept. The OECD Development Assistance Committee placed particular emphasis on participatory aspects [*OECD, 1992b, 1993, 1995bc*].

A concept like this cannot be isolated from its political environment, of which the neo-patrimonial state is a part. The concept cannot therefore avoid relating to both the public and the private sector and must include both the political leadership and the public administration as more narrowly defined.[49]

This third major component of post-Cold War political liberalism was to be pursued both through (second-generation) aid conditionality and through 'positive measures' – aid for purposes assumed to improve 'good governance'. Again, in principle there was little new about this concern: donors have always considered an effectively functioning government structure on the recipient side to be of crucial importance, both for a country's development in general and its ability to utilise the development assistance effectively. They have targeted foreign aid for this purpose, first through technical assistance and later by contributing to institution building and management training. However, the new concern with governance in the early 1990s transcended earlier efforts both in scope and emphasis.[50]

---

48. The World Bank definition was criticised from different perspectives; however, the initiative remained with the World Bank during the early period. Leftwich [*1994: 372, 381*] found its prescriptions for good governance naïve because they failed to recognise that good governance is a function of state character and capacity, which in turn are functions of politics: 'Neither sophisticated institutional innovations nor the best-trained or best-motivated public service will be able to withstand the withering effects of corruption or resist the developmentally-enervating pulls of special or favoured interests if the politics and authority of the state do not sustain and protect them.'

49. In a neo-patrimonial system, efforts to improve the competence of the civil service would, almost by definition, only marginally address the more fundamental problems of bad governance, and this also applies to formal rules and guidelines: the root of the problem rests with the governing elite [*Sørensen, 1993a, 1995; Leftwich, 1994*]. For an attempt to develop a concept of governance for comparative political analysis in line with this broader approach, and a stimulating discussion that also, at an early stage, offered a guided tour through the literature, see Hydén [*1992*].

50. The World Bank can claim ownership of the 'good governance' concept and made major efforts in exploring and defining it. This was not by coincidence: as noted, the Bank needed an explanation for the limited success of the structural adjustment programmes, particularly in Africa, and had, in the late 1980s, come to the con-

Most donors considered these three dimensions of political liberalism to be interwoven values to be pursued simultaneously: together they represented what was considered *good government* in a Western tradition. The emphasis of individual governments and multilateral agencies differed. However, when it came to human rights, the emphasis was arguably more on political and civil rights than on economic, social and cultural rights, if not almost exclusively so, although the many international conferences may convey a different picture. When it comes to democracy, the emphasis has been on the more formal side of democracy building – free and fair elections – rather than the broader aspects focused on in the discourse.[51] The concern for participatory development on the part of the OECD (DAC) may, however, indicate a different conclusion. In addition, the DAC found it useful to distinguish between the form of political regime, the processes by which authority is exercised in the management of a country's economic and social resources, and the capacity of government to formulate and implement policies and discharge government functions [*OECD, 1993: 14*].

Some ambiguous and controversial aspects of these policies have already been identified. The most controversial aspect is probably the intervention in the policies of the countries concerned, particularly when pursued by means of aid conditionality.[52] Development assistance has always represented an intervention in the domestic affairs of recipient countries: bilateral and multilateral donors have always been concerned with the effectiveness and efficiency of the administrative systems on the recipient side. In the 1980s, they even set up their own administrations in some countries to implement aid-financed activities, bypassing domestic administrative systems at national and local levels. As observed [*Mutahaba, 1990*], this may have improved efficiency but contributed negatively to development in terms of competence and capacity building in the relevant countries.

Even against such a background, the good government policy constituted a novelty in relations among sovereign states, particularly because the interference in internal political and administrative structures and processes represented such a broad and massive advance. As

---

clusion that in order to work, economic policy reforms had to be bolstered by reform of the governing system of the countries concerned [*World Bank, 1989b; Lancaster, 1993; Waller, 1992*].

51. For a comprehensive analysis of US policy in promoting democracy abroad – norms and practice (and effects) – see Carothers [*1999*]. See also Santiso [*2000*]. For an analysis of EU efforts in this field, see Santiso [*2003*].

52. This problem is discussed elsewhere. See, in particular, Stokke [*1995*] and Hewitt and Killick [*1996*].

argued by Martin Doornbos, the package of measures imposed represented a new phase in the processes of state formation in the countries concerned, placing them under external supervision. If measures are imposed from outside, the effects might well be contrary to those intended, he warned [*Doornbos, 1990, 1995*]. This identifies an inherent conflict between first-generation aid conditionality, as implemented, and one of the core objectives of second-generation conditionality.

In the 1990s, Amartya Sen carried the development discourse involving these values further through his entitlement approach, particularly in *Development as Freedom* [*Sen, 1999*]. The discussion on how to understand and deal with core development challenges, such as poverty, famine, population growth, unemployment and gender inequality, is organised from the philosophical position that the ultimate aim of development is to expand human freedom.

IV. SOME MAJOR THEMES ON THE DEVELOPMENT AGENDA OF THE 1990S AND BEYOND

Both the evolving international situation (outlined in section II) and the development thinking (outlined in section III) have influenced the development agenda of the 1990s. So have insights generated by research or learned the hard way from experience in implementing long-term development co-operation, as well as emerging crises crying out for immediate solutions.

At the global level, the *World Bank* has been particularly well placed to influence the development agenda through a combination of extensive resources for research and development and a capacity to finance development. The main themes chosen for its annual reports will both influence the agenda and reflect evolving priorities. In addition, the *United Nations* and its specialised agencies and programmes, the UNDP in particular, are well placed to influence the development agenda. The political legitimacy of the UN system is broader than that of the Bretton Woods institutions, and its 'intellectual' capacity has also been developed but not centralised to the same extent as that of the World Bank and not sustained with a similar financing capacity. In the late 1980s and 1990s, the annual UNDP human development reports exerted strong influence. More than anything else, the UN has been able to set the agenda through a series of high-level reports and global thematic conferences. As already noted, in the community of donors the *DAC* has also played an important role in setting and maintaining norms. Traditionally, the developing countries have chosen the UN system as their main arena for influencing the international development agenda.

1. Conflict Resolution and Aid for Relief and Reconstruction

The new conflict patterns emerging after the end of the Cold War (see above, section II.3) brought the security situation to the fore: peace and stability were increasingly seen as preconditions for development. In some instances (*inter alia*, in the former Yugoslavia, Kuwait and Somalia), 'international society' intervened with armed force, at extremely high economic cost; in other conflicts (for example, in the Great Lakes Region) such interventions were ambivalent and at high cost in human lives and suffering. This led to a *security first* concern in international councils, such as in the United Nations, involving conflict resolution and peacekeeping, and at times even peace-enforcing measures. Such man-made catastrophes led to *complex emergencies* for 'international society' (involving the UN system and/or regional organisations as well as a multitude of non-governmental international and national organisations), where the work of diplomats was supplemented by armed forces in various capacities and humanitarian relief operations.

New actors, often with little or no previous insight in local (social, political and cultural) conditions, were appearing on the scene; in many complex emergencies they were often the first to arrive. Co-ordination between the many different actors posed a necessity that was almost impossible to meet, resulting in incoherent policies and actions and gross waste of resources. At the conceptual level this led to the recognition that a more holistic perspective was needed, integrating more strongly security aspects in a wider development concept. Promotion of peace and prevention of violent conflict gained prominence in the discourse on development – and became objectives in their own right for development assistance.

Aid was increasingly conceived of as a tool to secure and maintain an enabling environment for development, and to contribute to preventing conflicts from flaring up, conflict resolution, peacekeeping and peacebuilding, including rehabilitation and reconstruction of the damage caused by violent conflicts. Humanitarian aid of various kinds, including relief aid and support for 'traditional' refugees and a growing number of 'internal' refugees, was part of this picture.

This 'new', high-profile role of aid was not uncontroversial. What was the particular role of aid *vis-à-vis* other traditional foreign policy instruments in conflict resolution, such as diplomacy, sanctions and peacekeeping – even humanitarian intervention? A distinction has been made between three approaches to the use of aid: (i) to gain greater insights into the problem areas (research), (ii) to improve such basic conditions that, if not improved, might create a fertile ground for violence or trigger war (the classic role of aid – *inter alia*, by fighting poverty), and (iii) as a lever to influence the policies of recipients in order to prevent a conflict from breaking out or to bring the parties to solve it by other means (aid conditionality as applied to new objectives). This triggered a discussion of the strength and limitation of aid as compared with other instruments in conflict prevention and peace promotion [*Stokke, 1997: 202ff.*].[53]

In actual practice, the main role played by aid in this setting has been to provide humanitarian assistance, particularly aid for relief and refugees, which took an increasing share of total aid as from the late 1980s and during the 1990s. This development involved changes in priorities, away from the 'traditional' concern for economic and social development. It also affected the geographical distribution of aid. The large and many small conflicts caused humanitarian aid to grow steeply, more with some donors than with others. The major conflicts that exploded in Central/South-East Europe distorted the established patterns for the distribution of aid: much assistance went to new recipients within Europe at the expense of traditional recipient countries in the South. The changed priorities reflected in such re-allocations of aid were not uncontroversial.

Humanitarian assistance itself, previously shielded from criticism because of the humanitarian imperative to assist people in ultimate need, came increasingly under fire. Based on evidence from humanitarian relief operations, some critics argued that, instead of being instrumental in preventing conflict, humanitarian aid might generate conflict and trigger conflicts into violence, and even prolong them.[54]

---

53. The DAC High-Level Meeting on 25–26 April 2001 addressed violent conflict prevention as a high-priority issue and considered it an integral part of the quest to reduce poverty. The 2001 DAC report summarises the cost of violent conflict as follows: 'Since the end of the Cold War, violent conflict has caused over 5 million casualties, 95% of whom are civilians. Out of the 38 poorest countries of the world, 20 are in conflict. ... Ten of the 24 most war-affected countries between 1980 and 1994 were African. In Rwanda alone, an estimated 800 000 people were killed in the 1994 genocide, 1.5 million people were internally displaced and a further 800 000 made refugees. The Carnegie Commission on Preventing Deadly Conflict has estimated that the costs to the international community – in addition to the costs to the countries actually at war – of the seven major wars in the 1990s, excluding the Kosovo conflict and calculated before the close of the decade, had been USD 199 billion' [*OECD, 2002: 26–8, 135 (quotation) ff.*]

54. See, *inter alia*, De Waal [*1997*], Esman [*1997*] and NAR [*1996*]. Prendergast [*1996: 17ff.*] argued that aid sustained conflicts in three major ways: it can be used directly as an instrument of war, it can be indirectly integrated into the dynamics of

This dilemma – that aid in a given context may run the risk of being misused and cause harm as well as good – is not new, nor limited to humanitarian aid.

2. A Revival of the Social Dimensions of Development

Towards the end of the 1980s, protests against the neo-liberal orthodoxy gained strength as the social costs of the structural adjustment policy became increasingly recognised: as already noted, the pendulum was about to swing back again. In this process, the UN system played a crucial role by giving emphasis, in various ways, to the humanitarian, social and ethical dimensions of development. In 1990, poverty alleviation was again given prominence in the UN strategy for the fourth development decade [*UN, 1990: 5, 22ff.*]. A series of global conferences contributed to direct the focus to the social dimension of development more generally or to particular aspects of this dimension.

The World Bank was not unaffected by the mounting critique against the structural adjustment policy coming not only from the UN system but also from some of its member governments (the critique from the UN system and prominent researchers served a 'legitimising' function in this regard). A clear indication of a change of emphasis – at the level of analysis and rhetoric – can be traced to the 1989 *World Development Report* [*World Bank, 1989a*], which emphasised institutional development and prescribed a more active role for the state in development. The subsequent report [*World Bank, 1990*] turned its focus towards poverty alleviation. The main focus of the World Bank in the 1990s – on good governance, in both the public and the private sector, where the state has a crucial role to play – also represented a policy readjustment away from a one-sided focus on macroeconomics.[55]

---

the conflict, and it can exacerbate the root causes of war and insecurity. With illustrations provided from several conflict areas, particularly in the Sudan, Ethiopia and the Great Lakes Region, he shows how the conflicting parties (and host governments, in particular) were able to manipulate and blackmail donor agencies to serve their own objectives, particularly through denial of food aid, manipulation of the distribution of this aid in order to gain favours and support, and even to maintain the military presence and the direct diversion of humanitarian aid to sustain their capacity to fight.

For a discussion of the moral dilemmas involved, see Slim [*1997: 16–17*]. Discussing the risks involved in humanitarian intervention, he noted, *inter alia*, that getting involved 'is bound to lead the helper into an encounter with humankind at its worst, as well as its best. In such situations, helpers soon find themselves dining with the devil. And in doing so, no matter how long the spoon, they will tend to find themselves feeding on moral compromise and getting dirty hands' [*ibid.: 3*].

55. For the follow-up, see, *inter alia*, the strategies established [*World Bank, 1991*], the manual for implementation [*World Bank, 1992a*] and reports in 1994 on the fol-

The UN's most important contribution was perhaps the UNDP search for alternative indicators of development, first presented in the first issue of the *Human Development Report* [*UNDP, 1990*], itself instrumental in turning the tide. The effort to develop a *human development index* continued, and was further elaborated in subsequent reports [*UNDP, 1991–95*].[56] The main effect was to direct attention more directly towards the social and 'human' aspects of development instead of the traditional ones (indicators of economic growth). From a conceptual point of view, it represented a silent revolution: a shift of focus with regard to what development was all about.

Other arms of the UN system followed up on the social and human dimensions of development. The *Social Summit* in Copenhagen in March 1995 was instrumental in placing the social dimension firmly on the development agenda. Fighting poverty became the overall theme. Poverty represents a multi-faceted problem, and development assistance may be used as both a lever and a targeted contribution. The main contribution of the summit in this regard was the so-called 20/20 agreement: developing countries were to allocate 20 per cent of their public expenditures for basic social programmes, with particular reference to health and education, and the donor countries, in turn, were to allocate 20 per cent of their ODA for these same purposes. Education and health were singled out as strategically important both for attaining 'development' and for fighting poverty [*UN, 1997: 1113ff.*].

The emphasis on health and education was not accidental: in the follow-up to the structural adjustment policy, expenditure on social services was the first to be cut. In several countries, the major advances in education and health in the 1960s and 1970s – including life expectancy – not only stagnated: in the 1980s, development trends reversed. The emphasis was also, at the 'practical' level, a follow-up to the new emphasis on human development, triggered by the work on the human development index. It was part of the wider perspective that came increasingly to the fore in the development discourse of the 1990s, emphasising competence building and institutional development.

---

low-up [*World Bank, 1994ab*]. However, rhetoric and policy implementation do not always coincide [*Emmerij, 1995*].
56. Mahbub ul Haq, for years situated at the centre of power both in his home country (Minister of Finance in Pakistan) and internationally (in the World Bank and the UN system), was the main architect of this silent revolution as far as the perception of development is concerned. See also his reflections on human development [*Haq, 1995*].

Other global summits in the 1990s and beyond focused on other, more specific, issues that are part of the social and human development agenda. Thus, the plan of action from the world summit for children in 1990 [*UN, 1999: 797ff.*] and a series of world conferences on women, including the fourth in Beijing, also contributed to putting the social dimension of development on the agenda in the 1990s.

3. The New Poverty Agenda

Poverty alleviation, increasingly expressed in more active terms as fighting poverty, has always been central in the aid and development rhetoric: improving conditions of the poor has been at the core of development assistance. In the early 1970s, the World Bank formulated a redistribution strategy and identified the poorest part of the population as a target group for aid [*McNamara, 1973*]. Poverty alleviation constituted a central objective in the basic needs strategy of the mid-1970s and was a major concern when the social and human dimensions of development were again in the forefront in the late 1980s and early 1990s. In the 1990s, reduction of poverty came to be the primary declared objective of most donors, multilateral as well as bilateral. The rhetoric – the justifications given for aid and the objectives set – has represented good intentions. However, the discrepancy between intentions and deeds has remained large for many governments, international aid agencies and development finance organisations. International agencies and some bilateral donors have only recently initiated the difficult process of systematically transforming lofty objectives into operative activities.

Poverty is a complex concept and processes leading into and out of poverty are even more complex. In the early days, economic growth was the simple answer – and it still remains part of the answer, although insufficient in itself. A much more holistic development concept has emerged with effects also for strategies to fight poverty involving a web of approaches including, *inter alia*, 'traditional' economic growth as well as social, institutional, political and cultural development, freedoms, the participation of those involved – and human dignity. Security, in the extended version of the concept, has increasingly been a part of the development concept. Such components are increasingly seen as interrelated.

The World Bank exercises a strong influence on both the international development discourse and development policy and its implementation, both as an instrument for other major actors and as a major actor in its own capacity. This gives added weight to conceptualisations and policies formulated in major World Bank publications. Although analyses and recommendations in the same direction had come from other quarters, even more pointed, the *World Development Report 1990* [*World Bank, 1990*] and its follow-up (see above, subsection 2) probably had a stronger signal effect. *WDR 2000/2001* – given the title *Attacking Poverty* [*World Bank, 2001*] – influenced both the conceptualisation of the problem and the ways to tackle it. The theme remained high on the agenda when the core development challenges were identified in the 2003 report [*World Bank, 2003:Ch. 1*].

The strategy departs from the millennium development goals (see below, subsection 9). It outlines a complex reality as the starting point for a strategy to fight poverty. Poverty is seen as more than inadequate income or human development; it is also vulnerability and a lack of voice, power and representation.[57]

The World Bank report relied on and summed up debates that, during the 1990s, had been driven even more forcefully by other international and some regional and national institutions, and by individual academics as well. During these years, the poverty issue and commitments to reduce poverty have been high on the international agenda and on the agenda of individual countries as well. However, the follow-up has been weak at all levels with regard to the conceptualisation of the problems involved, and even more so when it comes to the operationalisation of the commitment to fight poverty. It remains to be seen to what extent the proposed strategy (and instruments, such as the introduction of poverty reduction strategy papers (PRSPs) in the planning procedures for recipients of aid) will be instrumental in this process.

4. Development and the Environment

Stimulated by the 1972 UN conference in Stockholm, a concern for the environment came gradually to the fore in the second part of the 1970s and the early 1980s. With the Brundtland Report [*WCED, 1987*], whose main conclusion was that economic growth should be sustainable and should not affect negatively the opportunities of future generations to satisfy their needs, the issue emerged as a priority theme on the agenda. A precondition for both economic redistribution (social development) and ecologically sustainable development was a strong, intervening (developmental) state. Sound environmental development would also be dependent on an effective international

---

57. The three main pillars of the strategy outlined were promotion of opportunity, facilitating empowerment and enhancing security. The report also explored the instrumentalities of the approach.

regime. In the 1980s, the prevailing ideological climate was not favourable in this regard.

Nevertheless, the UN system took up the challenge and the World Bank joined in. The World Conference on Development and the Environment in Rio de Janeiro in 1992 brought the environmental issue forcefully on to the agenda and developed a link between the environment and development assistance, particularly through the mechanism of the Global Environmental Facility (GEF) to be administered by the World Bank. However, the environment has an institutional set-up of its own and a broader scope than that set for development assistance. The core concern of an environmental policy that identifies poverty as a major threat, the multi-faceted challenge of achieving sustainable development [Stokke, 1991], is also a primary objective for development assistance.[58]

---

58. The broader issue, the relation between 'development' and 'the environment', is controversial and hinges on the questions raised in several contexts related to aid: What kind of development, for whom, and at what costs? For environmental 'fundamentalists', almost any kind of 'development', particularly any related to Western development models of the post-World War II era, is seen as part of the problem, not of the solution. In the post-Rio discourse, this is illustrated in contributions by, *inter alia*, Sachs [1993), Hildyard [1993] and Finger [1993]. Sachs argued that 'the governments at Rio came around to recognizing the declining state of environment, but insisted on the relaunching of development. Indeed, most controversies arose from some party's heated defence of its "right to development" ... the rain dance around "development" kept the conflicting parties together and offered a common ritual which comforted them for the one or other sacrifice made in favour of the environment. At the end, the Rio Declaration ceremoniously emphasized the sacredness of "development" and invoked its significance throughout the document wherever possible' [Sachs, 1993: 3].

Who were the winners? According to Hildyard, the 'World Bank not only emerged with its development policies intact, but with control of an expanded Global Environmental Facility (GEF), a prize that it had worked for two years to achieve. The US got the biodiversity convention it sought simply by not signing the convention on offer. The corporate sector, which throughout the UNCED process enjoyed special access to the Secretariat, also got what it wanted: the final documents not only treated TNCs with kid gloves but extolled them as a key actor in the "battle of the planet". Free-market environmentalism ... has become the order of the day, uniting Southern and Northern leaders alike' [Hildyard, 1993: 22].

Although firmly placed on the development agenda, perspectives differ. At one end of the spectrum is the 'Limits to growth' school, at the other are technical optimists (associated with the World Bank), arguing that resource constraints can be overcome at modest cost provided the correct policies are put in place [Bhaskar and Glyn, 1995]. For critical perspectives, see, *inter alia*, several contributions in Bhaskar and Glyn (eds) [1995] and Sachs (ed.) [1993], Middleton *et al.* [1993] and Chatterjee and Finger [1994].

This is not the place for assessing what has been achieved within this policy area and what has not been attained since environmental issues appeared as a top priority on the international agenda. Perhaps the main achievement has been the broad recognition of the 'precaution first' principle – a concern for avoiding the possible damage to the environment of new activities before they are initiated. With particular reference to development-financed activities, this principle has wide implications, par-

The financing of activities geared toward improving the environment became an item in the discussion. Several governments argued that ODA budgets should be protected: some followed up on this argument, with particular reference to part of the contributions to UNEP and the GEF. For most governments, however, improvement of the environment became a prime development objective in its own right, to be pursued both through aid targeted on such projects and as an integral part of all development assistance. Some governments distinguished between global environmental challenges that were to be financed by all governments (outside the ODA budgets) and the solution of national and local environmental problems that might be assisted through aid programmes. Aid might also be used to help developing countries fulfil their international environmental commitments.

However, many of the challenges, particularly the global issues, needed an international consensus that was hard to attain – and even more so after the Republican Administration took over in Washington in 2001. When consensus is needed, the governments resisting reform tend to have the upper hand. Thus, although Agenda 21, the Rio commitments, served as an inspiration and remained high on the international agenda, little was achieved. In the course of the 1990s, a polarisation on this issue built up between the US and Europe – manifest also in the 2002 World Summit on development and the environment in Johannesburg. The lack of consensus affected negatively what could be achieved.

5. Gender and Development

The *Human Development Report 1995* and, even more strongly, the Fourth World Conference on Women in Beijing in September 1995, brought the gender perspective strongly to the fore, although it had been prominent on the human development agenda since 1976 and increasingly in the 1980s. In the context of aid policy, this cross-cutting issue has been addressed, almost since the mid-1970s, through a two-pillar strategy: supporting activities targeted on women and integrating a concern for women in all development activities – and gen-

---

ticularly for large infrastrucural projects (including huge energy projects that involve regulations of rivers, dams, etc.). The sensitivity of the international institutions of development finance (and bilateral donors) in this regard has increased during the 1990s, particularly when 'helped' by 'green' national and international protest movements. However, there is a need to differentiate between governments – on donor and recipient sides alike: attitudes to environmental issues have differed along ideological lines, national interests have been interpreted differently by governments, and national economic interests have played a more prominent role for some governments than for others, as demonstrated at the 2002 Johannesburg summit.

erally ensuring that development assistance does not negatively affect development from a gender perspective.

The Beijing conference addressed the interrelated issues of development, equity and peace from a gender perspective, as did the first world conference in Mexico City 20 years earlier, further developed during the UN Decade for Women (1976–85). An agenda for action was defined with the purpose of empowering women socially, politically and economically, in both private and public life, and eliminating all forms of discrimination against them. The connection between the advancement of women and progress for society as a whole was again emphasised. Societal issues had to be addressed from a gender perspective in order to ensure sustainable development. The Beijing Declaration committed the international community to advance the goals of equity, development and peace and to implement the Platform for Action ensuring that a gender perspective was reflected in all policies and programmes. The platform outlined 12 critical areas of concern: poverty, education, health, violence against women, armed conflict, economic structures, power sharing and decision-making, mechanisms to promote the advancement of women, human rights, the media, the environment and the girl child [*UN, 1997: 1168ff.*].

The 1995 conference set a broad agenda for gender and development, ranging from human rights to mechanisms to improve opportunities for women in the workforce and in business. It brought the issue strongly on to the agenda for the rest of the old millennium and into the new. The fact that women are over-represented among the poor – the 'feminisation of poverty' – has negative consequences both for societies and for families, affecting children's opportunities for development as well. Any strategy for poverty reduction involving development assistance will therefore of necessity include a gender perspective.

6. Debt and Development

In the second part of the 1970s, a debt problem affecting most developing countries was mounting and exploded in the 1980s. The long-term debt of developing countries had increased from US$154bn. in 1975 to US$424bn. in 1980 and more than doubled during the following decade to reach US$1024bn. in 1990. During the 1990s, the debt continued to rise, reaching US$1354bn. in 1995 [*UNCTAD, 1985, 1995*].[59] Its magnitude emerges when compared with total ODA transfers of about US$50bn. a year in the late 1990s.

---

59. Definitions vary – according to the IMF, the debt of developing countries in 1995 amounted to US$1714bn., increasing to US$2065bn. in 2000 [*IMF, 2001*].

The composition of this debt had changed profoundly during the late 1970s and early 1980s: while previously it consisted to a large extent of official loans from bilateral and multilateral institutions of development finance (ODA), private financial institutions – with petrodollars in abundance – were increasingly taking over as major creditors. In the early 1980s, when developing countries began to find it difficult to service these debts, the alarm bells started to ring in the commercial financial institutions: the debt problem became a highly prioritised political problem also in the North. In August 1982, Mexico declared its inability to service its debt.[60] From the perspective of the Northern commercial banks involved, supported by their governments, a vital concern was to find ways and means of enabling the indebted Third World governments to service and repay their loans.

The increasing debt burden had a negative effect on the development of the countries concerned. Debt servicing demanded an increasing share of their income from exports and reduced their ability to invest in development. In section III, the debt crisis was identified as one of the causes of the advance of neo-liberal orthodoxy in the 1980s. In their analysis of the crisis, the Bretton Woods institutions laid the responsibility one-sidedly on the doorsteps of the governments concerned: their economic policy was wrong and had to be reformed. As already noted, development assistance became the major instrument in this regard: first-generation aid conditionality, with structural adjustment lending as the tool, was used to force governments to adjust expenditures to earnings. But this went beyond economic housekeeping; the instrument was also used to enforce the liberalisation of trade and foreign-exchange restrictions, deregulation of the economy, privatisation and other elements of the 'less state, more market' strategy of the so-called Washington Consensus, reflecting the neo-liberal orthodoxy.

---

60. This is not the place for analysing the causes of the debt crisis. The governments incurring the debts bear their share of the responsibility – also for spending the credits in a way that was not economically viable. However, international developments outside their control also contributed to the misery: the stagflation in the industrial countries that reduced the demand for imports of the traditional exports of developing countries and brought prices down, while the prices of their imports (oil and capital goods) went up. A strict monetary policy in the main industrialised countries raised interest rates, making the loans more expensive. And a resurgence of protectionist and restrictive trade policies in the industrial countries affected the exports of developing countries negatively, further reducing their income. Changes in the world economy, beyond the control of these governments, therefore represent the major factors propelling the debt crisis.

The debt crisis has been high on the international agenda since it exploded in the early 1980s. Over the years, many ways of solving the problem have been put forward; it has even been suggested that the commercial banks themselves should write off part of the debt, thereby sharing the responsibility for the crisis. The main strategy, however, has been to enable the debtors to pay back their loans by loan rescheduling plus some waiving of debt involving public creditors.[61] Such arrangements have more often than not been covered by the aid budget of the donor country concerned. From an aid and development perspective, little has been achieved: ODA has been used to bail out unviable commercial projects, many of which had originally not been considered or justified as development assistance.

During recent years, the focus has increasingly been directed towards helping out heavily indebted poor countries (HIPCs) and adding a poverty-reduction dimension to the debt relief efforts. NGOs have played a decisive role in this regard, with particular reference to the Jubilee 2000 initiative.[62]

---

61. A landmark in the many efforts to find solutions to the debt problem was the so-called Baker Plan (1985), whereby the commercial banks agreed to extend their lending to developing countries that agreed to economic reforms under IMF control (US$20bn. over three years), provided the multilateral institutions of development finance did likewise (US$3bn. on a yearly basis). Under the so-called Brady Plan (1989), the focus shifted to debt reduction, voluntary exchanges of old debt for new bonds and lending by the multilateral development financial institutions to enable debtor countries to buy back debts at a lower price. These arrangements, while easing the situation for some middle-income countries, did little to help the low-income countries. In the late 1980s and the 1990s, NGOs appeared more strongly on the scene focusing especially on the debt situation of poor countries, with particular reference to the Jubilee 2000 initiative. They managed to persuade governments, including the G-7 countries, to improve schemes for debt reduction more generally and to consider the HIPC initiative in particular.

62. Prominent international actors dealing with the debt issue have been the World Bank, the IMF and the so-called Paris and London clubs, in which major bilateral creditors also played a decisive role. The debtors have been given little possibility to influence the policy: the power relations between North and South have seldom been more openly asymmetrical than in these negotiations.

Arrangements that had some form of burden sharing between debtors and creditors and that discriminated between debtors according to their economic ability with regard to debt waiving have, over time, increasingly come to the fore. Another concern – that debt waiving should be to the benefit of the developing countries concerned and not to other creditors – was part of this discussion. In addition, the idea of a multilateral debt-exchange mechanism, involving agreements between the creditor country and the developing country involved, whereby the debt relief was to be connected to environmental or other development-promoting efforts by the developing country, has been brought to the fore (debt swap for equity, environmental and social projects) – making the use of ODA for debt relief more acceptable from a 'development' perspective. Thus, the HIPC initiative, for example, tries to set the preparation of a Poverty Reduction Strategy Paper by the country concerned as a condition for the debt relief. However, as argued in UNCTAD's 2000 report on the least-developed

7. *A Conducive Environment for Development Co-operation: Participation and Ownership*

In the late 1980s and 1990s, the participation of the authorities of developing countries in development efforts was increasingly acknowledged as a precondition for development. With particular reference to the World Bank rhetoric, developing countries' 'ownership' was necessary for development assistance to be effective. The main lesson learned from decades of development co-operation was that the national authorities should be in the driving seat in setting national development priorities as well as in the planning, implementation and evaluation of the development assistance. The role of the donors was to assist the governments of developing countries in their efforts (see, in particular, OECD [*1992b, 1993, 1995abc, 2002: 79ff.*]).

However, these were not the only lessons learned. The neo-patrimonial state was also part of the 'new' knowledge with a strong impact on the thinking on development and aid: it was seen as equally important to create an enabling environment for development. In the 1990s, good government appeared as the main answer, involving fighting corruption as well as assistance for competence and institution building in the countries concerned.[63] With the fight against poverty emerging as task number one, strategies and guidelines for poverty reduction became part of this package – to be established and

---

countries, the HIPC initiative is considered inadequate to meet the needs of the poorest countries for sustained growth and poverty reduction, and there is a danger that debt relief will be substituted for development assistance. The addition of the poverty reduction conditionality to the already extensive list of formalities may complicate even further the situation of these countries, it is argued [*UNCTAD, 2000*].

63. A recent theme in this debate concerns the conditions under which aid will work. A World Bank research report, *Assessing Aid, What Works, What Doesn't, and Why* [*World Bank, 1998*], has exerted strong and widespread influence, not least in settings where aid policy is formed and implemented. The main message is that aid may work provided recipient governments pursue the 'right' policies (that is, policies adjusted to the 'Washington Consensus'; a great number of developing countries were found to do so), otherwise not. This argument has been pursued in a series of publications emerging from the Development Research Group of the World Bank (*inter alia*, Burnside and Dollar [*2000*], Collier and Dollar [*2001, 2002*], and Dollar and Kraay [*2001*]). The analysis has been criticised and the conclusions contested by other economists (see, *inter alia*, Hansen and Tarp [*2001*], Lensink and White [*2001*], Amann *et al.*[*2002*] and Morrissey *et al.* [*2002*]). Morrissey *et al.* argue, based on an econometric study of aid and growth in sub-Saharan Africa, that aid works regardless of government policies.

A paper prepared for the UN Monterrey conference on financing for development sums up lessons learned and policy positions of the World Bank with regard to the role and effectiveness of development assistance [*World Bank, 2002*].

implemented by the developing countries concerned and scrutinised by the donor 'community'.[64]

The donor policies are therefore full of contradictions and paradoxes. The same governments and multilateral organisations which designed and implemented first- and second-generation aid conditionality in an effort to 'convince' the governments of developing countries that reform of their economic policy and even their political systems was necessary if they wanted to receive development assistance, also insisted that the very same governments should claim 'ownership' of these policies. This illustrates both that 'lessons learned' may vary, and that such lessons and the aid rhetoric are not always alone in informing and forming policies.

8. Policy Coherence

In 1991, the high-level meeting of the OECD Development Assistance Committee recognised the need for effective participation by developing countries in the global economy and their co-operation in confronting global challenges and in solving regional security issues. Enlightened self-interest was made explicit, particularly the longer term interest in the effective functioning of the global economy and the solving of other problems that were global in scope, including the threat to environmental sustainability and threats associated with orderly political and economic transitions in many parts of the world. As perceived by the DAC, such conditions could not be obtained without substantial political, economic and social development in the South.

Development co-operation, in the form of financial and technical assistance to help the countries in building national capacities, remained essential. Because of their significant impact, it was also considered crucial that OECD macroeconomic, trade, finance and other policies towards the South should be part of an integrated approach in support of developing countries. The DAC meeting opted for broad coherence across this range of OECD country policies as a central feature of a larger economic and political framework for managing global challenges over the coming decades [*OECD, 1992a: 31*]. This common ground was further elaborated in the years that fol-

---

64. In the words of the 2001 report of DAC, 'the number of Poverty Reduction Strategy Papers (PRSPs) have increased rapidly, to about 40, over 25 of them dealing with the Initiative to reduce the debt of the most Heavily-Indebted Poor Countries (HIPCs) – even if in most cases the strategies are interim ones, only about ten "complete" papers having already been put into effect. These rapid developments justified a thorough review of the PRSP approach that was recently launched by the World Bank and the IMF, with active participation by the partner countries, but also by bilateral donors, via the DAC in particular' [*OECD, 2002:11–12*].

lowed, with particular reference to the strategy report adopted five years later, *Shaping the 21st Century: The Contribution of Development Co-operation* [*OECD, 1996a*].[65]

In an earlier EADI study, Forster and Stokke [*1999: 20–3*] distinguished between four systemic frameworks within which policy coherence (or incoherence) in the donor countries' relations with developing countries takes place. Two of these relate to policies generated within the national systems of donor countries and two to policies generated within various international systems, involving bilateral, regional and multilateral arenas, namely:

- *the aid policy of a country* whose development assistance is channelled through many different systems that may affect both objectives and strategies and the outcome: a multitude of governments, non-governmental organisations, and multilateral aid agencies and development finance institutions, each with framework conditions that may be system-specific and with different aspirations, objectives and priorities;
- *the (various) policies of a country towards developing countries*, where development co-operation constitutes only one policy area; several of its policies within other areas (trade, migration, arms sales, etc.) affect countries of the South directly or indirectly;
- *the policies of (a multitude of) donors towards developing countries*, their aid policy in particular; and
- *the policies of (one or all) donors* vis-à-vis *the policies of a developing country*, with particular reference to development policies, including bilateral state-to-state relations involving development assistance, such relations involving regional donor institutions and governments and regional institutions in the

---

65. For a systematic analysis, see Forster and Stokke (eds) [*1999*]. The notion of coherence is associated with a rational choice model related, in particular, to a defined administrative and/or political system. Within this system, the core features of the concept are that:
   – *objectives* should be formulated in clear terms and harmonised. Objectives within a policy framework, such as the aid policy, should be internally consistent, and objectives pursued within this and other policy frameworks of the system, such as trade, the environment, migration and security, should be attuned to each other or, as a minimum, not be conflicting;
   – *strategies and mechanisms* should be attuned to the objectives or, as a minimum, not in conflict with them or with the intentions and motives on which they rest; and
   – *the outcome* of administrative and political interventions should correspond with intentions and objectives or, as a minimum, not conflict with these [*Forster and Stokke, 1999: 20*].

South, and similar relations involving multilateral institutions, whatever the degree of integration.

The constraints on policy coherence are system-specific and will, accordingly, differ from one systemic framework to another and, within the four frameworks identified, from one subsystem to another. It follows that, within each of these frameworks, the objectives pursued within one subsystem may contradict objectives pursued within others, resulting in policy incoherence. The more stakeholders involved, the greater the probability of a conflict of interest, resulting in incoherent policies, and the greater the challenge if policy coherence is the ultimate objective. The focus here will be on coherence within international systemic frameworks, with particular reference to the coherence of donors' policies towards developing countries.

The justification for the concern within the OECD for increasing the coherence of member countries' policies towards countries in the South, broadening the perspective to include other policy areas than development co-operation (trade, finance, the environment, etc.), has already been indicated. The main rationale for increased coherence of the aid policy has all along been to increase the effectiveness of aid *vis-à-vis* the established development objectives.[66]

However, the stated aspirations went far beyond considerations of effectiveness and efficiency. The 1996 'vision' took as its point of departure that, in 2000, four-fifths of the world's population would be found in the developing countries. Although conditions would be improved for most of them, the numbers in absolute poverty and despair would be growing. The industrialised countries had a strong *moral imperative* to respond to the extreme poverty and human suffering that still afflicted more than one billion people, and a strong *self-interest* in fostering increased prosperity in developing countries.

---

66. The development assistance of donor governments (and multilateral agencies and institutions) should be better co-ordinated in order to be more effective and efficient. Waste through duplication should be avoided and the development assistance of one donor should not counteract that of another. This should not only apply to their aid policy, but also to their policies within areas such as trade, investments and security [*OECD, 1992a, 1996a, 1997*].

The commitment of the DAC members to policy coherence was summed up as follows [*OECD, 1996a: 18*]: 'We should aim at nothing less than to assure that the entire range of relevant industrialised country policies are consistent with and do not undermine development objectives. We will work with our colleagues in the broad collaborative effort now underway within the OECD to examine linkages between OECD Members and the developing countries, building on the promising work being done in 1994 (OECD, 1995). We are confident that we can do more than just to avoid policy conflict. We will work to assure that development co-operation and other linkages between industrialised and developing countries are mutually reinforcing.'

'Our solidarity with the people of all countries causes us to seek to expand the community of interests and values needed to manage the problems that respect no borders – from environmental degradation and migration, to drugs and epidemic diseases. All people are made less secure by the poverty and misery that exist in the world. Development matters' [*OECD, 1996a: 1*].

Three objectives were highlighted. Although expressed in general terms, normal for consensus documents of this kind, the direction of the objectives was fairly clear [*ibid.: 8–11*]:

(i) *economic well-being*: the proportion of people living in extreme poverty in developing countries should be reduced by at least one-half by 2015;
(ii) *social development*: there should be substantial progress in primary education (universal primary education in all countries by 2015), gender equality (*inter alia*, eliminating gender disparity in primary and secondary education by 2005), basic health care and family planning; and
(iii) *environmental sustainability*: there should be a national strategy for sustainable development in the process of implementation in every country by 2005, so as to ensure that current trends in the loss of environmental resources – forests, fisheries, fresh water, climate, soils, biodiversity, stratospheric ozone, the accumulation of hazardous substances and other major indicators – are effectively reversed at both global and national levels by 2015.

These objectives were linked to more qualitative ones, such as a commitment to promote social integration by fostering societies that were stable, safe and just, and based on the promotion and protection of all human rights, democratic accountability and the rule of law as key elements of integrated development strategies. 'Investment of development resources in democratic governance will contribute to more accountable, transparent and participatory societies conducive to development progress' [*ibid.: 11*].[67] As concluded earlier [*Forster and Stokke, 1999: 29*], the overall perspective of the policy established by the DAC countries as a basis for a coherent policy towards developing countries was that of development in the South, based on the objectives traditionally set for aid as refined by those coming to the fore in the late 1980s and early 1990s, poverty alleviation being

---

67. These objectives were not 'new', but agreed to by the international community at major UN conferences earlier in the 1990s and accumulated by the DAC [*OECD, 2001: 23*].

the major concern along with human rights, democracy (with an emphasis on participation) and good government. The predominant ideology of the day, market liberalism, was also part and parcel of the policy.

During subsequent years, the DAC followed up on this policy.[68] Within the European Union, policy coherence became part of its European project and was written into the constitution already in the Maastricht Treaty. This also affects its relations with developing countries, including the aid and trade policy, and 'internal' policies (subsidies, agricultural policy, fisheries policy, industrial policy). Objectives of a similar kind were identified, and acclaimed, by the so-called Millennium Summit of the UN [*UN, 2000*] and the DAC high-level meetings in 2000 and 2001 [*OECD, 2001, 2002*].

9. The Millennium Development Goals

The Millennium Summit of the United Nations in September 2000 attracted the participation of representatives of 191 nations, including 147 heads of state and government. The largest ever gathering of world leaders, taking place at this particular point in time, did not serve a ceremonial function only; they adopted the United Nations Millennium Declaration.[69] From our perspective here, the most important outcome was their commitment to the millennium development goals (MDGs). The more specific goals related to development and poverty eradication were integral parts of a broader perspective that included peace and security, protection of the

---

68. See, for instance, the overview of the DAC Chair in the 1996 report, subsequent to the 1996 strategy document, where the commitment to greater co-ordination and policy coherence is explicitly related to the objectives set out in that document – the vision [*OECD, 1997: 8–10*]. The following year, the DAC Chair gave high priority to more intensive efforts to improve co-ordination and coherence within the OECD [*OECD, 1998: 14*]. The DAC high-level meeting in May 2000 reconfirmed the commitment [*OECD, 2001: 23*]. And so did the 2001 DAC report [*OECD, 2002: 33ff.*]. They were also followed up on the 'ground' [*OECD, 1996b*].

69. The participants reaffirmed their faith in the UN and its charter and recognised that 'in addition to our separate responsibilities to our individual societies, we have a collective responsibility to uphold the principles of human dignity, equality and equity at the global level. As leaders we have a duty therefore to all the world's people, especially the most vulnerable and, in particular, the children of the world, to whom the future belongs.' Among the fundamental values considered to be essential to international relations in the twenty-first century were freedom, equity, solidarity, tolerance, respect for nature, and shared responsibility as defined (Part I). In order to translate these values into actions, key objectives were identified with regard to peace, security and disarmament (Part II), development and poverty eradication (Part III), protection of our own environment (Part IV), human rights, democracy and good governance (Part V), protection of the vulnerable (Part VI), meeting the needs of Africa (Part VII), and strengthening the United Nations (Part VIII) [*UN, 2000*].

environment, human rights, democracy and good governance, protection of the vulnerable and particular attention towards those worst off, with special reference to Africa. Improvements within such areas would also impact on poverty eradication and development; they have increasingly been defined as development objectives in their own right even in an aid context.

Some of the particular commitments to development and poverty eradication were general but still important because they highlighted the direction of the joint efforts.[70] Other commitments were quantified and to be attained within a fixed date. These call for special attention. Most of the MDGs had been formulated in similar terms before, as recommendations from more specialised global conferences and meetings during the 1990s (and integrated in the DAC 1996 strategy document). The Millennium Declaration gave them added authority. The summit resolved [*UN, 2000: point 19*]:

---

70. It was stated [*UN, 2000: points 11–18*] that:
    – 'We will spare no efforts to free our fellow men, women and children from the abject and dehumanizing condition of extreme poverty, to which more than a billion of them are currently subjected. We are committed to making the right to development a reality for everyone and to freeing the entire human race from want';
    – 'We resolve therefore to create an environment – at the national and global levels alike – which is conducive to development and to the elimination of poverty';
    – 'We are committed to an open, equitable, rule-based, predictable and non-discriminatory multilateral trading and financial system';
    – 'We will ... make every effort to ensure the success of the High-level International and Intergovernmental Event on Financing for Development, to be held in 2001';
    – 'We also undertake to address the special needs of the least developed countries. ... We call on the industrialized countries: To adopt ... a policy of duty- and quota-free access for essentially all exports from the least developed countries; To implement the enhanced programme of debt relief for the heavily indebted poor countries without further delay and to agree to cancel all official bilateral debts of those countries in return for their making demonstrable commitments to poverty reduction; and To grant more generous development assistance, especially to countries that are genuinely making an effort to apply their resources to poverty reduction';
    – 'We are also determined to deal comprehensively and effectively with the debt problems of low- and middle-income developing countries, through various national and international measures designed to make their debt sustainable in the long term';
    – 'We also resolve to address the special needs of small island developing States. ... We urge the international community to ensure that ... the special needs of small island developing States are taken into account';
    – 'We recognize the special needs and problems of the landlocked developing countries, and urge both bilateral and multilateral donors to increase financial and technical assistance to this group of countries to meet their special development needs and to help them overcome the impediments of geography by improving their transit transport systems.'

- to halve, by the year 2015, the proportion of the world's people whose income is less than one dollar a day and the proportion of people who suffer from hunger and, by the same date, to halve the proportion of people who are unable to reach or to afford safe drinking water;
- to ensure that, by the same date, children everywhere, boys and girls alike, will be able to complete a full course of primary schooling and that girls and boys will have equal access to all levels of education;
- by the same date, to have reduced maternal mortality by three-quarters, and under-five child mortality by two-thirds, of their current rates;
- to have by then halted, and begun to reverse, the spread of HIV/AIDS, the scourge of malaria and other major diseases that afflict humanity;
- to provide special assistance to children orphaned by HIV/AIDS;
- by 2020, to have achieved a significant improvement of the lives of at least 100 million slum dwellers as proposed in the 'Cities Without Slums' initiative.

In more general terms, the participants of the Millennium Assembly also resolved to promote gender equality 'and the empowerment of women as effective ways to combat poverty, hunger and disease and to stimulate development that is truly sustainable', and to develop and implement strategies 'that give young people everywhere a real chance to find decent and productive work [*ibid.: point 20*].[71]

In the declaration, a general commitment was made to increase the volume of international development assistance in the years to come, and several governments made more specific commitments in this regard, with or without a time schedule for attaining such targets (see the Introduction and subsequent chapters in this volume). However, development assistance, although important, represents only one of the means necessary to attain these targets.

The quantified targets set in the declaration focus on *outcomes* within a given time horizon instead of, as in the past, on the *input* side. Jan Pronk, an experienced politician in the field of development

---

71. The participants also committed themselves to 'encourage the pharmaceutical industry to make essential drugs more widely available and affordable by all who need them in developing countries'; to develop 'strong partnerships with the private sector and with civil society organizations in pursuit of development and poverty eradication'; and to ensure that 'the benefits of new technologies ... in conformity with recommendations contained in the ECOSOC 2000 Ministerial Declaration, are available to all' [*UN, 2000: point 20*].

co-operation and North–South policies, has found this new approach a promising one, given the many commitments to inputs in the past that have not been honoured – including the targets set for ODA [*Pronk, 2003*]. I would have found it more reassuring if outcomes of the kind identified had been combined with commitments involving inputs, both with regard to policies and resources, ODA included.

The commitments made in New York in September 2000 represent a great challenge to the world leaders, internationally and nationally. They represent a risk as well: what will happen – in terms of disillusionment and lost legitimacy – if the commitments are not followed up?

V. SOME CONCLUDING OBSERVATIONS

1. Towards an International Development Regime?

The complex environments in which aid policies are moulded – involving, *inter alia*, conflicting values and interests both within and between major actors at national and international levels, different worldviews with regard to the problems involved, and a variety of approaches to development – make a broad international convergence of development perspectives and aid policy an odd probability. However, the evolving development agenda of the 1990s, in particular the increasing demand for greater coherence of policies towards developing countries, triggers the question: Do trends in the last decade of the past millennium and the first years of the new one indicate such a convergence?

Agreement on the general aspirations, including aid objectives at a high level of generalisation, is a *sine qua non* for an international development *cum* aid regime to come into existence, but not sufficient in itself.[72] It would be necessary to have agreement among most actors, including the major ones, also on the more specific objectives and how they should be attained; procedures for their implementation, including mediation and conflict resolution; and institutions that were responsible for policy-making, monitoring and enforcement of the rules set.

Most of these components are absent or weakly developed when it comes to development co-operation. How weakly is probably best illustrated by the fate of international volume targets, repeatedly

---

72. For a mainstream definition, see Krasner [*1982: 185*]: 'International regimes are defined as principles, norms, rules and decision-making procedures around which actor expectations converge in a given issue-area.'

agreed to in the strategies for successive UN development decades and in other contexts: only a few governments have met the 0.7 per cent of GNI target, even fewer the 1 per cent of GNI target. Most governments have unilaterally decided the level of their ODA, up or down – in the 1990s, mostly down, but with an upward turn in 2002 and 2003 (tables IN.1 and IN.3) – following up, in a modest way, the more positive signals given at the UN millennium summit and more recently in Johannesburg.

Nevertheless, remaining within an international regime concept, a start has been made: international objectives, rules and norms (guidelines) have been established involving several aspects of development co-operation. The UN Millennium Assembly was instrumental in rallying international consensus around some general, but vital *objectives* that had emerged from the several, more specialised global conferences during the 1990s. As noted, most of them had been included in the 1996 DAC strategy document [*OECD, 1996a*], outlined above (see section IV).[73]

International agreement on development objectives is no novelty of the new millennium, nor of the 1990s; such objectives were part of the development strategies of successive development decades. A system for monitoring such norms and commitments has been in existence for more than 40 years, with an institution charged with this task (OECD DAC). However, it is a long way from agreements on general objectives and norms and a monitoring system to an effective implementation of the stated policy that would include, *inter alia*, enforcement mechanisms. Nevertheless, monitoring of agreed norms

---

73. The statement from the DAC high-level meeting in May 2000, entitled 'Partnership for Poverty Reduction: From Commitment to Implementation', reconfirmed its commitment to the objectives set out in the 1996 strategy document and claimed that major strides had been made 'towards the common ground of a more effective model of co-operation, including an agreed set of indicators to monitor progress towards achieving the international development goals' [*OECD, 2001: 23*].

The 2001 DAC report opened by stating: '*The goal has now been set*, by the entire international community. In the wake of the Millennium Declaration adopted by the United Nations General Assembly in September 2000, eight Millennium Development Goals (MDGs) were formulated in September 2001. Seven of these objectives take the same approach as the International Development Goals that the DAC had set forth in its 1996 report *Shaping the 21st Century: The Contribution of Development Co-operation*, and they use practically the same quantitative indicators, while supplementing them as well. An eighth objective is to "develop a global partnership for development". The result of all this, the culmination of close co-operation between the United Nations, the World Bank, the International Monetary Fund and the OECD, is the solemn institution of a partnership between the developed and the developing countries, all resolved "to create an environment – at the national and global levels alike – which is conducive to development and to the elimination of poverty"' [*OECD, 2002: 11 (emphasis in original)*].

THE CHANGING INTERNATIONAL AND CONCEPTUAL ENVIRONMENTS

represents a major achievement that may also influence the behaviour of governments.[74] In addition, mechanisms to facilitate and co-ordinate donor policies are also in the making, with particular reference to strategies (*inter alia*, for poverty reduction) and planning procedures such as the Comprehensive Development Framework (see below).

The initiative in establishing international norms and objectives has – during the 1990s, as before and since – been with the UN system and its many arms, with particular reference to the many specialised global conferences and the Millennium Assembly. When it comes to policy coherence, the initiative has from the start been with the OECD. As already concluded, the DAC is playing an important role as an arena for the donor 'community', including both bilateral and multilateral actors, to formulate and co-ordinate their policies towards developing countries and, beyond that, as an institution that initiates norms and monitors agreed commitments.

The most important contribution of the World Bank in this regard has been in designing instruments that may be effective in implementing a coherent policy *vis-à-vis* the developing countries. The major recent contribution has been the so-called *Comprehensive Development Framework* (CDF), involving a co-ordinated long-term development strategy for each recipient country jointly agreed to by all foreign actors on the scene (bilateral donors and international agencies) and the national authorities of the country concerned, with an expected involvement of the private sector and civil society also. It presupposes that the authorities of the country concerned are at the wheel.[75] In form, it is a technical design seemingly removed from

---

74. For a discussion of the coherence *problematique* from such a perspective, see Forster and Stokke [*1999*], outlining the rationale for policy coherence within the four systemic frameworks identified, but also the main obstacles. With regard to co-ordination among donors, competing interests among the donor governments are identified as a major stumbling block.

75. In the 1999/2000 *World Development Report*, the efforts of the World Bank, during the 1990s, in designing and promoting new directions in development thinking are systemised, with particular reference to the main themes selected for the annual *WDRs*. The key themes referred to are *macroeconomic policy and trade* (1991, 1997), *government, regulation and corruption* (1996, 1997), *social safety nets* (1990, 1995), *health* (1993), *education* (1998/99), *infrastructure* (1994), *the environment* (1992, 1998/99), *rural strategy* (1990), *private-sector strategy* (1996, 1997) and *gender* (1990, 1993, 1998/99). The bottom line is that conditions varied over time and for various countries and regions, and that efforts to promote development had to be adapted to these various conditions. A holistic approach to development is the main recommendation, and the comprehensive development framework is presented as a major instrument to achieve such a development and reduce poverty [*World Bank, 2000*].

'politics', characteristic of World Bank designs.[76] In actual practice, it may develop into a powerful steering instrument in ensuring donor policy coherence *vis-à-vis* the individual recipient country – but not limited to that. Given the asymmetrical power relationship – the one recipient government vs. a more or less co-ordinated donor line-up – the design may prove even more effective in ensuring donor–recipient policy coherence, based on the premises of the donors, particularly those exerting the strongest influence among them.[77]

Although still only a few steps up the ladder, something that is approaching an international development *cum* aid regime is in the making. At the start of the new millennium, the 'donor community' is definitely 'at the wheel', rhetoric aside.

2. Power Politics – Irrelevant in a Development Context in the New Millennium?

Against this background, has the course of development in the post-Cold War international system made theory on international politics related to development co-operation – with particular reference to the realist and the humane internationalism approaches – less relevant in the study of development co-operation?

In the study of polity and politics, paradigms are instruments that have to be adapted to existing and shifting environments. This applies also to the two approaches briefly discussed earlier (see section II). With the end of the Cold War, several of the conditions under which development co-operation operated changed. However, even if the environment changes, some basic concerns related to values and interests remain, although redefined and adapted to the new environment. The most salient questions relate to such redefinitions of interests and values: in the new international system, with only one superpower,

---

76. Designs, technical in scope, are not always technical only. Thus, when the UNDP designed the Human Development Index as an alternative to the well-established GNP, it contributed to a change of focus, away from indicators of economic growth towards indicators of human development, ultimately a new development concept. Similarly with the World Bank, it influences the criteria on which development, that is good development, is to be measured. A major contribution in this regard is *Assessing Aid, What Works, What Doesn't, and Why* [*World Bank, 1998*].

77. The World Bank has for years been at the wheel in co-ordinating donor countries' policies, particularly through the so-called consultative groups established for the individual recipient countries, composed of the major donors under the chairmanship of the World Bank (in Paris). The Bank prepared the basis (main documents) on which the dialogue and negotiations between the many donors and the one recipient government were to take place. Recommendations emerging from these meetings were also followed up by tacit or explicit aid conditionality. If the recipient country was small, poor, debt-ridden and aid-dependent, the asymmetrical power relationship was extreme: the recipient government was not left with much of a choice.

have they moved towards ensuring common global goods[78] rather than securing national interests? Towards greater political and cultural pluralism and tolerance or less? The answers to such questions may vary with the perspective chosen – from the centre or the periphery of the new power relationships. However, they may be ambivalent, whatever perspectives are chosen, because there are several processes under way and all are not moving in the same direction.

The reality with which developing countries, varying extensively when it comes to resources and predominant values, are confronted today is the globalisation process, involving all kinds of relations – economic (finance, trade) as well as culture and politics. For many developing countries and social groups this process, at the global or regional levels, also involves marginalisation. These processes have been going on for many years and do not belong to the post-Cold War era. However, with the end of the Cold War and with the ideology of neo-liberalism in an almost hegemonic position, much of the restraints were removed. The major remaining constraint relates to the parallel processes of regionalisation. However, for most practical purposes involving most developing countries, this is part and parcel of the globalisation process: the only regional process with a prospect, in the shorter term, to pose a challenge to globalisation is confined to the North, namely the ongoing process in Europe (the EU).

In the transformation of the international system that took place after the end of the Cold War, the former bipolar system turned gradually, during the 1990s, into a unilateral system where the US appeared as the hegemon. Hegemonic power may, however, be exercised in many different ways.

With the emergence of the new Republican administration in the remaining superpower, the features of the new international system became clearer than under the Democratic administration in the 1990s.[79] From the beginning, the new Bush administration

---

78. The reference to common global goods, here and earlier in the text, is to the broad concept of 'international common goods' (such as international stability and peace, economic stability, the environment, etc.). It may also include elements of the more specific current concept of international public goods (such as health, education, safe drinking water, etc.). The UNDP Human Development Report Office, in particular, has recently explored this concept and policies that might follow from such a perspective [*Kaul* et al. *(eds), 1999, 2003*].

79. Since World War II, the foreign policy of the United States has been torn between two main approaches, unilateralism and multilateralism, with the isolationalism of the past lurking in the background. Both main approaches have been based in a realist tradition, although multilateralism has been associated with idealism as well. Unilateralism seeks the attainment of foreign policy objectives, as defined within a domestic framework, by means of its own national instruments – the full battery, involving bilateral diplomacy as well as economic and military power. However,

demonstrated strong determination to use the powers at its disposal to safeguard US interests, as perceived in Washington, and a unilateral approach became visible almost immediately.

Then came the shocking events of 11 September 2001. For the first time since Pearl Harbor in December 1941, the US was hit at home, with thousands of casualties, by an enemy that was not easily identified or located. The horrors were transmitted to the whole world. This act of terror strengthened the trends towards unilateralism, already visible, and the determination of the new administration to hit back hard against the invisible enemy and those states that were assumed to provide support or cover to that enemy, directly or indirectly. The terror was presented as a military assault and the response designed in military terms: the war against terrorism was to be fought by military means.

The ideological basis of the new Bush administration and the modes in which power has been exerted indicate a dramatic shift of foreign policy paradigm, towards a unilateral approach, with fundamental consequences for international politics, including development policy. The crucial question is: How robust will this shift of paradigm be, confronted with the US electorate and costs incurred from its implementation?

The unilateral approach has been particularly strongly demonstrated when it comes to security policy, as prescribed and practised. International agreements that might represent a constraint to security concepts have been unilaterally disbanded (the ABM treaty being an example) and entry into others declined. The new national security strategy formalises this practice, emphasising that the United States will, if necessary, act unilaterally to protect itself against what it considers threats to its security, even by pre-emptive action [*White House, 2002*].

The war in Iraq illuminated several aspects of the new approach, in line with the new strategic concept. While US governments traditionally, from the end of World War II, have been in the forefront in

---

multilateral solutions are not *a priori* excluded from a unilateral approach if the primary objectives are not compromised; they will, however, be circumstantial and not sought on a permanent and committing basis. Multilateralism, in contrast, seeks to realise foreign policy objectives through international co-operation in combination with the use of national foreign policy instruments. This strategy can involve ad hoc solutions, but will have a preference for more long-term collective commitments and international 'regimes' associated with international law. During most of the post-World War II period, and particularly during its first decades, the multilateral approach had the upper hand in US foreign policy – particularly under Democratic administrations – and this strategy has served the foreign policy interests of the superpower well.

defence of liberal political ideas and in building international 'regimes' to fortify such ideas, including international law,[80] the Bush administration set itself above the 'law' by acting unilaterally without approval of the UN Security Council. It acted against strong protests coming from some of its major European security partners in NATO.

A unilateral approach is discernible also in other policy areas, involving such international common goods as those included in the Kyoto Protocol, strongly resisted by the US administration. The other side of the coin, of equal importance, is that 'the rest of the world' has committed itself to reform within this policy area, at least as regards its rhetoric – however modest the reform may be.

This brief analysis does not justify the conclusion that interests, values and power relations have become less relevant in the study of international relations. On the contrary, it points to the need to define interests more broadly than short-term national interests; they should be extended to global common goods. When it comes to actual politics, one implication of the new world order may well be that these 'international common goods' – involving security as well as economic concerns, political ideology as well as norms regulating interactions – will increasingly be defined and implemented by the centre of the new world order, further marginalising old and new peripheries.

However, this may not be the only future scenario. The unilateral behaviour of a superpower may provoke other major actors to join forces around competing policies and foreign policy instruments. The 'dialogue' during recent years between 'Europe' and the US over a wide spectrum of issues indicates the possibility of such a scenario; this applies to the development agenda as well, as demonstrated at the Johannesburg summit on development and the environment. This dialogue may well strengthen the efforts, within the European Union, to attain a common foreign and security policy, and nurture European identity as well – even if the main conflicting issues split European governments as well.

The Iraqi conflict has illuminated yet another paradox: the limitations of superiority in military power when it comes to securing peace and stability once a war is won. There are also limitations to how far military superiority can be converted into influence within other policy areas. To illustrate the point: an old, loyal ally like Germany may be strongly criticised for not falling into line in a high-priority issue driven by Washington (even in a highly sensitive and critical situation

---

80. Admittedly, there have been exceptions and lack of consistency in this regard, as alluded to earlier (see, *inter alia,* footnotes 3, 42 and 46), particularly involving Republican administrations and US relations with Latin America.

for the government, a general election immediately ahead), but it cannot be bombed into agreement. No one would even consider trying to influence France, traditionally a more independent player, by the use of military force. And tough criticism from a hegemonic power does not always have the intended effect; rather the contrary. If the political supremacy of today's hegemon was to be challenged, future development co-operation would be affected. It might even become an important foreign policy instrument – no novelty for developing countries.

The new world order, with the hegemon following a unilateral foreign policy, represents a threat to the multilateral system, particularly to parts in which the hegemonic power is not in almost full control. This applies, in the first place, to the UN system, including its specialised organisations and programmes in the field of development assistance. It may also influence priorities of individual donors that find themselves in a dependency relation with the superpower, particularly involving security. In the short term, this may again be illustrated by the US military involvement in Afghanistan and Iraq: development assistance will be needed for years to come (although, in Iraq, income from oil may help) and directed to accommodate immediate suffering (relief aid), and to repair and reconstruct damage caused by war, rather than focusing on long-term development. This will reinforce a trend established already in the early 1990s, resulting from major conflicts elsewhere (in the former Yugoslavia, in particular) and with implications both for the geographical distribution of aid and the purposes for which it is used. The more long-term perspective is that development assistance will increasingly emerge as an instrument in a foreign policy that has security (fighting terrorism) as the primary concern, and that the agenda and priorities will be decided by an administration that is guided by a unilateral approach to foreign policy.

Such a perspective represents a challenge to those governments within the European Union that have a development co-operation tradition geared towards long-term economic and social development – in fortifying and further developing the best elements of that tradition within the regional setting of the EU and, from there, in multilateral settings such as the UN and the international institutions for development finance, the World Bank in particular.

*

At this stage it is not possible to come to any firm conclusions with regard to the '11 September' effects on the foreign policy of the new hegemon. The unilateralism of the Bush administration, strengthened by these events, will soon have to stand the test of the US voters, and the outcome of that election may again turn the tide. The extreme unilateralism pursued at present has resulted in a polarisation between the major political parties on this issue. The Democrats have committed themselves to multilateralism and a policy of consensus-seeking. And developments in Iraq have pushed the present administration into retreat, opting for the UN to intervene.

The hegemon has not succeeded in obtaining general acceptance internationally for its security agenda, involving doctrine and means, not even from most of its closest allies. For several European powers, the fight against terrorism did not start in September 2001. Nor have they relied on military means in the first place. In other corners of the world, criticism of the US policy has been even more outspoken: the US security policy has even been seen as a greater threat than that emanating from organised terrorism – because of its violation of international law and human rights, and because it has contributed to a weakening of democratic processes that have been under way in many regions of the world, the very values it purports to defend, and has strengthened rather than weakened international terrorism.[81]

So far, these developments have affected the aid and development policy of most European countries only marginally. The effects may be stronger in future, particularly if the UN takes over responsibility for future developments in Iraq. This may direct humanitarian assistance as well as assistance for reconstruction to that region beyond what has been provided so far.

3. Development Paradigms and Aid Policy: Towards an International Convergence?

It has already been concluded that something like an international development regime is in the making, based, in the first place, on the objectives set out by the UN Millennium Assembly and what seems to be a convergence of ideas about 'development' and ways and means of attaining these objectives. The transformation of the international system in the early 1990s resulted in a more complex development

---

81. In 2004, the Center of International Cooperation in New York organised, in co-operation with regional partners including NUPI, a series of regional consultations on various approaches to and understandings of the security concept after 11 September 2001. The project is to be summed up at a meeting (at NUPI) in Oslo in July 2004, where all regional partners will be present ('The UN High Level Panel on Threats, Challenges and Change'), and then to be published. Although regional perspectives varied from one region to the other, the scepticism regarding the US policy was outspoken in most regions, if least so in Asia [*Eide, 2004; NUPI, 2004*]. Se also Cooper [*2003*] and Kagan [*2003*].

paradigm than before. The so-called Washington Consensus of the mid-1980s, on which the neo-liberal orthodoxy had put its strong stamp, was part of this mixture. In the 1990s, a modified prototype of this 'consensus' emerged, mellowed by the growing concern for human development of the early years of the decade, involving 'investments' in 'human capital' – education, health and institutions. As we have seen, the UN system initiated this process, and the World Bank jumped on the bandwagon. Then came the strong emphasis on liberal political values as a fresh component of the new consensus. The new ideological blend may well be named *the Washington–New York consensus*. A concern for the global environment was part of this 'consensus', with particular reference to Agenda 21, emerging from the 1992 World Conference on Development and the Environment in Rio de Janeiro. As we have seen, development assistance was to play an important role in all components of this mix.

Two aspects of the new consensus should be highlighted: (i) it represents a pragmatic blend that draws on many and different development paradigms and also on experiences from development co-operation that were high on the agenda during the past two decades – not totally isolated from the predominant ideology of the time, and (ii) it reflects values predominant at the centre of the new world order after the system transformation of the early 1990s had been consolidated. A third aspect should be added: the consensus has been expressed at a high level of generalisation that allows room for widely different interpretations when it comes to operationalisation and implementation. Even at the theoretical level the mix may open up for many flowers to blossom.[82] The implication is that the consensus is fragile, and has a sell-by date.

All of the millennium objectives are ambitious; some are also specific. That agreement has been reached on these objectives is important in its own right, setting priorities and norms for international efforts. However, experience has taught us to differentiate between lofty commitments and actual performance. National interests do not disappear into the blue, even if they are not highlighted in

---

82. Some ideas of the past do not seem to thrive within the confines of the new consensus, with particular reference to the 'utopian' development paradigms of the mid-1970s, synthesised in the normative 'alternative development paradigm' that was developed by the Dag Hammarskjöld Foundation, based in Uppsala (Sweden), and the International Foundation for Development Alternatives, based in Nyon (Switzerland) [*Dag Hammarskjöld Foundation, 1975; IFDA, 1978; Nerfin, 1977; Nerfin (ed.), 1977*]. Such ideas are still around to be picked up when the time becomes ripe. In the first years of the new millennium, they are most strongly represented in movements protesting against globalisation, such as ATTAC.

international consensus documents. This also applies to practices rooted in a variety of values and ideologies, and to value systems (ideologies, development paradigms) themselves, rooted in national and international traditions. When opportunity and national interests coincide, they may get new lives. Such interests and values will be decisive for the extent of achievement of the international targets agreed on and for the role development co-operation will play in this regard.

In what direction has the policy convergence of the 1990s brought development? We have seen that a series of global conferences have focused on social issues and poverty eradication. However, although these conferences influenced the international agenda and the perception of what development is about, they have only marginally affected the power balance between the main international actors, *inter alia* between the Bretton Woods institutions (BWIs) and the UN system. What may be termed hard political issues, involving finance and trade, have increasingly been transferred to bodies where the major industrial powers are in almost exclusive control, for example, the BWIs and the new international trade regime, the World Trade Organisation (WTO). The World Bank, in addition to providing development finance, has increasingly moved into technical assistance, the field originally assigned to the UN (particularly the UNDP) in the division of labour between these two institutions. The World Bank has assumed an important role within other areas as well, such as the 'new' area of the environment (GEF). The UN system, traditionally relied on by developing countries in pursuit of their interests and where they exerted influence over policy outcomes, has been losing ground, discredited among the major industrial powers for lack of effectiveness and efficiency. The emerging power balance is illustrated by the fate of UNCTAD, in the 1970s the major instrument of developing countries in their efforts to improve their trade conditions: in the 1980s and increasingly in the 1990s, it lost much of its previous influence, eventually even its main mandate, to the WTO.

4. Unresolved Matters: The Limits of Insights and Lessons Learned

The state of development co-operation has been attributed to the evolving international situation, including both power relations and values and norms, and to the conceptual development, with particular reference to the evolving development ideology (development paradigms). The emerging policy convergence – the Washington–New York Consensus, reflected also in the Millennium Declaration – has also been attributed to new insights generated by research and lessons learned from decades of development co-operation, both at national and international headquarters and in the 'field'.

We have indicated in this chapter that there are limits to what power (military supremacy) can possibly achieve. Limits also exist when it comes to the knowledge acquired within this policy area: our insights are limited and most probably also skewed, with a Western bias. Thus, the new approaches and mechanisms resulting from this learning process have not always produced the results aimed at – which should come as no surprise.

In the past, development assistance has often been ill-conceived from a development perspective, constructed to serve purposes quite different from those conveyed by the rhetoric, such as the strategic or economic self-interests of the donor country, as illustrated for instance in the extensive practice of tied aid. In Europe, the regional economic integration has created a setting that has been politically conducive to the untying of aid, although the practice still lingers on. Even when intentions were less self-centred, the mechanisms selected did not work as intended. This can be illustrated by first-generation aid conditionality: it was not effective in producing the intended results and, at times, produced results that reversed a positive development, particularly involving human development. It was therefore eventually modified.

Second-generation conditionality, in pursuit of liberal political ideas such as human rights, democracy and good government, made 'bad performers' risk a reduction in development assistance or an end to aid relations – to the extent that the logic of the mechanism worked. However, the logic does not always work, particularly when other donor interests are at stake, such as interests related to an expanding market. This has resulted in gross policy inconsistencies. And when the logic works, it creates new problems: the humanitarian dimensions aside, should a growing number of 'bad performers' be 'freed' from the influence that development assistance might possibly exert on their behaviour – resulting in increased terror and instability generated by state-failure and disastrous living conditions for the population?

Nevertheless, important strides ahead have been made – stimulated during the last decade by knowledge generation within the framework of national and international research institutions (including such efforts within the UN and the World Bank) and experience gained from co-operative frameworks such as that of the Special Partnership with Africa – giving stronger emphasis to mechanisms that reduce transaction costs and provide more ownership to the recipient country (such as direct or indirect budget support). After all, the lessons learned from first-generation aid conditionality and structural adjustment policy contributed to the new emphasis on the social and human dimensions of development and an emphasis on poverty alleviation in the 1990s, perhaps also in less tied aid.

Pointing to unresolved matters and limitations in our present knowledge does not deny the importance attributed to good government in the widest sense of the concept as a *sine qua non* for 'development' to take place – with or without foreign aid. The greater emphasis given to good governance, including accountability, transparency and the rule of law, and to democracy and popular participation in the development process, including civil society participation, represents important progress. However, rhetoric and reality may be different things. And mechanisms prescribed with the best of intentions do not always deliver the intended results: they may even produce unintended results. After so many years, our knowledge basis is more limited than the impression conveyed by the self-confident rhetoric of major donor agencies.

*

The evolving international political situation influences the framework conditions for development assistance, as illustrated by the effects on aid of the system transformation of the early 1990s. So do the predominant development paradigms, both 'old' ones that, in an earlier hegemonic position, shaped the development thinking of a generation of people engaged in formulating and implementing aid policy, and 'new' ones.

However, although part of the foreign policy of a donor country, and therefore adapting to the international discourse on development and development priorities, aid policy is also moulded by the domestic policy of the country concerned. Here, other important framework conditions influence the policy, varying from one donor to another. Donor countries have different traditions in their relations with developing countries, different histories and different value mixes. Their economic, strategic and cultural interests *vis-à-vis* developing countries vary. The economic situation, with particular reference to employment and the balance of payments, varies from one country to another and, within a country, from one period to another. The domestic political agenda will reflect such variations. In addition, the political balance between the political parties, each representing a different mix of values and interests, will influence the aid policy of the day.

The following chapters in this volume, focusing on the aid policy of individual European countries, will relate to such framework conditions, which may vary from one country to the next. In the present chapter on international framework conditions, important developments within the European region itself – affecting the aid policy of individual governments, particularly those that are members of the European Union – have not been addressed. The explanation is simple: the aid and development policy of the EU will be analysed in a separate chapter of this volume.

REFERENCES

Adams, Nassau A., 1993, *Worlds Apart. The North–South Divide and the International System*, London and Atlantic Heights, NJ: Zed Books.

Amann, Edmund, Nektarios Aslandis, Frederik Nixson and Bernhard Walters, 2002, 'Economic Growth and Poverty Alleviation: A Reconsideration of Dollar and Kraay', paper presented at the EADI General Conference, Ljubljana, 19–21 September.

Amin, Samir, 1974, *Accumulation on a World Scale: A Critique of the Theory of Underdevelopment*, New York: Monthly Review Press.

Amin, Samir, 1976, *Unequal Development*, Sussex: Harvester Press.

Amin, Samir, 1977, *Imperialism and Unequal Development*, Sussex: Harvester Press.

Amin, Samir, 1993a, 'The Issue of Democracy in the Contemporary Third World', in Gills *et al.* (eds).

Amin, Samir, 1993b, 'The Challenge of Globalization: Delinking', in South Centre.

Baehr, Peter, Hilde Hey, Jacqueline Smith and Theresa Swinehart (eds), 1994, *Human Rights in Developing Countries Yearbook 1994*, Deventer: Kluwer Law and Taxation Publishers.

Bauer, Peter T., 1971, *Dissent on Development*, London: Weidenfeld & Nicolson.

Bauer, Peter T., 1981, *Equality, the Third World and Economic Delusion*, Cambridge, MA.: Harvard University Press.

Bauer, Peter T., 1984, *Reality and Rhetoric*, London: Weidenfeld & Nicolson.

Bauer, P. T. and B. S. Yamey, 1972, 'The Pearson Report: A Review', in Byres (ed.).

Bhaskar, V. and Andrew Glyn, 1995, 'Introduction', in Bhaskar and Glyn (eds).

Bhaskar, V. and Andrew Glyn (eds), 1995, *The North, the South and the Environment*, Tokyo, New York, Paris: United Nations University Press/Earthscan Publications.

Bird, Graham, 2002, 'The Completion Rate of IMF Programmes: What We Know, Don't Know and Need to Know', *World Economy*, Vol. 25, No. 6 (June).

Burnside, C. and D. Dollar, 2000, 'Aid, Policies, and Growth', *American Economic Review*, Vol. 90.

Byres, T. J. (ed.), 1972, *Foreign Resources and Economic Development*, London: Frank Cass.

Cardoso, Fernando Henrique and E. Faletto, 1969 (English edition 1979), *Dependency and Development in Latin America*, Berkeley CA: University of California Press.

Carothers, Thomas, 1999, *Aiding Democracy Abroad. The Learning Curve*, Washington, DC: Carnegie Endowment for International Peace.

Carter, Gordon, 1977, 'The Implications of Basic Services', in *Basic-needs-oriented Development Strategies with Special Reference to UNICEF's Basic Services Approach*, UNICEF/EADI International Workshop, Vienna, 4–8 December.

Chagula, W.K. *et al.* (eds), 1977, *Pugwash on Self-Reliance*, New Delhi.

Chatterjee, Pratap and Matthias Finger, 1994, *The Earth Brokers. Power, Politics and World Development*, London and New York: Routledge.

Chenery, Hollis and Others, 1974, *Redistribution with Growth*, Oxford: Oxford University Press.

Clay, Edward, 1995, 'Conditionality and Programme Food Aid: From the Marshall Plan to Structural Adjustment', in Stokke (ed.).

Cocoyoc Declaration, 1974, Mexico, in Chagula *et al.* (eds), 1977.

Cohen, M.D., J.G. March and J.P. Olsen, 1972, 'A Garbage Can Model of Organisational Choice', *Administrative Science Quarterly*, Vol. 17, No. 1.

Collier, P. and D. Dollar, 2001, 'Can the World Put Poverty in Half? How Policy Reform and Effective Aid Can Meet International Development Goals', *World Development*, Vol. 29.

Collier, P. and D. Dollar, 2002, 'Aid Allocation and Poverty Reduction', *European Economic Review*, Vol. 46.

Cooper, 2003, *The Breaking of Nations: Order and Chaos in the Twenty-first Century*, New York: Atlantic Monthly Press.

Cornia, Giovanni Andrea, Richard Jolly and Frances Stewart (eds), 1987, *Adjustment with a Human Face*, Vol. I: *Protecting the Vulnerable and Promoting Growth*, A Study by UNICEF, Oxford: Oxford University Press.

Dag Hammarskjöld Foundation, 1975, *What Now – Another Development*, Uppsala.

Dahl, Robert A., 1971, *Polyarchy: Participation and Opposition*, New Haven, CT: Yale University Press.

De Waal, Alex, 1997, *Famine Crimes, Politics and the Disaster Relief Industry in Africa*, Oxford: James Currey for Africa Rights and the International African Institute and Indiana University Press and Bloomington and Indianapolis, IN: Indiana University Press.

Deng, Lulal, Markus Koster and Crawford Young (eds), 1991, *Democratization and Structural Adjustment in Africa in the 1990s*, Madison, WI: University of Wisconsin–Madison, African Studies Program.

Development Committee, 2004, Statement by Mr. Richard Manning, Chairman, OECD Development Assistance Committee (DAC) (Joint Ministerial Committee of the Boards of Governors of the Bank and the Fund, On the Transfer of Real Resources to Developing Countries), Sixty-ninth meeting, Washington, DC, 25 April (DC/S/2004–0001).

Diamond, Larry, 1988, 'Introduction: Roots of Failure, Seeds of Hope', in Diamond *et al.* (eds).

Diamond, Larry, Juan J. Linz and Seymour Martin Lipset (eds), 1988–90, *Democracy in Developing Countries*, Vol. I: *Persistence, Failure, and Renewal*; Vol. II: *Africa*; Vol. III: *Asia*; Vol. IV: *Latin America*, Boulder, CO: Lynne Rienner Publishers.

Djukanovic, V. and E.P. Mach (eds), 1975, *Alternative Approaches to Meeting Basic Needs in Developing Countries*, A Joint UNICEF/WHO Study, Geneva: WHO.

Dollar, D. and A. Kraay, 2001, *Growth is Good for the Poor*, Washington, DC: World Bank, Development Research Group.

Doornbos, Martin, 1990, 'The African State in Academic Debate: Retrospect and Prospect', *Journal of Modern African Studies*, Vol. 28, No. 2, Cambridge: Cambridge University Press.

Doornbos, Martin, 1995, 'State Formation Processes under External Supervision: Reflections on "Good Governance"', in Stokke (ed.).

Eide, Espen Barth, 1997, '"Conflict Entrepreneurship": On the "Art" of Civil War', in McDermott (ed.).

Eide, Espen Barth, 2004, 'CIC–NUPI Workshop on Europe's Evolving Role in Multilateral Security Institutions: A Few Preliminary Thoughts', Oslo: NUPI and CIC.

Emmerij, Louis, 1977, 'Facts and Fallacies Concerning the Basic Needs Approach', in *Bacic-needs-oriented Development Strategies with Special Reference to*

UNICEF's Basic Services Approach, UNICEF/EADI International Workshop, Vienna, 4–8 December.
Emmerij, Louis, 1995, 'A Critical Review of the World Bank Approach to the Social Sector Lending and Poverty Alleviation', in *International Monetary and Financial Issues for the 1990s: Research Papers for the Group of Twenty-four*, Vol. 5, New York and Geneva: UNCTAD.
Engberg-Pedersen, Poul, Peter Gibbon, Phil Raikes and Lars Udsholt (eds), 1996, *Limits of Adjustment in Africa: The Effects of Economic Liberalization, 1986–94*, Oxford: James Currey in association with the Centre for Development Research (Copenhagen); Portsmouth, NH: Heinemann.
Esman, Milton J., 1997, 'Can Foreign Aid Moderate Ethnic Conflict?', *Peaceworks* No. 13, Washington, DC: United States Institute of Peace.
Fahy, John C., 1984, 'IMF Conditionality: A Critical Assessment', *OPEC Review*, Autumn, Vienna.
Finger, Matthias, 1993, 'Politics of the UNCED Process', in Sachs (ed.).
Forster, Jacques, 1984, 'Swiss Aid: Policy and Performance', in Stokke (ed.).
Forster, Jacques and Olav Stokke, 1999, 'Coherence in Policies Towards Developing Countries: Approaching the Problematique', in Forster and Stokke (eds).
Forster, Jacques and Olav Stokke (eds), 1999, *Policy Coherence in Developing Co-operation*, London: Frank Cass, EADI Book Series 22.
Fortin, Carlos, 1988, 'International Commodities and Third World Development: Trends and Prospects', in Stokke (ed.).
Frank, André Gunder, 1967, *Capitalism and Underdevelopment in Latin America. Historical Studies of Chile and Brazil*, New York and London: Monthly Review Press.
Galtung, Johan, 1971, 'A Structural Theory of Imperialism', *Journal of Peace Research*, Vol. 8, Oslo: Institute of Peace Research.
Gasper, Des, 1999, 'Ethics and Conduct of International Development Aid: Charity and Obligation', *Forum for Development Studies*, Vol. 26, No. 1, Oslo: NUPI.
Gasper, Des, 2004, *The Ethics of Development*, Edinburgh: Edinburgh University Press.
German, Tony and Judith Randel, 1995, 'Overview', in *The Reality of Aid 1995*, London: Earthscan Publications Ltd/Actionaid.
Gills, Barry, Joel Rocamora and Richard Wilson, 1993, 'Low Intensity Democracy', in Gills *et al.*(eds).
Gills, Barry, Joel Rocamora and Richard Wilson (eds), 1993, *Low Intensity Democracy*, London: Pluto Press.
Griffin, Keith, 1991, 'Foreign Aid after the Cold War', *Development and Change*, Vol. 22, No. 4, London: Sage Publications.
Hansen, H. and F. Tarp, 2001, 'Aid and Growth Regressions', *Journal of Development Economics*, Vol. 64.
Haq, Mahbub ul, 1995, *Reflections on Human Development*, New York: Oxford University Press.
Healey, John, Richard Ketley and Mark Robinson, 1993, 'Will Political Reform Bring about Improved Management in Sub-Saharan Africa?', *IDS Bulletin*, Vol. 24, No. 2, Brighton: Institute of Development Studies, Sussex.
Hettne, Björn, 1990, 1995 (second edition), *Development Theory and the Three Worlds*, Harlow, Essex: Longman.
Hewitt, Adrian P. and Tony Killick, 1996, 'Bilateral Aid Conditionality and Policy Leverage', in Stokke (ed.).
Hildyard, Nicholas, 1993, 'Foxes in Charge of the Chickens', in Sachs (ed.).
Hirschman, Albert O., 1958, *The Strategy of Economic Development*, New Haven, CT: Yale University Press.
Holm, Hans-Henrik and Georg Sørensen, 1995, 'International Relations Theory in a World of Variation', in Holm and Sørensen (eds).

Holm, Hans-Henrik and Georg Sørensen (eds), 1995, *Whose World Order?*, Boulder, CO and Oxford: Westview Press.

Hydén, Göran, 1983, *No Shortcut to Progress. African Development Management in Perspective*, London: Heinemann.

Hydén, Göran, 1992, 'Governance and the Study of Politics', in Hydén and Bratton (eds).

Hydén, Göran, 1994, 'Shifting Perspectives on Development: Implications for Research', in Masst *et al.* (eds).

Hydén, Göran, 1999, 'The Shifting Grounds of Policy Coherence in Development Co-operation', in Forster and Stokke (eds).

Hydén, Göran and Michael Bratton (eds), 1992, *Governance and Politics in Africa*, Boulder CO and London: Lynne Rienner Publishers.

IFDA, 1978, A United Nations Development Strategy for the 80s and beyond. Participation of the 'Third System' in its Elaboration and Implementation. A project description, Nyon, Switzerland, January.

ILO, 1976, *Employment, Growth and Basic Needs: A One World Problem*, New York: Praeger.

IMF, 2001, *World Economic Outlook*, Washington, DC: IMF.

Jackson, Robert H. and Carl G. Rosberg, 1986, 'Sovereignty and Underdevelopment: Juridical Statehood in the African Crisis', *Journal of Modern African Studies*, Vol. 24, No. 1, Cambridge: Cambridge University Press.

Kagan, Robert, 2003, *America and Europe in the New World Order*, New York: Vintage Books.

Kaul, Inge, Isabelle Grunberg and Mark A. Stern (eds), 1999, *Global Public Goods: International Cooperation in the 21st Century*, New York: Oxford University Press.

Kaul, Inge, Pedro Conceico, Katell Le Gulven and Ronald U. Mendoza (eds), 2003, *Providing Global Public Goods: Managing Globalization*, New York: Oxford University Press.

Kay, Cristóbal, 1989, *Latin American Theories of Development and Underdevelopment*, London and New York: Routledge.

Kay, Cristóbal, 1991, 'Reflection on the Latin American Contribution to Development Theory', *Development and Change*, Vol. 22, London: Sage.

Keohane, Robert O., 1984, *After Hegemony: Cooperation and Discord in the World Political Economy*, Princeton, NJ: Princeton University Press.

Krasner, Stephen D., 1982, 'Structural Cause and Regime Consequences: Regimes as Intervening Variables', *International Organization*, Vol. 36, No. 2.

Krueger, Anne O., Constantine Michalopoulos and Vernon W. Ruttan with others, 1989, *Aid and Development*, Baltimore, MD and London: Johns Hopkins University Press.

Krueger, Anne O. and Vernon W. Ruttan, 1989a, 'Development Thought and Development Assistance', in Krueger *et al.*

Krueger, Anne O. and Vernon W. Ruttan, 1989b, 'Towards a Theory of Development Assistance', in Krueger *et al.*

Lancaster, Carol, 1993, 'Governance and Development: The Views from Washington', *IDS Bulletin*, Vol. 24, No. 1, Brighton: Institute of Development Studies, University of Sussex.

Lancaster, Carol, 2000a, *Transforming Foreign Aid. United States Assistance in the 21th Century*, Washington, DC: Institute for International Economics.

Lancaster, Carol, 2000b, 'Redesigning Foreign Aid', *Foreign Affairs*, Vol. 29, No. 5 (September/October).

Larrain, Jorge, 1989, *Theories of Development. Capitalism, Colonialism and Dependency*, Oxford: Polity Press in co-operation with Basil Blackwell.

Leftwich, Adrian, 1994, 'Governance, the State and the Politics of Development', *Development and Change*, Vol. 25, Oxford: Blackwell Publishers.

Leftwich, Adrian, 1995, 'Bringing Politics Back In: Towards a Model for the Developmental State', *Journal of Development Studies*, Vol. 31, No. 3 (February).

Lensink, R. and H. White, 2001, 'Are there Negative Returns to Aid?', *Journal of Development Studies*, Vol. 37.

Little, Ian M., 1982, *Economic Development, Theory, Policy and International Relations*, New York: Basic Books.

Lumsdaine, David Halloran, 1993, *Moral Vision in International Politics. The Foreign Aid Regime, 1949–1989*, Princeton NJ: Princeton University Press.

Løvbræk, Asbjørn, 1990, 'International Reform and the Like-minded Countries in the North–South Dialogue 1975–1985', in Pratt (ed.).

Machiavelli, Niccolò, 1532, *Il Principe*, Firenze (see, *inter alia*, *The Prince*, edited by Quentin Skinner and Russel Price, Cambridge: Cambridge University Press).

Masst, Mette, Thomas Hylland Eriksen and Jo Helle-Valle (eds), 1994, *State and Locality*, Oslo: Norwegian Association for Development Research and Centre for Development and the Environment.

McDermott, Anthony (ed.), 1997, *Humanitarian Force*, PRIO Report 4/97, Oslo: NUPI and PRIO.

McNamara, Robert S., 1973, Address to the Board of Governors, Nairobi, 24 September, World Bank.

Meier, Gerald M., 1984, 'The Formative Period', in Meier and Seers (eds).

Meier, Gerald M. and Dudley Seers (eds), 1984, *Pioneers in Development*, New York: Oxford University Press for the World Bank.

Middleton, Neil, Phil O'Keefe and Sam Moyo, 1993, *Tears of the Crocodile. From Rio to Reality in the Developing World*, London and Boulder, CO: Pluto Press.

Morgenthau, Hans, 1948, *Politics Among Nations*, New York: Knopf.

Morgenthau, Hans, 1962, 'A Political Theory of Foreign Aid', *The American Political Science Review*, Vol. LVI, No. 2, Menasha, Wisconsin: The American Political Science Association.

Morrissey, Oliver, André McKay and Chris Milner, 2002, 'Aid and Growth in Sub-Saharan Africa: New Evidence on Aid Effectiveness', paper presented at the EADI General Conference, Ljubljana, 19–21 September.

Mosley, Paul, Jane Harrigan and John Toye, 1991a, *Aid and Power. The World Bank and Policy-based Lending*, Vol. 1, *Analysis and Policy Proposals*, London and New York: Routledge.

Mosley, Paul, Jane Harrigan and John Toye, 1991b, *Aid and Power. The World Bank and Policy-based Lending*, Vol. 2, *Case Studies*, London and New York: Routledge.

Mutahaba, Gelase, 1990, 'Foreign Assistance and Local Capacity-Building: The Case of Swedish Aid to Tanzania's Rural Water Supply', *European Journal of Development Research*, Vol. 1, No. 1, London: Frank Cass.

Myrdal, Gunnar, 1957, *Economic Theory and Underdeveloped Regions*, London: Duckworth.

Myrdal, Gunnar, 1968, *Asian Drama. An Enquiry into the Poverty of Nations*, New York: Twentieth Century Fund and Pantheon.

Myrdal, Gunnar, 1970, *The Challenge of World Poverty*, New York: Pantheon.

Myrdal, Gunnar, 1984, 'International Inequality and Foreign Aid in Retrospect', in Meier and Seers (eds).

NAR, 1996, *Development Cooperation Between War and Peace*, No. 111, December, The Hague: National Advisory Council for Development Cooperation.

Nelson, Joan M. with Stephanie J. Eglinton, 1992, *Encouraging Democracy: What Role for Aid?*, Washington, DC: Overseas Development Council.

Nerfin, Marc, 1977, 'Introduction', in Nerfin (ed.).

Nerfin, Marc (ed.), 1977, *Another Development, Approaches and Strategies*, Uppsala: Dag Hammarskjöld Foundation.

Ners, Krzysztof T. and Ingrid T. Buxell (eds), *Assistance to Transition Survey 1995*, Warzaw: Institute for EastWest Studies (IEWS), Policy Education Centre on Assistance to Transition (PECAT).

NUPI, 2004, 'Regionale konsultasjoner om ulike tilnærminger og forståelser av sikkerhetsbegrepet' (Regional Consultations on Various Approaches and Perceptions of the Security Consept), Nyhetsbrev fra FN-Programmet (Newsletter from the UN Programme), Oslo: NUPI.

OECD, 1992a, *Development Co-operation 1992 Report*, Paris: OECD DAC.

OECD, 1992b, DAC and OECD Public Policy Statements on Participatory Development/Good Governance, Paris: DAC (OCDE/GD(92)67).

OECD, 1993, DAC Orientations on Participatory Development and Good Governance, Paris: OECD (OCDE/GD(93)191).

OECD, 1995a, *Development Co-operation 1995 Report*, Paris: OECD DAC.

OECD, 1995b, *Participatory Development and Good Governance*, Paris: DAC, Development Co-operation Guidelines Series.

OECD, 1995c, Ad Hoc Working Group on Participatory Development and Good Governance, Paris: OECD/DAC DCD/DAC/PDGG/M(94)2/PROV.

OECD, 1995d, Linkages: OECD and Major Developing Economies, Paris: OECD.

OECD, 1996a, *Shaping the 21st Century: The Contribution of Development Co-operation*, Paris: OECD DAC.

OECD, 1996b, Building Policy Coherence, Tools and Tensions, Public Management Occasional Papers No. 12, Paris: OECD.

OECD, 1997, *Development Co-operation 1996 Report*, Paris: OECD DAC.

OECD, 1998, *Development Co-operation 1997 Report*, Paris: OECD DAC.

OECD, 2001, *The DAC Journal Development Co-operation 2000 Report*, Vol. 2, No.1, Paris: OECD DAC.

OECD, 2002, *The DAC Journal Development Co-operation 2001 Report*, Vol. 3, No.1, Paris: OECD DAC.

OECD, 2003, *The DAC Journal Development Co-operation 2002 Report*, Vol. 4, No. 1, Paris: OECD DAC.

OECD, 2004, *The DAC Journal Development Co-operation 2003 Report*, Vol. 5, No. 1, Paris: OECD DAC.

Ofstad, Arve and Arne Wiig (eds), 1993, *Development Theory: New Trends*, Bergen: Chr. Michelsen Institute.

Pearson, Lester, 1969, *Partners in Development*. Report of the Commission on International Development, New York: Praeger.

PMO, 1991, *Common Responsibility in the 1990s*, The Stockholm Initiative on Global Security and Governance, 22 April, Stockholm: Prime Minister's Office.

Pratt, Cranford, 1989, 'Humane Internationalism: Its Significance and Variations', in Pratt (ed.).

Pratt, Cranford (ed.), 1989, *Internationalism under Strain*, Toronto: Toronto University Press.

Pratt, Cranford, 1990, 'Middle Power Internationalism and Global Poverty', in Pratt (ed.).

Pratt, Cranford (ed.), 1990, *Middle Power Internationalism: The North–South Dimension*, Kingston and Montreal: McGill Queen's University.

Prebisch, Raúl, 1962, 'The Economic Development of Latin America and its Principal Problem', *Economic Bulletin for Latin America*, Vol. 7, No. 1.

Prendergast, John, 1996, *Frontline Diplomacy. Humanitarian Aid and Conflict in Africa*, Boulder,CO and London: Lynne Rienner Publishers.

Pronk, Jan, 2003, 'Collateral Damage or Calculated Default? The Millennium Development Goals and the Politics of Globalisation', Inaugural Address as Professor of the Theory and Practice of International Development delivered on 11 December at the Institute of Social Studies, The Hague, The Netherlands (mimeo).

Riddell, Roger C., 1987, *Foreign Aid Reconsidered*, London and Baltimore, MD: James Currey and Johns Hopkins University Press, with the Overseas Development Institute.

Rodney, Walter, 1972, *How Europe Underdeveloped Africa*, London: Bogle-L'Ouverture Publications.

Rostow, Walt W., 1960, *The Stages of Economic Growth*, Cambridge: Cambridge University Press.

Ruttan, Vernon W., 1996, *United States Development Assistance Policy: The Domestic Politics of Foreign Economic Aid*, Baltimore, MD: Johns Hopkins University Press.

Sachs, Wolfgang, 1993, 'Global Ecology and the Shadow of "Development"', in Sachs (ed.).

Sachs, Wolfgang (ed.), 1993, *Global Ecology. A New Arena of Political Conflict*, London and Atlantic Heights, NJ: Zed Books.

Sandberg, Svante, 1995, 'Sweden', in *The Reality of Aid 1995*, London: Earthscan Publications Ltd/Actionaid.

Santiso, Carlos, 2000, 'Democracy Assistance and International Development Co-operation: What Have We Learned?', *Forum for Development Studies*, Vol. 27, No. 1, Oslo: NUPI.

Santiso, Carlos, 2003, 'Sisyphus in the Castle: Improving European Union Strategies for Democracy Promotion and Governance Conditionality', *European Journal of Development Research*, Vol. 15, No. 1 (June).

Sen, Amartya, 1999, *Development as Freedom*, Oxford: Oxford University Press.

Senghaas, Dieter, 1995, 'Assessing War, Violence and Peace Today', *Security Dialogue*, Vol. 26, No.3, London: Sage Publications/PRIO (Oslo).

Slim, Hugo, 1997, *Doing the Right Thing. Relief Agencies, Moral Dilemmas and Moral Responsibility in Political Emergencies and War*, Studies on Emergencies and Disaster Relief, Report No. 6, Uppsala: Nordic Africa Institute in co-operation with Sida.

South Centre, 1993, *Facing the Challenge. Responses to the Report of the South Commission*, London and Atlantic Heights, NJ: Zed Books in association with South Centre.

South Commission, 1990, *The Challenge to the South*. The Report of the South Commission, Oxford: Oxford University Press.

Stiglitz, Joseph E., 2002, *Globalization and its Discontents*, Harmondworth: Penguin Allen Lane.

Stokke, Olav, 1978, *Sveriges utvecklingsbistånd och biståndspolitik* (The Development Assistance and Aid Policy of Sweden), Uppsala: Scandinavian Institute of African Studies.

Stokke, Olav, 1984a, 'European Aid Policies: Some Emerging Trends', in Stokke, (ed.).

Stokke, Olav, 1984b, 'Norwegian Aid: Policy and Performance', in Stokke (ed.).

Stokke, Olav, 1989a, 'The Determinants of Aid Policies: General Introduction', in Stokke (ed.).

Stokke, Olav, 1989b, 'The Determinants of Norwegian Aid Policy', in Stokke (ed.).

Stokke, Olav, 1989c, 'The Determinants of Aid Policies: Some Propositions Emerging from a Comparative Analysis', in Stokke (ed.).

Stokke, Olav, 1991, 'Sustainable Development: A Multi-Faceted Challenge', in Stokke (ed.).

Stokke, Olav, 1995, 'Aid and Political Conditionality: Core Issues and State of the Art', in Stokke (ed.).

Stokke, Olav, 1996, 'Foreign Aid: What Now?', in Stokke (ed.).

Stokke, Olav, 1997, 'Violent Conflict Prevention and Development Co-operation: Coherent or Conflicting Perspectives?', *Forum for Development Studies*, Vol. 24, No. 2, Oslo: NUPI.

Stokke, Olav (forthcoming), *The Dilemmas of Policy Coherence. Aspirations and Realities in the Norwegian Policies Towards Developing Countries.*
Stokke, Olav (ed.), 1984, *European Development Assistance*, Vol. I: *Policies and Performance*, Tilburg: EADI Secretariat, EADI Book Series 4.
Stokke, Olav (ed.), 1988, *Trade and Development: Experiences and Challenges*, Tilburg: EADI Secretariat, EADI Book Series 7.
Stokke, Olav (ed.), 1989, *Western Middle Powers and Global Poverty*, Uppsala: The Scandinavian Institute of African Studies.
Stokke, Olav (ed.), 1991, *Sustainable Development*, London: Frank Cass.
Stokke, Olav (ed.), 1995, *Aid and Political Conditionality*, London: Frank Cass, EADI Book Series 16.
Stokke, Olav (ed.), 1996, *Foreign Aid Towards the Year 2000: Experiences and Challenges*, London: Frank Cass, EADI Book Series 18.
Sundman, Folke and Jorma T. Mattila, 1995, 'Finland', in *The Reality of Aid 1995*, London: Earthscan Publications Ltd/Actionaid.
Sørensen, Georg, 1991, *Democracy, Dictatorship and Development*, London: Macmillan.
Sørensen, Georg, 1993a, 'Democracy, Authoritarianism and State Strength', *The European Journal of Development Research*, Vol. 5, No. 1, London: Frank Cass.
Sørensen, Georg, 1993b, *Democracy and Democratisation*, Boulder, CO: Westview.
Sørensen, Georg, 1995, 'Conditionality, Democracy and Sevelopment', in Stokke (ed.).
Tetzlaff, Rainer (ed.), 1993, *Human Rights and Development*, Bonn: Eine Welt.
Toye, John, 1987, *Dilemmas of Development. Reflections on the Counter-revolution in Development Theory and Policy*, Oxford: Basil Blackwell.
Toye, John, 1993, 'Dilemmas of Development: New Challenges, New Theories', in Ofstad and Wiig (eds).
UN, 1990, Development and International Economic Co-operation: International Development Strategy for the Fourth United Nations Development Decade (1991–2000), A/C.2/45/L.72, 5 December, New York: UN, General Assembly.
UN, 1997, *Yearbook of the United Nations 1995*, The Hague/Boston, MA/London: Martinus Nijhoff for the Department of Public Information, United Nations, New York.
UN, 1999 (*sic*), *Yearbook of the United Nations 1990*, The Hague/Boston, MA/London: Martinus Nijhoff for the Department of Public Information, United Nations, New York.
UN, 2000, *United Nations Millennium Declaration* (Resolution A/RES/55/2 8 September), Millennium Summit, New York, 6–8 September, New York: UN Department of Public Information (DPI/2163).
UNCTAD, 1985, 1995, *Handbook of International Trade and Development Statistics*, New York: United Nations.
UNCTAD, 2000, *The Least Developed Countries 2000 Report*, Geneva: UNCTAD.
UNDP, 1990, *Human Development Report 1990*, New York/Oxford: Oxford University Press.
UNDP, 1991, *Human Development Report 1991*, New York/Oxford: Oxford University Press.
UNDP, 1992, *Human Development Report 1992*, New York/Oxford: Oxford University Press.
UNDP, 1994, *Human Development Report 1994*, New York/Oxford: Oxford University Press.
UNDP, 1995, *Human Development Report 1995*, New York/Oxford: Oxford University Press.
UNICEF, 1977, *A Strategy for Basic Services*, New York: UNICEF.
Waller, Peter P., 1992, 'After East–West Détente: Towards a Human Rights Orientation in North–South Development Cooperation?', *Development*, No. 1, Rome: SID.

Waller, Peter P., 1993, 'Human Rights Orientation in Development Co-operation', in Tetzlaff (ed.).

Waltz, Kenneth, 1959, *Man, the State and War*, New York: Columbia University Press.

Waltz, Kenneth, 1979, *Theory of International Politics*, Reading, MA: Addison Westey.

Warburg, Anne Melte Rahbæk, 1995, 'Denmark', in *The Reality of Aid 1995*, London: Earthscan Publishers Ltd/Actionaid.

WCED, 1987, *Our Common Future*, Oxford and New York: Oxford University Press.

White, Gordon, 1995, 'Towards a Democratic Developmental State', *IDS Bulletin*, Vol. 26, No. 2, Brighton: Institute of Development Studies, Sussex.

White House, 2002, *The National Security Strategy of the United States of America*, Washington, DC: The White House. September.

Wood, Robert, 1986, *From Marshall Plan to Debt Crisis: Foreign Aid and Development Choices in the World Economy*, Berkeley, CA: University of California Press.

World Bank, 1989a, *World Development Report 1989*, Washington, DC: World Bank.

World Bank, 1989b, *Sub-Saharan Africa: From Crisis to Sustainable Growth*, Washington, DC: World Bank.

World Bank, 1990, *World Development Report 1990*, Washington, DC: World Bank.

World Bank, 1991, *Assistance to Strategies to Reduce Poverty*, A World Bank Policy Paper, Washington, DC: World Bank.

World Bank, 1992a, *Poverty Reduction Handbook*, Washington, DC: World Bank.

World Bank, 1992b, *Governance and Development*, Washington, DC: World Bank.

World Bank, 1994a, 'Poverty Reduction and the World Bank. Progress in Fiscal 1993', Washington, DC: World Bank.

World Bank, 1994b, 'The World Bank and the Poorest Countries. Support for Development in the 1990s', Washington, DC: World Bank.

World Bank, 1998, *Assessing Aid, What Works, What Doesn't, and Why*, A World Bank Policy Research Report, New York: Oxford University Press and the World Bank.

World Bank, 2000, *World Development Report 1999/2000, Entering the 21st Century*, New York: Oxford University Press and the World Bank.

World Bank, 2001, *World Development Report 2000/2001, Attacking Poverty*, New York: Oxford University Press and the World Bank.

World Bank, 2002, *The Role and Effectiveness of Development Assistance. Lessons from World Bank Experience*, A Research Paper from the Development Economics Vice Presidency of the World Bank, Washington, DC: World Bank.

World Bank, 2003, *World Development Report 2003, Sustainable Development in a Dynamic World*, New York: Oxford University Press and the World Bank.

Young, Crawford, 1991, 'Democratization and Structural Adjustment: A Political Overview', in Deng *et al.* (eds).

2

Austrian Aid Policy

MICHAEL OBROVSKY

I. THE GLOBAL FRAMEWORK

Political, economic and social developments on the international scene during the past two decades have had a significant impact on Austria's development co-operation policy. These effects cannot always be related directly to specific events but may be interpreted as necessary reactions or gradual adaptations to the international mainstream in the development debate. The various institutions of the international donor community (UN, OECD, EU, etc.) – and Austria as a member of them – have responded to political and economic changes with world summits (for example, Rio, Copenhagen, Beijing, Vienna, Monterrey) and new concepts.

Therefore we have to draw attention to the question of how the implementation of the decisions and goals adopted at international summits is reflected in Austria's official development co-operation.

The fall of the Berlin Wall in 1989 and, in its wake, the disintegration of the Soviet Union had a major and lasting impact on international politics in general, the global economy and relations between states. The disintegration of Yugoslavia and the crises in the Balkans led to new orientations in foreign policy and to high expenditures in humanitarian aid. As a consequence of the triumph of the free market economy over the state-planned systems, the opinion that many government functions could be performed more efficiently and economically by the private sector began to prevail the world over in the 1990s. Under the catch-phrase 'less state, more market', nationalised industries, public utilities, health care facilities and parts of general government administration were being outsourced and privatised in industrialised and developing countries. This trend also found its way into development co-operation and influenced the national development co-operation programme.

In the 1990s, the international donor community (UN, OECD, World Bank, IMF) reconsidered its strategy in an attempt to respond to the stalling academic debate on development issues and the aid fatigue, which had led to a decline in ODA flows. In 1996, the DAC published a strategy for the 21st century [*OECD, 1996*] and later the Millennium Development Goals [*OECD, 2000b*]. This reorientation at the international level was to leave marks on Austria's development policy.

The present analysis aims to draw a detailed picture of Austria's development policy framework, in order to explain the present state of Austria's development co-operation and assess its perspectives. Considering Austria's development policy and its co-operation, it is useful to start with a look at its long-term performance regarding the volume of aid.

II. THE VOLUME OF AID

During the most recent peer review of Austria's aid policy in 1999, the Development Assistance Committee (DAC) of the OECD recommended that Austria should increase its official development assistance to a level that would more appropriately reflect its economic performance and capacity [*OECD, 2000a*]. In a comparison of the economic strengths of EU member states in 2000, Austria ranked sixth, whilst a comparison of members' ODA flows showed it in twelfth place on a list of 15. As a percentage of gross national income (GNI), Austria's ODA amounted to 0.23 per cent in 2000, compared with a DAC average of 0.22 per cent and an EU average of 0.32 per cent.

The relatively small volume of Austria's ODA has long been a core point of criticism by the DAC (see Figure 2.1). During the 1980s, Austria's ODA was generally below the DAC average (with the exception of 1985), and during the 1990s it reached the DAC average, as ODA flows declined for all donors. A detailed analysis of the reasons behind Austria's ODA levels and their fluctuations over the past two decades would offer no great insights, since these fluctuations do not reflect political decision-making and concrete planning but rather have been the result of peculiarities in Austria's reporting practice or the respective dates when contributions to replenish the funds of international financial institutions fell due. In terms of total ODA, no interpretation of trends during the past two decades has been possible, because the Austrian ODA performance has been the result of many coincidences and peculiarities of statistical reporting.

TABLE 2.1
AUSTRIA'S ODA, 1981–2001 (US$m. AND % OF GNP/GNI)

	1981	1982	1983	1984	1985	1986	1987
Bilateral aid	161.00	165.00	126.50	137.19	174.35	141.48	156.87
Multilateral aid	59.00	71.00	31.12	44.15	74.12	56.18	44.19
Total ODA	220.00	236.00	157.62	181.34	248.47	197.66	201.06
In % of GNP	0.33	0.36	0.24	0.28	0.38	0.21	0.17
DAC average	0.34	0.38	0.36	0.36	0.35	0.35	0.35

	1988	1989	1990	1991	1992	1993	1994
Bilateral aid	162.36	200.94	299.37	434.83	420.39	410.75	535.72
Multilateral aid	139.03	81.53	94.39	113.65	136.03	133.28	119.56
Total ODA	301.39	282.47	393.76	548.48	556.42	544.03	655.28
In % of GNP	0.24	0.23	0.25	0.34	0.30	0.30	0.33
DAC average	0.36	0.32	0.33	0.33	0.34	0.31	0.30

	1995	1996	1997	1998	1999	2000	2001
Bilateral aid	559.74	412.14	306.20	291.54	343.87	256.76	341.61
Multilateral aid	207.07	144.89	220.98	164.20	182.82	166.51	191.35
Total ODA	766.81	557.02	527.18	455.74	526.69	423.27	532.95
In % of GNI	0.33	0.24	0.26	0.22	0.26	0.23	0.29
DAC average	0.27	0.25	0.22	0.23	0.24	0.22	0.22
EU average	0.38	0.37	0.33	0.33	0.31	0.32	0.33

Source: DDC/ÖFSE.

As Table 2.1 shows, the figures for 1982, 1985, 1991 and 1994 are relatively high for Austria. These correlate with high shares of officially supported export credits. Bilateral aid accounted for 70 to 80 per cent of total ODA. Only after Austria's accession to the EU in 1995 did the share of multilateral aid increase as a result of obligatory contributions to EU development co-operation programmes.

FIGURE 2.1
AUSTRIA'S ODA 1980–2001

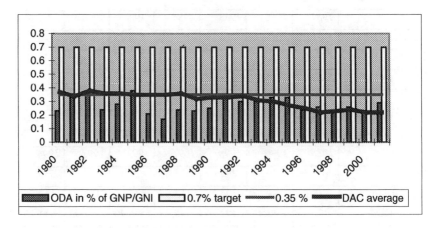

The strong fluctuations in Austria's ODA are in part due to the fact that 'Austria does not have a comprehensive aid budget' [*OECD, 2000a: I–35*], the DAC concluded. It recommended that 'Austria should have an overall aid strategy that links all its components to a clear set of development objectives' [*ibid.: I–11*].

In the 1980s and 1990s the main DAC criticism concerned the composition of the reported ODA expenditures and proportions of the individual components. The shares of government-supported export credits, the relatively high proportions of imputed student costs and the high expenditures for refugees in Austria were determining factors of the quantity and quality of bilateral ODA. Furthermore, there was criticism that these three activities were not primarily pursuing development goals and that the policies and programmes of the Department for Development Co-operation (DDC) of the Ministry of Foreign Affairs were not binding on the other ministries and agencies entrusted with their implementation. Austria has responded to this criticism to the extent that, as of 1999, its reporting practice for imputed student costs and aid to refugees was modified to meet the requirements of the DAC Statistical Reporting Directives. The share of concessional export credits started to decline significantly in 1996 (11 per cent of ODA). In 2000, however, for the first time repayments of loans, the disbursements of which had been reported as ODA back in the 1970s and 1980s, exceeded current disbursements for new loans. In ODA statistics, this resulted in a negative flow for this component. Since loan repayments are expected to increase in the future,

Austria's reporting practice employed thus far – and criticised by the DAC for not being in conformity with its Statistical Reporting Directives – has the effect of reducing its ODA. Austria is therefore negotiating with the DAC to find ways of changing its reporting on concessional export credits. Although the volume of these three predominant components started to decline as of 1996, it has not been possible to expand further the areas where aid policy governs the activities reported as ODA.

Austria had to increase the volume of its multilateral aid significantly with its accession to the EU in 1995. In 2000, this aid accounted for about 39 per cent of total ODA, whereas, in the early 1990s, it had been only about 25 per cent. Contributions to the EU accounted for 52 per cent of multilateral ODA flows. Contributions to the international financial institutions are subject to fluctuations determined by the cycles and negotiations regarding the replenishment of funds so that, as already noted, no clear trend may be seen here. Voluntary contributions to organisations of the United Nations were reduced in 2000 as a result of cuts aimed at budgetary consolidation.

The Ministry of Foreign Affairs (MFA) and the Ministry of Finance (MF) are both optimistic in their ODA forecasts to achieve an increase to 0.36 per cent of GNI in 2003 [*MFA, 2001b*]. This confidence stems from plans to implement debt forgiveness within the framework of the HIPC Initiative on debt resulting from export credit agreements with payments due in 2001, 2002 and 2003. As to the extent of eligibility of these expenditures as ODA (Austria reported part of the loans in question as ODA flows already at the time of their disbursement), Austria negotiated with the DAC in 2002 the way of correct reporting. For the year 2004, the forecast indicates a decline again. No proper budgetary provisions – considering total ODA – have been made as yet to compensate for this drop [*ibid.*].

The goal adopted by the EU during the Barcelona Council Meeting, in preparation for the Monterrey Conference on Financing for Development, of increasing the volume of member states' ODA to achieve an EU average of 0.39 per cent of GNI by the year 2006 was also endorsed by Austria. Austria is thus being encouraged to undertake concrete measures to increase its ODA volume in order to make the minimum contribution of 0.3 per cent of GNI to attain the European target, and the Minister for Foreign Affairs, Benita Ferrero-Waldner, announced the government's pledge to contribute the recent EU average of 0.33 per cent by 2006 [*MFA, Press Release, Vienna, 29 April 2002*]. The Ministry of Finance participated in the debate with the EU on this objective and is therefore well aware of the necessity of budgetary action. Achievement of the target set would require Austria to expand its aid funds progressively to reach an annual volume of approximately $800m., which means it would have to virtually double its 2000 volume of aid (see Table 2.2).

Furthermore. there is – in terms of volume – the more successful story of official aid (OA) to the Central and Eastern European countries and the new independent states of the former Soviet Union (CEECs/NIS). The lifting of the Iron Curtain meant for Austria a shift from a geographically peripheral position to being at the very centre of Europe and, as a result, the opening up of new business opportunities. Austria therefore had a vital interest in the rapid economic and political transformation of Eastern Europe, a fact clearly reflected in its commitment to help the states of Central and Eastern Europe and the former Soviet Union [*MFA, Referat VII/2e, 2001c; ÖFSE, 2001*]. Austria's historical role as a bridge to the Central and Eastern European countries has been used by the government as an argument *vis-à-vis* the EU to stress the country's special role and importance within the European Union, and by the Austrian business community to enlarge its activities in Eastern Europe. The disintegration of Yugoslavia and the crisis in the Balkans led Austria to step up its efforts in the field of humanitarian aid (aid to refugees, intake of and assistance to asylum seekers, disaster relief) and its participation in peace-building activities under the auspices of the United Nations.

Austria was in the lead in the international ranking of donors regarding OA, with 0.10 per cent of GNI in 2000 and 0.11 per cent of GNI in 2001. This reflects Austrian priorities as regards Eastern Europe and the Balkans. It also reflects its former economic interests since half the Austrian OA consists of a reduction in debt service for Poland, with reference to outstanding export credits [*MFA, 2001c; ÖFSE, 2001*]. The administration of OA falls within the responsibility of the Ministry of Foreign Affairs; the policy and the aid programme follow neither the programme of the Department for Development Co-operation nor the international mainstream of development co-operation.

TABLE 2.2
FINANCIAL REQUIREMENTS (US$m.)

	2003	2004	2005	2006
0.33% of GNI	703	732	761	792
0.35% of GNI	745	777	808	840
0.39% of GNI	830	865	900	936

Source: ÖFSE.

The main reason for the traditionally small development budget is the low importance attributed to international development policy in the Austrian Parliament and in Austrian society. Although the government has, in principle, confirmed the importance of international solidarity at international summits and meetings, the commitments in national budgets have never been implemented to any significant extent. Little attention is paid to international criticism by the DAC and national criticism from the small Austrian development community, because no international or national political consequences are expected.

III. THE ADMINISTRATION OF DEVELOPMENT CO-OPERATION

A major problem for Austrian development co-operation continues to be that the Department for Development Co-operation (DDC) of the Ministry of Foreign Affairs, based on its mandate as defined by the government, can apply its programmatic principles in the allocation of funds to only a mere 15 to 20 per cent of total ODA or 20 to 30 per cent of bilateral ODA (Table 2.3). While the budget for the DDC's programmes and projects was increased significantly in the early 1990s, reaching just US$100m. in 1996, the last two years of the decade saw a reduction following general budgetary cuts. In 2001 and 2002 the budget dropped again to an amount of $50m. An increase up to roughly $70m. in this budget line has been decided in Parliament for 2004. A further source of funding is Austria's European Recovery Programme, which provides approximately $7m. a year to the MFA for development programmes and projects. If these funds are added to the DDC budget the result is that the DDC can spend roughly $60m. annually according to its priority principles regarding its development programmes and projects. During the 1999 peer review of Austria's development co-operation, the DAC recommended that this 'core aid budget' be increased to a level comparable with other DAC members, but in the budget projections for 2003 and 2004, this recommendation has not been taken into account thus far. The announced increase in 2004 will not change the volume of Austria's ODA significantly.

TABLE 2.3
DDC PROGRAMME AND PROJECT AID 1991–2002 (US$m.)

	1991	1992	1993	1994	1995	1996
DDC budget	43.26	66.05	11.63	87.22	86.89	99.44
ERP fund	8.57	6.00	5.76	5.11	10.02	4.06
Total	51.82	72.06	93.89	92.30	96.91	103.41
% of bilateral ODA	12	17	23	17	17	25
% of ODA	9	13	17	14	13	19

	1997	1998	1999	2000	2001	2002
DDC budget	72.91	76.76	64.34	52.04	49.15	53.35
ERP fund	5.25	13.66	12.31	15.20	7.65	5.37
Total	78.21	90.42	76.66	67.24	56.80	58.73
% of bilateral ODA	26	31	22	26	17	n.a.
% of ODA	15	20	15	16	11	n.a.

Source: DDC/ÖFSE.

In 1992, by defining priority and co-operation countries for development programmes and projects financed by the core aid budget, a decisive move was made to end the focus of the 1970s and 1980s on sectors and the practice of dispersing funds over a wide range of recipient countries. The selection of priority countries was based on historical relations, on the presence and experience of Austrian NGOs and companies in these countries, on specialised know-how (for example, Austria's knowledge of forestry in high mountain regions), and on diplomatic relations (the presence of an embassy or other permanent representation). By 2000 about 70 per cent of the DDC funds was concentrated on all priority and co-operation countries (Table 2.4). The regional focus is on Africa, to which almost 50 per cent of DDC funds are allocated [*MFA, 2001a; ÖFSE, 2001*].

The *priority countries* (eight) and co-operation countries (12) of the DDC of the Ministry of Foreign Affairs are as follows: in West Africa: *Cape Verde, Burkina Faso*, Senegal; in East Africa: *Ethiopia, Uganda, Rwanda,* Tanzania, Kenya and Burundi; in Southern Africa: *Mozambique*, Namibia, Zimbabwe and the Republic of South Africa; in Central America: *Nicaragua*, Costa Rica, Guatemala and El Salvador; and in the Himalaya/Hindu Kush region: *Bhutan*, Pakistan and Nepal. In addition, Austria has a development co-operation programme with the Palestinian Administered Areas, which was incorporated in the geographical concentration concept as a special programme.

TABLE 2.4
DDC PROGRAMME AND PROJECT AID BY REGION, 1992–2002 (%)

	1992	1993	1994	1995	1996	1997	1998	1999	2000	2001	2002
Europe	0	9	0	0	0	0	0	1	2	1	1
Africa	56	45	41	48	56	53	56	45	48	46	45
Asia	12	13	15	10	13	14	12	21	20	14	21
America	7	17	21	17	16	14	18	19	15	20	17
Oceania	0	0	0	0	0	0	1	1	1	1	1
Unallocable	25	16	23	25	15	19	13	13	15	18	16
Total in US$m.	72.06	93.89	92.30	96.91	103.41	78.21	90.42	76.66	67.24	56.80	58.73

Source: DDC/ÖFSE.

Priority countries are those partner countries with which a comprehensive co-operation programme is jointly worked out and where long-term involvement is envisaged. Co-operation countries are those in which programmes and projects are implemented in a limited number of sectors. It is the aim to focus most of the DDC aid budget on priority countries, while a reduction of the aid volume or even a phasing out in co-operation countries is intended. Based on the relatively small amount of DDC aid available, each country of the priority group receives annual shares of between $2m. and 6m. Austria thus plays a small role even in its priority countries and excels only by providing specific know-how of specially tailored solutions. An exception is Bhutan, where in contrast to a few per cent of aid in the case of African countries the Austrian contribution to total ODA flows is 10 to 14 per cent [*OECD, 2003*].

A geographical distribution of Austria's total bilateral ODA – as published by the DAC – shows the recipient countries of concessional export credits and/or humanitarian aid and assistance to the CEECs/NIS, in the top ranks of recipients during the 1990s. In 2000, for instance, these were: Indonesia, Yugoslavia, Egypt and Bosnia-Herzegovina. This is why projects in the CEEC/NIS or export credits belong to different budget lines from those of development co-operation. Therefore these components do not follow the geographical concentration and the development objectives of the DDC, but efforts are, of course, reported in the ODA statistics.

Sectoral breakdowns of DDC programme and project aid show a shift of focus during the 1990s. Whereas, in the early part of the decade, the sectoral focuses were education and agriculture as well as other social infrastructure/services, transport, industry/SME, energy and water supply/sanitation, in 2000 projects in the sectors government/civil society, water supply/sanitation, and power generation/distribution started to gain in share. Education and agriculture continue to be key focus sectors, however. The cross-cutting issues of poverty, gender equality, the environment, democracy and human rights have been incorporated gradually since the mid-1990s. In addition to elaborating concepts and guidelines for project assessment, the government has been striving to mainstream these cross-cutting issues. The stronger emphasis on cross-cutting issues is a result of the involvement in international debates in the DAC and the EU.

The national debate on bilateral ODA is confined to the programme administered by the Department for Development Co-operation of the Ministry of Foreign Affairs, reflecting the division of administrative authority. Most of the other components of bilateral ODA spent by ministries other than the MFA do not follow the main priorities of the development policy of the MFA. This is an obstacle to overcoming the problem of lack of coherence in Austria's development policies, caused by the fragmentation of responsibilities among the various ministries.

The scope of the Ministry of Foreign Affairs' responsibility basically comprises the DDC aid budget, co-operation with the United Nations, EU development co-operation and, since 2000, the assistance programme for the CEECs/NIS. (In 2001, about 18 per cent of ODA came under the MFA.) The Ministry of Finance administers the largest share of ODA-relevant budgets. With regard to bilateral expenditures, the MF is responsible for credit financing and measures of debt relief. As to multilateral development co-operation, it administers the contributions to international financial institutions and Austria's payments to the EU. In 2001, for instance, a total of $323m. (approximately 61 per cent of ODA) was administered by the MF.

The Federal Ministry for Education, Science and Culture is responsible for 'imputed student costs'. In 2001, ODA expenditures by this ministry amounted to approximately 9 per cent of ODA. The Federal Ministry of the Interior manages the assistance to refugees, amounting in 2001 to about 5 per cent of ODA. Other, smaller shares of ODA are contributed by the budgets of the Federal Ministries of Agriculture, Forestry, Environment and Water Management, Defence, and Economics and Labour.

Co-ordination of the different aid activities is limited to persuasion. In the 1999 peer review of Austria, the DAC came to the conclusion: 'However, greater coherence could be achieved if the MFA [Ministry of Foreign Affairs] had a more extensive and explicit leadership role, backed by a clear political mandate from the government' [*OECD, 2000a*]. No effective change to upgrade the role of the MFA in aid co-ordination has been made since then. Reorganisation of the structures of Austrian development co-operation and official aid was announced in autumn 2002 [*MFA, MF, 2002*] and a fresh debate on outsourcing the management of development co-operation has since then got under way. The MFA and the MF argue that reorganisation – that means establishing a private agency to manage the budget of the DDC on behalf of the MFA – is necessary in order to take part more effectively in the international development co-operation of the European Union. In 2003 the MFA set up the new agency – the Austrian Development Agency (ADA) – which is supposed to start working in 2004.

IV. THE INSTITUTIONAL SETTING

At the beginning of the 1980s, Austrian development co-operation formed part of Department IV (Nationalised Industries) of Group A in the Federal Chancellery. Development co-operation, or rather development policy, at that time was perceived as a part of Federal Chancellor Kreisky's foreign policy.

In 1985, the responsibility was transferred to the MFA and a separate department for development co-operation (Section VI) was established. The reasons behind this transfer were seen partly as lack of interest in development co-operation on the part of Chancellor Sinowatz and partly as hopes, based on the co-ordinating function laid down in the law defining ministerial responsibilities, that the creation of a separate department for development co-operation would lead to 'a qualitative concentration of willingness to actively apply aid policy' and, as a consequence, to an increase in aid volume [*Höll, 1986: 148*]. An expansion of the administrative infrastructure and the transfer of all responsibilities regarding aid policy and co-operation were seen as prerequisites for this. In 1987, during the negotiations between the Social-Democratic Party (SPÖ) and the Austrian People's Party (ÖVP) to form a coalition government, foreign affairs was claimed by the ÖVP. Since then, the position of Foreign Minister has been held by an ÖVP representative. The SPÖ's abandonment of the MFA was criticised publicly by the then former Chancellor Bruno Kreisky and led to a breach between him and his party. By transfer to a Secretary of State (Jankowitsch/Ederer) linked to the Federal Chancellery, responsibilities for development co-operation were returned to and remained under Social-Democratic influence from December 1990 to November 1994 (Vranitzky III). Responsibilities for the assistance programme for CEECs/NIS fell within the authority of the Federal Chancellery, where they remained until 2000.

Since November 1994, development co-operation has again been under the authority of the MFA, first, as of May 1995, managed by a Secretary of State (Ferrero-Waldner) and later, as of April 2000, under the direct authority of the Federal Minister for Foreign Affairs (Ferrero-Waldner) together with the responsibilities for the assistance programme for CEECs/NIS.

Elections in 2002 and the new government (Schüssel II) did not change the former institutional responsibility.

These transfers of authority back and forth between the Federal Chancellery and the MFA, or rather between SPÖ and ÖVP during the period of their coalition governments, were not the result of a long-term development policy or party strategy but rather seemed to represent the surrender of a political field deemed less important for the sake of securing the respective party's own priority issues in the coalition negotiations. Austria's development co-operation thus became a political plaything testifying to its declining importance in domestic politics since the 1970s.

Despite these frequent changes during the past decades, most of the staff (experts, civil servants, even top- and middle-tier diplomats) remained in charge. This explains a certain continuity in the priorities of the co-operation programme and the priority countries, without large-scale changes taking place every time political responsibility was transferred.

V. THE LEGAL BASIS

For some time, the legal basis for Austria's development co-operation was the Development Co-operation Law of 1974. During the early 1990s, first initiatives were started on the part of the aid administration to amend this law [*Federal Chancellery, 1992, Preface by Secretary of State for Integration and Development Brigitte Ederer*]. Since, on the one hand, no agreement on the details of the planned revision could be reached with the non-governmental organisations involved [*Hartmeyer and Jäggle, 1992: 51*], and since Austria's accession to the European Union was approaching and thus a reorientation of its development policy framework was pending, on the other, the revision was postponed. The issue was taken up again in 1999, after Austria's Presidency of the European Council in 1998, and was included in the government proclamation of the ÖVP/FPÖ coalition government in 2000 as an item on the government's agenda in the area of development co-operation [*Österreichische Bundesregierung, 2000*].

The MFA presented a draft that, in general terms, took into account the new development goals and principles adopted by the international community, the revised terminology and the modified rules for implementation. Development NGOs and the political opposition also demanded that the quantitative target of 0.7 per cent of GNI be laid down by law, that the role of Austrian NGOs in the implementation of Austrian aid policy be taken into account to a greater extent, and that development education and public awareness-raising be expressly mentioned as development co-operation activities. This task was included in the wording of the new legislation, but an increase in ODA volume and the importance of Austrian NGOs was not incorporated. The new Bill on development co-operation was passed in February 2002 by majority vote of the government parties [*Federal Law Gazette for the Republic of Austria, Vol. 2002, issued on 29 March 2002, 49th Federal Law on Development Co-operation*]. The intention to pass the new legislation by a unanimous vote of all the parties failed on account of both the opposition's demand for a motion for a parliamentary resolution containing the stipulation of the ODA target of 0.7 per cent of GNI and the demand for concrete measures to ensure better co-ordination of development policies.

Although the legislation does include provisions for a co-ordinating role of the MFA in all activities reported as ODA, along the guidelines laid down in the Three-Year Programme, the Minister of Foreign Affairs is not in a position to enforce these guidelines in other ministries, since their responsibilities have not been modified by the new law. The new legislation definitely amounts to a legal framework for implementing Austria's development co-operation in which necessary adjustments in response to international changes have been made. However, despite the high expectations of NGOs and demands for clarification in the area of practical implementation, the government settled for pragmatic legislation that would not limit the scope of its action at the international level. Clarifications and definitions, for instance, regarding the function and role of the co-ordination offices, do not need to be laid down by law, but they should at least be made public in the Three-Year Programme or other MFA publications.

In July 2003, Parliament adjusted the development co-operation law in order to get a new legal base for establishing the new development agency (Austrian Development Agency – ADA) in 2004. Some articles had to be enlarged as to the way to establish the agency and regarding the new tasks and responsibilities. The relationship between the government, the MFA, the new agency, companies and NGOs were to be arranged by 2004. The MFA was convinced that establishing a new agency would improve Austrian development co-operation [*MFA, 2003a*]. NGOs were doubtful about it.

The new legislation reflects the international development rhetoric and provides new, adapted definitions, but it remains a piece of paper with no significant budgetary means to back it.

VI. OBJECTIVES AND INTERNATIONAL DEVELOPMENT RHETORIC

There was a shift in the objectives of Austrian development co-operation by the DDC of the Ministry of Foreign Affairs during the 1990s in accordance with international reorientations, their formulation being adapted to the objectives endorsed by the international community. The Three-Year Programme 1991–1993, in the chapter 'Guiding Principles and Criteria of Austrian Development Policy', mentioned a contribution by Austria to solving key economic and political problems in partner countries as the principal objective. In the Three-Year Programme 1993–1995, combating poverty already had priority over economic considerations, which were regressed to a certain extent. In the following period, the targets were directed towards the overall goals spelled out in the OECD's 'Strategy for the 21st Century' [*OECD, 1996*] and were integrated into Austrian development co-operation in accordance with the focus areas of DDC programme and project aid.

The overall goals laid down in Paragraph 1 of the Development Co-operation Law of 2002 are: combating poverty; securing peace and human security; and preservation of the environment and protection of natural resources. In pursuing these goals, development co-operation is guided by the following principles: ownership and partnership; integration into the socio-cultural environment – the use of appropriate technology; and equal participation of women and men. The arguments underpinning these objectives and the forms of their implementation are given in the Three-Year Programme [*MFA, 2001b*]:

Poverty reduction is to be achieved primarily by:
– selecting least developed countries as partner countries
– concentrating the available DDC aid budget on priority and co-operation countries up to a level of 70 per cent
– focusing on especially needy target regions, provinces and districts within partner countries

- adopting approaches which make an immediate impact on poverty, and
- selecting target groups which have to contend with being structurally especially disadvantaged.

The objective of *peace and human security* is closely linked with the issue of conflict prevention, a thematic focus of Austrian development co-operation. In addition to helping to build a network of institutions in Africa that are active in the field of conflict prevention, Austria supported the peace process in Burundi by participating actively in the peace negotiations.

The goal of *maintaining and supporting the natural environment* has its origins in the Rio Conference of 1992 and is being pursued in Austria's aid policy as a cross-cutting issue. Before approval, Austrian development projects are subjected to a standardised environmental impact assessment so that ecological aspects can be taken into consideration in good time. In funding projects and measures specifically focused on environmental protection, there has been a concentration in recent years on sustainable use of forest resources.

Implementation of these objectives had already started in 1992. But only after detailed criteria for assessing the cross-cutting themes of gender [*MFA, 1998*], the environment and poverty reduction had been developed – and corresponding markers for projects introduced – during the second half of the 1990s, did it become possible, in the area of DDC programme and project aid, to screen activities with regard to their targeting and derive data for analysis. Before that, criteria for evaluation and statistical instruments had not been sufficiently advanced to be able to supply reliable information.

Austria's accession to the European Union in 1995 had consequences not only in the shape of a compulsory financial contribution to the EU's development co-operation, but, above all, brought about an internationalisation of its aid policy following the adoption of the EU goals of harmonising development policies and co-ordinating concrete measures. Holding the Presidency of the European Council in 1998 gave Austria the opportunity to indicate directions in specific fields of EU development co-operation (such as conflict prevention, tourism, operational co-ordination) and led to increased efforts to integrate its development co-operation into the EU's common development policy framework. This is reflected in particular in those priority countries where Austria carries out a co-ordinating function of EU operations (for example, Mozambique and Burkina Faso).

VII. IMPLEMENTATION OF AUSTRIAN DEVELOPMENT CO-OPERATION

The structure for implementing Austria's development co-operation has grown historically. Austrian NGOs had been used by the DDC to implement ODA since the beginning of development co-operation in Austria. NGOs provided a cheap implementing instrument, specific know-how, and committed young volunteers, so it was not necessary to establish a government agency. Most of them were church-related NGOs. Some associations stemmed from solidarity movements and some NGOs were specialised in awareness-raising and Third World education. Project proposals submitted by NGOs were funded by the DDC in the spirit of the subsidiarity principle. NGOs and sometimes business companies and other institutions were commissioned to carry out projects. The core budget for development NGOs and projects channelled through NGOs was higher than 50 per cent of the DDC budget. During the 1980s and early 1990s, plans to create a government agency similar to the German 'GTZ' were being seriously considered but were then not realised for financial reasons [*Hartmeyer and Jäggle, 1992: 51*]. In the course of preparations for Austria's accession to the EU, it became evident that the structures for implementing DDC aid were tailored too specifically to NGOs and that other instruments had to be developed.

The Development Assistance Committee noted the following changes in the role of NGOs:

- more competitive regulations and guidelines (bidding for projects)
- stricter budget and administrative control
- progressive introduction of country and sector programmes
- stronger monitoring of NGO work by the DDC via its co-ordination offices
- new incentive schemes favouring private sector activities in development co-operation, and
- a stronger involvement of local NGOs [*OECD, 2000a: I–30*].

In the opinion of the DAC reviewers, these gradual changes in the general framework of relations with NGOs have reduced the privileged access that NGOs once had in project financing as well as their degree of autonomy in project design and implementation [*ibid.*].

Prior to Austria's accession to the EU, NGOs were being required to improve their efficiency and professionalise their work in order to meet the demands of the development community within the EU. The creation of a number of non-profit enterprises to carry out the implementation of government projects and some NGO mergers, for example, the *Kofinanzierungsstelle* (KFS), are evidence of the NGOs' attempt at greater professionalism. Insufficient equipment and, in most cases, organisational and management structures typical of non-profit entities remain the main reasons for their poor competitiveness. In the field of EU projects, NGOs were in fact successful at tapping co-financing funds from the budget line intended for NGO activities (B-6000). However, neither NGOs nor business companies from Austria were very successful in gaining access to funds from other budget lines, let alone from sources within the framework of the European Development Fund.

With Austria's accession to the EU, the Ministry of Foreign Affairs gradually introduced changes in the funding modalities of aid activities. NGOs no longer receive 100 per cent funding but have to contribute private means to a co-financing scheme with the state. Furthermore, the MFA has started to conclude framework agreements with larger NGOs allowing them to implement projects co-financed by them in priority countries. These framework agreements are intended to facilitate longer term planning and continuity in the work of NGOs. When contracting out the execution of projects a tendering process is required. On the small Austrian market, these tendering processes ultimately lead to an increase in costs for NGOs, which are insufficiently equipped with capital resources in the first instance, because they have to finance their participation in the bidding process. Moreover, the competitiveness situation among NGOs does not foster co-operation, co-ordination and exchange of information between actors in the priority countries.

Private enterprises have been commissioned to manage and implement various tasks such as educational, cultural and public awareness work, co-financing programmes, or recruitment and assignment of the Ministry's experts and consultants, and structures resembling partial agencies have been established for this purpose (for example, Trimedia, GPR, Kommunalkredit). By contracting out public administration tasks the operative DDC aid budget has been reduced again, because administrative costs were no longer a part of the general budget of the MFA but came out of the DDC budget.

The outsourcing process of MFA activities, criticised by NGOs as a severe deterioration in general conditions, is seen as a necessary reform, both by the MFA and the DAC, above all in the light of the envisaged intensification of co-operation with government authorities in the priority countries, in the form of budgetary assistance within the framework of Poverty Reduction Strategy Papers and a stronger involvement of the private sector. The small total volume of the DDC aid budget, together with the general budgetary cuts in DDC project and programme funds in the course of efforts at budgetary consolidation, has impeded the funding, on a regular basis, of investment programmes and budgetary or general programme aid for priority countries. Austria's bilateral development co-operation can therefore hardly be considered an effective instrument of its foreign policy.

Progress was made during the 1990s in the field of evaluation. During the 1980s, it was barely possible to derive analytical information about impacts and the implementation of Austrian development co-operation projects. However, in the 1990s evaluation became more important within DDC projects and programmes. The progress mainly consists of the general feeling that evaluation has become an accepted fact and is regarded no longer as a threat but rather as a necessary tool for corrections in the course of further development.

Since the move towards geographical concentration in 1992, co-ordination of the DDC's development co-operation with other donor countries and with recipient governments has been organised in a decentralised form by establishing regional co-ordination offices in priority countries. This means that, based on its field presence, the DDC is in a position to harmonise and negotiate its contributions to country and sector programmes with the respective governments. Moreover, the regional offices have a special function in Austria's participation in the PRSP approaches of the World Bank. Since the role of the regional offices is not clearly defined, the work of some offices in specific priority countries is being criticised by Austria's development NGOs, which are under the impression that their concerns are not being adequately represented by the aid administration. Austria's participation in general donor co-ordination in bodies of the OECD, the EU and other multilateral organisations is managed by the Department for Development Co-operation of the Ministry of Foreign Affairs.

Multilateral development co-operation is managed by the Ministry of Finance (with the international financial institutions and the regional development banks) and the Ministry of Foreign Affairs. According to the MFA's Annual Report for 2000, Austria lobbies for the principles and objectives of its development co-operation *vis-à-vis* the multilateral institutions and supports all efforts to arrive at a common EU position and thus give European policies more weight [*MFA, 2001a: 65*]. Its contributions to policy debates within the multilateral institutions are described in very general terms in the Three-

Year Programme and the annual reports; they do not seem to be characterised by any outstanding initiatives. This inconspicuous policy corresponds with Austria's recognition of its limited powers to influence the policies of multilateral institutions and also with its somewhat negligible financial input – although Vienna does host some major UN organisations. This minor influence may be the reason why multilateral contributions are not an important issue in parliamentary or public debates. In support of a European approach to development co-operation, Austria will, however, have to increase its multilateral contributions to ensure equitable burden-sharing among member states in accordance with their economic strength.

VIII. CONCLUDING REMARKS

The importance of development policy in Austrian international relations has declined considerably since the 1970s and is no longer of great political or public interest. Given the size of Austria's total ODA, governments since the 1980s have not provided budgetary resources for development co-operation that can be planned and that reflect the country's economic strength. Although the Austrian public express a mainly positive attitude towards development assistance, the role of international development is not sufficiently acknowledged when it comes to federal budget allocations.

During the past 20 years, a variety of motives for development co-operation, ranging from international solidarity by means of strengthening the economy and trade in developing countries and support for self-help initiatives to combating poverty and preventing migration, have been brought forward as arguments in the political debate. Used mainly to underpin the principal necessity of development co-operation, they were, however, an insufficient incentive to guide political action and consequently increase the volume of ODA. Although none of the four parties represented in Parliament is against development co-operation, there is no consensus on its method of implementation and/or its volume.

Although the SPÖ pleads for an increase in Austria's ODA volume, it did not put this into practice when it was in power. While the ÖVP stresses the importance of Austrian development co-operation and sees it in a positive light, the FPÖ uses the catch-phrase of 'new poverty in Austria' as an argument against increasing the aid volume [*Der Standard, 30 April 2002*]. During the flood disaster in Austria in the summer of 2002, for example, the FPÖ demanded that the development co-operation budget be spent on the flood victims.

Although public support for development co-operation has been polled at regular intervals since 1974 and, in 2001, 83 per cent of the Austrian population agreed to engaging in development co-operation (48 per cent were even in favour of a significant increase in the aid volume) [*Andrlik, 2001*], this public support is not enough to achieve a concrete increase in the budget. Public opinion stands here in clear contradiction to the political decision-making level. Nevertheless, it seems that SPÖ and ÖVP politicians do not want to risk a public debate on spending more on development co-operation while taxes are being raised and structural and economic reforms discussed.

In this context special mention should be made of a fund-raising campaign (Neighbour in Need), initiated and run by the Austrian Broadcasting Corporation (ORF) and private aid organisations, which has raised a total of €125 million in private funds since 1992 for humanitarian aid to the former Yugoslavia. This campaign created public awareness about solidarity and humanitarian assistance in the 1990s, and has supported the collective national conception that Austria is the champion in funding humanitarian aid. It seems that politicians prefer to participate in the publicity of the mass media instead of implementing international goals in domestic policies.

The criticisms of Austria's ODA performance have been well known for many years, having been expressed straightforwardly and to the point, above all, by the DAC. That these shortcomings have not yet been overcome is due mainly to the fact that the civil servants in charge are hindered by political decisions regarding the improvements they propose and that the Foreign Ministry's co-ordinating function in aid activities barely works in practice [*Raffer, 1995*].

The political authorities have neither provided the financial means nor created the structural and administrative conditions necessary to implement a development policy that would at least bring Austria near to fulfilling its international commitments and financial pledges. Since there is no prospect in the near future of an overall aid budget or a development policy framework which is binding on all activities reported as ODA, no sustainable quantitative or qualitative improvements can be expected, even if HIPC debt reduction efforts bring Austrian ODA close to the average EU contribution for the next couple of years.

Looking at DDC development co-operation as an isolated component and as the area of Austrian ODA where policy and programmatic principles are applied, significant qualitative improvements are noticeable, especially since a geographical concentration was initiated in 1992. Since Austria's accession to the EU, DDC programmes and projects have been progressively focused on the goals adopted by the international community and, through the establishment of regional co-ordination offices in priority countries, have improved considerably in qualitative terms.

Austria's accession to the EU has led to an internationalisation and, as a consequence, to modifications in the general framework and conditions for implementing development co-operation activities. Austrian NGOs see this as a threat to their work, because they have to contribute their own funds and are no longer the preferred implementors of official development co-operation projects. The cross-cutting issues of gender, the environment, poverty reduction and participatory development/good governance are gradually being incorporated and implemented in DDC aid, but they play a minor role in overall ODA. In the formulation of country and sector programmes there has been visible improvement. Attempts towards new projects and other forms of assistance such as general programme or budgetary aid have been constrained by the restrictive budgetary policy.

Neither the gradual qualitative and programmatic modifications in the field of DDC aid since 1992 nor the creation of a new development agency in 2003 can obscure the fact that there is a lack of coherence in overall Austrian ODA and that the aid volume is far below the EU average. During the past two decades, the political authorities in Austria contented themselves with announcements and pledges without providing the budgetary scope.

REFERENCES

AGEZ (ed.), 1992, *Soll und Haben, Konturen der österreichischen Entwicklungspolitik 1* (Debits and Credits, Outlines of Austrian Development Policy 1), Vienna: Südwind.
AGEZ (ed.), 1993, *Zusammenarbeit, Konturen der österreichischen Entwicklungspolitik 2* (Co-operation, Outlines of Austrian Development Policy 2), Vienna: Sandkorn, Südwind.
Andrlik, E., 2001, *Ein Zwischenbericht über die bisherigen Aktivitäten des Arbeitskreises, Public support for international co-operation* (An interim report about the activities of the working group Public Support for International Co-operation), Vienna: VIDC.
Bartsch, J.,1997, 'Die österreichische Entwicklungszusammenarbeit am Prüfstand der OECD – die Ergebnisse der Österreichprüfung' (Review of Austria's Development Co-operation – Results), in Liebmann and Amon (eds), pp.351–60.
Dachs, H. and P. Gerlich (eds), 1991, *Handbuch des politischen Systems Österreich* (Handbook of the Political System in Austria), Vienna: Manz.
Der Standard, 30 April 2002, 'Österreich hilft' (article in the Austrian newspaper *Standard* – Austria helps), Vienna.
Federal Chancellery, 1992, *Dreijahresprogramm der österreichischen Entwicklungshilfe 1993 bis 1995* (Three-Year Programme of Austria's Development

Co-operation 1993–1995), Vienna: BKA, Bundeskanzleramt, Sektion VI – Entwicklungszusammenarbeit.
Hartmeyer, H. and M. Jäggle, 1992, 'Kein Profil. Zur Situation der österreichischen Politik im Bereich der Entwicklungszusammenarbeit' (No Profile – the Situation of Austrian Development Policy), in AGEZ (ed.), pp.36–55.
Höll, O., 1986, *Österreichische Entwicklungshilfe 1970-1983*: Kritische Analyse und internationaler Vergleich (Austrian Aid 1970–83, Critical Analysis and International Comparison), Vienna: Braumüller.
Lennkh, G., 1997, 'Die öffentliche Entwicklungszusammenarbeit Österreichs' (Official Development Assistance of Austria), in Liebmann and Amon, pp.361–91.
Lennkh, G., 1998, 'Neue Tendenzen und Überlegungen in der Entwicklungszusammenarbeit – ihre Bedeutung für Österreich' (Trends in Development Co-operation – Relevance for Austria), *Journal für Entwicklungspolitk*, Vol. XIV/1, pp.7–22, Vienna: Südwind.
Liebmann, A., 1993, *Leitfaden zur europäischen Entwicklungszusammenarbeit* (Guidelines to European Development Co-operation), Schriftenreihe Europa des Bundeskanzleramtes, Band 3, Vienna: Verlag der Österreichischen Staatsdruckerei.
Liebmann, A., 1994, *Handbuch der österreichischen Entwicklungszusammenarbeit*, (Handbook of Austrian Development Co-operation), Vienna: Österreichische Staatsdruckerei.
Liebmann, A., 1997, 'Die Finanzierung der Entwicklungszusammenarbeit' (Financing Development Assistance), in Liebmann and Amon (eds), pp.393–403.
Liebmann, A. and W. Amon (eds), 1997, *Dimensionen 2000: Umwelt, Friede und Entwicklung* (Dimensions 2000: Environment, Peace and Development), Vienna: Holzhausen.
Luger, K.,1997, 'Das Bild der Dritten Welt in Österreichs Öffentlichkeit' (Third World Public Image in Austria.), in Liebmann and Amon (eds), pp.291–300.
MFA, 1998, *Gender und Entwicklung, Grundlagen für die Gleichstellung von Frauen und Männern in der Entwicklungszusammenarbeit* (Gender and Development, Basics for Development Co-operation), Vienna: BMaA, Bundesministerium für auswärtige Angelegenheiten, Sektion Entwicklungszusammenarbeit.
MFA, 2001a, *Jahresbericht 2000* (Annual Report 2000), Vienna: BMaA, Bundesministerium für auswärtige Angelegenheiten, Sektion Entwicklungszusammenarbeit.
MFA, 2001b, *Dreijahresprogramm der österreichischen Entwicklungszusammenarbeit 2002 bis 2004* (nicht veröffentlichter Entwurf) (Three-Year Programme of Austria's Development Co-operation 2002–2004 – draft), Vienna: BMaA, Bundesministerium für auswärtige Angelegenheiten, Sektion Entwicklungszusammenarbeit.
MFA 2001c, *Bericht über die österreichischen Unterstützungsmaßnahmen 2000 für Mittel- und Osteuropa sowie die Neuen Unabhängigen Staaten* (Report on Official Aid 2000 to CEEC/NIS), Vienna: BMaA, Bundesministerium für auswärtige Angelegenheiten, Referat VII.2e.
MFA, 2003a, *Weltnachrichten 2/2003 Informationen der Österreichischen Entwicklungszusammenarbeit im Außenministerium* (World News, Information on Development Co-operation in the MFA), Vienna: BMaA, Bundesministerium für auswärtige Angelegenheiten, Sektion Entwicklungs- und Ostzusammenarbeit.
MFA, 2003b, http://www.bmaa.gv.at/,Vienna, BMaA, Bundesministerium für auswärtige Angelegenheiten.
MFA, MF, 2002, *Vortrag an den Ministerrat betreffend Reorganisation der österreichischen Entwicklungs- und Ostzusammenarbeit* (Report to the Council of Ministers regarding the reorganisation of development co-operation and co-operation with Eastern Europe), Vienna: BMaA, Bundesministerium für auswärtige Angelegenheiten, BMF, Bundesministerium für Finanzen.

Nier-Fischer, F., 2002, 'Österreich: Eine politische Wende in Europa – auch in der Entwicklungspolitik' (Austria – A Political Turnabout in Europe – also in Development Co-operation), *epd Entwicklungspolitik*, Vol. 19/20, Frankfurt.

Obrovsky, M., 1993, 'Strukturanpassung. Zu den politischen und administrativen Rahmenbedingen der österreichischen Etwicklungszusammenarbeit' (Structural Adjustment. The Political and Administrative Frame of Austrian Development Co-operation), in AGEZ (ed.), pp.79–95.

Obrovsky, M., 1998, 'Auswirkungen der Mitgliedschaft Österreichs bei der EU auf die österreichische Entwicklungspolitik und Entwicklungszusammenarbeit' (Impacts of the Austrian EU Membership on Austrian Development Co-operation and Policy), *Journal für Entwicklungspolitik*, Vol. 1, pp. 93–102, Vienna.

Obrovsky, M., 2001, 'Standortbestimmung der entwicklungspolitischen NROs in Österreich' (Position-fixing of Austrian NGDOs), *Journal für Entwicklungspolitik*, Vol.17, pp. 139–42, Vienna.

OECD, 1996, *Shaping the 21st Century: The Contribution of Development Co-operation*, Paris: OECD.

OECD, 2000a, 'Peer Review of Austria', *The DAC Journal, International Development*, Vol. 1, No. 2, Paris: OECD.

OECD, 2000b, *A Better World for All: Progress Towards the International Development Goals*, Paris; see also: http://www.paris21.org/betterworld/home.htm.

OECD, 2003, *Geographical Distributions of Financial Flows to Aid Receipients, 1997–2001*, Paris: OECD.

ÖFSE, 1999, *Österreichische Entwicklungspolitik, Berichte, Informationen, Analaysen* (Development Co-operation of Austria, Reports, Information, Analyses), Vienna.

ÖFSE, 2001, *Österreichische Entwicklungspolitik, Berichte, Informationen, Analysen* (Development Co-operation of Austria, Reports, Information, Analyses), Vienna.

ÖFSE, 2003, www.eza.at,Vienna.

Österreichische Bundesregierung, 2000, *Regierungsprogramm* (Government Programme 2000), Vienna.

Österreichische Gesellschaft für Außenpolitik und Internationale Beziehungen, 1995, *Österreichisches Jahrbuch für Internationale Politik 1995* (Austrian Yearbook of International Politics), Vienna: Böhlau.

Raffer, K., 1995, 'Österreichs Entwicklungshilfe: Ein trauriges Kapitel' (Austrian Aid – A Sad Chapter), in Österreichische Gesellschaft für Außenpolitik und Internationale Beziehungen, pp.21–47, Vienna.

Republic of Austria, 2002, *Bundesgesetzblatt der Republik Österreich*, 49. Bundesgesetz: Entwicklungszusammenarbeit, 29. März 2002 (Federal Law Gazette, Law for Development Co-operation), Vienna, Republik Österreich.

Republic of Austria, 2003, *Bundesgesetzblatt der Republik Österreich*, 65. Bundesgesetz: EZA- Gesetz-Novelle 2003, 14. August 2003 (Federal Law Gazette, Law for Development Co-operation), Vienna, Republik Österreich.

3

Belgian Aid Policies in the 1990s

NATHALIE HOLVOET AND ROBRECHT RENARD

I. INTRODUCTION

The 1990s were a challenging decade for development co-operation in Belgium. In Africa the former colony (Democratic Republic of Congo[1]) and the territories formerly administered by Belgium (Burundi and Rwanda) witnessed *coups d'états*, civil war, regional conflicts, genocide and immense suffering of the population. Hundreds of thousands of people lost their lives in all three countries. Aid to this region constituted the core of Belgium's African policies, in terms of both volume of aid and geopolitical attention up to the late 1980s. The crises affecting the three countries wiped away decades of development aid and constituted a major challenge for Belgian policy-makers. In the view of some observers, they were a testimony to the failure of Belgian aid policies.

At a political level, the reactions to the crises of the 1990s by successive Belgian governments varied. Belgium suspended aid to DR Congo in 1991, following the reported killing of students by Mobutu's paratroopers on the Lubumbashi university campus, and from then on took a tough stand with the regime. In the ensuing period of regime change, instability and war, Belgium was active on the diplomatic front and supported regional efforts to find a peaceful solution. It also provided humanitarian and emergency aid to DR Congo (making it the foremost bilateral donor to DR Congo in 1999 to 2000). In Burundi, Belgium supported the transition to a democratic system in the early 1990s and used aid to this effect. But in October 1993 the first democratically elected president Ndadaye was murdered by the Tutsi military after only a few weeks in office,

---

Note of acknowledgement. In addition to the participants in the Nijmegen seminar, the authors wish to thank Johan Debar for useful comments on this chapter. A companion paper containing more detailed statistical information is Holvoet and Renard [2002].
1. We shall use the new name Democratic Republic (DR) of Congo even when referring to the period when the country was still called Zaïre.

and the country plunged into political and military instability from which it had yet to recover at the end of the decade. In Rwanda the early 1990s witnessed the onset of a major internal conflict between the regime in place in Kigali and the invading Rwandan Patriotic Front troops from Uganda made up mainly of former Tutsi refugees. Belgium was involved in the search for a peaceful solution here, using both aid and diplomatic means, and putting pressure on Rwandan President Habyarimana to enter into negotiations with the rebels and to sign the 1992 Arusha peace agreement. Belgian troops, present in the country as part of the deal to supervise the implementation of the agreement, were caught up in the fighting that started on 6 April 1994 when the plane carrying President Habyarimana and his colleague from Burundi was shot down. Ten Belgian paratroopers were killed by the Rwandan army, and the Belgian government pulled out its troops and contributed to the non-intervention of Western governments in the dreadful weeks that followed.[2] The government contributed humanitarian aid following the genocide, and then gradually resumed its structural aid to Rwanda.

During the course of 1995 the Belgian political world was shaken by a series of highly critical articles published in the Flemish daily *De Morgen*.[3] They painted a picture of failed Belgian aid projects worldwide. A major charge was that commercial interests had the upper hand in decision-making, resulting in wasteful 'white elephants' which did more harm than good. The projects singled out for attack had received most of their funding through the Ministry of Finance rather than the aid department, yet the former was never seriously questioned and the latter got all the blame. That a series of articles could have such an influence may seem at first sight remarkable. But over the years official aid policies had acquired a poor reputation with political commentators and many rank-and-file members of Parliament. The criticism thus struck a chord with these disenchanted observers, and was amplified in the media and in Parliament. A special parliamentary commission was set up that formulated a set of recommendations to improve the working of the aid department [*Chambre des Représentants de Belgique, 1997*]. Several senior officials were removed, and the government embarked on a major overhaul of the administration. The Belgian Administration for Development Co-operation (BADC), a semi-autonomous unit within the Ministry of Foreign Affairs, was restructured and integrated,

---

2. An all-party parliamentary commission was later set up to investigate the role of Belgium in the 1994 events in Rwanda [*Sénat de Belgique, 1997*].
3. The journalist who wrote the articles later produced a book on the same topic. See de Coninck [*1996*].

although not fully merged, into the Foreign Ministry and renamed the Directorate General for Development Co-operation (DGDC). At the same time, an independent operational branch, Belgian Technical Co-operation (BTC), was created, modelled on the German GTZ.[4] This reform has been criticised for being overhasty and badly conceived, but it certainly answered an urgent political need.

During the 1990s Belgian government discourse shifted towards political and civil rights, and 'soft' issues such as the environment and culture. In an important break with the past, it was decided that aid administered by the DGDC (and implemented by BTC) would no longer be tied to procurement in Belgium. Development aid became, so to speak, purer. But it also became leaner. Aid flows diminished as a share of GNP, following a trend in other countries, although the fall was less dramatic than for DAC countries as a whole.

The influence of NGOs on official policies grew considerably. The staff of personal advisers of the Secretaries of State[5] for development co-operation holding office during the 1990s consisted of collaborators with an NGO background, at the expense of officials from the aid administration or experts from universities, consultancy firms, or private business. Underlying this trend was the fact that, in the view of public opinion and the political class, NGOs were the only major actors untainted by the scandal surrounding aid policies, notwithstanding the more critical assessment of the role of NGOs that began to appear in the press [see *Achterhuis* et al., *1993*]. The ascendancy of NGOs was helped by a number of reforms in co-financing modalities that strengthened their financial and analytical muscle.

All three Secretaries of State serving during the 1990s came from Flanders, the Dutch-speaking part of the country. This continued a trend of the previous two decades, in which development co-operation attracted more attention in the Flemish than in the francophone part of the country. For instance, the scandals pertaining to the aid department were exposed by a Flemish daily paper, were widely discussed by Flemish politicians and NGOs, but were less noticed in the French-speaking part of the country. What was not foreseen was that the linguistic issue would all of a sudden become a major factor in the aid debate. In discussions among the coalition partners during the course of 2000 concerning a wide range of policy reforms providing more federal subsidies to the cash-stricken Walloon region and francophone community and more autonomy to Flanders, the possibility of handing the responsibility for major parts of development co-operation over to

---

4. By Law of 21 December 1998, published in *Moniteur Belge*, 30 December 1998.
5. A Secretary of State is a junior minister without a seat in the Cabinet.

the regions became a burning issue. It was thrust on to the political scene by a small Flemish political party (Volksunie or Flemish Union), but there was increasing support for such transfers of resources and political power on the francophone side, notably from the Socialist Party. The principle of decentralising development was negotiated during an all-night marathon session and became part of the reform package, to the dismay of many of the development experts in the main political parties, the NGO movement, and the academic world.

Quantitatively, Belgium did not achieve the 0.7 per cent of GNP target during the 1990s. On the contrary, whereas it stood slightly higher than 0.4 per cent at the beginning of the 1990s, Official Development Assistance (ODA) declined markedly towards the mid-1990s. This was the period when European economic and monetary union was being launched and the Maastricht criteria had been set to assess the eligibility of the EU member states. Belgium had to make a convincing effort to bring down its colossal public debt ratio and to reduce its fiscal deficit to the target rate of 3 per cent of GNP. This factor certainly helps explain the poor ODA performance. A contributing factor was the crises in the three traditional aid-recipient countries in Central Africa. The low ODA percentages continued until 1999. In 2000 ODA rose again to 0.36 per cent (see Table 3.2).

II. MOTIVES AND OVERALL OBJECTIVES

In Belgium, the official rhetoric on aid has evolved during the 1990s. The end of the Cold War and the political and military instability that followed, not only in Eastern Europe but also in Africa, brought new issues into focus. More than in the past, aid began to be seen as an instrument to bring peace and stability to developing regions and to protect European borders from floods of immigrants. AIDS had also alerted public opinion that there was no effective way to cut the Western world off from the developing countries and their woes. Such changes in the justification for international aid mark the brochures and policy documents produced by successive Secretaries of State in this period [*BADC, 1996, 1998; DGDC, 2000*]. The new rhetoric may well be just that: a way of phrasing policy intentions in fashionable jargon for sale to the general public, without necessarily any effect on actual decision-making. But this would be an unfair assessment of the considerable efforts by successive Belgian governments to renovate development co-operation policies.

TABLE 3.1
GOVERNMENTS IN POWER IN THE 1990s

Government Coalition	Secretary of State/Minister of Development Co-operation (Political Party)	Minister of Foreign Affairs (Political Party)
May 1988–September 1991 Christian Democrats + Socialists + Flemish Nationalists	Minister A. Geens *(Flemish Nationalists)*	M. Eyskens *(Flemish Christian Democratic Party)*
September 1991–June 1995 Christian Democrats + Socialists	Secretary of State E. Derycke *(Flemish Socialist)*	W. Claes *(Flemish Socialist Party)* From October 1994 onwards: F. Vandenbroucke *(Flemish Socialist Party)*
June 1995–September 1999 Christian Democrats + Socialists	Secretary of State R. Moreels *(Flemish Christian Democrat)*	E. Derycke *(Flemish Socialist)* Exceptionally, Moreels reported to the Prime Minister, J.-L. Dehaene, *(Flemish Christian Democrats)*
September 1999–July 2003 Liberals + Socialists + Greens	Secretary of State E. Boutmans *(Flemish Green Party)*	L. Michel *(Francophone Liberal Party)*

Note: A new government (Liberals+Socialists) came into power from mid-July 2003 onwards. Secretary of State E. Boutmans was replaced by Minister M. Verwilghen (Flemish Liberal Party).

Table 3.1 gives an overview of the coalition governments and the politicians in charge of development co-operation in the 1990s. From the end of 1991 onwards, three successive Secretaries of State[6] pursued overall objectives that were similar in significant ways and gave Belgian aid policies a new sense of direction. The adoption of the first law on International Co-operation by Parliament [*Moniteur Belge, 1 July 1999*] constituted a consolidation of these changes. The law sets out a new framework for Belgian international aid policy. The global objective is to achieve sustainable human development by combating poverty, and poverty reduction is presented as the prime objective of the aid programme, the concept of poverty being seen as multi-dimensional, including economic, social and political aspects, and opening up room for including the concept of human rights. In line with recommendations contained in the DAC strategy paper [*OECD,*

---

6. Secretaries of State, as junior ministers, report to a designated senior minister when they have to bring decisions to Cabinet level. Two of the Secretaries of State for development co-operation reported to the Minister of Foreign Affairs, the third one directly to the Prime Minister.

1996], the law also reinforced the importance of sector policy papers and country strategy papers [*OECD, 1997*]. These documents are meant to translate the broad aid policy objectives into more specific objectives and concrete activities for a particular sector and country, and provide an objective basis for subsequent monitoring and evaluation.

III. BALANCE BETWEEN OLD AND NEW OBJECTIVES

The Law on International Development of 1999 stipulated that five sectors – public health, education and training, agriculture and food security, basic infrastructure, and conflict control and reconstruction – and three cross-cutting issues – the environment, gender and the 'social economy' – should constitute the core of official Belgian aid efforts. Although all the Secretaries of State who were in charge of development co-operation during the 1990s emphasised similar sectors and themes, there are also specific priorities distinguishing the three politicians. During the first half of the decade, mostly coinciding with the period in office of Secretary of State Derycke, a Flemish Socialist, the importance attached to human rights and democracy was striking. At that time, Belgium was fine-tuning its policy of political conditionality. In 1993, for instance, direct bilateral aid to Rwanda was made conditional on democratisation and improvements in the human rights situation, and negotiations with Indonesia on a new general agreement broke down because the Indonesian authorities refused to include a human rights clause. Government officials stressed repeatedly that official development assistance relations with DR Congo could only be re-established when human rights and other aspects of governance improved. In February 1995, an important policy paper was published by the Minister of Foreign Affairs Vandenbroucke, and the Secretary of State for development co-operation Derycke, which established new priorities in Belgium's Africa policy [*Vandenbroucke and Derycke, 1995*]. While the paper stressed Belgium's willingness to provide continued aid to Central Africa, it added that this would no longer be based on honouring any 'privileged relationship'. The paper also proposed a reorientation from a paternalistic approach towards a 'genuine partnership' building on notions of 'shared responsibility', 'local capacity-building', 'sustainable development' and the untying of aid.

During the second half of the decade, Secretary of State Moreels, a Christian Democrat, drew special attention to humanitarian concerns. A medical doctor who prided himself on his experience as a 'war surgeon', and who had presided over the Belgian section of the international NGO Médecins Sans Frontières before entering politics, Moreels brought some experience with him in the field of humanitarian aid, and a passionate belief in linking aid to good governance and human rights. In practice, food aid, emergency aid, preventive aid and short-term rehabilitation aid became more important aid categories in Belgium. Conflict prevention came to the forefront, and internationally Belgium led the lobby for anti-personnel mine clearing and against the small arms trade. Within DGDC a special unit responsible for conflict prevention activities was installed and in 1997 two new financing instruments were created in order to be able to react rapidly and flexibly in cases of need.

Secretary of State Boutmans (Flemish Green Party) who took office in 1999 had no problems endorsing the priority sectors and themes of the Law on International Development adopted in 1999 under his predecessor. As a politician from a Green Party, he put extra emphasis on environmental issues, both in rural and also increasingly in urban areas, on the medical, social and economic problems caused by AIDS, and on culture. As a lawyer he was also keen to continue the emphasis laid by Derycke and Moreels on human rights, conflict prevention and peace-building. But he also felt that Belgium was not in a moral position to lecture recipient governments on how to run their own countries, and consequently was more reluctant to apply political conditionality.

IV. RELATIONS WITH RECIPIENTS AND OTHER DONORS

In 1989 Belgium launched an experiment in '*co-gestion*' or joint management in its project funding to some recipient countries. Under 'co-gestion' the responsibilities of donor and recipient were spelled out for the whole project cycle. The recipient was responsible for project identification, formulation and implementation. Belgium would have the final say in accepting the identification and formulation reports, and together with the recipient would take charge of evaluation. The system was meant to ensure that, while still very recognisably 'Belgian', projects would to a larger extent than before be integrated into the spending procedures of the recipient country. A further important advantage was that projects would be at least partially untied, with tendering for goods and services coming under the supervision of the recipient and being based on the principle of free competition. Some weaknesses of project aid remained, however, such as the use of Belgian rather than recipient government monitoring and control procedures, and the de facto imposition of

Belgian technical assistance during project formulation and implementation. Unfortunately the countries selected for implementation of the new management system, Rwanda and Burundi, were caught in a spiral of institutional instability and violence, and the experiment ran only very partially. It was applied tentatively to some other beneficiary countries, such as Bolivia, Tanzania and Vietnam, but was limited to certain non-project transactions, such as debt relief operations [*OECD, 1995*]. Most of the features of 'co-gestion' have been adopted in the new management system put in place in 1998 to 2000. However, the system is now more complicated and unwieldy, with two partners on the Belgian side: the DGDC for planning and BTC for the supervision of implementation.

Belgium has been reluctant to provide general *budget support* to developing countries. Budget support alters relations not only with the recipient but also with other donors, with whom co-ordination is desirable. Although Belgium was sceptical about granting budget support during most of the 1990s, there were some counter-indications towards the end of the decade, notably in the case of Niger and Mozambique.[7] A factor complicating the use of this instrument was the opposition coming from the *Inspecteur des Finances*, the Budget Department Official stationed at the DGDC who has to approve all spending prior to execution, and who opposed budget support on legal grounds. The scepticism with regard to budget support prevailing among Belgian aid officials and policy-makers can be explained in several ways. One is the political dimension: as a small donor Belgium cannot expect to play a significant role during negotiations with the recipient which are dominated by more powerful donors. Another is the dominance of traditional project experts within the aid department, and the relative scarcity of macroeconomists or other macro experts.

V. STRUCTURE AND MAIN COMPONENTS OF THE AID PROGRAMME

The aid under the responsibility of the DGDC may be broken down into four categories: *'direct' bilateral aid*,[8] negotiated by DGDC on behalf of the Belgian government, the execution of which is entrusted to BTC; *'indirect' bilateral aid*, i.e. funds entrusted to NGOs and

---

7. Both deals became operational in the period after 2000 and are not discussed further here.
8. The distinction between 'direct' and 'indirect' aid is distinctive to Belgium. It will be used in the rest of the chapter, as it is of some importance for a proper understanding of some of the issues dealt with.

other Belgian non-public sector actors who have the right of initiative in allocating these funds, *multi-bi*; and finally *multilateral aid*. The aid administered by other ministries and official bodies is discussed later in this chapter. Table 3.2 provides a breakdown for the period under review. One of the remarkable features of the 1990s is the considerable fall in the importance of direct bilateral aid. From around 40 per cent of all DGDC aid at the beginning of the decade it fell to around 20 per cent in the closing years of the same decade. This represents a fall not only in relative, but also in absolute terms. The slack has been taken up by 'indirect' bilateral aid, especially through the NGOs and the universities, and by multilateral aid.

The reduction in direct bilateral aid, for which the DGDC has effective responsibility, was not explicitly planned by any of the three Secretaries of State. It can probably be best explained as a consequence of the continued failure of the official aid agencies to deliver aid in a timely and effective manner, together with the introduction of programme financing for the indirect actors, which greatly reduced transaction costs for the public sector and allowed vastly increased spending. What is striking in this respect is the absence of a serious effort to develop a consistent global policy in which the different aid channels are weighted against each other and complementarities between them exploited.

Technical assistance no longer commands the important place in direct bilateral aid it held during previous decades. The number of long-term expatriate experts declined from several thousands in the 1970s to some 1,200 at the end of the 1980s. By the mid-1990s the number had fallen below 500, and at the end of the decade to less than 300. The breakdown or at least suspension of aid relations with the three main recipients of Belgian technical assistance during the 1990s, DR Congo, Burundi and Rwanda, is a major cause of this striking trend. Belgian aid officials were probably also not insensitive to the criticisms voiced against long-term expatriate experts [*OECD, 1995, 1997*].

Project aid remained the key instrument of Belgian direct bilateral aid, but its importance in total aid declined. It was further characterised by a strong lack of co-ordination. In Rwanda, for instance, Belgium was funding some 12 interventions in the health sector at the turn of the century, including some institutional technical assistance at the level of the health ministry. But no explicit overall sector strategy seemed to underscore the different interventions, which were managed as isolated events.

TABLE 3.2
BREAKDOWN OF TOTAL BELGIAN ODA AND THE EXPENDITURES OF THE DGDC/BADC BY TYPE OF ACTION (1990–2000, US$m.,[a] CURRENT PRICES)

	1990	1991	1992	1993	1994	1995	1996	1997	1998	1999	2000
I. DGDC administration costs	33.14	22.53	29.53	32.81	33.47	36.44	40.07	34.83	39.68	35.51	27.10
II. Bilateral (total)	383.84	327.99	409.76	322.76	276.10	328.72	309.47	295.27	307.83	313.12	281.35
Direct	246.08	213.95	248.85	164.77	157.89	139.64	153.76	119.93	120.70	129.23	120.28
Indirect	137.76	114.05	160.9	157.99	118.20	189.07	155.7	175.34	187.12	183.89	161.06
NGO co-financed	64.15	62.47	88.26	90.79	66.26	108.08	90.61	100.13	107.63	98.31	88.15
University co-operation	31.01	24.52	32.86	28.02	16.12	50.44	32.15	33.92	40.02	42.60	37.24
III. Bi-Multi DGDC (incl. Belgian Survival Fund, emergency aid)	69.65	49.41	41.73	56.27	43.77	73.43	98.58	58.29	65.87	50.35	78.92
IV. Multilateral	126.58	103.05	113.66	204.25[c]	115.10	135.16	100.19	154.76	152.51	162.46	165.51
UN Group						31.69	34.23	26.62	19.41	21.58	33.45
EU Group						84.08	47.66	53.96	73.40	93.29	67.27
IDA/World Bank Group						0.75	0.325	57.79	47.79	47.05	52.26
TOTAL DGDC/BADC	613.23	502.98	594.66	615.28	468.40	573.13	548.31	543.15	565.89	561.45	552.86
Ministry of Finance	232.15	290.42	229.88	153.45	180.62	367.85	251.08	134.45	162.00	120.31	152.84
Bilateral[b]	33.03	77.79	33.93	27.9	24.02	6.56	10.97	−17.00[d]	−4.81	−12.78	3.63
Multilateral	199.12	212.64	195.95	125.54	156.60	360.45	240.10	151.44	166.81	133.10	149.21
Office National du Ducroire (OND)	0	0	52.7	21.25	37.98	38.31	69.11	36.10	102.83	19.87	31.49
Foreign Affairs (except for DGDC)	10.36	14.42	18.38	21.01	13.53	33.55	33.02	27.54	27.08	24.60	37.37
Other Federal Ministries	21.09	10.05	6.45	6.22	8.86	5.71	8.54	6.33	6.82	6.98	6.82
Other Regional Ministries, Provinces and Communes	17.22	16.21	18.63	17.97	20.67	26.06	30.99	24.34	25.12	29.98	36.90
TOTAL (ODA)	894.06	834.07	920.68	836.00	747.56	1044.5	941.23	772.04	889.75	766.71	818.28
ODA/GNP (%)	0.46	0.42	0.42	0.40	0.32	0.37	0.35	0.31	0.35	0.30	0.36

Sources: 1990–1994: DGDC [2001b]; 1995–2000: DGDC [2002d]

Notes: [a]: using DAC exchange rates; [b]: including, among others, loans from government to government; [c]: exceptionally large due to the fact that in 1993 BADC instead of the Ministry of Finance paid Belgium's contribution to IDA. From 1997 onwards this became the rule. It explains the steep increase in the multilateral contribution from BADC/DGDC to the World Bank Group from 1997 onwards; [d]: negative due to a decline in amount of intergovernmental loans, in combination with an increased payment of the latter.

Belgium was active in the field of *debt relief*. In the early 1990s the official export credit agency, the *Office National du Ducroire* (OND), was in dire financial straits. Many of the export credit and insurance contracts it had accepted in earlier years had fallen through and the agency had collected a huge amount of unrecoverable claims on developing countries, while, on the other hand, it had to pay out large sums of money to the Belgian firms which were parties to the contract. In 1991 it was decided that the Treasury would buy some of the OND's outstanding claims on developing countries for half their face value, far above their market value, and thus provide needed liquidity to the organisation. As part of the deal, the development co-operation administration was asked[9] to contribute approximately US$15.5m. a year over a period of 12 years, in exchange for which it would receive from the Treasury some of the debt claims which it could then in turn cancel. Whether this is indeed genuine aid is a debatable point, but the DAC accepted it as ODA under its generous accounting rules. A striking feature of Belgian debt relief operations during the 1990s is how often aid officials insisted that the recipient government should set aside the debt relief in a special counterpart fund managed jointly with Belgium. In this way budget support was averted and replaced by the more familiar instrument of project aid. Aid officials remained largely oblivious to the criticisms voiced in this respect by academics and other experts.

The 1990s were characterised by sturdy growth in the share of indirect bilateral aid, as indicated above. This was made possible by the introduction of programme funding, which increased the autonomy and responsibility of the indirect partners. The Belgian administration distinguishes between three partners in indirect co-operation: recognised non-governmental organisations (NGOs), Belgian universities, and specialised non-profit and research organisations.

During the 1990s, the relationship between the government and NGOs was characterised by a gradual move from project-by-project funding to programme funding under which NGOs submit five-yearly programmes and annual so-called 'action plans'. In 1991 programme funding was introduced for the larger and more experienced NGOs, while the traditional project co-financing scheme was kept in place for the bulk of NGOs. Programme aid was generalised to all NGOs in 1997. The main objective of these reforms was to reduce transaction costs for both NGOs and the administration through a simplification of administrative procedures. Incentives were created to increase pro-

---

9. It is more correct to say that the deal was imposed on reluctant aid officials by colleagues from the Treasury.

fessionalism and far-reaching clustering among NGOs. The work of the aid administration shifted from laborious bureaucratic *ex-ante* control of individual projects to monitoring and *ex-post* evaluation. Whereas DGDC officials had difficulty adjusting to the new type of aid management and while they still tended to assess the yearly action plans as packages of projects, they nevertheless switched increasingly to the new programme logic, as seen by the increase in field inspections. In addition, external experts have been appointed to assist the administration in its policy dialogue with the NGOs and the evaluation of programmes. Co-ordination among NGOs was encouraged through the formal recognition and financing of two federations, one for the Flemish NGOs and one for the francophone (and German-speaking) NGOs, which represented NGOs in their relations with the government, thus institutionalising consultation between the NGOs and the government. Consultation with individual NGOs was strengthened through policy dialogues on the occasion of the submission of the programme and the yearly action plans. An NGO consultative committee was also installed.

Although generally recognised as a major breakthrough, NGO programme funding suffers from some weaknesses. First and foremost, the government failed to weed out the large number of small NGOs which previously had access to project financing. Many such NGOs are run by a few well-intentioned individuals, but with no professional staff to speak of, and are barely equipped to perform their assigned tasks adequately. As a consequence, programme funding was extended to up to 100 NGOs, most of them too small and institutionally unstable to warrant the considerable trust put in them under this form of funding. A much smaller number of, say, 10 to 15 would have been preferable from a public sector management perspective, with the rest either receiving project funding directly or through consortia of larger NGOs. The financial incentive for NGOs to merge or form alliances did not generate the expected results, as subsequent Secretaries of State considered that small emerging NGOs ought to be given a chance to get direct access to government co-financing, and in this way contributed to the dispersion rather than the concentration of funding.

Second, the opportunity offered by the new consultation mechanisms to enter into a policy dialogue with the NGO community, in light of a co-ordination of activities among themselves and with the public sector, was not taken up, due in part to the resistance of the NGOs, but also because the DGDC did not bring a coherent policy strategy to the negotiating table. Third, evaluation of the functioning of the NGOs and their field achievements, crucial for a system of programme funding to work fairly and efficiently, was not undertaken systematically. The administration thus lacked the authority to assess individual NGOs credibly and reward good performance.

Reform of the *indirect university co-operation* was governed by agreements between the government and the Inter-university Councils of the two major communities, namely the Flemish Inter-university Council (VLIR, representing the Flemish universities) and the Inter-university Council of the French Community (CIUF, representing the francophone universities). The main focus of the reform was to give the universities greater responsibility, while allowing the administration to concentrate on the important functions of policy dialogue, monitoring and evaluation. Although the programme framework has undoubtedly allowed the universities to gain experience as development actors and to exert considerable peer control, the DGDC again failed to exploit the advantages offered by the policy dialogue to co-ordinate university co-operation with its own bilateral programme. On the positive side, evaluations are being undertaken on a regular basis.

Multilateral aid constitutes a considerable part of Belgian aid (see Table 3.2). There is an almost complete separation in the management of multilateral and bilateral aid. Complementarities are not pursued nor are synergies sought. The major item is the contribution to the European Development Fund (EDF) of the European Union, which rose to 41 per cent of the multilateral budget in 2000. Successive Belgian governments have been enthusiastic supporters of the EU, and contributions to the EDF, negotiated at the EU level, were consequently hardly questioned, although the quality of EU development aid was not in general highly regarded within the DGDC. Another part of the multilateral budget is taken up by compulsory contributions or internationally negotiated contributions, such as to the International Development Association (IDA) of the World Bank and the Global Environmental Facility (GEF). The remainder consists of a large number of highly dispersed and often relatively small voluntary contributions to dozens of organisations, often earmarked for particular interventions. The Law on International Co-operation of 25 May 1999 envisaged the concentration of multilateral aid to about 20 multilateral organisations. In 2002, an innovative follow-up envisaged pluri-annual commitments within a programme framework for some of these organisations.

The Belgian Survival Fund (BSF) is the major vehicle for *multi-bi co-operation*. It was set up in 1983 as a result of a parliamentary initiative, with the aim of providing additional aid funds, collected from the state-run lottery, to improve the food security of households in 18 selected countries of sub-Saharan Africa (not necessarily programme countries of Belgium development co-operation), affected by chronic food shortages, poor access to basic services and a high infant mortality rate. An innovative feature of the programme was the adoption of an integrated approach towards food security and the simultaneous co-operation with a number of international development organisations (IFAD, FAO, UNCDF, UNICEF). Some 15 larger Belgian NGOs also received funds from the BSF. In a later phase some of the BSF funds were also used for direct bilateral projects.

The major criticism of the BSF has been its slow pace of implementation. During its first phase, covering a period of ten years, there was an authorisation from Parliament to commit about US$330m., but by the beginning of the 1990s only half of this had effectively been spent. This is maybe not so unusual, given that such complex projects took a long time to formulate and were implemented over a considerable number of years, but the criticism remained. Over the years the BSF modalities have been further adapted [*DGDC, 2002c*].

VI. COUNTRY PROGRAMMING

Statistics on geographical allocation indicate that Belgian aid is highly dispersed. One reason for this is that several ministries and other public agencies are managing parts of the aid programme, and there is very little co-ordination between them. But this is certainly not the main reason. The situation is in fact only slightly better if we turn to the part of the aid budget managed by the DGDC. During the first half of the 1990s, the 20 major recipients of aid from the DGDC received on average US$10 million a year, more than 40 per cent of which was delivered through multilateral channels, NGOs, universities and other 'indirect' actors who all have the right of initiative and a large degree of autonomy from the DGDC in their respective fields.[10] During the second half of the decade, the corresponding average dropped to US$8m., more than half of it delivered through 'indirect' actors. The budgetary resources that the DGDC can negotiate with the individual programme countries are thus fairly modest, and decreasing over time. Yet, fixed costs are high in terms of geographical desks in Brussels and field offices in the recipient country. Even with only a few staff manning the desks, it adds up to an important sum compared to the small amount of aid that is being allocated.

---

10. This may be illustrated by looking at geographical priorities. If one compares the ten most important recipients of aid through NGO financing and through 'direct' DGDC financing during 1995 to 2000, only three countries overlap: DR Congo, Rwanda and Bolivia.

TABLE 3.3
RELATIVE IMPORTANCE OF BELGIUM IN ITS TEN MAJOR ODA RECIPIENTS (1999–2000)

		Rank^a	*Importance^b*
1	Vietnam	9	2%
2	DR Congo	2	96%
3	Rwanda	8	8%
4	Tanzania	>10	na^c
5	Bolivia	>10	na
6	Cameroon	10	2%
7	Niger	6	6%
8	Ivory Coast	>10	na
9	Burkina Faso	>10	na
10	Tunisia	6	2%

Source: On the basis of OECD [*2002a, 2002b*].
Notes: ^{a} Belgium's rank among multilateral and bilateral donors; ^{b} Belgium's contribution as a percentage of those donors preceding it in the ranking list; ^{c} data not available in DAC table.

One consequence of the geographical dispersion of aid may be seen from Table 3.3. The ten countries listed are 'major' recipients from a Belgian perspective and all but two (Cameroon, Tunisia) have the status of programme countries in Brussels. Yet from a recipient perspective Belgium is hardly an important donor, with the exception of DR Congo. In view of the fact that Belgium contributes only between 1 and 1.5 per cent of total DAC bilateral aid, this result is not surprising, but it puts the notion of programme countries into perspective. Until the 1980s, the geographical distribution of Belgian aid was highly skewed towards DR Congo, Rwanda and Burundi. In the five years preceding our period of study (1985 to 1989) these three recipients obtained more than half of bilateral ODA. In these countries Belgium was a major donor. The remaining bilateral aid was widely dispersed over a large number of countries, many of them receiving funding for one or two projects. During the 1960s and 1970s the concentration was even more pronounced.

But during the 1990s, when aid to the three traditional recipients dwindled, Belgium became a small donor with little clout in almost all the countries in which it was active. In the past, the issues thrown up by the marginal status of Belgian aid in comparison with dozens of larger multilateral and bilateral donors were mainly addressed by putting the emphasis on project aid as a major instrument and by evading policy dialogue at sector or macro levels. But if it is believed, as is at present fashionable, that project aid as a form of conditionality is an illusion [*World Bank, 1998*], and if it is further argued that in aid-dependent countries project aid, with all its donor-imposed bureaucracy, undermines the institutional capacity of the recipient government to plan, budget and implement its own development, the question must be faced as to what the best strategy is for small donors. The consideration of what, if any, the contribution might be of small bilateral donors in a post-project area, except as providers of funds, has not been squarely faced by Belgian aid officials. With regard to Central Africa, there is the feeling that there is some role for an overall foreign policy in which aid has an important place. For the other countries, it is not clear that the notion of a programme country makes much sense.

Although the above results would suggest otherwise, successive governments congratulated themselves on the efforts at concentrating bilateral aid. All three Secretaries of State in charge of development co-operation during the 1990s produced lists of priority countries. The lists were far too long, however, to have any effect on the concentration of aid. A maximum of ten countries instead of up to 25 would be more reasonable from this perspective. Second, the list changed from one government to the next. There seemed to have been no agreement across political parties to keep the list the same for a period of time bridging several legislations. As a consequence, 35 countries were at one time or another programme countries during the 1990s; out of these only six were on the list for the whole decade. Furthermore, non-programme countries also continued to receive aid, sometimes more than the programme countries.

Belgium had a reasonably good track record on *poverty orientation* during the 1970s and 1980s. This has to do with the geographical concentration on Central Africa. The situation in the 1990s was much less impressive. If we consider the percentage of aid going to low-income countries, there was not much difference between Belgium and the DAC average. The increasing share going to lower middle-income countries is striking. This was not the consequence of some determined effort at reallocating aid resources, but rather of the crises in the three Central African countries which eased the constraint on budget allocations to the other aid recipients. Low middle-income recipients, such as Bolivia and Vietnam, had a greater absorption capacity than the low-income countries, and they were accordingly rewarded with more aid.

VII. INSTITUTIONAL SET-UP: CHANGES AND PROBLEMS

At the end of the 1980s and during the early 1990s several attempts were made to reform aid administration. These reforms had in common that they were based on the idea that the same administrative division should be responsible for as large a part of the cycle of planning and implementation as possible – what came to be called the principle of A to Z responsibility. A corollary was that heads of department could also be held accountable for the final outcome. During the second half of the decade another, much more drastic overhaul of the administration was prepared under Secretary of State Moreels, who also oversaw the beginning of the implementation of this reform. That Moreels was able to go much further in his reform than his predecessors was due to a number of factors. There were the scandals that had rocked the administration at the beginning of his mandate. Moreels also had the advantage of reporting directly to a forceful Prime Minister Dehaene, who supported his fellow Flemish Christian Democrat.

The reform was carried out in the period 1998 to 2000 and consisted of three parts. First and foremost, a new unit, Belgian Technical Co-operation (BTC), was set up as a public company. This had the advantage of allowing a flexible personnel policy, and also addressing the problem of a cruel shortage of high-level personnel that had plagued the administration for a long time. BTC was governed by a board of experts appointed by the government. Moreels, with his background as president of a prestigious humanitarian NGO, Médecins Sans Frontières, considered the major weakness to be the dilatoriness and ineffectiveness of the implementation of operations in the field; BTC was the new tool that would remove this constraint. The remaining administrative and management tasks, mainly strategic planning and policy preparation, would be left to the BADC. As already noted in section I, the BADC was part of the Ministry of Foreign Affairs, but had enjoyed substantial autonomy. Moreels integrated this department more fully in Foreign Affairs without fully merging it. What this meant was that the BADC was renamed the Directorate General of Development Co-operation (DGDC), but with a separate budget and personnel that did not rotate within the Ministry. A third part of the reform was the creation of a separate evaluation unit, with an independent head to be recruited from outside the administration.

It is too early (mid-2003) to judge this major reform by its outcome, but a number of critical comments are in order. First, the reform goes in the opposite direction from previous efforts. Instead of A to Z responsibility, the functions of policy preparation and implementation were radically separated, with all the risks of poor co-ordination and administrative in-fighting that had been experienced during the 1980s. At the very least, this testifies to a lack of institutional memory in the administration's modernisation. This is linked to the fact that in Belgium such reforms are very much imposed by the government on the administration, with minimal participation from the latter. Every new minister or secretary of state, with his inner circle of hand-picked personal advisers, starts afresh the process of analysing the weaknesses of the administration, driven by a political will to make a difference, to plant the flag of the new man in charge, even if it means swinging back the pendulum of reform in the opposite direction from his predecessor.

Second, the reform was not in line with the new thinking that was emerging at the time in international circles. BTC, like the German GTZ on which it was loosely modelled, was set up to leave the donor in charge of implementation, and seems to have been conceived with old-style donor-driven projects and technical assistance in mind. Moreels, with his background in humanitarian aid, was either unaware or oblivious to the increasing calls for more recipient-country ownership during implementation. To put it bluntly, if the tendency is towards budget support under recipient-formulated and -owned strategies for development and attacking poverty, as is now attempted with the PRSP approach, then BTC seems an odd choice for preparing the administration for these new tasks.

Third, the reform did not acknowledge the important weaknesses in terms of strategy preparation and policy analysis. The part of the BADC that remained after the implementation tasks were removed was structurally weak and unprepared for its important residual tasks. The change of name to the DGDC and the partial integration into Foreign Affairs did not solve any of this. In one way, the integration did not go far enough to benefit from a pooling of human resources with Foreign Affairs. But it could not go much further than it actually did without risking a wholesale takeover by the much larger and conservative ministry.

Fourth, the new evaluation structure emphasised the accountability dimension of evaluation to the detriment of its learning function. In effect, the removal of the evaluation function from the DGDC suppressed the critical reflection on current practices as a tool of internal quality control within the organisation.[11] Fifth, the new set-up was administratively very complex and bound to run into co-ordination problems. For instance, both BTC and the DGDC set up field offices in the 25 programme countries, each of which has to communicate with the geographical desk in the respective headquarters in Brussels.

Another institutional aspect we wish to comment on is the relationship between the aid administration and other ministries that are also involved in aid policies. We have already noted how the DGDC was partially integrated into the Ministry of Foreign Affairs without being fully merged with it. Notwithstanding this closer administrative and physical proximity, co-ordination between the DGDC and the rest of the Foreign Ministry did not work very well during the first few years. This was most striking in the field of human rights, humanitarian aid and conflict prevention, all of which were taken up by development co-operation as new themes during the 1990s, while at the same time remaining areas of interest to the Ministry of Foreign Affairs. There has been surprisingly little exchange and collaboration between the departments, leading to a loss of efficiency and effectiveness in Belgian actions. It seemed at times that the new relationship was too close for comfort, not only for the administration, but also for the politicians. There have been frequent clashes between Secretary of State for Development Co-operation Boutmans and Foreign Minister Michel, who belonged to different political parties and language groups, and had very different styles of operating.

Relations between Development Co-operation and the Ministry of Finance are more distant but equally difficult. As already indicated, people at Finance are very jealous of their role of representing Belgium at the international financial institutions.[12] Development Co-operation has not been allowed to post any staff to the delegation to the World Bank in Washington, for instance, in contrast with the practice of the larger bilateral donors such as the US, France, Germany, the UK, and also Scandinavian donors and the Netherlands. The Ministry of Finance did not relent, even though the DGDC suggested that it would bear all the costs, and even after it had started paying the contributions to IDA from its budget. This has led to lack of communication between Development Co-operation and the IFIs. One of the consequences is that the Belgian aid administration has

---

11. A similar critical assessment is made in the DAC Peer Review of Belgian aid [*OECD, 2001*].
12. The following incident illustrates this. In 2001 a Treasury official, stationed at the Belgian delegation to the World Bank in Washington, was sent back to Brussels, apparently as a penalty for having consulted with a colleague from the DGDC over preparations for the Monterrey Conference on financing development.

largely missed out on the lively debate going on in Washington on such diverse matters as debt relief, the need to integrate poverty reduction into macroeconomic strategies, or on new aid instruments such as budget support. During much of the 1990s the Belgian Minister of Finance, Philippe Maystadt, was chairman of the influential Interim Committee of the Board of Governors at the IMF. In this capacity he was involved in some important discussions on debt relief. Yet, remarkably, almost nothing of this was communicated internally between the two departments; aid officials would find out indirectly what Maystadt was saying on behalf of the Belgian Government, and also indirectly on their behalf. There were no important debriefings, and little feedback, let alone dialogue.

Another difficult relationship was that between the federal government in Brussels and the many complex layers of local government that had been given more and more responsibility after each round of constitutional reform in Belgium. This has some bearing on our topic, as the 1993 reform devolves considerable rights to the Regions (Flanders, Wallonia, Brussels) and the Communities (French-, Flemish- and German-speaking) in the domain of foreign policy. There is probably no other country in the world where decentralised political authorities have such far-reaching powers. This has, for instance, led to intense consultation between the federal and lower governments during Belgium's presidency of the European Union in the second half of 2001, including EU ministerial meetings being chaired by ministers from the decentralised governments. The same overlap of authority exists in the area of development co-operation. The Flemish government, for instance, has signed international co-operation treaties with South Africa and Chile and extended aid to them.

The above may help to explain why an independent development policy at the federal level, without any involvement of the decentralised levels of government, is becoming increasingly difficult politically. Not only have the decentralised governments, using their own resources, begun to deploy their own development policies, but they became more and more outspoken in their claims for a bigger share of the federal aid budget. During the 1990s, all the main parties in the Flemish Parliament, for instance, publicly supported the principle that large parts of development co-operation should be handed over to the Flemish government.

All this came to a head in 2000, during the so-called Lambermont negotiations, in preparation for a series of complex reforms that would yet again reshuffle some of the responsibilities between different layers of government. It was decided that parts of development co-operation would be handed over to the decentralised governments. Following the principle 'in foro interno, in foro externo', every layer of government represents Belgium outside the country in areas for which it has responsibility within the country. This is the case, for instance, with regard to education, agriculture, the infrastructure and the environment. On this basis the Flemish and francophone governments had asked that development co-operation should be split along sector lines, and the federal government had accepted this in principle. In the most ambitious version of this plan, the Flemish and francophone governments would select their own partner countries, set up their own aid agencies, and fund their own projects in areas under their responsibility. The federal government would still be in charge of multilateral aid, for instance, and the tasks traditionally entrusted to the Ministry of Finance. The final Lambermont agreement, accepted by the parties of the ruling coalition and some of the opposition, accepted the principle that part of development co-operation would be handed over to decentralised levels of government by 2004, but left the details to be worked out later.

The agreement was met with dismay by the majority of NGOs and development experts. Staff at the DGDC and BTC were shaken. The reform initiated by Moreels had not yet been completely worked out in all its details, and already another, even more massive reform of development co-operation was hanging in the air. Secretary of State Boutmans publicly expressed his opposition, but did not resign over the issue. The Development Assistance Committee [*OECD, 2001*] was openly critical in its 2001 peer review of Belgium. A Senate commission discussed the issue in the same year, and invited experts and representatives of NGOs and the academic world to give their views. Significantly, in the end the commission could not summon up a majority in favour of the so-called regionalisation of development co-operation.

VIII. MAIN ACTORS AND THE PUBLIC SUPPORT OF THE AID PROGRAMME

All the major political parties in Belgium have remained committed to development co-operation. If anything, a stronger consensus on the importance of aid prevailed at the end of the decade than at the beginning. Leading politicians of the Liberal, Christian-Democratic, Socialist and Green parties pursued active policies in the areas of aid and foreign policy, especially towards Central Africa. The Liberals, traditionally the most sceptical and the most likely to emphasise Belgian commercial and economic interests over development issues, became major defenders of a foreign policy 'of the heart'. Of the government in office at the end of the decade, both Prime Minister Guy Verhofstadt (Flemish Liberal) and Foreign Minister Louis Michel (francophone Liberal) were keenly interested in development policies and active on the diplomatic front. There were no signs of disagreements within that coalition between the progressive Greens and the conservative Liberals. When tensions arose, they were related more to conflicts of personality or territory than to ideological differences. The only political party to adopt an outspoken anti-aid stand is the 'Vlaams Blok', a right-wing Flemish nationalist party that gained ascendancy during the 1990s but remained excluded from power by an agreement between the other parties.

The press has also remained largely positive towards development aid. In the 1990s, it provided a more critical analysis than during previous decades, but largely in a constructive spirit, with the exception of the almost vitriolic attacks that appeared over several months in the Flemish *De Morgen*, as noted earlier. NGOs continued to be very forceful actors, influencing both Parliament and the press. During the mid-1990s a series of scandals led to a worrying loss of credibility of official aid policies. Secretary of State Moreels, who was highly regarded at the time by public opinion and whose moral integrity was not in doubt, began to redress this situation. He also paid extra attention to informing and educating the general public on development issues, and increased the budget considerably [*OECD, 1997*].

Opinion polls did not always reveal a clear picture. A major poll in Flanders in 1999 indicated that one out of every three people regarded poverty in the Third World as among the three most important societal problems. This was in contrast with some earlier results, and suggested that there might be considerable support in the country for official aid. At the end of the decade the government decided to increase aid spending gradually over the coming years. This decision has not met with any major criticism in the media.

IX. CONCLUSION

Aid policies in the 1990s were being pursued in a climate of major upheavals, some of them externally caused, such as the dramatic events in Central Africa. These events, in particular, marked the end of the post-colonial period in which Belgium had been providing the major part of its bilateral aid to DR Congo, Rwanda and Burundi. In the past Belgium emphasised its 'special ties' with Central Africa, leading to unclear and unarticulated aid policies. All that has changed. The three countries went through dramatic crises, and the whole region remained highly unstable during most of the period. Belgian aid dwindled, and diplomacy became as important, if not more important, than aid. The aid that no longer went to the three traditional beneficiaries was spread very widely, without any clear sense of direction or a convincing effort at imposing some geographical concentration. At the level of motives and overall objectives, aid became more complex, with all the major political parties involved in government putting emphasis on human rights and governance issues. In a significant break with past practice, aid tying was, if not abandoned, then at least greatly attenuated, except for the bilateral aid administered by the Ministry of Finance. Indirect actors, especially NGOs and the universities, became major channels for official aid, to the detriment of direct bilateral aid.

The Achilles' heel of aid policies remained administrative management, despite, or maybe because of, several efforts at restructuring. A major reform at the end of the decade, splitting the administration into two entities, the DGDC for planning and BTC for implementation, has not been in operation long enough for us to make a balanced judgement, but there is some reason to be sceptical about the good it may bring. The major cloud hanging over development policies at the turn of the new millennium, however, was the proposal to hand over important parts to the regional governments. This would, in practice, result in three governments in Belgium being in charge of development co-operation, one federal and two regional, and would call for yet another major rethinking of management structures.

REFERENCES

Achterhuis H., D. Barrez, Y. Tandon *et al.*,1993, *Het orkest van de Titanic: werken aan andere Noord-Zuid verhoudingen* (The orchestra of the Titanic: working towards other North–South relations), Brussels: VUB Press.

BADC, 1995, *Répartition de l'APD par Organisme de Financement et de l'Aide AGCD par Forme de Coopération*, Vol. 1 (50 premiers pays) (Composition of ODA by contributing government department and composition of ODA by BADC over different aid components (data for the 50 most important recipients), Brussels.

BADC, 1996, *From Quantity to Quality* (Policy Note of E. Derycke), Brussels.

BADC, 1998, *Showing our Colours* (Policy Note of R. Moreels), Brussels.

BADC, various issues, *Annual Report*, Brussels.

BTC, various issues, *Annual report*, Brussels.

Chambre des Représentants de Belgique, 1997, *Suivi des Problèmes de l'Administration Générale de la Coopération au Développement (AGCD), Rapport (8 juillet 1997) fait au nom de la Commission Spéciale* (Follow-up of the problems of the aid administration AGCD, report of 8 July 1997 made for the Special Commission), Brussels.

De Coninck, D., 1996, *Witte olifanten: de miljardenschandalen van de Belgische Ontwikkelingssamenwerking* (White elephants: scandals of Belgian Development Co-operation), Leuven: Van Halewyck.

DGDC, 2000, *Quality in Solidarity. Partnership for Sustainable Development* (Policy Note of E. Boutmans), Brussels.

DGDC, 2001a, *La Coopération Belge en chiffres de 1995 à 2000* (Belgian Development Co-operation in figures from 1995 to 2000), Brussels.

DGDC, 2001b, *Mémorandum de la Belgique présenté au Comité d'Aide au Développement. Années 1997–1999* (Belgium's Memorandum presented to the Development Assistance Committee 1997–1999), Brussels.

DGDC, 2001c, *Ontwerp strategienota onderwijs en vorming (versie van 24/04/2001)* (Draft sector policy note on education and training (version of 24 April 2001)), Brussels.

DGDC, 2002a, *Strategienota Basisgezondheidszorg (versie van 21/02/2002)* (Sector Policy Note on Basic Health Care (version of 21 February 2002)), Brussels.

DGDC, 2002b, *Parlementaire nota over de Multilaterale Samenwerking* (Parliamentary Note on Multilateral Co-operation), Brussels (http://www.DGDC.be/nl/txt acteurs/multilateral/multilateral.html).

DGDC, 2002c, *Het Belgische Overlevingsfonds* (Belgian Survival Fund), Brussels (http://www.DGDC.be/nl/txt/acteurs/fondssurvie/general.html).

DGDC, 2002d, *Statistieken Officiële Belgische Ontwikkelingshulp* (Statistics on Belgium's ODA), Brussels.

DGDC, various issues, *Annual report,* Brussels.

DGDC, various issues, *INCOM: Newsletter for International Communication and Co-operation,* Brussels.

Holvoet, N. and R. Renard, 2002, *Breaking with the Past? Belgian Development Co-operation at the Turn of the Century,* IOB Working Paper No. 8, Antwerp: University of Antwerp.

Jennes, G. and M. Schellens, 1997, 'Belgium', pp. 35–41, in J. Randel and T. German (eds), *The Reality of Aid 1997/8: An Independent Review of Development Co-operation,* London: Earthscan Publications Ltd.

Moniteur Belge, 30 December 1998, *Loi du 21 décembre 1998 portant création de la 'Coopération Technique Belge' sous la forme d'une société de droit public* (Law of 21 December 1998 on the creation of the 'Belgian Technical Co-operation' under the form of a public-law company with social purposes), Brussels.

Moniteur Belge, 1 July 1999, *Loi relative à la Coopération Internationale belge* (Law on Belgian International Co-operation), Brussels.

Moniteur Belge, 14 June 2000, *Arrêté royal déterminant les organisations internationales partenaires de la coopération multilaterale* (Royal Decree determining the international organisations that will be the partners of Belgian multilateral co-operation), Brussels.

Moniteur Belge, 17 November, 2001, *Loi du 3 novembre 2001 relative à la création de la Société belge d'Investissement pour les Pays en Développement et modifiant la loi du 21 décembre 1998 portant création de la 'Coopération technique belge' sous la forme d'une société de droit public (3 novembre 2001)* (Law of 3 November 2001 on the creation of a 'Belgian Investment Corporation', modifying the Law of 21 December 1998 on the creation of the 'Belgian Technical Co-operation' as a public corporation), Brussels.

NCOS, various issues, *De Belgische Ontwikkelingssamenwerking* (Belgian Development Co-operation), *NCOS Jaarrapport* (NCOS Annual Report), Brussels.

OECD, 1995, *Belgium. Development Co-operation Review Series*, No. 7, Paris: OECD DAC.

OECD, 1996, *Shaping the 21st Century: The Contribution of Development Co-operation,* Paris: OECD DAC.
OECD, 1997, *Belgium. Development Co-operation Review Series,* No. 23, Paris, OECD DAC.
OECD, 2001, *Belgium: Development Co-operation Review. Main Findings and Recommendations, DAC Journal,* Vol. 2, Paris: OECD DAC.
OECD, 2002a, *Belgian Aid at a Glance Chart (Aid and Debt Statistics: Donor Aid Charts),* Paris: OECD, DAC (http:www.oecd.org/EN/countrylist).
OECD, 2002b, *Aid at a Glance. Aid Charts for Recipient Countries and Territories (Aid and Debt Statistics: Recipient Aid Charts),* Paris: OECD DAC (http:www1.oecd.org/dac/htm/aid_recipients.htm).
OECD, various issues, *Development Co-operation. Efforts and Policies of the Members of the Development Assistance Committee,* Paris: OECD DAC.
Schellens, M., 2000, 'Belgium', pp. 97–103, in J. Randel, T. German and D. Ewing (eds), *The Reality of Aid 2000: An Independent Review of Poverty Reduction and Development Assistance,* London: Earthscan Publications Ltd.
Sénat de Belgique, 1997, *Commission d'Enquête parlementaire concernant les événements du Rwanda* (Parliamentary Inquiry Commission on the events in Rwanda), Brussels.
Vandenbroucke, F. and E. Derycke, 1995, *La Belgique et l'Afrique subsaharienne: agenda pour un nouveau dialogue et une nouvelle cooperation: Note Politique* (Belgium and sub-Sahara Africa: agenda for a new dialogue and a new co-operation: Policy Note), Brussels: Ministère des affaires étrangères et Secrétariat d'état pour la coopération au développement.
World Bank, 1998, *Assessing Aid – What Works, What doesn't, and Why,* Oxford: Oxford University Press for the World Bank.

4

British Aid Policy in the 'Short–Blair' Years

OLIVER MORRISSEY

I. INTRODUCTION

The (New) Labour government of Tony Blair was elected on 1 May 1997, following some 18 years of Conservative government. Changes to British aid policy were among the first measures implemented by the new government. The Overseas Development Administration (ODA) became the Department for International Development (DFID). While the political head of the ODA was a Minister without a seat in the Cabinet, Clare Short was appointed Secretary of State for International Development with a seat in Cabinet. This was not a cosmetic change: Short not only provided DFID with clear and strong leadership, she was also a powerful advocate of development policy within the government. DFID was soon involved actively in policy formulation and a White Paper was produced as early as November 1997.

The enhanced political status of development was matched by funding. The British aid budget was increased by 40 per cent from $3.2 billion to $4.5bn., representing 0.26 and 0.32 per cent of gross national income, in 1997 and 2000 respectively [*IDC, 2002: 30*]. The Chancellor of the Exchequer, Gordon Brown, frequently expressed his commitment to provide funds to support developing countries and reduce poverty, and DFID fared well in the Comprehensive Spending Review announced on 15 July 2002. Aid spending is to increase to some $7 billion, equivalent to 0.4 per cent of gross national income, by 2005/6. In broad terms, the real value of British aid will have doubled over ten years. While aid advocates would have liked more (Britain will remain some way short of the 0.7 per cent target), this is

---

Note of acknowledgement. Helpful comments were received from Olav Stokke and Paul Hoebink, neither of whom are implicated in the final version.

a significant achievement. It has been applauded by most NGOs and commentators, and is evidence of what can happen when the Prime Minister, the Chancellor of the Exchequer and the Secretary of State share the same basic commitment to assisting the poorest countries.

The orientation of aid policy has also changed, away from the concern with promoting British commercial and political interests that prevailed in ODA towards a focus on reducing poverty in the poorest countries. DFID has been concerned that aid is used effectively and targeted towards reducing poverty, especially if the volume of flows is to be increased. This concern has been manifested in a number of ways, notably through coherent policy-making and analysis, reflected in the tenor of the two international development White Papers [*DFID, 1997, 2000*], and a 'results-oriented' perspective that has guided the way the department operates in practice.

The assessment of DFID in this chapter concentrates on three issues – organisational change, aid practice, and international development policy. Section II briefly reviews the organisational change in the transformation from ODA to DFID. Section III considers trends in aid allocation and composition, on the principle that how much of different types of aid is spent where is a good indicator of what DFID is doing. Section IV then turns to policy evolution by considering the two White Papers. We conclude in section V with a discussion of the direction and identity of British aid policy under DFID.

II. ORGANISATIONAL CHANGE AND IDENTITY

The change from ODA to DFID was more than a change in name; the orientation and focus of British aid policy has also changed since 1997. John Vereker was Permanent Secretary of DFID from its establishment until 2002, having previously been Permanent Secretary of the ODA since 1994. He thus had unique insight into the institutional change in establishing DFID. He described this as having three components: 'the strategic transformation, encapsulated as the shift from Aid to Development; the institutional transformation, from an Agency to a Department; and the transformation in the expectations of our political leaders, which I would summarise as being from Administration to Delivery' [*Vereker, 2002: 134*]. These three elements structure the following discussion of the emergence of DFID as a leading aid agency.

The first of the three elements related to the change in policy focus (discussed in section IV below). Aid agencies traditionally were largely concerned with spending their money, and often valued the recognition associated with a visible project (the 'we built that road' mentality, often ignoring whether the road was maintained or even went anywhere useful) more than the long-term contribution of aid to growth and development. Under the leadership of Clare Short, DFID espoused a different approach. The purpose of aid was to promote growth, development and poverty reduction, and DFID officials were expected to devote more attention to analysing how British aid could better contribute to such an objective. Whereas the ODA appeared largely to accept the lead of the World Bank on development policy, such as the appropriateness of structural adjustment programmes, DFID has been more inclined to have its own people study the policy issues. For example, DFID was among the leaders in getting economists and social development advisers to consider the effects of macroeconomic stabilisation on the poor, and DFID staff were seconded to the IMF to advise on this issue.

One immediate effect of this shift in 'institutional orientation' was an increase in the number of young professionals employed in DFID, especially but not exclusively economists. DFID increased its in-house 'analysis and policy' capacity, and also engaged with academics and NGOs. This takes time to implement but after five years it had become a feature of DFID. While the 1997 White Paper was largely an in-house document, for almost a year in the run-up to the 2000 White Paper DFID held consultations and workshops with interested parties and commissioned a number of background papers. As a result, DFID became a voice of donor policy and developed an in-house research capacity, even offering staff research sabbaticals – free from official duties to work on a particular policy issue for a few months. For example, at a DAC 'Informal Experts Meeting on Aid Effectiveness' in 2001 the two presentations on policy issues were by DFID staff. Senior DFID officials found themselves addressing 'complex policy issues ... designed to improve the economy and governance of poor countries' [*Vereker, 2002: 135*] rather than considering which projects to spend their budget on.

The second element, the institutional transformation, had a number of core elements – notably preparation, purpose and expertise – identified by Vereker [*2002: 137–8*]. The change in government was not a big surprise, and senior ODA officials were aware that Labour would establish DFID as a department independent of the Foreign and Commonwealth Office (renamed Department of Foreign Affairs). Contingency plans were in place for the transition. Clare Short, from the beginning, provided purposeful leadership that acted as motivation to

DFID. In many respects it was a new department concerned with global poverty reduction, rather than the old ODA concerned with spending aid money and supporting British commercial interests. Strong leadership is a powerful force in policy change and DFID has acquired an identity that many would associate with Short. While she might argue, publicly and privately, with Blair or others in Cabinet, there was a strong political commitment to international development from the top in the Labour government. This will have contributed to the successful institutional development of DFID.

The objectives of DFID, in terms of aiming for development targets, were to be achieved by building partnerships with recipients and pursuing the development 'targets in partnership with poorer countries who are committed to them' [*DFID, 1997: 22*]. As the donor has the financial resources needed by the recipient, there is an inherent inequality, and partnership is difficult. It is evident, and was stated at the time, that DFID set itself a demanding challenge in aspiring to partnerships, especially if these were to be of the 'strong' variety with jointly agreed transparent country programmes and a financial commitment over many years [*Maxwell and Riddell, 1998*]. To some extent the partnerships will be exclusive, as donors will at the least look for countries willing to express shared values and objectives, but may still be a positive move forward 'from the confrontational approach that characterised the era of structural adjustment' [*Kayizzi-Mugerwa, 1998: 224*]. With hindsight, DFID appears to have been quite successful, at least with selected countries.

It appears that DFID negotiated relatively openly with recipient countries, in the sense that the latter knew where they stood. Partnership was intended to be more than a byword and, while the respective parties may not have been equal partners, DFID reached agreement on objectives with recipients, and resulting DFID country strategies have been made publicly available. These strategies state the policy commitments and responsibilities of both DFID and the recipient government. Arguably, DFID interpreted partnership as building a strong relationship based on mutual respect. In terms of an operational aid programme, this is a reasonable approach and it does seem to be taking effect in those countries where DFID is building a strong programme (see section III for a discussion of how bilateral aid is becoming increasingly focused on selected countries).

The approach to partnership has been reinforced by the increase in the number of DFID local offices. In 1999 only three countries had their own local office (Bangladesh, India and Nepal); no African country was listed as having a local office, although there were overseas offices serving Southern, Eastern and Central Africa [*DFID, 1999: 130*]. Two years later local offices had been established in Uganda, Nigeria and Tanzania [*DFID, 2001: 128*], such that the Eastern office became in effect the Kenya office, and by 2002 there were offices in Ghana and Rwanda. This understates the overseas staffing levels, as it is a moot point whether a number of DFID staff in the British High Commission constitutes a local office or DFID requires a separate (part of the) building. The important point is that, relative to the ODA, the number of DFID employees based overseas increased. At the end of 2000, DFID had 2,257 staff, 965 (43 per cent) were overseas and 424 of these (19 per cent of the total) had been appointed in-country. By the end of 2002, DFID had 2,807 staff, 47 per cent were overseas and 31 per cent had been appointed in-country.[1] There was also an increase in the number of DFID staff working essentially on policy issues, whether international economics, social development or more recently governance. Consequently, DFID combined a greater policy orientation in head office with greater local engagement and experience overseas. One result was increased respect for DFID as a donor, both among other donors and recipient countries.

The third element, political expectations of delivery rather than administration, was a feature of 'New Labour' rather than specific to DFID. There was a technocratic element, and strong central control overtones, in the focus of government on measurable results. Britain saw a proliferation of 'public sector league tables' ranging from performance of primary schools to university research, from response times of ambulance crews to the number of deaths in surgery in different hospitals, and DFID was not exempt. Acronyms proliferate and we shall be brief [see e.g. *DFID, 1999: 26–8*]. A Performance Reporting and Information System for Managers (PRISM) was introduced in 1998, since when all projects above £500,000 are evaluated and given a score. The aid programme has stated objectives and targets, with progress monitored against both the Policy Information Marker System (PIMS) and the Policy Objective Marker (POM). The former assesses bilateral spending against objectives and the latter links spending to specific objectives or targets. One should give credit for transparency; much of the information on performance and assessment is on the DFID website and the annual Departmental Reports are informative.

This target- or result-oriented approach to evaluating civil service performance has affected the way in which DFID operates, and in

---

1. Data from the DFID website (www.gov.dfid.uk).

some respects this appears to have been an improvement. While DFID officials cannot guarantee that any specific target will be met, they know what they are working towards. Targets do help to focus the mind and encourage one to think about the most effective way to move forward. In the context of partnership, agreed targets can have a positive effect as both parties know where they should be heading. If targets are not met, judgements will have to be made regarding who, if anybody, was at fault, or whether external circumstances changed, and how severe the achievement deficit is. Agreed targets are nevertheless preferable to imposed conditions – the traditional approach to conditionality – and DFID is providing leadership on donor–recipient relations. We return to the importance of targets and partnerships in the final section of this chapter.

Vereker [*2002: 139*] suggests three characteristics that capture the 'national identity' of British civil servants: horizontal co-ordination (officials talk to each other and share information relatively freely across and within departments), collegiality and professionalism. While these may be characteristics of British officials, they do not fully capture the identity DFID has established as a bilateral donor. Two distinguishing features are a policy focus on poverty reduction and a local commitment to strong partnership (even if the parties are inherently unequal). DFID is a donor that has contributed to thinking about aid effectiveness in development policy, and that has supported its own approach to partnership by establishing local offices to sustain relationships. However, professionalism has been essential to implementing its new purpose, and collegiality may have smoothed the transitional experience for those involved. Horizontal co-ordination has been important in determining the relationship between DFID and other departments. The relationship with Her Majesty's Treasury (HMT) has become close, partly because the Chancellor of the Exchequer has been committed to debt relief and helping the poorest countries (and DFID's budget has been increasing), and partly because HMT established its own International Poverty Reduction Team that worked with DFID on policy issues. The relationship with the Department of Foreign Affairs may have been made easier when DFID became an independent department. It is likely, however, that the relationship with the Department of Trade and Industry (DTI) became more strained as DFID was strongly committed to untying aid and not using it to support British exporters (see below).

III. TRENDS IN AID ALLOCATION

The 1990s were a 'bad' decade for aid, real volumes fell and most donors allowed the share of national income devoted to aid to decline. For DAC donors in general, aid as a percentage of national income fell on average from about 0.33 per cent in 1990 to about 0.22 per cent by 1997, remaining below 0.25 per cent in 2000. The corresponding figures for Britain were 0.27, 0.26 and 0.31 per cent [*IDC, 2002: 30–1*]. While the May 2002 commitment to increase aid to 0.4 per cent of income in 2006 is positive, it is worth noting that this falls a long way short of the 0.5 per cent share of the late 1970s, which was more than halved during 18 years of Conservative government. The DFID years have witnessed a significant increase in aid and, perhaps more importantly, a significant shift in the pattern of its allocation. In the early 1990s about 70 per cent of aid went to the poorest countries, but by 2002 some 80 per cent went to low-income countries and this is projected to rise to 90 per cent by 2006.

Table 4.1 provides data to compare the allocation of aid in 1997/98 against 2000/1 (although the latter figures are estimates, they capture the pattern); data for 1990/91 are included for comparison to capture the situation at the end of the Thatcher era. A number of trends are evident. First, more aid was allocated directly to countries. While total aid increased by 40 per cent between 1997 and 2001, the bilateral country programme increased by 47 per cent. However, this was lower than the share of bilateral aid in 1990 (largely because contributions to the EU only grew in the 1990s). Second, within the country programme, there were dramatic increases to parts of sub-Saharan Africa (SSA), and Africa's share of country aid rose from 40 to 55 per cent (compared with some 33 per cent in 1990). This is discussed in more detail below. There were significant declines in nominal allocations to East Asia and the Overseas Territories, and a decline in aid to Russia (accounting for most of the reduction in aid to former Soviet Union states).[2] Third, while the share of multilateral aid in total rose dramatically from 1990, it remained fairly constant after 1997 although the share going to the IFIs fell significantly. Finally, the share of 'Bilateral other' fell slightly after 1997, largely accounted for

---

2. The major reduction to East Asia was a more than halving of spending on the Aid and Trade Provision (ATP), although the allocation to China more than doubled; these account for almost 90 per cent of the regional total. Almost all of the reduced spending on Overseas Territories is attributable to the aid given to Monserrat following the earthquake there. Note that in terms of DAC definitions, aid to Russia is Official Aid rather than Official Development Assistance.

by the Commonwealth Development Corporation being made independent of DFID and Chevening Scholarships being taken out of the DFID budget (from 1999 these were paid from the Department of Foreign Affairs budget).

TABLE 4.1
ALLOCATION OF BRITISH AID, 1997/98 AND 2000/1

	1990/91	1997/98		2000/01		change
	%*	£m.	%	£m.	%	97–1 %
Country Programme						
Africa Greater Horn		34.6	4.7	66.9	6.2	93.4
Central Africa		67.2	9.2	156.1	14.4	132.3
Eastern Africa		111.1	15.1	194.0	17.9	74.6
Southern Africa		39.2	5.3	41.0	3.8	4.6
West and North Africa		42.0	5.7	132.2	12.2	214.8
Total Africa [a]	35.2	297.3	40.5	594.5	55.0	100.0
India	12.9	75.1	10.2	104.0	9.6	38.5
Bangladesh	5.4	36.5	5.0	74.2	6.9	103.3
Nepal	1.7	16.4	2.2	17.0	1.6	3.7
South East Asia	+	24.2	3.3	35.5	3.3	46.7
Eastern Asia & Pacific	6.3	66.6	9.1	48.7	4.5	-26.9
Western Asia	3.4	37.4	5.1	34.9	3.2	-6.7
Total Asia [b]	33.1	256.1	34.9	319.1	29.5	24.6
Caribbean	4.2	16.4	2.2	20.5	1.9	25.0
Latin America	2.0	23.0	3.1	23.4	2.2	1.7
Overseas Territories		54.0	7.4	37.7	3.5	-30.2
C&SE Europe [c]	+	38.2	5.2	39.1	3.6	2.4
Former Soviet Union [d]	1.9	48.7	6.6	46.8	4.3	-3.9
Bilateral [e]	1020	733.8		1081.0		47.3
	%					
Total Aid (nominal)	1883	1996.5		2800.6		40.3
Bilateral country	54.2	733.8	36.8	1081.0	38.6	47.3
Bilateral other	12.8	286.5	14.4	365.0	13.0	27.4
Multilateral	33.0	976.2	48.9	1354.6	48.4	38.8
European Union	16.8	542.5	27.2	760.0	27.1	40.1
IFIs [f]	10.2	264.1	13.2	276.3	9.9	4.6

Notes: Figures are out-turn for 1990/91 and 1997/98 and estimates for 2000/1, and percentages are region's share of Bilateral Country Programme. Although out-turn for 2000/1 is available in DFID [2003], some classifications are different from those used here. Lower panel is composition of total aid budget. The percentage change in nominal aid value between 1997/98 and 2000/1 is expressed relative to the 1997/98 (starting) value.
* classification in 1990/91 is slightly different: Africa divided into north and south of Sahara, combined Asia Far East (+) and combined Europe (+).
[a] includes Africa Policy and Economics Department.
[b] includes Asia Regional Economics and Policy Department.
[c] Central and South East Europe includes Baltics and Balkans.
[d] includes Russia, Ukraine and Central Asia.
[e] total for Bilateral Country Programmes.
[f] international financial institutions includes regional development banks.

Sources: DFID [2001: 159–65]; for 1990/91 ODA [1992: Tables A3 and A12].

It is clear that the major changes in aid allocation were the increased flows to the SSA regions. Total aid to Africa doubled in (nominal) value terms between 1997 and 2001, while Africa's share in bilateral aid increased by 56 per cent between 1991 and 2001. Often, this reflected increased focus on specific countries, in particular those moving in the right policy direction (e.g. Ghana) and/ or emerging from conflict (e.g. Rwanda). The largest proportional increase was the more than 200 per cent increase in nominal aid to West and North Africa. This was accounted for by three countries: aid to Ghana increased from £15m. to £67m., aid to Sierra Leone increased from £3m. to almost £39m. (reflecting conflict-prevention activities), and aid to Nigeria increased from £7m. to £17m. The only significant allocations to North Africa, for Egypt and the Aid and Trade Provision (see below), actually fell.

The next largest increase in allocation was to Central Africa, a more than doubling of aid so that the region's share of the country programme increased from less than 10 per cent to almost 15 per cent. Aid to Zambia more than quadrupled to £51.5m., aid to Malawi almost trebled to £62m., aid to Mozambique increased by some 50 per cent to £29m. while the allocation to Zimbabwe remained at about £11m. When Mozambique adopted economic policy reform to promote reconstruction DFID was one of the donors to provide commitment, both financial and in terms of technical support, and the relationship has been sustained. Similarly, DFID has shown commitment to the (faltering) attempts of Malawi to reform and stabilise its economy, with an especially strong focus on support for the health sector.

Eastern Africa is the region that gets the largest share of bilateral aid and has seen an increase from 15 to 18 per cent of the country programme. Although precise shares vary from year to year, the broad pattern is that Kenya gets a quarter of the regional allocation and Tanzania and Uganda each get just under 40 per cent. All now have sizeable local DFID offices. Uganda, since the mid-1990s, and Tanzania more recently are reaping the benefits of demonstrating commitment to economic reform. It is fair to say that Kenya has exhibited no such commitment, but it is politically important to Britain and the aid relationship is as important to Britain as it is to Kenya [*see Cumming, 2001*].

The other region that saw a large increase in aid – almost double – was the rather misleadingly named 'Africa Greater Horn' Department that includes Angola, Indian Ocean islands and Eastern Africa Regional. Almost all of the increase and half of the aid under this department went to Rwanda, a country that received no British aid in 1997 but almost £33m. in 2000. After the genocide, Britain was one of the first donors to commit to supporting the regime of Paul Kagame and is one of the major donors in Kigali. It has increased the aid allocation quickly, with a special focus on the education sector, and reinforced the relationship by publishing a country strategy that is essentially an agreed partnership commitment between the two governments. The only other significant country allocations for this department in 2000/1 were £6m. to Ethiopia and just under £3m. to Angola.

Overall, DFID's aid to Africa is quite concentrated, with only ten countries planned to receive more than £25m. in 2001/2 (and accounting for almost 80 per cent of all aid to Africa); of these, South Africa is the only one that is not low-income. In order of importance they are: Malawi, Tanzania, Uganda, Ghana, Zambia, Kenya, Mozambique, Rwanda, South Africa and Sierra Leone.[3] According to the 2001/2 out-turn, India (£180m.), Bangladesh (£60m.), Pakistan (£43m.), China (£23m.) and Nepal (£20m.) are the five major recipients outside of SSA [*DFID, 2003: 126*].[4] While some reallocation among countries will occur, South Asia is likely to account for just under 20 per cent of British aid. Table 4.1 shows a clear trend away from the middle-income countries of East Asia that have seen not only a reduction in their share but also in the level of aid. There seems also to be a general tendency for the share of aid going to Latin America and the Caribbean and to the economies in transition to decline, from a combined share of about 25 per cent in 1997/98 to 15 per cent in 2000/1.

Along with these shifts in the pattern of allocation have been some observable shifts in the type of aid given. Morrissey [*1999*] shows that in the early 1990s project aid and technical co-operation each accounted for about 45 per cent of bilateral aid. The mixed credits scheme, the Aid and Trade Provision (ATP), accounted for almost 9 per cent of bilateral aid in 1992 but had declined to 5 per cent by 1996. The ATP was a scheme where aid was specifically used to support British exporters; it was mostly to middle-income countries, and was heavily criticised [*see Morrissey, 1991*]. The 1997 White Paper

---

3. Seven of these were also among the 15 largest recipients in 1990/91. The three that were not were Rwanda, South Africa and Sierra Leone (whereas Nigeria, Zimbabwe and Sudan were among the 'top 15'). All ten did receive at least £25m. in aid in 2001/2, although the actual ranking changed. Uganda, Tanzania and Ghana received the most and Kenya the least (just below £25m.) of the ten [*DFID, 2003*].
4. Compared to 1990/91, Nepal could be said to have 'replaced' Indonesia among the major recipients. Russia also emerged as a major recipient of (official) aid (£22.5m.).

announced that the ATP would be abolished (see discussion in the next section), and it was for new projects from November 1997. As may be seen in Table 4.2, however, ongoing ATP commitments accounted for almost 3 per cent of bilateral aid in 1999/2000, although this was highly concentrated in Far East Asia (almost all in China and Indonesia).

TABLE 4.2
BRITISH BILATERAL AID BY TYPE, 1997/98 AND 1999/2000

	Aid £m.	Prog %	Proj %	TC %	ATP %	H&E %
1990/91						
Bilateral aid *	945.5	16.1	18.4	51.5	9.9	4.0
1997/98						
Africa total	344.7	21.5	9.3	57.2	2.2	6.9
South of Sahara	332.4	22.3	9.6	56.1	2.2	7.1
America total	100.0	10.0	7.0	62.4	0.7	13.5
Central	7.9	-	-	88.6	5.1	2.5
Caribbean	69.3	14.4	10.1	49.2	-	17.9
South	21.5	-	-	93.0	1.4	4.2
Asia total	288.4	0.3	16.5	60.0	17.4	3.3
South	178.9	-	25.9	69.9	0.4	2.5
Far East	82.7	-	1.3	36.2	59.5	1.8
Central	3.5	-	-	98.6	-	1.4
Total aid	1045.9	5.9	8.3	64.6	5.9	8.6
1999/2000	Aid £m.	Prog %	Proj %	TC %	ATP %	H&E %
Africa total	498.9	26.5	18.2	32.3	0.5	6.4
South of Sahara	484.7	27.3	18.7	34.2	0.5	6.5
America total	96.6	8.5	10.9	45.6	0.5	7.3
Central	6.8	-	-	42.6	2.9	14.7
Caribbean	65.9	12.4	15.9	35.7	-	7.7
South	21.6	-	-	72.7	1.4	5.1
Asia total	325.5	-	22.9	42.4	8.7	9.5
South	224.0	-	31.9	44.1	0.3	5.0
Far East	70.4	-	2.6	32.8	39.3	17.1
Central	3.0	-	-	93.0	-	-
Total	1326.2	10.6	13.3	40.1	2.6	16.7

Notes: Total is total bilateral aid. Other categories are: Programme (Prog), Financial Aid, Project and Sector (Proj), Technical Co-operation (TC), Aid and Trade Provision (ATP) and humanitarian assistance and emergency relief (H&E). The omitted category is debt relief (in 1999/2000 a category of 'Grants and other Aid in Kind' is listed, accounting for 12% of the total, but that is omitted). Regional totals include some aid unallocated to subregions.
* refers to the bilateral aid programme; project aid refers to financial aid.
Sources: DFID [*1999: 153–6*], DFID [*2001: 169–72*]; for 1990/91 ODA [*1992: Table B1*].

Table 4.2 compares the composition of bilateral aid in 1997/98 and 1999/2000 (with comparative data for 1990/91). A number of trends are evident. First, as already noted, the ATP has been abolished and its share of aid halved over the two years (this also accounts for much of the reduction in aid to middle-income countries), declining significantly from its peak of 10 per cent in 1990. Second, technical assistance (TC) is a declining share of the total (although it increased in the early 1990s), falling from almost 66 to 40 per cent. In SSA, TC has fallen from over half to about one-third of bilateral aid, and in South Asia from 70 to 40 per cent. In 1997/98, Far East Asia (where the ATP made up a large share) was the only region where TC accounted for less than half of bilateral aid. In 1999/2000, Central Asia and South America were the only regions in which TC accounted for more than half (almost all in the former case), and both were minor recipients in terms of aid volume.

Third, there was an increase in programme aid and a reduction in aid allocated to projects (although both diminished in importance compared with 1990). This shift is partly understated in the table as sector aid is included under projects. As sector aid is mostly (dedicated) budget support, it is closer to programme aid than to conventional projects. Britain gives an increasing share of its bilateral aid as budget support, usually linked to the existence of IMF- or World Bank-approved structural adjustment programmes. With sector aid counting as budget support, this may account for 15 per cent of bilateral aid in 2000. It is most pronounced in the low-income countries in SSA where aid is being concentrated. Programme and project aid accounts for 46 per cent of bilateral aid to SSA (up from 32 per cent in 1997), reflecting the fact that DFID's commitment to a country is conditional on the government reaching agreement with the IFIs. Although programme aid also appears important in the Caribbean in 1999, all of this refers to aid to Monserrat.

Finally, the share of humanitarian aid in the total has almost doubled over the two years, although the trend varies by region (and, by definition, varies from year to year). The large increase in humanitarian and emergency (H&E) aid as a share of aid to Central America is accounted for by Honduras and Nicaragua (hit by severe hurricanes), and that to the Far East was mostly for East Timor. In both regions, also, the value of aid fell. The reduced share to the Caribbean reflects reduced needs in Monserrat. While bilateral aid to SSA increased by almost half, the H&E share was maintained. Similarly, aid to South Asia increased by about a quarter but the H&E share doubled.

The DFID years have witnessed an increasing concentration of British aid. To some extent this is following a trend initiated by Lynda Chalker, the last Conservative ODA Minister, but her motivation was how best to use a diminishing budget. The Fundamental Expenditure Review published in 1996 pledged to 'ensure that all but a fraction of bilateral assistance would be devoted to 20 [countries] primarily in Africa' [*Whaites, 1998: 205*]. DFID has retained the concentration, but has a larger budget and uses different considerations to determine allocations. Bilateral country aid is absorbing an increasing share of the total and is being allocated specifically towards selected poor countries in SSA and South Asia. Excluding South Africa, more than half of British aid in 2002 will go to nine SSA and three South Asian countries, all low-income. India is the single most important in volume terms but, like China, it poses a peculiar problem: both are low-income countries with many millions of poor people, but are also large important economies. In terms of the importance of British aid to the country and the influence of DFID, the SSA recipients are of greater significance, and SSA is at the top of the British aid agenda. Furthermore, the composition of aid is changing, with a lower share going to technical assistance and traditional projects, and an increasing share for programmes and sector-wide projects. The share of aid for humanitarian purposes is also increasing.

IV. POLICY EVOLUTION THROUGH TWO WHITE PAPERS

Aid policy emerges from a policy-making process where, typically, development interests are weak relative to donor political (strategic) and economic (commercial) interests. Business lobbies have been quite effective in capturing aid as an instrument of trade policy, and Britain was no exception [*Morrissey, 1993*]. Historically, British aid policy has been incremental, with no major changes in the allocation or composition of aid following changes of government since the 1960s [*Morrissey* et al., *1992*]. The major legacy of the Conservative governments of the 1980s and 1990s was a significant reduction in the volume of aid. On two of the main 'policy innovations' regarding aid – economic conditionality in the 1980s and political conditionality in the 1990s [*see Stokke, 1995*] – the British tended to adopt and follow the lead of the World Bank [*Cumming, 2001*]. The previous two sections have detailed how the organisational structure of the aid agency and the allocation of aid did change under the Blair government (arguably, the latter continues an incremental trend). Policy has also seen some significant changes under the new government, and these are considered by reference to the two White Papers. We limit attention to three issues: tied aid, the global economy, and partnerships.

1. The Untying of British Aid

The 1997 White Paper [*DFID, 1997*] was the incoming government's statement of international development policy, with the aspiration in the title of 'eliminating world poverty'. The policy objective was to improve the quality of life of the majority of people in developing countries. Aid should be guided by the interests of the poor in developing countries, and should be targeted at the poorest countries with the objective of promoting development. It was interesting that an aid agency should accept, so early and so openly, that donor practices, especially tied aid, affected aid allocation and undermined the potential to promote development. In many respects the White Paper aimed to present a clear break with Conservative policy which, despite the rhetoric, had been strongly influenced by business interests [see *Morrissey* et al., *1992*].

The proposal to abolish the Aid and Trade Provision was unambiguous: 'no more applications will be accepted for ATP assistance, and the scheme will be closed' [*DFID, 1997: 45*]. Morrissey [*1998*] argues that the actual commitment to untying aid was less clear-cut as the commitment to abolish the ATP 'does not preclude deploying development assistance in association with private finance, including in the form of mixed credits' [*ibid.: 45*]. The section of the White Paper on 'Partnerships in Britain' appeared something of a euphemism for business interests [*Morrissey, 1998*]. In fact, Britain has gone further than almost any other donor in untying its aid, to the extent that by 2002 Clare Short could safely claim that British aid was effectively fully untied. This compares with about 60 per cent of bilateral aid being tied in the early 1980s and just over a quarter by the mid-1990s [*Morrissey, 1999*].

An anecdote demonstrates the change in policy – the case of the Tanzanian decision to purchase an air traffic control system from British Aerospace (BAe). The BAe system was a military specification, twice as expensive as the civil system Tanzania actually required. In the 1980s few would have raised an eyebrow if this had been funded by tied aid. However, this project had not been supported by British aid, and DFID was not able to stop it by withdrawing funding. Clare Short believed strongly that Tanzania should not spend its money in this way, and argued, loudly and publicly, that the British government should refuse an export licence to BAe. The Cabinet approved the licence, on the basis that the exports supported British jobs (the standard argument for tied aid, led as usual by the DTI). Short then argued that the Tanzanians should terminate the contract (they argued it was legally binding), and that if they did not she would withhold aid to the value of the money she believed Tanzania was wasting. This is compelling evidence of the sincerity of her views against tied aid.

Tanzania could have been financially punished, but the reason was not so much the implied fungible use of aid as the lack of transparency. Tanzania had not allocated the contract in a publicly transparent manner (in Britain, one can observe that BAe was not asked to explain its behaviour). When, later in 2002, President Mkapa announced his intention to purchase a jet for government use, Short defended Tanzania, against objections from the World Bank among others, *because* this decision had been justified to Parliament. There is a consistency in the DFID approach: the cornerstone of partnership is transparency, it is explaining and justifying spending rather than what (with reasonable limits) the spending is on that is important.

2. The Global Economic Environment

Both White Papers contain substantial sections on international trade and capital flows, with the general premise that trade and private capital are ultimately more important than aid for ensuring development and poverty reduction. Comparing the two, one notices variations in tone and nuance rather than any change in substance. The 1997 paper reflected the time: the Uruguay Round had been completed, the WTO was in full flow and the East Asian crisis had not fully struck (at the time of writing). It is not too surprising that the tone was buoyant regarding the benefits of trade and capital flows. By 2000 reality had set in: the East Asian crisis was still fresh in the mind and confidence in private capital had been undermined, and serious questions were being asked of the global trading system in the run-up to the Seattle Ministerial meeting. The influence of these circumstances can be seen, between the lines at least, in the second paper.

Section 3 of the first White Paper [*DFID, 1997: 58–67*] considers international development policy in relation to economic policies, specifically trade, agriculture and investment. In many respects the views expressed appear naïve, especially in respect of multilateral liberalisation of investment (see below). The views expressed are strongly in favour of liberalisation. Further multilateral liberalisation can 'encourage and assist developing countries to become more fully integrated into the multilateral system ... [and help] developing countries build their own capacity to take advantage of globalisation' [*DFID, 1997: 58*]. There is some recognition that liberalisation can impose costs on developing countries, but usually this is noted only in respect of agriculture or loss of trade preferences for certain countries. The adjustment costs of trade liberalisation can be quite high, and some developing countries can reasonably expect to be net losers from multilateral trade liberalisation [*Morrissey, 2000*]. At the least, the White Paper failed to appreciate the inequities of global trade.

Thinking had clearly evolved by the second White Paper: 'support for open trade is not to be confused with unregulated trade [and] there are substantial inequities in the existing international trading system' [*DFID, 2000: 69*]. The whole tone of this paper was, as stated in the title, 'making globalisation work for the poor'. Rather than simply advocating liberalisation, attention was given to the constraints facing poor countries and how donors could help in removing these constraints. For example, high transport costs, poor infrastructure and limited access to markets were all recognised as major barriers to increasing exports from poor countries. Rich countries had to do more in opening up their markets; they should also assist poor countries (through aid and technical assistance) by improving infrastructure and providing information on marketing and the product standards required. This shift has had practical influences on aid decisions. For example, in 2002 DFID initiated a substantial programme to assist trade policy capacity-building in SSA. The Africa Trade and Poverty Programme and related projects encompassed Botswana, Ghana, Kenya, Malawi, Tanzania and Uganda. They amount to a new form of technical co-operation where the project is demand-led (i.e. it should be requested by the recipient government as an active participant) and the emphasis is on establishing local capacity to design and implement trade policy (especially the legal and administrative requirements of meeting WTO commitments).

Agricultural trade, despite being the single most important issue in multilateral trade liberalisation for the poorest countries, gets surprisingly little attention – about one page in DFID [*1997*], mostly on EU reform, and scattered paragraphs amounting to perhaps a page in DFID [*2000*]. The first White Paper recognises that developed countries must remove protection of their agricultural markets, but says very little beyond this; if there is multilateral liberalisation, developing countries will benefit. The second White Paper reiterates the same points but goes further and addresses what developing countries should do in order to benefit from trade opportunities –

liberalise agricultural markets, improve transport and infrastructure, and diversify their exports. The major difference in DFID policy as revealed by the two White Papers is a realisation that the poorest countries, for a variety of reasons, are the least able to benefit from liberalisation. As both papers are statements of development policy, a major criticism is that they did not contain more explicit and stronger statements on the absolute necessity of genuine liberalisation of trade in agricultural commodities, with the removal of price and production subsidies in developed countries. On a more positive note, DFID [*2000*] showed a greater awareness of the realities of the international trading environment.

There are a number of respects in which DFID [*1997*] reflects its timing. Most of the discussion of trade, standards, intellectual property and even environment is in the context of the WTO and related international negotiations. This is most apparent in the almost unconditional support for liberalisation of foreign direct investment (FDI) through advocating the Multilateral Agreement on Investment (MAI) that was being negotiated within the OECD at the time. The MAI is supported as being 'based on the principles of non-discrimination against foreign investors, open investment regimes and investor protection' [*DFID, 1997: 62*]. There seems to be a complete lack of recognition that FDI is not always beneficial to developing countries and that multinational companies tend to engage in anti-competitive practices. Morrissey [1998] is very critical of the assessment of liberalisation of investment and the MAI in DFID [*1997*]. As it turned out, the MAI failed, partly because of orchestrated public opposition but largely because the developed (investor) countries could not reach agreement among themselves [*Morrissey, 1998*].

A different approach to investment is evident in DFID [*2000*]. It is still considered important that developing countries attract FDI and private inflows, and the need for appropriate domestic policies (sound macroeconomic management, financial sector liberalisation and competition policy) is stressed. As in the case of trade liberalisation, the emphasis is on how countries can ensure that they benefit from FDI and encourage domestic investment. 'The challenge is to maximise the benefits of increased foreign investment, including that from transnational corporations, by creating strong links to the domestic economy' [*DFID, 2000: 49*]. Similarly, private capital is to be encouraged, but the Asian crisis has 'shown that there are risks associated with rapid capital account liberalisation' [*ibid.: 51*]. Considerable attention is devoted to the role of the international financial system in ensuring financial stability. Overall, the White Paper shows considerable awareness that achieving development and poverty reduction requires action at a global level rather than simply appropriate policies by developing countries themselves. What is absent, however, is proposals on the kinds of instruments DFID envisages to foster investment and private sector development.

Another feature of the second White Paper is that considerable attention is devoted to the role of multilateral agencies, including the WTO and the OECD, and especially the need to reform the aid practices of the European Commission, the World Bank and the United Nations. The European Commission aid programme is heavily criticised as very inefficient and not focused on poor countries, while the World Bank is urged to improve its co-operation with regional development banks and bilateral donors [*DFID, 2000: 95–7*]. Britain is dissatisfied with the performance of the European Commission on aid policy, and is prepared to say so publicly (see below).

3. Development Partnerships

Partnership was the big phrase in the 1997 White Paper: section 2 (some 28 pages) was entitled 'Building Partnerships', although only four pages were devoted to 'partnerships for development' and almost twice as many pages to 'partnerships in Britain'. The sentiment is clear: 'genuine partnerships between poorer countries and the donor community are needed if poverty is to be addressed effectively and in a coherent way' [*DFID, 1997: 37*]. What this may mean in practice is also set out. To be partners, recipients have to demonstrate commitment to poverty reduction by pursuing policies that promote pro-poor economic growth and 'responsive and accountable government' and have to 'want' Britain as a partner [*ibid.: 39*]. Britain is then committed to providing 'an enhanced level of resources' with some flexibility in how they are used in a long-term agreement.

Commentators on the White Paper welcomed this commitment to partnership while expressing concerns and reservations about how effectively it would be implemented. The obvious concern is how donors can be transparent in demonstrating the way recipient governments can qualify as partners [*Maxwell and Riddell, 1998*]. It is not sufficient that partners are revealed *ex post* as being those with whom an aid agreement is signed. There must be a mechanism for announcing in advance how recipients can demonstrate their suitability as partners, and this implies empowering recipient governments. Whaites [*1998*] sees the emphasis on partnership as 'enabling the state', because recipient governments that satisfy the requirements for partnership will be given greater control over their budgets and decisions. Kayizzi-Mugerwa [*1998*] draws an interesting comparison between the White Paper and a Swedish proposal on 'partnership with Africa' that also appeared in 1997. He also sees an element of enabling the state in both proposals. 'African governments should, at a minimum, be able to defend their programmes before their domestic constituencies (as indeed the donor governments do theirs) on the basis of their own convictions' [*Kayizzi-Mugerwa, 1998: 222*]. What emerged is that recipients, by revealing commitment to good government (including reasonable economic policies), could earn the entitlement to a partnership.

The second White Paper does not use the language of partnership (the word does not even appear in any of the chapter headings or subheadings), but the sentiment remains. The emphasis of this paper is on what Britain will do to help achieve the international development targets, not only as an aid donor but also by exerting influence on other donors and multilateral agencies. There is also considerable emphasis on the responsibilities of developed countries to 'make globalisation work for the poor' and Britain reveals a willingness to take an international leadership role. Developing countries too are given responsibility. 'Progress is dependent on developing country leadership but some of the resources needed will have to be provided by the international community' [*DFID, 2000: 14*]. In this White Paper, developing countries can reveal their commitment by adopting poverty reduction strategies, and DFID can in effect declare an objective criterion to determine which countries are eligible for a partnership agreement. This is the current guiding principle for DFID: chapter 2 of DFID [*2003*] is entitled 'Partnerships to Reduce Poverty'. In current DFID thinking, budget support is seen as the instrument of partnership and is recommended as the most appropriate means to deliver aid in countries with high levels of poverty that demonstrate their commitment to poverty reduction. To target aid on the poor, emphasis is given especially to budget support for health and education spending.

V. CONCLUSION: BRITISH AID POLICY AT THE START OF THE MILLENNIUM

The New Labour government, in power in Britain for just over seven years, has implemented significant changes in aid policy. The aid agency is now a department led by a Secretary of State with a seat in the Cabinet. The real value of the aid budget (as a share of gross national income) has been increased by about one-third. The allocation of aid has become increasingly concentrated on the poorest countries, especially in Africa. Most importantly, however, the guiding purpose of aid has changed. British aid is no longer part guided by supporting commercial interests and given only to countries that implement specified economic policy reforms. Under DFID, aid is directed at serving the development needs of the poorest countries, provided their governments demonstrate a commitment to poverty reduction, typically by initiating a poverty reduction strategy. DFID has acquired an identity as a donor committed to providing the resources to tackle poverty in poor countries that show the willingness to enter into a transparent partnership. The signs are that it will continue to have the resources to fulfil this role.

In assessing how DFID is likely to evolve over the next few years, the targets recently set out provide some guidance. The Comprehensive Spending Review announced in July 2002, which included the dramatic increase in planned aid spending to 2006 alongside large increases in public spending on other sectors (especially transport, health and education), was tied to a Public Services Agreement [*PSA, 2002*]. All departments signed up to the PSA and this document set out the objectives and performance targets for each department. In principle, these targets have to be met if the department is to secure in full the planned increase in spending. In practice, 'satisfactory attainment' would be sufficient. The important point is that DFID has specific targets to aim for and these are set out in the public domain. These targets are linked to one of five objectives for DFID [*PSA, 2002: 23–4*]:

- *Reduce poverty in SSA*. The main target is to achieve a sustained reduction in the poverty headcount to below 48 per cent across the region. Specific targets (to be demonstrated for 16 key countries) include increasing primary school enrolment from 58 to 72 per cent and reducing the proportion of 15–24-year-old pregnant women with HIV to below 16 per cent on average.
- *Reduce poverty in Asia*. In this case targets should be demonstrated for four key countries, and were similar to, but more ambitious than, those for SSA. Headcount poverty is to be reduced to 10 per cent in East Asia and 32 per cent in South Asia. Examples include lifting primary school enrolment from 95 to 100 per cent and reducing the prevalence rates of HIV among vulnerable groups to below 5 per cent on average.
- *Reduce poverty in other regions* (no specific targets).
- *Increase the effectiveness of multilateral agencies in poverty reduction and responding to humanitarian crises*. Specific targets

were to increase the proportion of EC aid going to low-income countries, and to ensure that 75 per cent of the Heavily Indebted Poor Countries that show a commitment to reducing poverty do receive debt relief by 2006.
– *To develop innovative approaches to international development that are based on evidence* (no specific targets).

Obtaining value for money in the aid budget is also listed, although not as a specific objective, and the associated targets are raising the proportion of bilateral aid going to the poorest countries to 90 per cent by 2006 and improving the proportion of projects evaluated as successful. In identifying DFID aid policy from this, two issues can be highlighted. First, regarding British bilateral aid, the targets are all specified in terms of reducing poverty and improving indicators of well-being (such as health and education), and targeting aid to the poorest countries. Clearly, recipient commitment is essential if these targets are to be met, so effective partnerships are important. Second, the complementary role of multilateral agencies in ensuring poverty reduction is recognised.

Arguably, too little emphasis is given to the role of multilateral agencies, and there are no targets identifying specific agencies other than the target on the allocation of EC aid. Perhaps DFID is clever in recognising that, while the overall aim may be to increase the poverty reduction effectiveness of the international system, targets should be limited to the EC allocation and the HIPC Initiative over which Britain has some influence. On both of these issues DFID works quite closely with the Treasury, and the HIPC target is joint for the two departments. A few days after the announcement of the results of the spending review, Clare Short published a piece in the *Guardian* (29 July) arguing that reform of the EC aid programme was essential. She maintained that the reforms being implemented were, rather like the disbursement of EU aid, too slow and cumbersome and that far too small a share of EC aid was being allocated to the poorest countries. This may well have been the first broadside in what will become a campaign by Britain (DFID and HMT) to ensure a reallocation of EC aid away from Central and Eastern Europe and the Mediterranean states and towards the poorest countries, especially in SSA and South Asia. It seems likely that Britain will adopt a strong line and, on criteria of recipient needs allocation of aid, it should get support from NGOs and development lobbies and some of the other member states.

The wording of one target is particularly interesting, especially in the context of the two White Papers. A joint target with the DTI is to 'secure agreement by 2005 to a significant reduction in trade barriers leading to improved trading opportunities for the UK and developing countries' [*PSA, 2002: 24*]. It would not be easy to identify specific trade barriers that, if removed, would improve the trading opportunities of *both* the UK and developing countries. One could even argue, mischievously perhaps, that tied aid is a trade barrier and DFID's untying reduced opportunities for the UK. The two departments may not have to co-operate. Presumably the DTI will advocate the reduction of barriers where there are gains to the UK, and DFID will lobby on behalf of developing countries; together, they will promote multilateral liberalisation. In fact, DFID is assisting African countries in trade policy through the capacity-building programme mentioned earlier, which includes assistance in participating in trade negotiations. This is an example of the DFID approach to partnership that is based on helping developing countries to make their own policy decisions.

British aid policy has changed in the past five years, and Clare Short is the single most important reason that DFID has become the organisation it now is. The existence of financial and policy support, from the Chancellor of the Exchequer and the Prime Minister, for the goals of international development gave Short the base from which to drive change in DFID. However, all will acknowledge that she provided strong leadership, intellectual and moral, and in negotiations with recipient governments personified DFID's approach to partnership. Donors and recipients can never be equal partners, and donors will set out requirements of the types of economic and governance policies they want to see. DFID has done this without being overprescriptive, relying on demonstrations of commitment to poverty reduction rather than specific economic liberalisation policies, and has shown transparency by publishing country strategy papers and fully documenting DFID policy. The department is likely to continue in this way for the foreseeable future, and will encourage other donors, especially the European Commission, to follow a similar policy of using aid to assist the poor in the poorest countries.

Clare Short's resignation as Secretary of State for International Development in early 2003 deserves mention, because DFID has lost the forceful personality that drove British development policy since 1997. Her replacement, Baroness Amos, who was Spokesperson on International Development in the House of Lords, was likely to follow a broadly similar policy, as may be seen in DFID [*2003*]. However, Baroness Amos was a less outspoken person (it would be difficult to be more outspoken than Clare Short) and it remains to be seen if

DFID retains its high profile under her successor, Hilary Benn. To Clare Short's credit, among other achievements, she increased the status of international development issues on the British political agenda, and contributed to greater public awareness of the importance of reducing global poverty.

REFERENCES

Cumming, G., 2001, *Aid to Africa: French and British Policies from the Cold War to the New Millennium*, Aldershot: Ashgate.
DFID, 1997, *Eliminating World Poverty: A Challenge for the 21st Century*, White Paper on International Development, Cmnd 3789, London: The Stationery Office.
DFID, 1999, *DFID Departmental Report 1999*, London: The Stationery Office.
DFID, 2000, *Eliminating World Poverty: Making Globalisation Work for the Poor*, White Paper on International Development, London: The Stationery Office.
DFID, 2001, *DFID Departmental Report 2001*, London: The Stationery Office.
DFID, 2003, *DFID Departmental Report 2003*, London: The Stationery Office.
IDC, 2002, *Financing for Development: Finding the Money to Eliminate World Poverty*, 2 volumes, House of Commons International Development Committee, Fifth Report of Session 2001–02, HC 785-I and II, London: The Stationery Office.
Kayizzi-Mugerwa, S., 1998, 'Africa and the Donor Community: From Conditionality to Partnership', *Journal of International Development*, Vol. 10, No. 2, pp. 219–26.
Maxwell, S. and R. Riddell, 1998, 'Conditionality or Contract: Perspectives on Partnership for Development', *Journal of International Development*, Vol. 10, No. 2, pp. 257–68.
Morrissey, O., 1991, 'An Evaluation of the Economic Effects of the Aid and Trade Provision', *Journal of Development Studies*, Vol. 28, No. 1, pp. 104–29.
Morrissey, O., 1993, 'The Mixing of Aid and Trade Policies', *The World Economy*, Vol. 16, No. 1, pp. 69–84.
Morrissey, O., 1998, 'ATP is Dead. Long Live Mixed Credits', *Journal of International Development*, Vol. 10, No. 2, pp. 247–56.
Morrissey, O., 1999, 'Aid and Trade Policy (In)Coherence', in Forster and Stokke, (eds), *Policy Coherence in Development Cooperation*, London: Frank Cass, pp. 373–88.
Morrissey, O. 2000, 'Foreign Aid in the Emerging Global Trade Environment', chapter 16 in F. Tarp (ed.), *Foreign Aid and Development: Lessons Learned and Directions for the Future,* London: Routledge, pp. 375–91.
Morrissey, O., B. Smith and E. Horesh, 1992, *British Aid and International Trade*, Buckingham: Open University Press.
ODA, 1992, *British Aid Statistics 1987/88–1991/92*, Government Statistical Service: Overseas Development Administration.
PSA, 2002, *Public Services Agreement White Paper*, published by The Stationery Office and at http://www.hm-treasury.gov.uk/mediastore/otherfiles/psa02.
Stokke, O., 1995, 'Aid and Political Conditionality: Core Issues and State of the Art', in O. Stokke (ed.), *Aid and Political Conditionality*, London: Frank Cass for EADI, pp. 1–87.
Vereker, J., 2002, 'Blazing the Trail: Eight Years of Change in Handling International Development', *Development Policy Review*, Vol. 20, No. 2, pp. 133–40.
Whaites, A., 1998, 'The New UK White Paper on International Development: An NGO Perspective', *Journal of International Development*, Vol. 10, No. 2, pp. 203–14.

5

Danish Aid Policy in the Post-Cold War Period: Increasing Resources and Minor Adjustments

GORM RYE OLSEN

This chapter considers some of the main features of Danish aid policy in the period following the end of the Cold War. The exclusive focus on this period is not based on an assumption that there was a radical break in policy between the Cold War era and the years after the fall of the Berlin Wall. Rather, to the contrary, Danish aid policy is characterised by a continuity which is quite remarkable, bearing in mind the dramatic effects that the end of bipolarity had on a number of other international issues and fields of policy.

I. WHAT IS CHARACTERISTIC AND UNIQUE ABOUT DANISH DEVELOPMENT AID POLICY?

Danish aid policy in the post-Cold War period is unique in at least two respects. First and perhaps most significant is the fact that by 1992 Denmark had achieved the target, laid down by Parliament in 1985, of donating 1 per cent of its GDP to development aid, making it the top donor according to this measure (Table 5.1). At the same time, Denmark seems to be a unique donor as far as the quality of its aid is concerned. On several occasions, Danish aid has been evaluated positively by the Development Assistance Committee of the OECD, for example, in the 1999 DAC report [*OECD, 1999*]. Such positive evaluations are interesting, but consideration of them falls outside the

---

Note of acknowledgement. Some of the information and evaluations presented in this chapter have been collected during the past ten years via personal observations, informal conversations and formal interviews with civil servants in Danida, none of whom are responsible for any of the interpretations presented in the chapter.

TABLE 5.1
DANISH DEVELOPMENT ASSISTANCE 1965–2003 (DKm. (CURRENT PRICES) AND %)

	1965	1975	1985[1)]	1996	2000	2001	2002	2003
Bilateral assistance	22	606	2.380	5.155	6.161	6.440	5.876	5.953
As a share of total aid	*35*	*58*	*53*	*49*	*51*	*47*	*46*	*46*
Multilateral assistance	42	437	1.982	4.790	5.266	4.320	4.087	4.935
As a share of total aid	*65*	*42*	*44*	*46*	*43*	*32*	*32*	*31*
Environmental aid, ass. to refugees (2001ff.)						2.196	2.242	2.183
As a share of total aid						*16*	*17*	*17*
Running costs			140	476	624	644	658	656
As a share of total aid			*3*	*5*	*5*	*5*	*5*	*5*
Gross aid (DKm.)	63	1,043	4,502	10,437	12,051	13,560	12,864	12,828
Total aid as a percentage of BNP/GDI [a]	0.1	0.53	0.77	1.06	1.0	1.03	0.96	0.96 [b]

Source: Yearly reports from Danida plus information from the Information Office of Danida. The method of calculations has been changes several times. Therefore, comparisons have to be treated with caution. The figures 2001–2003 are calculated on identical basis.
Notes: [a] Since 1985 the share of aid has been calculated on GDI.
 [b] Preliminary figure.

scope of this chapter. The second notable feature of Danish aid policy is the fact that, in spite of the increasing budget appropriations during the 1990s, its basic features did not change until the beginning of 2002. The potential changes now are the result of the liberal-conservative government which came to power at the end of 2001. Some of the most important of these changes will be dealt with in the concluding section of this chapter.

The core argument of the chapter is that the remarkable increases in budget allocations during the 1990s as well as the basic continuity of the aid policy must be explained by a combination of two circumstances that characterise policy-making on aid in Denmark. The first is the existence of a tight-knit policy community actively engaged in designing Danish development aid. A policy community is defined by its 'stability and the high degree of institutionalisation of the relations between administration and organisations [that] is primarily a consequence of common perceptions that moreover promote a value community with a common view of what the problems are and how they should be solved. Radical policy changes that are not accepted by all actors will therefore rarely be feasible' [*Christensen and Christiansen, 1992: 94; Smith, 1993: 56–66*].

The second is related to the very nature of the policy field itself. At the beginning of the 1980s, Knud Erik Svendsen argued that Danish aid policy is a particular policy field because it is 'characterised by idealism and therefore more open to argumentation, meaning much less tied to the traditional clash of economic interests' [*Svendsen, 1981: 11*]. The strength of that idealism and thus the weight of the moral arguments in favour of giving aid are usually considered the most important motives of Danish aid. The prominence of moral arguments has caused Olav Stokke to categorise Denmark, together with Norway, Sweden and possibly the Netherlands, as 'humane internationalist', defined by their 'acceptance of the principle that citizens of the industrial nations have moral obligations towards peoples and events beyond their borders; it implies a sensitivity to certain cosmopolitan values such as the obligation to refrain from the use of force in the pursuit of national interests and respect for human rights' [*Stokke, 1989: 10–11; Pratt, 1990: 5*].

In summary therefore it is the core assumption of this chapter that the two main characteristics of Danish aid policy, the increasing budgets and the limited changes in policy, may be explained as the result of two basic circumstances. First, that for many years only a limited number of actors have been engaged in policy-making, thus making it almost impossible to make radical changes in the basic features of the aid policy. Second, the strong moral arguments in favour of development aid have made it difficult to argue publicly against aid, and also to suggest radical changes in the fundamental features of the policy.

The structure of the chapter is as follows. The next section argues that a number of different approaches may be used to study Danish development aid policy in the 1990s, posing different questions and leading to different answers. A public administration approach and an international relations approach are briefly presented. Then follows a presentation of the main approach used in the chapter which is one of comparative politics. This approach is also used when selecting the topics to be dealt with in analysing the main features of Danish aid policy in the post-Cold War era. The final sections examine the chief actors in the Danish aid policy community starting with the most important, namely the aid agency 'Danida' and its relationship with the other actors, namely Parliament, the NGOs and the business community.

II. DIFFERENT APPROACHES, DIFFERENT THEORIES

The use of a public administration approach inevitably brings Danida into focus as a separate actor. During the 1990s, Danida faced two crucial challenges: on the one hand, how to spend the considerable amount of money which continued to increase from year to year, bearing in mind that the spending should not involve too many public scandals. On the other hand, the administration was also the target of numerous and changing demands about how to design the best and most adequate aid policy. Such demands came from both the international donor community and the local Danish environment. It is remarkable that the administration was to a very large extent able to face up to all these different challenges without running into scandals or particular problems [*Olsen, 2003a).* The conclusion that Danida was quite efficient at meeting a whole range of demands, while at the same time administering increasing budget appropriations, is both interesting and remarkable. It contradicts the traditional thesis in the public administration literature, namely that large public resource appropriations result in 'laziness' in government institutions [*Downs, 1967; Niskanen, 1971*].

That Danida did not incur the expected 'laziness' is to be explained by a unique political-bureaucratic capacity within the agency both to formulate policy and to transform its ideas into operational guidelines. Danida has also had the capacity to implement the more specific guidelines without running into big problems [*Olsen, 2003a*]. However, this apparent success is to be explained not only by Danida's capacity. It has to be seen against the background of the special character of aid as a public policy issue in Denmark. The strong moral sentiments involved set a strict limit on how tough the critique of Danida and its policy could be. Moreover, the strong moral element in Danish aid has probably also been both a driving force and an incentive for the daily work of the civil servants involved. The highly career-focused culture of Danida and the Danish Ministry of Foreign Affairs in general probably also contributes to explaining the considerable capacity of the aid administration. This issue will be touched on later in the chapter.

An international relations approach throws light on how Danish development aid policy may be understood in a broader foreign policy context. As a starting point, it is the official position that 'Danish development policy is an integral part of Danish foreign policy' [*Danida, 2000b: 4*]. This is also reflected in the internal organisation of the Ministry of Foreign Affairs, which is a unified service divided into two sections based on geographical rather than functional criteria. One section, the North group, is responsible for all relations with the OECD plus the former Communist countries, whereas the South group has the responsibility for relations with the developing countries as regards not only development aid, but also 'traditional' foreign policy.

The ending of the Cold War led to a reformulation of Danish foreign policy away from the traditional policy of adaptation towards a new and much more pro-active 'active internationalism' [*Udenrigskommissionen, 1990; Due-Nielsen and Petersen, 1995; Holm, 1997, 2002: 22–3*]. During the 1990s, two policy instruments in particular appeared to be crucial in the government's policy towards developing countries. First and maybe foremost, the size and quality of the development aid budget appear to have been crucial means to achieving international influence in Denmark's deliberate attempt to 'punch above its weight', which is the essence of active internationalism. The Danish armed forces increasingly became the second instrument used for pursuing the policy of active internationalism [*Jakobsen, 2001; Holm, 2002: 28ff.*]. Danish soldiers were deployed in Eritrea and Afghanistan, and Danes were also deployed as military observers in the West Sahara and Sierra Leone.

Moreover, with the new aid plan *Strategy 2000* support for the building of peacekeeping forces in Third World countries was made part of official aid policy [*Danida, 2000b: 97–100*]. Conflict prevention in developing countries was also made an official priority of Danish development policy [*Danida, 1997*]. Among other things, this has resulted in support for the building of conflict management skills in Southern Africa by stationing a Danish major in the headquarters of the Southern African peacekeeping force in Harare, Zimbabwe. Based on these changes, it may be argued that, viewed from an international relations perspective, Danish aid policy has achieved another rationale than simply promoting economic and social development in the Third World. A high-profile aid policy has become a crucial foreign policy instrument, and as such has a value in itself, irrespective of its impact on development in poor countries.

Finally, an international relations perspective makes it necessary to mention one obvious but nevertheless important fact about Danish aid policy, namely that Denmark is a 'small state' [*Holm, 1982: 20–4*]. The basic international condition for small states has always been that they must seek influence by non-military means, while, at the same time, having a fundamental interest in the development of an international society which is not anarchic but is governed by norms and international agreements. For precisely these reasons, Denmark has traditionally supported the establishment of an international legal system and has therefore always been a strong supporter of international organisations, not least the United Nations. It is this simple fact that explains why it has stuck so vehemently to the principle of a fifty–fifty division between multilateral and bilateral aid, which appears in Table 5.2. The fact that Denmark is a small state, combined with an interesting policy inertia, explains why it continues to adhere to the principle of an equal division of Danish aid, in spite of the increasing problems that the UN system has faced during the 1990s. This continued support of the United Nations may be all the more striking, since it is publicly acknowledged that membership of the European Union is now the most important determinant of Danish foreign policy [*Holm, 2002: 33–7*].

TABLE 5.2
OFFICIAL DANISH DEVELOPMENT AID. SELECTED BUDGET LINES
(DKm. (CURRENT PRICES) AND %)

	2001	2002	2003
Total bilateral assistance	6,439.61	5,876.42	5,953.10
Total sector programme and project aid	5,381.23	4,803.08	4,474.40
Programme and project aid to the main recipient countries in Africa, Asia, Latin America as a share of total bilateral aid	74.4	75.4	75.0
Aid through NGOs, share of total bilateral aid	15.3	15.2	15.1
Human rights and democracy assistance as a share of total bilateral aid	1.2	1.5	1.8
Total multilateral assistance (DKm.)	4,320.21	4,087.30	4,035.90
UNDP as a share of total multilateral aid	12.2	10.8	10.9
World Bank group as a share of total multilateral aid	13.0	12.8	10.8
Aid through the EU, relative share of total multilateral aid	2.7	5.7	8.1
Aid through humanitarian organisations, relative share	6.0	5.6	6.0
Other expenditures financed by the aid budget: reception of refugees (DKm.)	949.10	863.80	750.00
Other expenditures financed by the aid budget: community financed EU development aid (DKm.)	612.33	623.96	650.00
Total Danish aid as reported to DAC (DKm.)	13,599.77	12,863.76	12,827.60

Source: www.um.dk/danida/bistandsstoerrelse.

III. THE DOMESTIC POLICY ENVIRONMENT

1. The Motives of Danish Aid

In applying a comparative politics approach to the study of Danish aid policy in the 1990s, the first and crucial question to be addressed is: What are the motives for Denmark's aid policy? The so-called aid motivations debate touches on this issue in general terms with its distinction between, on the one hand, 'recipient needs' and, on the other hand, 'donor interests' with their emphasis on trade and investments, security and finally political interests such as influence [*McKinlay and Little, 1977, 1979; Maizels and Nissanke, 1984*]. It has already been mentioned that Denmark is assumed to fall into the category of humane internationalists, assumed to be mainly motivated by so-called 'recipient needs'.

Despite the fact that Denmark is usually listed as one of the few countries that are sensitive to the needs of recipient countries, this does not automatically mean that there are only idealistic motives behind its aid policy. Quite considerable business interests are involved as well. Among other things, this is clearly illustrated by the international debate on the untying of aid, to which the Danish Government and the entire Danish policy community on aid have been strongly opposed. In that case, no attempt was made to hide the fact that very important motives for the Danes are commercial interests. It was clearly stated that the untying of aid in general would cause considerable losses for Danish contractors, sub-contractors and so on [*Development Today, 2002: 6*]. Nor, in this context, is it to be ignored that an estimated 6,000 to 8,000 jobs are directly dependent on the existence of a Danish 'development aid industry'.

2. 'Political Peace as Political Goal'

Turning to the question of 'how' Danish aid policy is made, it has already been mentioned that the policy is to a large extent influenced by the existence of a very narrow policy community. In order to locate that community within the larger Danish political system, it is pertinent to refer to the academic literature which argues that policy-making in Western political systems in general takes place within separate sectors of the national system, and is departmentalised and differentiated along functional lines [*Rhodes and March, 1992; Smith, 1993*]. This implies that decision-makers in aid policy are to a large extent separated from those in other policy fields such as agriculture, environmental policy, etc. As a consequence, it is highly probable that individuals and institutions involved in decision-making on foreign policy issues in general may be separated from those involved in development aid.

This statement is true for Denmark as far as all the actors involved are concerned, with the exception of the civil servants in the Ministry of Foreign Affairs and a limited number of politicians. Because development aid policy and foreign policy in the Danish case are to a large extent managed by the same individuals located in a single organisation, it is not possible to maintain that the departmentalisation exists when the focus is exclusively on the Ministry of Foreign Affairs. The same argument applies for a limited number of MPs who are members of small political parties. However, the departmentalisation is very conspicuous in most political parties that operate with a division of labour between the spokespersons for development and foreign policy issues.

This departmentalisation has two consequences. The first is that aid policy is a sector remarkably separate and thus isolated from basically all other sectors of Danish policy. This is partly to be explained by the unique financial situation of continuously rising budget allocations that has characterised this particular policy sector since 1985. The in-built automatic increase in the budget has, so to speak, shielded the sector from political debate. At least this was the case until the general elections of November 2001 brought a liberal-conservative government to power. One of the main pledges made by the liberal and conservative politicians during the election campaign was to cut the aid budget quite dramatically and use the funds saved to finance the public health sector and other domestic welfare services. The liberal-conservative government did in fact reduce the aid appropriation in 2002 by some 10 per cent.

The second consequence of the policy departmentalisation touches upon the issue of what is characteristic of the aid policy sector and thus of the policy community in Denmark. It has already been indicated that the members of the aid policy community to a large extent share a common ideology, and in particular a self-perception that 'we are all in this because of idealistic and unselfish motives'. Precisely this self-perception has contributed to lift aid out of the general political debate and turn it into an issue that can hardly be contested. This is very much in agreement with Smith's argument that it is characteristic of policy communities in general that 'the policy options are limited by the ideology of the policy community itself since ideology also defines the policy problems as well as their possible solutions' [*Smith, 1993: 226*].

But why did aid policy stop being a contested issue? No doubt, the main actors within the policy community share a basic interest in avoiding too much political debate on the use and value of aid both for the recipients as well as for Danish society as such. This has even led to the formulation of a hypothesis stating that the 'primary goal for Danish aid policy is political peace' [*Boel, 1986*]. Maybe it is possible to argue that every society needs a political refuge which is not subject to the everyday political battles that tend to reduce politics to a question of dollars or pounds. Thus, it may be maintained that the generous aid policy has had a special value in itself, because it has served to show that Denmark is a unique actor in the international context. Nor is it to be neglected that giving aid, and in particular giving a lot of aid, makes many Danes 'feel good'. A similar point with regard to Europeans in general has previously been put forward by John Ravenhill in his excellent book on the Lomé Conventions [*Ravenhill, 1985*]. Finally, it cannot be dismissed that the reason why the aid policy sector has been allowed to live its own more or less independent life can be explained by a general popular indifference towards Third World issues, which at the same time implies that involvement in development issues and in aid policy is basically an elite phenomenon, reserved for the members of the policy community.

This state of affairs is not surprising. Surveys indicate that the electorate is largely indifferent to international issues and that the voters are preoccupied with domestic problems such as welfare issues, immigration, law and order, etc. [*Andersen* et al., *1999*]. It is therefore possible to claim that Danish efforts at 'development education' have been both remarkably unsuccessful and remarkably successful. Every Danish young person knows about development issues and is knowledgeable about the massive poverty in the Third World. On the other hand, in the last decade or so, this knowledge has not led to a strong personal involvement in development issues. Allegedly Danish young people think it is all right to give aid, but also that it is not really their concern. They, so to speak, 'feel solidarity' with the poor of the world but in a very uncommitted way [*Pittelkow, 2001*].

3. *Public Opinion*

A final question that could be asked within a comparative politics approach would be: What is the role of public opinion in relation to Danish development aid? Since the mid-1970s, public support for development aid has amounted to over 60 per cent of the adult population. In 1997, 73 per cent were positive about the current level of 1 per cent of GDP [*Udvikling, 1997*] which was reached in 1992. This finding is confirmed by a Eurobarometer report: both in 1996 and 1998, 84 per cent of Danes thought that 'aid to developing countries' was either 'very important' or at least 'important' [*INRI, 1999: 2*].

Two reservations are pertinent here. First, these attitudes may not be consistent. Second, when we use another type of question, we get different answers. This implies that the 'traditional' surveys showing strong popular backing may equally well reflect what the respondents considered to be the 'politically correct' answer. At least, when asked to prioritise between different public issues, 'supporting the poor in the Third World' came far down the list. In a 1994 survey, 'help to developing countries' was one of the most unpopular fields of government spending [*Borre and Andersen, 1997: 200*]. In the latter half of the 1990s, it is obvious that 'aid to developing countries' did not figure on the list of priorities of Danish voters [*Andersen et al., 1999*]. Nor was development aid on the so-called 'political agenda of the voters'. Most important, aid was in no way a decisive issue for the voters, when casting their votes in favour of one particular party. Moreover, when respondents were asked to point out where the government was spending too much money, more than half indicated development aid.

Having established this state of affairs, it is necessary to warn against an interpretation that public opinion in general is important for designing Danish aid policy. Public opinion is hardly important, apart from pushing the policy community into compromising with explicit reference to the existence of strong public backing for the existing aid policy [*Olsen, 2001*]. But reference to such popular backing is not necessarily an indication that mass opinion mattered in the 1990s. It may just as well be argued that public opinion served to legitimise decisions already agreed on by the elite participants in the policy community. At least, this was the situation up to the general elections in November 2001, if the findings are based on an opinion survey from the late 1990s [*Andersen et al., 1999*].

IV. DANISH AID POLICY IN THE 1990S: THE MAIN FEATURES

After a cautious start in the early 1960s, Danish aid policy has become more and more visible both internationally and locally [*Holm, 1982*]. In 1971 the objective was laid down in Law No. 297 on international development co-operation, which states that 'the goal for Denmark's public aid to the developing countries is through co-operation with the governments and authorities to support their efforts to achieve economic growth and thereby to contribute to securing their social progress and political independence in accordance with the United Nations Charter'.

In 1989 political agreement was reached that Denmark's bilateral aid should in future be concentrated on 20 so-called co-operation countries. However, it was not until 1994 that the process of selecting the 20 countries was concluded, resulting in 64 per cent of total bilateral aid that year being directed to those countries. During the selection process, Danida initiated a new strategic planning procedure in1992, which included the preparation of four to six country strategies per year. With Parliament's approval of the 1994 strategy, it became a clear priority to involve both the authorities in the recipient countries and the relevant segments of the resource base in Denmark in the entire process of formulating and later revising the individual country strategies. In practice, this meant that private companies, NGOs and research institutions with special knowledge of a country were invited to discuss and make inputs into the planning of future Danish aid efforts. For each of the 20 co-operation countries, it was assumed that the strategies would mark out the framework for the total allocation of bilateral aid within a limited number of sectors and in the cross-cutting concerns in the official aid strategy [*Danida, 1994: 12–13*]. Table 5.3 shows the geographical distribution of Danish bilateral aid by the end of the selection process. It is obvious that Africa had the highest priority with 50 per cent of total bilateral aid in 2002. It was followed by Asia with 24 per cent, whereas Latin America received only 20 per cent of the total amount.

The country strategies represent the basis for long-term co-operation with the individual main recipient countries. Officially, a 'country strategy is to reflect the development needs of the partner country and its political priorities and it will to an increasing extent be

TABLE 5.3
DANISH BILATERAL AID 2002. GEOGRAPHICAL DISTRIBUTION AND TYPES OF AID
(PERCENTAGE DISTRIBUTION)

	Project and programme aid	*NGO aid*	*Other bilateral aid*	*Total*
Africa	62	33	23	50
Asia	24	24	25	24
Latin America	10	18	5	10
Balkans	4	1	8	4
Total including non disbursed	100	100	100	100

Source: Danidas årsberetning 2002, København: Danida [*2003*].

based on the national policy or national strategy for poverty reduction' [*Udenrigsministeriet, 2000: 15*]. At the same time, there is no doubt that the individual country strategies contain a clear signal to the Danish resource base of what Danida emphasises and prioritises. The respect for the national priorities of the recipient country does not mean that Danida plays an unobtrusive role in the dialogue on future development policy. On the contrary, the preparation of the country strategies requires 'an in-depth poverty analysis including an analysis of the expected consequences for the areas chosen for Danish efforts. The poverty analysis also has to make sure that representatives of the civil society including the poor groups in society are involved' [*ibid.*].

1. Increasing Resources

The ending of the Cold War led to a number of fundamental changes in the international system which in general terms had only limited impact on international aid policy. The most significant has been the reduction in the aid budgets of most donors, both bilateral and multilateral [*OECD, 2002*]. In this context, Denmark is a significant exception from the general pattern, since its development aid budget continued to increase through the 1990s up to 2001.

This has been the most important precondition for Danish aid policy in the post-Cold War era. The growth started to speed up with a decision taken in the Danish Parliament in 1985 to increase the level of aid from 0.77 to 1 per cent of GDP by 1992. This decision also meant an automatic increase built into the budget from 1992; if GDP increased, then the aid budget would automatically increase by the same percentage. The liberal government of the time was a minority government and it was strongly opposed to the decision to raise the aid budget. But, due to a conspicuous change in Danish parliamentary practice, it did not resign but agreed to administer the legislation it openly disliked [*Damgaard, 1992: 31–6; Olsen, 1998: 611–12*].

Moreover, in the wake of the international summit on the global environment in Rio in 1992, Parliament, once again against the will of the government, decided to establish a special fund aimed at environmental initiatives and emergencies. According to plans followed up to the general elections in November 2001, this special appropriation was forecast to increase to 0.5 of GDP by 2002 [*Danida/Danced, 1998: 5ff.*]. With this 0.5 per cent added to the 1 per cent of GDP in 'traditional' development assistance, this would mean that by 2002 Denmark would be contributing 1.5 per cent of GDP in different forms of transfers abroad, mainly to developing countries.

2. Sector-wide Approaches

The drastic changes in the international system caused by the end of the Cold War were the direct reason for the launching in 1994 of a new strategy for Danish aid policy. Entitled *A World in Development*, this strategy is regarded as representing an 'adjustment of Danish aid policy to the situation after the end of the cold war' [*Degnbol-Martinussen and Engberg-Pedersen, 1999: 135*]. The strategy made it clear that poverty reduction was the overall aim of Danish policy. Moreover, it explicitly launched three so-called 'cross-cutting' concerns: the participation of women in the development process, protection of the global environment and sustainable exploitation of natural resources, and, third, promotion of democracy and respect for human rights. Added to these concerns were a number of thematic priorities concerning population, trade with developing countries and debt relief [*Danida, 1994*].

On the one hand, with its emphasis on the fight against poverty and with its cross-cutting concerns, *A World in Development* was in a number of ways a continuation of the policy up until then. On the other hand, the plan aimed to describe the goals a little more clearly and in more detail than before. Nevertheless, the bottom line was the concrete policy and its implementation in individual projects and country programmes, till now left to the individual desks and individual civil servants in Danida. In summary, despite its many elements of continuity, the 1994 strategy introduced some new initiatives and concepts, the most far-reaching of which were doubtless the reflections on 'sector-wide approaches' and the so-called 'active multilateralism'.

The concept of sector-wide approaches means that the donor approaches development problems at the level of the sector instead of, as up until now, the project level. Sector-wide approaches to aid imply, for example, that not only agriculture as a sector becomes a target for Danish aid efforts; the same is true for transport, health, education, water and sanitary installations, etc. The new concept presupposes a considerable effort on the part of Danida to get involved in an intensified dialogue with the recipient country in order to determine the goals and resources for each individual sector. On the other hand, implementation of this form of development assistance requires a considerable expansion of administrative capacity in the recipient country, because the strategy presupposes precisely the channelling of donor money through the state. In order to implement the new concept, the donor is in reality forced to assist in building up administrative capacity in the recipient's civil service.

In 1997, Danida undertook an evaluation of the results of the new aid strategy up until that date. This so-called mid-term evaluation commented that the change from project aid to the new programme aid appeared to be much more complicated and time-consuming than originally envisaged [*Udenrigsministeriet, 1997: 15*]. Danida therefore drew attention to the need for a pragmatic approach to this form of aid in which new working routines had to be developed within the aid administration in line with the development of the sector-wide approaches. It was suggested that the management should be concentrated on the yearly examination of the programmes [*Danida, 2000a: 70*]. In the Spring of 2001, the auditor of the public accounts issued his first report on a Danish sector programme. It concerned support for a health sector programme in Ghana [*Statsrevisorer, 2001*]. The report showed considerable understanding of the difficulties with the implementation of the programme; there were a number of minor points of criticism, but in general it was quite positive. According to the report, the most important problems lay with the Ghanaian partner.

TABLE 5.4
SECTORAL DISTRIBUTION OF DANISH BILATERAL AID, 2001 AND 2002

Sector	2001 DKm.	2002 DKm.	2001 percentage	2002 percentage
Social infrastructure	2,808	2,593	43.8	44.1
– education	576	474	9.0	8.1
– health	685	634	10.7	10.8
– water & sanitation	596	472	9.3	8.0
– public administration	167	128	2.6	2.2
– development planning	36	23	0.6	0.4
– 'other' social infrastructure	478	863	11.7	14.7
Economic infrastructure	1,076	1,001	16.8	17.0
– transport	670	704	10.5	12.0
– energy	348	258	5.4	4.4
– misc.			0.9	0.7
Productive sectors	899	877	14.0	14.9
– agriculture	738	676	11.5	11.5
– manufacture, trade, etc.	162	200	2.5	3.4
Multi-sector integrated projects	581	486	9.1	8.3
Programme aid	470	348	7.3	5.9
Non categorised	575	572	9.0	9.7
Total bilateral aid	6,409	5,876	100	100

Source: Danidas årsberetning 2002, København: Danida [*2003*].

By the end of the 1990s, the reorganisation of bilateral aid from project aid to sector-wide approaches was concluded. Table 5.4 shows the sectoral distribution of the bilateral aid in the following years, i.e. 2001 and 2002. The table clearly proves the high priority given to 'social infrastructure', which received around 45 per cent of total bilateral aid, whereas 'economic infrastructure' received 17 per cent. Because of increased demands both on Danida and on the public administration in the recipient countries as a consequence of the shift to programme aid, an unexpected increase in the use of advisers was observed. By the end of 1999, 236 advisers on long-term contracts were employed by the bilateral programme. The number on short-term contracts has likewise increased and the numbers in both categories are expected to increase in future [*Udenrigsministeriet, 2000: 121–2*].

3. 'Active Multilateralism'

The concept of 'active multilateralism' was launched as part of the 1994 strategy and was subsequently put into operation in a plan of action in the Spring of 1996. The purpose is to influence and strengthen the international system in agreement with the aims of Danish development policy, but at the same time respecting the mandate of the individual international organisations. The plan implies that Denmark has to make demands on the administrative efficiency of the organisations; it has to require that a clearer division of labour be established among them and also that a better co-ordination of the international aid system be pursued. As a new and remarkable goal, the 'Action Plan for active multilateralism' aims at getting more Danes employed in the different international organisations, plus increasing the Danish share of deliveries to the multilateral organisations [*Udenrigsministeriet, 1996*]. In order to achieve these goals, the 'Action Plan' operates in a number of different ways, such as forming alliances with other countries, getting involved in political and professional dialogue with the organisations, earmarking of contributions, and establishing a much better interaction between bilateral and multilateral aid.

In the 1997 mid-term evaluation, it was established that 'more resources have been used in order to promote Danish aims and goals'. Because of this active policy, Denmark has been able to secure greater influence in those organisations where it is a significant contributor [*Udenrigsministeriet, 1997: 18–19*]. In an analysis in October 2000, it is emphasised that 'it is the general evaluation that Denmark through active multilateralism has obtained increased influence on the decision-making processes in the governing boards of the international organisations and thereby on the general policy decisions' [*Danida, 2000a: 91*]. In relation to the UN system, Denmark has emphasised in particular monitoring the efforts of selected UN organisations operating in some of the Danish co-operation countries. Criticism from the Danish side has been aimed at FAO and WHO in particular, and this has resulted in reductions in Danish contributions to these organisations from the mid-1990s.

4. 'Partnership 2000'

After no more than five years, a revision of *A World in Development* was launched. The new strategy, called *Partnership 2000*, contained an analysis of the international changes and developments that had taken place. In itself, this is not an innovation, since in its yearly 'rolling 5-year plan' Danida has continuously reflected on trends in international development. Characteristically, the focus has been on 'globalisation', 'trade', 'debt' and 'donor co-ordination'. In *Partnership 2000*, the agency presented a number of so-called 'strategic priorities' [*Danida, 2000b: 24–5, 63–4*], a number of which were already well known, while others must be described as new. Among the new priorities, violent conflicts may be mentioned as significant and also HIV/AIDS plus the consequences of 'the increasing urbanisation' in the poor world.

In summary, the new Danish aid strategy again stressed that 'Danish development policy is an integral part of Danish foreign policy, in which the promotion of common security, the promotion of democratic governance and human rights plus the promotion of economic, social and environmental sustainable development are the main goals' [*Danida, 2000b: 4*]. Nor was it an innovation that 'partnership' was stressed in the 2000 strategy. It had been a fundamental precondition in the 1994 plan that development may be promoted only in co-operation with the recipient countries, which must have the possibility of taking on the responsibility for implementing sustainable development programmes. 'Danish development policy has to be based on a partnership with the developing countries and their populations and on a partnership with the groups in Danish society which participate in development work', it was stressed officially [*ibid.: 5*].

V. THE CRUCIAL ROLE OF DANIDA

The Danish aid agency has played a crucial role not only in formulating but also in implementing the adjustments in development policy that have taken place since the 1980s. A (centrally) located observer in 1989 characterised the Danish development agency as follows: 'For many years, Danida has exerted considerable independent influence, not only concerning the actual adaptation of the principles for aid policy. It has also been the case as far as the policy formulations are concerned. Basically, the management of Danida has decided whether a file or a decision was characterised as part of the executive function or whether it fell into the realm of policy formulations' [*Martinussen, 1989: 245–6*]. This evaluation is based on an analysis of the course of events related to the preparation of the so-called *Action Plan of 1988*. Martinussen concludes that 'the administration has played a much more influential and independent role compared with the policy formulation process up to now' [*ibid.: 247, 248, 256*]. So it is the argument that the influence of Danida has not only manifested itself in daily administration; it has also shown itself in relation to policy formulation and to mapping out more concrete strategies for the implementation of overall policy goals.

In addition, the civil servants in Danida played a crucial role in relation to the preparation of the two aid plans which appeared in the years following the end of the Cold War, both *A World in Development* and *Partnership 2000*. It is symptomatic that in both cases the agency was responsible for drafting the analysis of the international situation which was assumed to require new policy initiatives. In both, Danida's analysis of the international changes pointed to the need for greater political and administrative flexibility. It also referred to the necessity of including new areas where Denmark ought in future to play a role. Later on, both the request for flexibility and the new policy areas became part of overall Danish aid strategy.

Based on discussions regarding both plans with the so-called 'resource base', which in the Danish context includes the business community, the non-governmental development organisations plus a few research institutions, Danida selected the comments and the type of critique that were included in the subsequent revisions of the first drafts. Via this method, it achieved quite considerable influence on policy formulation in the final plans. If the experience of the 1988 'Plan of Action' is also included, the discussions on the three strategies taken together provide a picture of a ministry and thus of an aid agency that in reality formulated what were the problems and what were the solutions. However, the debates on the 1994 and the 2000 strategies differed from the situation related to the preparation of the 1988 plan. During the preparation of the two post-Cold War plans, Danida received some input from the resource base. Nevertheless, in all three situations the agency itself formulated the strategic consequences of its own analyses.

The introduction of the two central innovations in Danish aid policy in the 1990s can probably contribute to illustrating Danida's very special capacity and also the degree of its influence on that policy. The introduction of the concept of sector-wide approaches was inspired by the international aid environment; in particular, it was the interaction with the World Bank and selected bilateral donors that led Danida to suggest that the idea be integrated into Danish aid policy. In contrast, 'active multilateralism' seems to have been the product mainly of one single individual in Danida, namely the then head of the department, Ole Lønsman Poulsen, who was a person with years of professional experience of multilateral aid work. Inspired by a number of critical consultancy analyses of selected UN organisations [*Engberg-Pedersen, 1991a, 1991b*] and the attempt in 1992 by the Minister of Foreign Affairs, Uffe Elleman-Jensen, to make Parliament change the balance between bilateral and multilateral aid, the head of Danida had reached the conclusion that something radical had to occur if the large Danish multilateral aid component was to be maintained in the future. The idea of active multilateralism has precisely those qualities that emphasise an increase in Danish influence and also make it possible to pay attention to specific Danish interests. In conclusion, irrespective of whether the demands come from the international aid community or from within the local Danish policy community, during the 1990s Danida was able to adapt and integrate the new demands into its general policy framework without any appreciable problems.

1. A Special Culture in Danida?

The remarkable capacity of the aid agency to react both to international as well as to local influence is probably buttressed by the existence of a special culture within the Ministry of Foreign Affairs. This particular culture is encouraged by a number of mechanisms, one of which is the highly obvious career structure that is probably its most decisive feature. The Ministry of Foreign Affairs is one of the few ministries in the Danish central administration that still operates with a subdivision of individual administrative units. Thus, the typical unit in the South group is subdivided into four sections, each with a head of section. To this is added something peculiar to the central government administration, namely each administrative unit contains a formal position as deputy head for the Permanent Under-Secretary. This career structure is probably further strengthened by the fact that there is also a quite clear hierarchy among the individual embassies and representations abroad and that all employees are forced to serve on a rotation basis.

The special aura surrounding diplomatic work drives in the same direction of strengthening the existence of a special culture that is very much related to the fact that the Ministry of Foreign Affairs has a remarkably limited exchange of personnel with other ministries in the central administration. It is literally unthinkable that the Ministry of Foreign Affairs would recruit an executive from another ministry [*Jensen and Olsen, 2001: 172f.*]. Nor are younger employees encouraged to accumulate experience by working in other ministries. Such a change in career would in any event set the individual back in the internal struggle for promotion [*ibid.: 172*]. This fact implies that the idea that it is positive to rotate between ministries is not followed to any extent worth mentioning by the Ministry of Foreign Affairs. However, at the level of Permanent Under-Secretary there has been a certain inclination to 'lend' individuals to other ministries, in particular to the Prime Minister's Office.

In summary, the Ministry of Foreign Affairs is highly career focused, meaning moving 'upwards' in an obvious hierarchical organisation which, in addition, is closed in comparison with other departments and ministries in the central administration. Moreover, its staff are characterised by a considerable arrogance which presumably has to do with the idea that diplomatic work is not 'just' administration of Denmark's relations with other powers and other countries. Diplomacy is to a large extent a unique profession. The considerable institutional capacity, combined with the strong internal culture of the Ministry, no doubt explain to a large extent the prominent role and not least the strong influence of Danida within the aid policy community [*Olsen, 2003a*].

VI. THE MAIN ACTORS WITHIN THE AID POLICY COMMUNITY

1. Parliament

At regular intervals, Parliament has discussed Danish development aid and the goals of Danish efforts in the Third World. The debates have often resulted in decisions that required the government to include one or more new goals in the future version and implementation of the aid programme. Traditionally, Parliament has been influenced by the new trends in the international debate on development co-operation where there has been a quite remarkable attention to current debates within the United Nations. Over the years, a considerable majority in Parliament have pointed to issues such as 'rural development', 'basic human needs', 'women', 'environment', 'democracy and human rights', 'indigenous people' plus 'population' that ought to become important goals for Danish aid policy.

Subsequently, the general pattern has been that civil servants in the Ministry of Foreign Affairs have drawn up strategies for the individual thematic areas often in collaboration with the relevant sections of the local Danish aid environment. Thus, since the mid-1980s, Danida has produced a number of policy papers that have tried to outline Danish attitudes in such fields as 'Women in development', 'the environment' and 'democracy and human rights'. Finally, ideas from the current minister have also been included in Danida's work without much difficulty. For example, the adoption of HIV/AIDS as yet another goal for Danish development policy was included in *Partnership 2000* and was (also) used by the then Minister for Development Co-operation to raise his own profile.

In spite of these scattered examples, it needs to be stressed that in general the influence of Parliament and of the MPs has to be described as limited. This is largely to be explained by the fact that very few politicians during the 1990s took much interest in issues related to development aid. During the earlier decades of Danish aid policy, the situation was different; at least a number of individual MPs took a strong interest in development issues. What was interesting at that time was that this active minority contributed to giving their parties a strong profile on development issues [*Svendsen, 1989: 93*], despite the fact that this profile did not necessarily evoke a positive response from the electorate.

During the 1990s, this situation seems to have changed. First, it is very difficult to identify individual politicians who take a strong interest in aid and development issues. It is striking that this is true even of the so-called left wing in Parliament, which traditionally has been strongly in favour of 'international solidarity', with development aid as one its high-profile flagships. In the 1990s, the left wing seems to have lost not only the capacity but also the initiative within this particular policy field. Judging from opinion surveys from the late 1990s, it is possible to argue that the low profile of all political parties including the left wing has moved in a direction where they are more in agreement with the general voter's attitudes in downgrading the significance of development aid [*Andersen* et al., *1999*].

2. Non-governmental Organisations

In the early 1980s, Knud Erik Svendsen argued that development aid policy 'is more influenced by idealism and thus is more open to argumentation, i.e. it is less narrowly tied to ordinary economic conflicts of interest' [*Svendsen, 1981: 11*]. The strong idealistic element is an important explanation of the existence of a whole network of voluntary organisations, the so-called NGOs which include both religious but mainly secular organisations, which is a characteristic feature of the Danish aid community.

There is no doubt that before the 1990s the Danish NGOs in particular exerted considerable influence on selected parts of Danish aid policy, being active in influencing the agenda concerning issues such as the necessity for popular participation in the development process and the need to respect human rights. The same is true of arguments in favour of paying attention to the special role of women and concerns about the environment. All these issues have become part of the official profile of Danish aid policy. A paper produced by NGO representatives emphasises that they have also influenced the choice of co-operation countries, the maintenance of the fifty–fifty division between bilateral and multilateral aid and the decision in 1992 regarding the establishment of a special budget line for aid to the environment and for emergencies [*Jørgensen et al., 1993*].

During the latter half of the 1980s and the first half of the 1990s, the NGOs succeeded in attracting an even increasing share of public aid funds, from around 7 per cent of total bilateral aid in 1988 to 13 per cent in 1992 [*Danida/NGO, 1994: 5*]. In the second half of the 1990s, the percentage reached almost 17 per cent [*Udenrigsministeriet, 1997: 24*]. It is recognised by NGO representatives that this significant increase is not only to be explained by their skills and influence on the aid policy process. To a large extent it is an expression of the general need of the aid agency to find alternatives to the traditional forms of aid, including finding new outlets for the increasing aid budget.

It was a central element of *World in Development* of 1994 that there should be a wider involvement of the resource base in designing Danish aid policy. It has been characteristic for the NGOs to be involved both in the current policy process and in the administration of an increasing percentage of Danish aid resources, as already mentioned. They are also represented on the Board for International Development Co-operation and are active participants in the Council for International Development Co-operation. Membership of the Board is quite significant, among other things because the basic law on Danish development co-operation established that it is the Minister who appoints the nine members of the Board, and that the Board should have a crucial role in 'advising the Minister' on his/her implementation of the tasks ascribed by the law.

The members of the Council for International Development Co-operation are also appointed by the Minister. But this is a much less important task as the law establishes that the Council is only to 'monitor the work of the board and to receive reports from it'. Membership of the Board, the Council and not least the administration of large public funds has meant that, in reality, these popular organisations have been co-opted, with the result that the NGOs have ended up being more or less similar to Danida in working methods, structure and product. Closely related to this, it has to be recognised that the NGOs in general have become less critical of official development aid policy [*Jørgensen et al., 1993*].

According to Danida, two developments took place in its co-operation with the NGOs during the 1990s. First with regard to the five big organisations, a consolidation took place, while at the same time a platform was established for co-operation with the smaller organisations [*Udenrigsministeriet, 1997: 24*]. In 1991 and 1992, Danida agreed to a multi-year framework of agreement with the four biggest NGOs – DanChurchaid, the Danish Red Cross, Ibis and the Secretariat for development within the Labour Movement – thereby securing them a budget frame which made planning much easier for the organisations. In 1996, a similar agreement was made with CARE-Denmark. It is worth noting that a number of bilateral donor organisations in other OECD countries had made such agreements with their national NGOs several years before Danida. When Danida did so, it also established a special secretariat within the ministry in order to improve its co-operation with the smaller NGOs.

The core of the practical co-operation between the big NGOs and Danida consists of two main elements: Danida's delegation of responsibility for planning, implementation and evaluation of projects to the organisations, and, second, the financial framework derived from the development aid appropriation in the Finance Bill [*Danida/NGO, 1994: 16*]. Within the framework of the agreed budgets, the organisations can initiate projects and activities without having first to apply for Danida's endorsement. In practical terms, there is a similar freedom and predictability in relation to the public financing of the personnel assistance implemented by two NGOs, the Danish Volunteer Service and Ibis.

Danida's non-financial demands on the NGOs have primarily been concentrated on the increasing demand for focusing and professionalism in their development work. The agreement framework is not considered solely an advantage for the NGOs. They have pointed out that Danida makes too many demands on their internal administrative procedures, budget procedures and forms of financing. They are therefore asking for more flexibility in budgeting which would give them greater freedom to move resources from one budget line to another. Such flexibility would mean that they would not all the time have to be applying for additional funds to meet unforeseen expenses. Finally, application procedures and debriefings are also considered by the NGOs to be too burdensome [*Danida 18, 2000: 17*].

3. The Business Community

Knud Erik Svendsen's emphasis on the significance of moral values and idealistic attitudes for the design of Danish aid policy has not prohibited commercial and narrow political interests from influencing both the form and the content of the policy. There is general agreement among observers that, over the years, quite important considerations have been taken into account as far as domestic economic interests are concerned [*Svendsen, 1981*; *Boel, 1986*; *Martinussen, 1989*]. The increasing significance of commercial interests in Danish development aid has to be seen against the background of the very limited role that business motives played in relation to the initial design of Danish development policy [*Betænkning, 1970: 21*]. Thus, the Federation of Industries had to be persuaded to become a member of the Board of Danida when it was set up [*Holm, 1982: 137*]. During the 1970s, this sceptical attitude gradually changed to a more positive one which doubtless has to be seen in connection with the fact that Danish development aid had by then reached a considerable level [*ibid.: 138*].

In the *Action Plan* of 1988, the interests and viewpoints of the business community were taken into account in vital areas. Nevertheless, the efforts to take care of the business interests were not necessarily a success. Among other things, this may be seen from the fact that a drop in the so-called 'return percentage' took place in the years following 1989 [*Danida, 1992: 110*]. As some form of compensation, Danida launched a number of new initiatives. First, a special administrative unit for business involvement in development aid was set up with the aim of securing contact between the business community and the development administration. Second, a special programme in support of the development of a private sector in developing countries, the so-called Private Sector Programme, was launched, aimed at developing a much more long-term and committed involvement in development work on the part of the Danish business community. Through the important elements of support for the establishment of viable commercial partnerships, the PS programme aimed at strengthening the private sector in recipient countries where Danida supplies policy advice and grants to a large number of activities [*Danida, 1997*]. It is considered an attractive new form of aid both by the Danish business community and by business executives in the recipient countries [*Degnbol-Martinussen and Engberg-Pedersen, 1999: 137*].

The 1994 strategy stressed that the business community is considered an important part of the resource base for Danish development aid within a number of sectors that are central to bilateral aid. The aim was therefore an intensification of the co-operation with the business community, which is now consulted both on the design of country strategies and on sector policies. Moreover, there are numerous contacts between the business community and the country officers and the Technical Advisory Service (TSA) [*Udenrigsministeriet, 1997: 22*]. As with other parts of the resource base, Danida holds meetings and seminars with business organisations and individual companies on a regular basis.

In line with the increasing internationalisation, a change has taken place in the views of the business community on the markets in developing countries. Danish business interests have shifted from being only about the sale of Danish products towards an increasing focus on establishing co-operative relations and forming strategic alliances. In this context, the Fund for Industrialisation in Developing Countries (IFU) has proved to be a useful instrument since it is allowed to make investments in the more wealthy developing countries.

Finally, the change from project to sector programme aid has opened up new opportunities for the business sector. Because Denmark continues to adhere to the principle of tied aid, reinforced efforts focused within a limited number of sectors where Denmark and Danish private companies have an obvious expertise offer new business opportunities. On the other hand, the focus on capacity-building within the public administration in the recipient countries has resulted in a smaller Danish share of the deliveries of goods to some individual sector programmes. In return, as already mentioned, a conspicuous growth has taken place in the number of private advisers to the programmes. The aggregate result of developments during the 1990s, however, is that the relationship between Danida and the business community has become much less conflict-ridden than before.

VII. DENMARK AND THE EU

Denmark has an ambiguous attitude towards the European Union. To make things even more complicated, it is possible to argue that the (political) elite is in favour of an ever closer European co-operation while the population in general is far more sceptical. The ambiguous attitude manifests itself in a generally positive attitude towards the European project while, at the same time, the politicians send mixed messages about how far Denmark wants to participate in the European integration process. The ambiguous position has resulted in the four so-called reservations which were accepted by the European Council in Edinburgh in December 1992. One of them is about Danish participation in the development of the European defence dimension, another is about the common currency.

Within the realm of foreign affairs, Denmark is in principle in favour of developing a common foreign policy, but when decisions touch upon defence matters, Denmark opts out and does not participate in the debates or the decisions on defence issues. Therefore, when the EU took over the responsibility for security in Macedonia on 1 April 2003, Denmark pulled out its troops.

Denmark is positive towards common foreign policy initiatives, and towards the common development policy. Having established a positive attitude towards common development initiatives, it has to be stressed that the European Union is not important to Danish development policy in general and neither is European development policy an important political issue within the Danish development debate. Among other things, it has to do with the simple fact that even though around 45 per cent of the total Danish development budget is multilateral aid, only a remarkably small percentage of this money is channelled to the European Union. In 2001, for example, around 5.4 per cent of the total Danish aid budget and 14.6 per cent of the total multilateral budget equivalent to US$88 million were transferred to Brussels [*OECD, 2003: Table 14: 247*].

Because the Danish aid transfers to the European Union are thus relatively insignificant *vis-à-vis* the total budget, the EU does not receive much attention either politically or bureaucratically. As of March 2003, only two employees in the Ministry of Foreign Affairs in Copenhagen were working full time with what is called the 'The EU. The Development and Aid Policy of the Community'. In comparison, three staff members worked full time with the regional development banks while at least ten persons were engaged in the management of 'The development programmes of the UN'.

Within the limited Danish priority given to the EU's development and aid policy, it is possible to identify two broad core issues. First, Denmark has in its bilateral aid policy as well as internationally lobbied in favour of making poverty eradication the leading principle for its development performance. Second, and as a natural consequence, Denmark has given high priority to promote development in sub-Saharan Africa. Because of this high priority, Denmark has in recent years stressed the need for conflict prevention in that region in particular. However, when it comes to deploying troops, the Danish position has until now been that the United Nations was the adequate organisation for carrying out peacekeeping operations. Here, it is necessary to call to mind the Danish reservations towards the common defence policy of the EU. And the reservations also include 'soft' Petersberg operations.

Moreover, it has to be emphasised that Denmark, in spite of the official positive attitudes in general, is hesitant about delegating too much power to the EU also when it comes to foreign policy. That is probably the explanation of why Denmark (always) supports the United Nations' efforts in Africa to promote peace and stability. It may also explain why Denmark is also clearly in favour of strengthening the local (that is, the African Union) or the regional (for example, ECOWAS) capacity for peacekeeping. It is either because such initiatives can be seen to be within the framework of the UN or because the initiatives delegate or 'decentralise' the responsibility to the regional/local forces. The latter is, on the one hand, in agreement with traditional Danish policy positions and, on the other hand, delegation removes the military option from the EU to the local authorities. The remarkable Danish position during the Iraqi campaign in the spring of 2003 does not change the traditional positive Danish position towards the UN.

Concerning the administration of European development aid, Denmark has traditionally been critical of the lack of efficiency, the lack of transparency and the general lack of coherence in European development aid policy. However, the Danish position has not been very outspoken. This low-key position probably has to be explained by the fear of the decision-makers (the political elite) of adding fuel to the sceptical popular attitudes towards the European integration process. When the former Danish Minister of Development, Poul Nielson, became EU Commissioner for Development Aid, he exposed the traditional Danish criticism of the common development policy. There is no doubt that the Danish administrative system/civil service strongly supported Poul Nielson in his efforts to streamline and improve the efficiency of the development administration.

The new Danish liberal–conservative government that came into power in late 2001 reduced the Danish development aid budget quite remarkably. The reductions hit both bilateral and multilateral aid. The transfers to the UN system were reduced quite significantly. In that context it was striking that the transfers to the common European aid system were not touched at all. The amount of money transfers from Copenhagen to Brussels was maintained at the existing level in spite of the fact that the process of reforming the aid administration in Brussels was far from concluded. The apparent lack of coherence in Danish aid policy on this particular point probably has to be explained by the aforementioned general positive attitudes among decision-makers towards the European integration project.

VIII. CONCLUSION

In the post-Cold War era, Danish aid policy has been characterised by two main features: on the one hand, the increasing budget allocations and, on the other hand, a remarkable continuity as far as goals and policy are concerned. This has been due to two main circumstances. First, the existence of a small, closed policy community explains why it has been so difficult to make radical changes of policies, priorities and goals. Second, the prominence of the moral arguments involved lifted aid out of the general public debate, making it almost impossible to question the basic features of the policy including the remarkable automaticity of the financial allocations.

When the general elections of November 2001 brought a liberal–conservative government to power, it seemed as if the long-standing political truce on development was over. One of the openly declared goals of the new government was to cut the development aid budget. The government succeeded in obtaining a large majority behind its first Finance Bill, for 2002, which contained dramatic reductions of both the bilateral and the multilateral aid budgets. The most conspicuous feature of the reductions in the bilateral budget was the closing down of three programmes with co-operation countries – Zimbabwe, Malawi and Eritrea – within a few months. Reductions were also implemented in a number of other partner countries such as Vietnam, Nepal, Bangladesh and Uganda. In the multilateral budget, the biggest reduction came in the contributions to UNDP, based on the argument that the implementation of the necessary reforms within the organisation was slow.

The radical 'adjustment' of the policy was criticised by the former ruling parties, the Social Democrats and the small Radical Party plus the two left-wing parties in Parliament. NGO representatives also criticised the reductions quite severely, which may be seen against the background of the fact that the budget allocations to the NGOs were also hit. The Finance Bill for 2003 maintains the level from the previous year, meaning a reduction compared with 2001, but a stabilisation if compared with 2002 [*Danidavisen, No.1, 2002*].

What appears as a very dramatic change in Danish development aid policy, of course, raises the question as to whether the interpretation put forward in this chapter remains valid after November 2001. It is obvious that the new liberal–conservative government has changed one fundamental feature of the policy up until then, namely the automatic growth in the aid budget. But, apart from the dramatic reduction in financial contributions, it is open to debate how dramatic were the other changes carried out by the new government [*Olsen, 2003b*]. Moreover, it is not possible within the short time horizon to conclude anything definite as far as the policy community is concerned. However, it does not seem to have changed in fundamental respects. Nor are there strong indications that the moral and ethical imperatives in giving aid have been reduced.

So, in summary, Danish development aid policy is still characterised by the same basic features as before 2002. On the other hand, it is still worth noting that the in-built financial inertia has been broken by a new government coming to power. Seen from a democratic point of view, this is positive. From the point of view of the development aid community it may not be so good, even though the possibilities for implementing changes of policy may be much better now than they were before. This statement, of course, presupposes that changes in the existing policy are necessary or desirable.

REFERENCES

Andersen, Johannes *et al.*, 1999, *Vælgere med omtanke. En analyse af folketingsvalget 1998* (Thoughtful Voters. An Analysis of the 1998 General Election), Aarhus: Systime.

Betænkning No. 565, 1970, *Betænkning om Danmarks samarbejde med udviklingslandene. Udvalget til revision af bistandsloven* (White Paper on Denmark's Cooperation with Developing Countries), Copenhagen: Statens Trykningskontor.

Boel, Erik, 1986, 'Politisk ro som politisk mål' (Political Peace as a Political Objective), *Politica*, Vol. 18, No. 2.

Borre, Ole and J.G. Andersen, 1997, *Voting and Political Attitudes in Denmark. A Study of the 1994 Election*, Aarhus: Aarhus University Press.

Carlsen, P. and H. Mouritzen (eds), 2003, *Danish Foreign Policy Yearbook 2003*, Copenhagen: Danish Institute for International Studies.
Christensen, J. Grønnegaard and P. Munk Christiansen, 1992, *Forvaltning og omgivelser* (Civil Service and Social Environment), Herning: Systime.
Damgaard, Erik, 1992, 'Denmark: Experiments in Parliamentary Government', in Damgaard (ed.), pp. 19–49.
Damgaard, Erik (ed.), 1992, *Parliamentary Change in the Nordic Countries*, Oslo: Scandinavian University Press.
Danida, 1992, *Danidas årsberetning 1992* (Danida Annual Report 1992), Copenhagen: Udenrigsministeriet.
Danida, 1994, *En verden i udvikling. Strategi for dansk udviklingspolitik frem mod år 2000* (A World in Development: Strategy for Danish Development Policies Towards the Year 2000), Copenhagen: Udenrigsministriet, Danida.
Danida, 1997, *Privat Sektor-programmet. Vejledning til virksomheden 3ac* (Private Sector Programme. Guidelines for Private Companies), Copenhagen: Udenrigsministeriet, Danida.
Danida, 2000a, *Tendenser i international udviklingsbistand. Arbejdspapir 11, Partnerskab 2000* (Tendencies in International Development Aid. Working Paper 11, Partnership 2000), Copenhagen: Udenrigsministeriet, Danida.
Danida, 2000b, *Danmarks udviklingspolitik. Analyse. Partnerskab 2000* (Denmark's Development Policy. Analysis. Partnership 2000), Copenhagen: Udenrigsministeriet, Danida.
Danida, 2003, *Danidas årsberetning 02* (*Danida Annual Report 2002*), Copenhagen: Udenrigsministeriet: Danida.
Danida 18, 2000, *Hovedproblemstillinger i den danske NGO-bistand. Kvalitet og fornyelse, folkelig forankring* (The Main Issues of Danish NGO Aid. Quality and Innovation. Popular Support), Copenhagen: Udenrigsministeriet, Danida.
Danida/Danced, 1998, *Miljøbistand til udviklingslandene. Årsberetning 1998* (Environmental Development Aid. Annual Report 1998), Copenhagen: Udenrigsministeriet and Miljø- og Energiministeriet.
Danida/NGO, 1994, *Strategi for Danidas NGO-samarbejde. Situations- og perspektivanalyse* (Strategy for Danida's Co-operation with the NGOs), Copenhagen: Udenrigsministeriet.
Danidavisen, No. 1, 2002, 'Danmarks u-landsbistand efter regeringsskiftet' (Denmark's Development Aid after the Change of Government), Copenhagen: Udenrigsministeriet, Danida.
Degnbol-Martinussen, John and P. Engberg-Pedersen, 1999, *Bistand. Udvikling eller afvikling. En analyse af internationalt bistandssamarbejde*. Copenhagen: Mellemfolkeligt Samvirke (translated and published as: Aid. Understanding International Development Cooperation, MS & Zed Press, 2003).
Development Today, Vol. XII, No.7, 31 May 2002.
Downs, Anthony, 1967, *An Economic Theory of Democracy*, New York: Harper & Row.
Due-Nielsen, Carsten and N. Petersen, 1995, 'Denmark's Foreign Policy since 1967: An Introduction', in *Adaptation and Activism. The Foreign Policy of Denmark 1967–1993*, Copenhagen: DJØF Publishing, pp. 11–54.
Engberg-Pedersen, Poul, 1991a, *Effectiveness of Multilateral Agencies at Country Level. Case Study of 11 Agencies in Kenya, Nepal, Sudan and Thailand*, Copenhagen: Udenrigsministeriet, Danida.
Engberg-Pedersen, Poul, 1991b, *Effectiveness of Multilateral Agencies at Country Level. European Community in Kenya and Sudan*, Copenhagen: Udenrigsministeriet, Danida.
Heurlin, B. and H. Mouritzen (eds), 1997, *Danish Foreign Policy Yearbook 1997*, Copenhagen: Danish Institute of International Affairs/DUPI.
Heurlin, B. and H. Mouritzen (eds), 2002, *Danish Foreign Policy Yearbook 2002*, Copenhagen: Danish Institute of International Affairs/DUPI.

Heurlin, B. and C. Thune (eds), 1989, *Danmark og det internationale system* (Denmark and the International System), Copenhagen: Politiske Studier.
Holm, Hans-Henrik, 1982, *Hvad Danmark gør... En analyse af dansk ulandspolitik* (What Denmark Does... An Analysis of Danish Development Policy), Aarhus: Forlaget Politica.
Holm, Hans-Henrik, 1997, 'Denmark's Active Internationalism: Advocating International Norms with Domestic Contraints', in Heurlin and Mouritzen (eds), pp. 52–80.
Holm, Hans-Henrik, 2002, 'Danish Foreign Policy Activism: The Rise and Decline', in Heurlin and Mouritzen (eds), pp. 19–45.
INRI, 1999, *European and Development Aid. Eurobarometer 50.1*, INRI (Europe), Brussels: European Coordination Office S.A., 8 February.
Jakobsen, Peter Viggo, 2001, 'FN's fredsoperationer i Afrika i dag og i morgen' (UN Peace Operations in Africa Today and Tomorrow), *Militært Tidsskrift*, Vol. 130, No. 2.
Jensen, Hanne Nexø and P. Lind Olsen, 2001, 'De nye topembedsmænd: Topchefers karriereforløb og ministeriers rekrutteringsmønstre' (The New Top Civil Servants; Executive Career and Patterns of Ministerial Recruitment), in Knudsen (ed.), pp. 151–78.
Jørgensen, Hans et al., 1993, *Debatoplæg om danske NGOers rolle i ulandspolitikken* (Discussion Paper on Danish NGOs' Role in Development Policy), Copenhagen: Mellemfolkeligt Samvirke.
Knudsen, Tim (ed.), 2001, *Regering og embedsmænd. Om magt og demokrati i staten* (Government and Civil Service. On State Power and Democracy), Aarhus: Systime.
Maizels, A. and M.K. Nissanke, 1984, 'Motivations for Aid to Developing Countries', *World Development*, Vol. 12, No. 9.
Marcussen, Martin and K. Ronit (eds), 2003, *Internationaliseringen af den offentlige forvaltning i Danmark. Forandring og kontinuitet* (Internationalisation of Danish Civil Service. Change and Continuity), Aarhus: Aarhus University Press.
Martinussen, John, 1989, 'Danidas handlingsplan. Et essay om administrationens rolle i formuleringen af dansk bistandspolitik' (Danida's Plan of Action. An Essay on the Administration's Role in the Formulation of Danish Aid Policy), in Heurlin and Thune (eds), pp. 244–58.
McKinlay, R.D. and R. Little, 1977, 'A Foreign Policy Model of US Bilateral Aid Allocations', *World Politics*, Vol. XXX, No.1.
McKinlay, R.D. and R. Little, 1979, 'The US Aid Relationship: A Test of the Recipient Need and the Donor Interest Models', *Political Studies*, Vol. XXVII, No. 2.
Niskanen, William A., 1971, *Bureaucracy and Representative Government*, Chicago: Aldine-Atherton.
OECD, 1999, *Development Co-operation Review of Denmark*, Paris: OECD.
OECD, 2002, *Development Co-operation 2001 Report*, Paris: OECD.
OECD, 2003, *Development Co-operation 2002 Report*, Paris: OECD.
Olsen, Gorm Rye, 1998, 'The Aid Policy Process of a "Humane Internationalist": The Danish Example', *Journal of International Development*, Vol. 10.
Olsen, Gorm Rye, 2001, 'European Public Opinion and Aid to Africa: Is There a Link?', *The Journal of Modern African Studies*, Vol. 39, No.4.
Olsen, Gorm Rye, 2003a, 'Forvaltningen af bistanden til den tredje verden og det internationale politikfællesskab' (The Administration of Development Aid to the Third World and the International Policy Community), in Marcussen and Ronit (eds).
Olsen, Gorm Rye, 2003b, '"Annus Horibilis" for Danish Development Aid: Has Denmark's Influence been Reduced?', in Carlsen and Mouritzen (eds).
Pittelkow, Ralf, 2001, *Det personlige samfund: et portræt af den politiske tidsånd* (The Personal Society: A Portrait of the Political Zeitgeist), Copenhagen: Lindhardt og Ringhof.

Pratt, Cranford (ed.), 1990, *Middle Power Internationalism. The North–South Dimension*, Kingston and Montreal, London, Buffalo, New York: McGill-Queen's University Press.

Ravenhill, John, 1985, *Collective Clientelism. The Lomé Conventions and North–South Relations,* New York: Columbia University Press.

Rhodes, R.A. and D. March, 1992, 'New Directions in the Study of Policy Networks', *European Journal of Political Research*, Vol. 21, Nos.1–2.

Smith, M., 1993, *Pressure, Power and Policy. State Autonomy and Policy Networks in Britain and the United States*, Hemel Hempstead: Harvester-Wheatsheaf.

Statsrevisorer, De af Folketinget Valgte, 2001, *Beretning om Danmarks støtte til sundhedssektorprogrammet i Ghana* (Denmark's Assistance to the Health Sector Programme in Ghana), Copenhagen.

Stokke, Olav, 1989, 'The Determinants of Aid Policies: General Introduction', in Stokke (ed.), pp. 9–23.

Stokke, Olav (ed.), 1989, *Western Middle Powers and Global Poverty*, Uppsala: The Scandinavian Institute of African Studies.

Svendsen, Knud Erik, 1981, 'Dansk bistandspolitik: Problemer og synspunkter' (Danish Aid Policy: Problems and Points of View), *Den Ny Verden*, Vol. 15, No. 3.

Svendsen, Knud Erik, 1989, 'Danish Aid : Old Bottles', in Stokke (ed.), pp. 91–115.

Udenrigskommissionen, 1990, *Udenrigstjenesten mod år 2000* (The Foreign Service Towards the Year 2000), Betænkning nr.1209, Udenrigskommissionen af 1 April 1989, Copenhagen.

Udenrigsministeriet, Danida, 1996, *Aktiv Multilateralisme – handlingsplan. Mål og midler i den aktive multilateralisme* (The Plan of Action for Active Multilateralism. Goals and Means of Active Multilateralism), Copenhagen: Udenrigsministeriet, Danida.

Udenrigsministeriet, Danida, 1997, *Vurdering af gennemførelse af hovedelementerne i Strategi for dansk udviklingspolitik mod år 2000 'En verden i udvikling'* (Evaluation of the Main Elements of Strategy for Danish Development Policy Towards the Year 2000 'A World in Development'), Copenhagen: Udenrigsministeriet, Danida.

Udenrigsministeriet, Sydgruppen, 2000, *Den rullende 5-årsplan 2001–2005* (The Rolling Five-Year Plan 2001–2005), Aarhus: Phønix-trykkeriet.

Udvikling, 1997, 'Solid opbakning' (Solid Support), *Udvikling*, No. 10.

6

Finland: Aid and Identity

JUHANI KOPONEN WITH LAURI SIITONEN

Finnish aid policies have been characterised by sharp discontinuities within basic underlying continuities. Finnish aid today looks very different from what it was in its heyday in the late 1980s and early 1990s. Aid volumes have shrunk, and although the growth has revived, it is proceeding at a rather slow pace. The internationally agreed target of 0.7 per cent of Gross National Income (GNI), once reached and then immediately abandoned, remains the official goal; the actual volumes are barely half this. At the same time, the whole approach to development co-operation has changed. Big projects flying the Finnish flag and building up infrastructure and industries, run by Finnish consultants with the maximum of Finnish deliveries, have given way to smaller, culturally more sensitive undertakings that are much better integrated into the structures of the recipients. A new concern for the effectiveness of aid encompasses demands for sustainability and ownership. Yet the transition from donor-driven aid modalities to genuine partnerships remains far from complete on the ground, to say the least, and very little is known about the actual impact of Finnish aid.[1]

In this chapter, the contradictory and idiosyncratic performance of Finnish aid is discussed via a constructivist approach. The argument takes as its point of departure the idea that aid policy may be seen as shaped and informed by different considerations, which we call developmentalist and instrumentalist. The important point is that these considerations are to be understood as comprehensive and dynamic, accounting not only for conscious motivations but also for the unacknowledged conditions of action.[2] Developmentalism refers to development thinking and development action – the whole complex of

---

The bulk of this chapter has been written by Juhani Koponen. It incorporates ideas and observations found in Siitonen [*forthcoming*].
1. For a bibliography on Finnish ODA, see Siitonen [*1996*].
2. For some elaboration, see Koponen [*1999*].

ideas, discourses, ways of action, institutions and other structures that has grown up around the notion of development during the last 50 years or so. Whatever the prevailing definition of the development ideal – modernisation or poverty reduction – the underlying assumption remains that development is desirable and beneficial to all and that a well-intentioned, rationally constructed social intervention will lead to ideal development. Without such morally grounded belief, there would be no development aid. But aid has been guided by other considerations as well. Here they are called instrumentalist, since they involve the use of resources released under the banner of developmentalism for other purposes, political or commercial. Aid policies are informed by the interplay of all such considerations. The bulk of this chapter traces how this has unfolded to produce the twists and turns in Finnish aid.

How these different notions are used and combined depends upon the way the states involved see and construct their identities and their interests. This follows the argument, most forcefully pursued by Alexander Wendt [*2000*], that whereas states may be seen as collective intentional actors, they do not have fixed interests and needs. Instead, they keep on constructing and reconstructing them. How this happens depends to a large extent on how states see and construct their identities. What you want depends on who you are, or who you want to be, and that applies to states as well. Identity is thus a property of intentional actors that generates motivational and behavioural dispositions. It is constituted by both internal and external structures. Accordingly, actors may have both intrinsic and social identities, each with associated needs of reproduction, or objective interests. Actors' understanding of these in turn constitutes the subjective interests that motivate their action. This often instructs them on how to define their interests. This approach allows us to consider the concept of national interest in a more flexible way. It refers to the reproduction requirements or security of state–society complexes, but it is far from fixed. Rather, variations or changes in state identity affect the national interests and policies of states [*Wendt, 2000: 198, 224, 234*]. Such an approach seems to provide fruitful insight into the formation of Finnish policies, including aid policy.

I. CONTRADICTORY GOALS AND MEANS

In a developmentalist discourse, aid is perceived in a rationalistic fashion. First, certain goals are set and then appropriate means will be sought and employed to achieve them. In setting the goals, shared values and political considerations are acknowledged to carry considerable weight, but discovering the means is assumed to be a more technical process in which instrumental rationality prevails. But this is hardly how the world works. Aid is not only a well-intentioned planned intervention through which resources are transferred from the more prosperous countries to the less well-to-do ones for the latter's well-being and development, but it serves other purposes as well. Aid practices consist of a continuous struggle and negotiation over the use of the resources that are made available for developmental purposes. Aid must be seen as a dual phenomenon: it is *both* a well-intentioned exercise in the rational transfer of resources, *and* a process of social negotiation and even struggle over the resources themselves. In addition to the explicitly articulated goals there are more hidden ones, a situation in which ends and means easily mix and what are represented as goals may in fact become means and vice versa.

Finnish aid currently has a fairly elaborate pattern of goals and means to guide and inform its activities. The overall aims of Finnish aid policies were set in a succession of major White Papers: the aid strategy paper of 1993 [*MFA, 1993*]; the Decision-in-principle of 1996 [*MFA, 1996*]; the paper on relations with developing countries of 1998 [*MFA, 1998*] which was 'operationalised' in an implementation plan in 2001 [*MFA, 2001a*]; and the White Paper of the new centre–left government [*MFA, 2004*]. Despite the changing provenance of the documents, the major lines appear to be clear. Development co-operation has been established as one 'instrument' in Finnish policy towards developing countries. The goals of development aid are defined, in slightly varying terms, as the reduction of world poverty, combating global environmental threats, and promoting human rights, democracy and equality in developing countries. These are to contribute to the ultimate aims of Finland's development policy: the increase of global security in the broad sense of the word, including 'human and ecological security', and the 'increase in economic interaction'. The partner countries are expected to show the will for development, that is, a commitment to share these goals and themselves to take the responsibility for finding the means to achieve them. Finland will only support them in achieving the mutually agreed goals. The policy goals are underwritten by a negative conditionality: if a partner lacks 'commitment' to the common goals, co-operation with it needs to be 'reconsidered'.

Taken together, these documents introduce a new doctrine of Finnish development policy. With globalisation proceeding inexorably, Finland as a political actor will not stand idly by but wants to optimise the positive and minimise the undesirable effects of globalisation – both for itself and for developing countries. The old 'Nordic development co-operation thinking' is discredited as old-fashioned global social policy and the quest for 'Finnish value-added' is introduced instead.[3] There is no point in failing to admit that development co-operation can also benefit the donor. In an interdependent world, Finnish national interests are seen as to a large extent compatible with the interests of developing countries and both may be pursued simultaneously. In the South, problems and crises must be prevented in advance, before they spill over and produce threats to Europe such as climate change, increased crime, the spread of drug abuse, epidemics and religious fundamentalism. A preventative influence can be exercised by supporting developing countries in implementing the agreed development targets and helping them to tackle the underlying reasons for their problems. Efforts must be made on a broad front. Development co-operation – which basically means aid – is one instrument among many. Other means to promote Finland's broader goals include political dialogue, trade and other economic interaction. All these means must be brought into line with each other. It no longer suffices to maintain an aid policy coupled with traditional diplomatic representation and international trade relations. What Finland needs, and has now managed to put together, is one holistic policy on relations with developing countries in which the links between foreign policy, security policy, trade policy and aid policy are tight and the whole set works coherently towards a common goal. The demand for coherence is especially pronounced in the 2004 White Paper.

However lofty this may sound, and however much a proactive and comprehensive approach is to be welcomed in principle, a closer look suggests that the bases of the policy are far from flawless. The goals are highly ambitious, encompassing every major problem facing the globe; but trying to pursue them through means related to developing countries is set to run into a number of inconsistencies. Even leaving aside the fundamental question as to the extent to which it is the developing countries that are primarily responsible for the global woes from climate change onwards, the goals themselves are internally inconsistent. As noted in a recent major evaluation commissioned by the Ministry for Foreign Affairs (MFA), 'potential conflict and trades-off... may arise' among the major goals [*Telford, 2002: 14*] – a fact that is implicitly acknowledged in the strong demand for coherence.

---

3. This is explicitly stated in the 1993 strategy [*MFA, 1993: 9–10*], but not repeated afterwards.

The policy papers represent a political compromise and they have been deliberately left open to different interpretations. At least two competing discourses rub shoulders in them. The thrust is strongly and unmistakably pro-market and even neo-liberal, but this is repeatedly punctuated by more sceptical notions about the challenges of globalisation. Exhortations to integration and free trade are mixed with welfarist and interventionist demands to look after the special needs of a diverse set of groups, ranging from the poorest countries and poorest people to women and girls and indigenous peoples, not forgetting Finnish companies. Imperatives of partnership and 'ownership' are stressed throughout. Yet it is made clear that Finland will support only partners whose strategies and programmes converge with its own development goals and who are committed to carry them out. Human rights are seen as universal by nature and thus their violations give 'the international community ... a legitimate right to impose commercial or development cooperation sanctions' [MFA, 1998: 6].

Further undermining the potential of the new policy to provide firm direction for action in practice is the tendency to blur ends and means, something which is most striking in the 1998 policy paper. In the earlier documents, the goals were loosely formulated without tangible, let alone quantifiable, benchmarks. In the most recent White Paper the tone changes. The Millennium Development Goals, with all their detailed indicators, are presented as the goals of Finnish development co-operation and the whole Finnish development policy more generally. Yet, when looking at the more general declarations, especially in the 1998 paper, it is difficult to get a grasp of the intended internal hierarchy among the goals and the mechanisms supposedly connecting the means to the goals. The implied relationship between the two is easily blurred, and what are represented as goals begin to appear more like means and vice versa. Development co-operation is clearly meant as a means to global security, but what about economic growth and what is called economic interaction? The document may also be read as advocating the export of a market economy and the institutions supporting it to developing countries as a major goal of Finnish policy. Such a reading would leave the rhetoric on poverty reduction and the special attention devoted to the poor, women and minorities to serve as means towards this end.

The overall impression arising from the documents is that Finland is now offering 'our' dominant Western institutions, values and models for the solution of the problems of developing countries and the world as a whole. The stress on democracy, human rights and gender considerations notwithstanding, the highest status is given to the market economy, which appears as the ideal path to development, ignoring the existence of the many varieties of capitalism and the differing combinations in which markets and states are always intertwined in modern societies. Though not presented as a straightforward goal, the market economy is ubiquitous throughout the policy documents under labels such as 'increasing economic interaction'. Even development co-operation is defined as a policy tool for supporting the integration of developing countries into the global economy. In other words, for poorer countries and people, integration into the world economy is presented as the solution: the real danger is seen in their exclusion.

Whatever the fair interpretation of the present goals and means and the relationship between them in Finnish development co-operation – and obviously the differing readings will be tested in practice – one thing is clear: the policy is quite different from what it used to be, both in developmentalist and in instrumentalist respects. For a decade and a half before the policy papers of the 1990s, during which time Finnish aid volumes grew very rapidly, no normative guidelines were issued. They came only after the crash in appropriations of the early 1990s had forced the Ministry for Foreign Affairs to rethink the use of diminishing resources. The guiding norms of earlier times were informal and unwritten, shaped within the higher echelons of the ministry and to a great extent shared by key decision-makers. The developmentalist goals were broader, if at all defined, reflecting a cautious quest for modernisation peppered with occasional exhortations to respect the recipients' own development plans, self-reliance and social equality.[4] Development was unproblematically understood as economic growth and modernisation, which were supposed to bring prosperity to the Third World while at the same time allowing for procurements from Finland. Democracy and human rights were not entirely overlooked, since aid was not given where the powerful political lobbies felt they were grossly violated. As for an early example, Chile was dropped from the list of recipients after the Pinochet coup of 1973. But their institutional interpretation was much more liberal, stressing the recipient's sovereignty in its internal affairs and allowing for one-party states and outright oligarchies to proliferate among the recipients. Corruption was a taboo word in official dealings.

Such thinking became suspect in the 1980s when environmental concerns came to the fore after the energy crises, while the debt crisis and structural adjustment revealed the hollowness of the

---

4. For the latter, see e.g. *Kehitysyhteistyökomitean mietintö* (Report of the Committee on Development Co-operation) [*1978: 136–7*].

modernisation process in most of Finland's partner countries. With the changes in the outside world, the ideas of development changed. Poverty reduction replaced modernisation as the developmentalist ideal in the early 1990s in the wake of the *World Development Report 1990* [*World Bank, 1990*], something which was hardly unrelated to the growing recognition of the ecological limits to growth. Democracy and human rights came along after the collapse of the Berlin Wall. The Nordic countries in particular emphasised them: it gave them a reason to abandon their old non-interventionist stance and provided them with a morally acceptable basis to exercise conditionality. Finland was at first less inclined to follow suit but it soon realised its advantages. Yet it took the drastic cuts in aid appropriations before the new insights were codified into a policy doctrine.

Neither the explicit nor the implicit goals have been subject to much open political discussion before or after they were adopted. The formulation of aid policies may be seen as a particularly striking example of the Finnish consensus-based way of policy-making. Three very different government coalitions have been involved, with remarkably similar outcomes. The 1993 strategy paper was issued by a centre–right government. It was prepared almost exclusively within the narrow confines of the Ministry for Foreign Affairs and only rubberstamped by Parliament, although the politically appointed Advisory Board for Relations with Developing Countries managed to intervene and leave its mark. The initiative for the later documents came from the red-blue-green 'rainbow' government, and in the drafting of some of them the Advisory Board was actively involved, particularly with regard to the 1998 policy paper. The papers were discussed and agreed by the whole Cabinet. This process allowed more room for political inputs and provided the relevant lobbies with an outlet to get their voices heard, especially as the responsible ministers were from the Green Party and were thus more receptive to the voices from civil society. The preparation of the last policy paper by the new centre–left government was even opened for a broader public participation and a new Development Policy Committee. Yet in all cases the actual drafting work was orchestrated by the Foreign Ministry bureaucracy.

The basic policy thrust of these documents has not been much challenged by outsiders in spite of all the contradictions involved. A few occasional voices have pointed out some of the inconsistencies and some low-key discussion has lingered on the legitimacy of instruments such as the pre-mixed concessional credits, which channel aid funds to shore up the exports of Finnish companies to emerging markets. Otherwise the principles have been surrounded by general complacency.

Yet one may ask whether the satisfaction is more apparent than real and is due to the internal inconsistencies of the policy, thanks to which something can be given to everybody. Senior officials and ministers responsible for aid are happy about what they regard as the mainstreaming of developmentalist considerations into the bulk of foreign policy: they see their traditional concerns and areas of competence having moved from the periphery of power to its very core. The normative emphasis on amorphous poverty reduction satisfies those who follow international developmentalist discussion as well as those with a more straightforward morally based humanitarian bent. Even Finnish commercial interests can be content that they have not been forgotten. It may be that the only politically feasible way to reach such a *modus vivendi* is to deliberately leave the conflicting discourses and multiple objectives where they are. But it is also a potential source of trouble, as the question remains as to the extent to which such an eclectic compilation is able to provide practically useful guidance for action, at least if no main criterion is singled out around which the policies should cohere. One may also wonder what is the point of the elaborate conceptual planning framework if confidence is ultimately placed in market solutions which by their very nature defy planning.

II. AID APPROPRIATIONS: SLOW MOVEMENT

What has met with somewhat less satisfaction and where the whole Finnish performance seems most erratic concerns the volume of funds allocated to aid. Given the size of its economy, Finland remains a small donor in terms of the absolute volume of aid. At its peak, the Finnish share of total world ODA was about 1.5 per cent, and currently it corresponds to about 0.7 per cent. In relative terms, the performance has changed dramatically from one decade to another. After a slow start, Finnish aid appropriations grew at a record rate in the 1980s and reached the peak of 0.8 per cent of GNI in 1991. Thereafter, with the onslaught of a severe economic depression, aid funds were slashed and plummeted to some 0.3 per cent in 1994. The drastic cuts were initially justified as a temporary measure. Although the depression passed and the Finnish economy boomed again in the late 1990s, the level of development aid appropriations for years stagnated at around a little over 0.3 per cent.

But the recent trends are a little more encouraging. The actual disbursements turned into an appreciable rise already in the late 1990s. At first this did not affect the ODA/GNI ratio due to the sustained growth of the GNI, but it has now begun to be seen there as economic growth has slowed down (see Figure 6.1 and Table 6.1). Meanwhile, some steps have even been taken to restore the 0.7 target.

FIGURE 6.1
FINLAND'S ODA DISBURSEMENTS 1961–2002

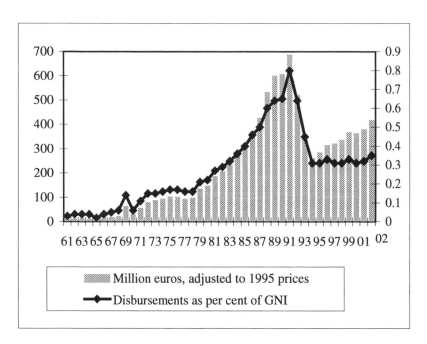

Source: http://global.finland.fi (12 December 2003)

TABLE 6.1
FINLAND'S ODA DISBURSEMENTS 1961-2002[a]

Year	€m.	% of GNI	Year	€m.	% of GNI
61	11.1	0.03	82	206.6	0.29
62	12.4	0.04	83	233.5	0.32
63	15.3	0.04	84	271.8	0.36
64	13.2	0.04	85	315.0	0.40
65	9.7	0.02	86	369.9	0.46
66	15.6	0.04	87	427.2	0.50
67	18.8	0.05	88	534.5	0.60
68	23.1	0.06	89	601.2	0.64
69	64.4	0.14	90	607.8	0.65
70	31.8	0.06	91	687.5	0.80
71	57.3	0.11	92	521.7	0.64
72	79.3	0.15	93	356.3	0.45
73	88.7	0.15	94	261.4	0.31
74	96.6	0.16	95	285.2	0.31
75	104.8	0.17	96	315.6	0.33
76	102.4	0.17	97	323.0	0.31
77	93.7	0.16	98	337.7	0.31
78	98.7	0.16	99	368.1	0.33
79	135.1	0.21	00	364.2	0.31
80	148.2	0.22	01	380.9	0.32
81	187.2	0.27	02	419.1	0.35

Source: http://global.finland.fi (12 December 2003).
[a] €m., adjusted to 1995 prices.

The debate on the volume of aid appropriations has been livelier than that on goals and means. The fluctuations in the levels of funds reveal some of the major forces affecting decision-making. Granting or withholding state funds is the point where the political machinery is most closely involved in aid affairs. The annual appropriation vote is the prerogative of Parliament, but in reality the greatest influence is wielded by the Ministry of Finance which traditionally has been very reluctant to allocate funds to ODA. As an indication of their thinking, every time a recession looms and the representative of the ministry produces a list of appropriate objects for fiscal cuts, the list invariably includes the reduction of aid funds to 0.25 per cent of GNI. All the major parties in the government, on the other hand, have commitments to raise the appropriation back to 0.7 per cent of GNI, and the parliamentary Foreign Affairs Committee routinely demands a timetable to be set for this goal. A political compromise has to be sought and the most consequential part of the negotiation takes place within the main parties, not between them. Each major party has its pro-aid factions and its 'realists', and the former tend to be far fewer and weaker. In order to understand how this works we need to take a closer look at the Finnish political context.

The best explanation for the permanently reduced level of Finnish aid appropriations has to do with the changes in the political functions of Finnish aid, and thereby with changes in Finland's identity. As long as Finland was outside the European Union and tied to the special relationship with the Soviet Union, development co-operation had a very different political function from the present one. It was seen as one of few means to maintain the international profile of Finland as a Nordic and thus basically a Western country. To do this with any credibility, the level of aid also had at least to approach the levels of the other Nordic states which had already attained and surpassed the 0.7 per cent of GNI target. With Finland's accession to the EU in 1995, being a Nordic country adds nothing in this regard, and the need to buy such an entry ticket to the Nordic club has evaporated.

As in so many other respects, Finland has assumed the role of model pupil within the EU and has adopted the 'EU average' as the main yardstick for its aid efforts, an EU average which happened to be conveniently close to the level to which the Finnish funds crashed in the early 1990s. The old kind of political motive to recover the aid level was therefore lacking even when the recession came to an end and the economy was booming again. There was also less room to do so, because the public debt had soared and the stipulations of the monetary union made it impossible to run large budget deficits. Politically effective movement towards raising the aid ratio again began only when the pressures building up within the EU had become too evident to ignore. After the Union, bracing itself for the UN Development Financing Conference in Monterrey in March 2002, had set an average target of 0.39 per cent for 2006, the Finnish government set up a special committee which recommended restoring the ratio 0.7 per cent by 2010. The new centre–left government ostensibly endorsed the proposal. However, it immediately watered it down by failing to commit itself to a rigorous timetable adding the clause that 'the overall economic situation' has to be 'taken into account' when increasing funds.[5]

The role of public opinion in all this has been muted, although such major figures as the President of the Republic, the Archbishop and the Chairman of the Trade Union Central Association have thrown their weight behind the demand for the return of aid funds to the 0.7 per cent target. Successive opinion polls show that the idea of development co-operation enjoys fairly wide and even increasing support among the population. Nevertheless, what goodwill there is remains passive and fails to turn into public action. Campaigns organised by development NGOs in favour of increasing the aid appropriations have met with a lukewarm reception, in a striking contrast to the 1980s when the grass-roots 'One Per Cent Movement' was instrumental in mustering public support for raising the aid level. Few have taken the government to task for contradictions such as whereas the official goal is 0.7 per cent by 2010, economic situation permitting, the actual resources which the government itself has reserved for aid will at their very best raise the ODA/GNI ratio to some 0.44 per cent when its mandate expires in 2007.

TABLE 6.2
FINLAND'S ODA COMPOSITION (%)

Year	Bilateral	Multilateral
1991	63.0	37.0
1992	65.4	34.6
1993	68.2	31.8
1994	73.6	26.4
1995	56.7	43.3
1996	52.5	47.5
1997	52.8	47.2
1998	52.7	47.3
1999	57.8	42.2
2000	58.5	41.5
2001	57.7	42.3
2002	54.4	45.6

Source: http://global.finland.fi (12 December 2003).

---

5. The Programme of Prime Minister Matti Vanhanen's Government on 24 June 2003. http://www.valtioneuvosto.fi/tiedostot/pdf/en/39357.pdf.

While much of the domestic discussion has been focused on bilateral aid, only roughly half of the appropriations go to bilateral activities. Almost another 50 per cent is for multilateral purposes (see Table 6.2). Thus the multi-funds have made a remarkable recovery from their heaviest cuts. However, new popular media themes such as debt relief have been allocated fewer funds than one might expect from their visibility in policy declarations. The principal United Nations development organisations have again been given preference as channels of multilateral funds, whereas the share going to the international financial institutions fluctuates depending on the timing of replenishments. On average, the UN's share in recent years has been some 40 per cent and that of the IFIs around a quarter of all multilateral aid. Yet, for both, Finland's subscriptions have remained below their pre-cut levels. Only a few favourites such as the Global Environment Facility (GEF) receive a higher contribution than before. The major change that has swelled and continues to swell multilateral aid as a whole comes from the EU, whose development co-operation Finland has financed since it joined in 1995, to the extent of roughly one-third of the total multilateral aid over the last years. Even this level will soon be inadequate: the commitments to the European Development Fund (EDF), which Finland has been financing since 1998, are ticking away like a time bomb. These have been kept very modest only because of the EDF's administrative lags in using up the funds previously committed. A marked increase in Finnish contributions to the EDF is in the pipeline. Or, as the operationalisation paper remarked, 'it may be difficult for Finland to respond to the growing obligations ... without any effect on bilateral and other multilateral activities' [*MFA, 2001a: 20*]. That is, unless Finnish aid appropriations in general are raised.

III. TRENDS IN THE 1990S: FRAGMENTATION AND INEFFECTIVENESS

One factor keeping the level of aid funds down is the widespread disillusionment about the effects and impacts of Finnish aid, both in developmentalist and instrumentalist respects. This became evident in the early 1990s, when only a few voices spoke up to defend the funds against the drastic cuts during the depression. With unemployment exploding, the general public had more urgent worries, while many activists who had lobbied for increased aid in the 1980s were disappointed by the commercial use to which the funds had been put. A fundamental critique denouncing all development aid and, by the same token, development itself gained ground in some Finnish NGO circles. And the scepticism about the effects of Finnish aid continues to linger. More recent aid has been widely criticised for extensive fragmentation leading to poor effectiveness.[6] Evaluations attest that, whereas Finnish projects and programmes may have been relatively effective in terms of achieving their stated short-term objectives, next to nothing is known about their long-term impact. Very little development, whether in terms of economic growth or of poverty reduction, seems to take place in such traditional major partner countries as Tanzania, Zambia and Nepal.

A major factor making it possible for Finnish aid to grow so fast in the 1980s was that much of it was channelled to Finnish deliveries – both of manpower and of goods. From an early stage, Finnish projects were entrusted to Finnish consultancy firms and were commonly planned with a view to maximising Finnish deliveries. During the time of ample funds, this led to large and expensive infrastructural and industrial projects with top-down planning and parallel implementation structures. They may have been effective in a short-term immediate sense but their longer term results left much to be desired, especially as regards sustainability and long-term impact [*Koponen and Mattila-Wiro, 1996*]. Even the commercial benefits turned out to be only modest and short-lived. After the deliveries paid for by aid dried up, trade on commercial conditions failed to emerge. Few if any major Finnish companies got a decisive boost for their most profitable operations from aid procurements. Few could foresee the future of Nokia when the company was sponsored by aid funds for rural electrification.

Many historical ironies flavour Finnish aid in the 1990s. The heavy cuts turned out to be a blessing in disguise in that they provided the Ministry for Foreign Affairs with an opportunity to enhance the quality of the projects. They strengthened developmentalist trends under way within the MFA and offered an excuse to get rid of the most dubious projects and activities of the old type. The new bilateral projects embarked upon represented a different kind of development thinking. As a rule they were much smaller in scale, aware of the demands of sustainability and local ownership, sensitive to gender and environmental considerations, and directed towards local development, good governance and democracy. Technical assistance teams are nowadays made up of, say, two rather than 20 Finns, and local expertise is increasingly included. Consultancy firms continue to run the majority of projects but they are fewer, leaner, and more experienced. Their role has changed, too. The consultants are no longer

---

6. Not least by the other donors, as in OECD peer reviews [*1999 and 2003*]. For domestic criticism, see e.g. *Kumppani*, 2001.

'implementing agencies' but 'provide support services' for locally recruited implementers. (For these changes, see Koponen and Mattila-Wiro [*1996*] and Moore *et al.* [*1996*]. Planning guidelines have been rewritten and the evaluation machinery strengthened.

Yet these changes did not do away with commercial elements in the Finnish aid programme; they were just made to change places. One item not only survived the cuts unscathed but even grew somewhat, namely, the funds earmarked for purposes furthering Finnish exports, notably pre-mixed concessional credits. The disbursements for interest subsidies for concessional credits rose up to 1995, after which restrictions introduced by the OECD in 1992 for 'commercially viable' deals (the so-called Helsinki package) began to take effect. Concessional credits have given rise to some debate, and criticism has been levelled at them from differing viewpoints. In the Decision-in-principle of 1996 they were battered by neo-liberal arguments and labelled as an 'unsound ... form of support distorting free trade and competition'. Finland promised to try to put an end to them by means of a common agreement within the OECD; meanwhile it would decrease their share and restrict their use to environmental technology and the social sector. With such restrictions, the demand for new loans was greatly reduced [*Osterbaan and Kajaste, 1999: 38*]. At one stage they almost seemed to be outliving their usefulness, but the MFA continued to push them by relaxing their conditions.

For a while, a new political and diplomatic instrumentalism invaded Finnish aid. Finland's direct relations with developing countries had been confined to a restricted, historically determined sample of a few countries mainly in Africa. EU membership confronted Finland with demands to support the EU's policies materially wherever the need arose. Aid funds were the obvious source for this. Earlier, under different names and slightly varying composition, Finland had focused on a group of 10 to 15 hand-picked major partner countries. For a time lists of preferred co-operation countries were not needed. A 'more flexible policy in the choice of partners' was announced. A different type of assistance was to be extended to different types of countries. Several new, mainly post-conflict recipients were added, with South Africa, Kosovo, Bosnia-Herzegovina and East Timor as the leading ones. In addition, a host of other countries were included in the name of 'widening the contact surface'. Diminishing funds were spread among an increasing number of recipients in increasingly small instalments. It was calculated that by the end of the 1990s almost 100 countries were enjoying Finnish aid.[7]

Such trends removed much of the clout from what had traditionally been among the most debated and politicised issues in Finnish aid – the selection of recipients. After all, alongside the level of aid appropriations, the choice of recipients had been the subject on which the donor had held the greatest leeway. Now much of the job was handed over to market forces and the 'international community'. Concessional credits are market-led and go where there is purchasing power and demand for suitable Finnish products. Selection of the 'flexibility countries' is dictated by the exigencies of world politics and the stance of the EU. As a result, the composition of the major recipients of Finnish aid was transformed. Concessional credits brought commercially more promising emerging markets high up among the recipients. China was catapulted to number one for several years, as most of the subsidised credits went there. Other recipients such as Thailand rose temporarily for the same reason. Traditionally privileged partners such as Tanzania and Zambia have fallen below the newcomers.

Although little hard evidence exists, what is available seems to confirm fears that such policies are bound to lead to ineffective aid. After an early evaluation had come up with highly critical conclusions, no serious attempt was made for a decade to re-evaluate the developmental effects and impacts of Finnish concessional credits and, when one was undertaken, it did not bring much illumination. How the 'flexibility countries' have recovered may be followed at a general level, but the contribution of foreign, let alone Finnish, aid has not been systematically assessed. Traditional long-term partners have been subjected to more assessment. A major round of evaluations was recently undertaken on the Finnish programmes in most of them. The resulting judgements were, however, rather superficial and inconclusive on the whole, apart from a few exceptions, and the synthesis study summing them up was very brief. Finnish programmes are mostly made up of scattered projects, ill connected with each other, the study concluded. Project objectives have been achieved in a number of cases, but the overall impact and sustainability were largely unknown or unimpressive, especially as regards poverty reduction. Nothing was said about the costs and opportunity costs, as most of the country evaluations had failed to undertake analysis on

---

7. There are different calculations around. According to the DAC 1999 review, the total number of recipients was 92 in 1997 [*OECD, 1999: 18*]. The number of official bilateral partners was 33 in 2000 and 44 in 2001 [*MFA, 2000: 54, 2001b: 50*].

cost-effectiveness. Such findings are disturbingly similar to those of the previous evaluation synthesis concerning Finnish projects in the 1980s and early 1990s.[8] This is disturbing, particularly in view of the fact that Finnish aid modalities did change in the 1990s. If it is widely accepted that the change was for the better, then why is that not reflected in the effects and impacts?

IV. PARTNERS AND INSTRUMENTS: PICKING UP THE WINNERS

A determined attempt to reclaim the policy initiative and lay down clear and transparent criteria for country selection has now been made in what is called the operationalisation of development policy objectives [*MFA, 2001*]. The category of politically chosen main recipients has returned, under the name of 'long-term partner countries'. The 'majority' of bilateral grant assistance is to be concentrated on them with poverty reduction as the main objective. This is to be achieved primarily through policy dialogue and the provision of resources to bilateral government-to-government development programmes and projects, although other means (in particular 'commercial and economic co-operation') are not excluded. Alongside the long-term partners, several other country categories will be used. The category 'other partnerships' is likely to be highly important, although the plans for them are still vague. Co-operation with them is thematically targeted and its duration is determined on a case-by-case basis. Themes may include whatever is fashionable in developmentalist thinking and is included among the Finnish priorities, from the environment and good governance to combating HIV/AIDS and furthering international trade. The forms of co-operation will obviously vary, as well, from supporting personnel and other capacity-building in public administration to working through NGOs and propping up the private sector. 'Economic and commercial co-operation' will be included as well. Even bilateral government programmes can in some cases be considered. In addition, there will be some countries that are seen to be in transition between these two major categories, long-term and other partnerships.

As recipient need – poverty – is declared to be the supreme criterion here, an obvious implication is that Finnish aid, or at least bilateral grant aid, should mostly be channelled to the poorer countries. What has been decided is that 60 per cent of the bilateral aid is to be allocated to eight poor primary partner countries. To make room for

---

8. Cf. Telford's synthesis [*2002*] with that of Koponen and Mattila-Wiro [*1996*].

this, what has been decided is that the existing middle-income recipients will be removed from the long-term partner category. Bilateral grant assistance to such countries as Egypt, Namibia and Peru will be brought to an end in a few years' time. To be sure, this by no means spells an end to all co-operation with these countries as 'other co-operation instruments' may well be adopted. Finland would no doubt like to intensify 'economic and commercial co-operation' with them, but the preferred forms of this are not very clear. The instrument closest to the heart of the MFA (if not necessarily its Department for International Development Co-operation) – concessional credits – has not been in great demand even in these more prosperous countries, no doubt because of their fairly strict conditions up until now.

For the poorer countries, a host of further criteria have been imposed, in addition to the need for assistance. These are related to two sets of demands, those of what is called commitment to development and those of the prerequisites for aid effectiveness. 'Commitment to development' means compliance with the stated goals of Finland's new development policy. A country is thus seen as committed if it 'systematically harnesses available resources to poverty reduction by promoting economic growth based on principles of market economy [and] endeavours to become integrated into the world economy'. A number of other expectations are added to this, from access to basic education to reduction of corruption and protection of the environment. The prerequisites for aid effectiveness include conditions such as that conflicts do not hamper co-operation, the administrative capacity of the partner is sufficient, and the experiences of the effectiveness of previous co-operation are positive. Finland is also expected to have expertise and experience of work in the long-term partner country and preferably to maintain a diplomatic mission there.

An insider joke in the MFA dryly quipped that if such criteria were taken seriously the only possible partner country would turn out to be Sweden. In any event, the criteria were applied in the operationalisation exercise to produce an altered mix of Finnish partnerships. Two apparently needy and promising countries were identified where Finland will significantly expand its co-operation programme: Mozambique and Vietnam. In contrast to these prospective success stories, three former long-term partners were singled out where commitment and performance was deemed so wanting that compliance with the criteria of co-operation based on grant assistance was to be actively monitored: Kenya, Nicaragua and Zambia. In other words, they were taken as naughty and put under surveillance. Three of the existing long-term partners – Ethiopia, Nepal and Tanzania – were not deemed affected. However, after elections and changes of government in Kenya, Nicaragua and Zambia these seem to have regained their former status, while the future of the Finnish aid to Nepal has become more unsecure due to continuing internal violence and political instability in the country.

To what extent and in which ways these decisions will be modified in implementation is another matter. The developmentalist line which emphasises co-operation with 'poor but promising' partners seems now to have prevailed over the diplomatic 'flexible choice' line which wishes to switch funds from one target to another according to political and diplomatic expediency. Obviously there are still differing lines of thought within the MFA. But the prospects for breakthrough of the new policy line are good; changes between ministers have not affected the underlying pattern. In any event the composition of Finnish development partners will not be changed out of all recognition. The degree of concentration in the new configuration will significantly depend on the proportion of aid funds allocated to the long-term partners, concerning which the emphasis will continue to be on Africa.

These choices have given rise to relatively little public discussion, especially in comparison with the 1980s when the previous, deliberately political country choices evoked loud debates. To some extent this must be due to the way decisions have been arrived at within the MFA, which until quite recently did not leave much room for public discussion. But deeper reasons may also have been involved. It appears as if the selection of aid partners, and development aid as an issue, has been largely depoliticised. No longer are vociferous lobbies speaking up for their 'own' recipient countries or against those of their political adversaries. Partly this reflects the belated recognition that the realities of the Third World are too complex to be confined beneath donor countries' domestic political labels. Then again, this should not lead one to believe that the decisions were taken without any consideration of domestic politics.

When seen from the outside, it appears that the country selections were based on a mixture of the perceptions of the Finnish officials dealing with the countries concerned and ideas about the preconditions of aid effectiveness circulating among the donor community. The review on which the choices were based was undertaken by desk officers and their superiors in the Helsinki headquarters; the round of country evaluations was commissioned by the MFA only after the operationalisation exercise had been practically concluded. The latter was therefore based on contributions from Finnish missions in the field. The decisive criteria appear to have been related to the prospects for success in aid effectiveness. After having for decades extended aid to countries that one after another turned out to be disappointing basket cases, Finland now wants to pick up the winners. According to any conventional developmental criteria, no country among traditional Finnish partners has been as successful during the last decade as Vietnam. Not only has its economic growth been brisk and its poverty reduction impressive but a structural transformation is also under way. Vietnam's vision of turning into an industrialised country by 2020 may not be that far-fetched. In Africa, where success stories are desperately scarce, Mozambique is widely seen as an exception full of promise. Its post-conflict economic growth rates have been nothing short of phenomenal and donors have found its officials congenial and easy to deal with.

Such choices resonate well with the prevailing views which tie aid effectiveness to the policies and institutions of the recipient. A new conventional wisdom has emerged, popularised by the World Bank and spreading fast among the rest of the donor community, which argues that aid works but only if the policy and institutional environment of the partner is good, meaning that the country respects the virtues promoted by the Bank.[9] Then, naturally, aid should be directed to such countries, among which Vietnam and Mozambique are now included (alongside China, India, Uganda and Poland).

For a small donor like Finland, however, this raises a few awkward questions. One can ask to what extent it is a wise policy for a small donor to follow the winning bandwagon. If aid is crowded into countries which have received a good policy certificate from the Bank and consequently are heavily aided, the Finnish share will be a drop in the ocean. Vietnam has become one of the largest aid recipients in the world in absolute terms, although much of it is in the form of loans from Japan – a far cry from the 1980s when Finland and Sweden were the only Western donors left in the country. In Mozambique, aid per capita – US$50 in 2000 – is among the highest in Africa and growing. At some point the 'absorptive capacity' will have been exhausted and further aid will turn out to be counter-productive. Furthermore, both countries are known to be relatively corrupt. The huge sums of aid money flowing in and oiling relations with national and local administrative machineries fit uneasily with donors' anti-cor-

---

9. The argument was originally put forward in the report *Assessing Aid* [*World Bank, 1998*], developed and amended in a string of publications thereafter (e.g. Collier and Dollar [*1999, 2000*], to be taken to a triumphalist conclusion in *The Role and Effectiveness of Development Assistance* [*World Bank, 2002*]).

ruption rhetoric. Underlying these is the basic question as to the probability that the Holy Grail has finally been found and the present thinking of the World Bank and the Finnish MFA holds the key to development. It is not difficult to appreciate that, if the state has collapsed to the extent that 'nothing works', aid also is unlikely to work. But it is much less certain that, if a country fails to develop, it is mainly because it refuses to respect good policy advice from the donors. The donors, including Finland, increasingly profess to have learnt that the realities of the developing world are complex and that there are no 'one-size-fits-all' solutions. But this is difficult to square with the heavy push for good policies interpreted as open economies and market-friendly approaches.

Finnish aid has been scattered not only in terms of recipient countries but also in terms of sectors of co-operation. Perhaps the only sector to which sustained attention has been devoted throughout is forestry, and even this sector has not been present in all partner countries and its approach has been transformed, following the dominant ideas of the day [*Koponen and Siitonen, 2001*]. Otherwise almost all the sectors have been in one way or another included in one country or another: industry and infrastructure, agriculture and rural development, water in rural and in urban areas, education and health at different levels. A major change took place in the early 1990s when the drastic cuts in funding forced the abandonment of the large infrastructure and industrial projects in favour of smaller and softer activities involving more local resources. The environment and basic education and health took on a grander stance, and smaller sums were channelled to politically visible aims related to the furtherance of democracy and human rights. The operationalisation decision points in two directions. On the one hand, it wants to further de-emphasise the traditional sectors and themes and direct attention to 'issues that have become highly topical in the development discussions of recent years', such as HIV/AIDS, globalisation, trade and development, support for the development of partner countries' private sectors, and information and communications technology. On the other hand, it states that the focus will be on fewer and larger efforts, and that in each long-term partner country no more than three sectors or three government programmes will be included. How this will evolve is unclear; country evaluations were rather unhelpful. Environment-related activities, from forestry to water, are apparently to be prioritised in the future as well.

The modalities of Finnish aid are also expected to change. The bulk of it has been, and continues to be, implemented through projects run by Finnish consultancy firms. Finland has been engaged in programme aid to a far less degree than, for instance, the other Nordic countries. Most of what the Finns have called programmes have been projects in disguise. There is now definite movement towards other modalities, especially the currently so fashionable sector-wide approaches (Swaps) and direct budget support under World Bank- and IMF-sponsored Poverty Reduction Strategy processes. Finland is planning to shift increasingly over to the sector mode in many of its major partner countries and also to increase its budget support. But this line is far from unanimously accepted within the MFA. The policy papers make it clear that Finland will participate only in strategies and programmes where the policies converge with its own development goals and the partner country is considered to possess adequate capacity and to be committed to good governance. Where such conditions are not present, direct sectoral or budgetary support is 'not yet possible' and Finland will support programmes that are 'in line with the partners' priorities'. Even the Swaps are seen to involve projects within their frameworks and Finnish consultants have been reassured that there will be room for them in their execution.

These trends raise the crucial issue of the role of Finnish personnel and the administrative limitations of the MFA. One thing on which all evaluations have been unanimous is that the Finnish aid administration suffers from a number of weaknesses. The decision-making is heavily Helsinki-centred and the field administration is starved of resources. Normally, Finland runs its programmes in the major developing countries with far fewer administrative personnel on the spot than, for instance, the other Nordic countries. Desk officers in Helsinki change frequently and the institutionalised memory at headquarters suffers from frequent cuts. Now, if Swaps and budget support are to be seriously embarked upon, this will shift much of the responsibility for planning and implementation to the partner, but at the same time will allow more policy responsibility and monitoring demands to fall on the donors. The MFA has learned to rely on consultants up to the point of outsourcing the monitoring of sectoral programmes to them in some major partner countries. One would expect the MFA to take much greater responsibility here, both for the reason that policy matters belong to it and not to private consultants and also for the sake of keeping the accumulating knowledge in-house. As a first step in this direction, the embassies in major partner countries have been strengthened with sectoral or other advisers both recruited locally and dispatched from Helsinki.

While the guidelines for operation in long-term partner countries are in place and fairly well elaborated as far they go, plans concerning the 'instruments' and other modalities to be used in 'other partnerships' are much more vague. What the instruments will be other than government-to-government co-operation is not specified. NGOs and civil society are mentioned a few times and praised for their potential but little has been said about how they will be employed except as 'alternative channels'. The commitment in the 1996 Decision-in-principle that NGOs will be allocated 10 to 15 per cent of the budget for operational development co-operation is specified in the 2004 White Paper to mean 14 per cent. 'Economic and commercial instruments', for their part, are repeated throughout but it remains unclear what else this could involve beyond the push to concessional credits. Concessional credits are no longer condemned but encouraged. It is promised that 'the selection of countries will be widened' beyond China, hinting at the possibility of introducing special arrangements for countries that rank low on creditworthiness. The legislation on credits has been amended so that their conditions have been made more palatable to exporters and debtors alike. Conditions concerned with the domestic contents and grace and repayment periods have also been loosened.[10] In the 2004 White Paper the restrictions channelling the credits into the environmental and social sectors were done away with, something which, of course, will allow much freer use of them.

V. AID AND THE CONSTRUCTION OF STATE IDENTITY

A popular argument among responsible MFA officials is that lessons have been learnt from past mistakes and Finland is now mature enough to pursue an effective aid policy. In the late 1980s a great deal of money was available but the policies were poor and the effects remained unsustainable. Now good policies are in place and if sufficient resources are granted the results will be effective aid and poverty reduction and development in partner countries, and thereby increased global security for all. In this chapter we have agreed that many lessons have indeed been learnt and Finnish aid policy is today in many respects much better grounded than it used to be. But the issue is more complex than the advocates of the official line maintain. As we have attempted to show, the agreed policies are themselves inconsistent and cannot be taken as guaranteeing success. The record

---

10. For details, see MFA [n.d.].

of the effectiveness of Finnish aid is not convincing, and uncertainty surrounds the development of its modalities.

Our argument is that such inconsistencies in aid policy result from the interplay between developmentalist and instrumentalist considerations. There would be hardly any aid at all without the acknowledgement of the basic developmentalist premises. But within them there is room for many sorts of policies depending on instrumentalist considerations. In Finnish aid policies the latter have predominated. Developmentalism has remained a weak influence on the aid policies until quite recently. The underlying idea has been the legitimacy of multiple motivations and the compatibility of interests as a basic starting point. From the first committee report onwards it has been maintained that aid can be undertaken from various motivations, none of which is inherently superior. The main thing is that resources are kept moving, not why they are moved. Consequently, aid has been used for various purposes, from raising the profile of Finland as a Western country to propping up Finnish companies to widening the 'contact surface' of Finland.

Many explanations may be offered for this weight given to non-developmentalist considerations. An obvious one is that, historically, developmentalism has been a shallow force and mainly imported into Finland. It had practically no indigenous roots when Finland started its aid activities. Public opinion, political forces and 'civil society' woke up only later to such considerations. Due to its geographical position and small size, Finland had few traditional contacts with Southern countries and peoples. There were a few shops for 'colonial goods' such as coffee and spices, and some missionaries were known to work in faraway Africa. Aid was not much better known. Finland was not itself a recipient of the Marshall Plan aid extended by the United States for the reconstruction of Europe; during the aftermath of the Second World War Finland's position was still uncertain and it therefore preferred not to damage its relations with the neighbouring Soviet Union. Yet Finland received considerable Western support via the World Bank. If World Bank infrastructural loans are included, Finland was until 1968 a net receiver of concessional finance. But that was not perceived as aid. Developmentalism came in only when Finland was bracing itself for its role as a donor at the United Nations. This took the diplomatic form of commitment to internationally agreed aid targets, the most consequential of which was the target of 0.7 per cent of GDP to be allocated for official development assistance, given at the 25th UN General Assembly in 1970.

This brings us to the second part of the argument of this chapter: aid serves the construction of state identity. The Finnish 'national habitus', to borrow a highly suggestive phrase of Norbert Elias, has been shaped by the uncertainty stemming from the country's ambivalent position between two great European civilisations, the Western/Germanic and the Eastern/Slavonic. The Finns have been unsure where they belong. Ethnically different, situated on the northern fringe of Europe, and sandwiched between two powerful neighbours (and former colonial powers), Russia and Sweden, Finland has for centuries straddled the more contingent line that separates 'Western' from 'Eastern' Europe and which in a cultural sense runs right across the country. Politically, the Finnish state has sought to emphasise the 'Western' elements, since it gained its independence from what had just been turned into Bolshevik Russia in 1917. This endeavour was greatly facilitated by the identification of the West with rising capitalism. The geopolitical realities of the post-Second World War situation brought a dilemma for Finnish foreign policy when it became necessary to actively maintain good relations with the big eastern neighbour. The solution was found in remaining cautious internationally and concentrating mainly on Finland's own development. However, when it became possible Finland started to search for the status of an internationally active neutral, Nordic country. That search became one of the hidden anchors of Finnish foreign policy, and an element of its external identity. With the end of the Cold War and the collapse of the Soviet Union, Finland, together with Austria and Sweden, abandoned neutrality and joined the European Union. This was a much bigger change for Finland than for the others, because it meant a change of side in the security system and brought the long-term political identity project to completion. Within the Union the quest for national identity has again assumed new elements. Finland is eagerly penetrating what is called the hard core of Europe and enjoys sitting as a master at the table of the masters in Brussels (in ex-Prime Minister Paavo Lipponen's words).

When viewed in this light, changes in aid policy provide a revealing picture of how Finland prefers to be seen and how it sees its national interests being constructed. When the primary identification was with the Nordic group, the role model was Sweden, an internationally active and highly developed 'modern' state. The implicit message of the aid policy was written between the lines: by engaging in development co-operation Finland was underlining its independence and Western orientation and demonstrating that it was not a Soviet satellite. Developmentalism in such a set-up carried little weight. Instead, the door was open for the commercial use of aid funds, which was understood as the direct national interest. After joining the EU in 1995, Finland positioned itself less as an independent actor and more as a member of a new political unit in the making. The earlier identification with the Nordic countries resurfaces occasionally, but the emphasis has definitely shifted. The United Nations, which used to be given great prominence in the early decades, is reduced to providing development ideals such as the Millennium Goals. For operational purposes, ideas and outlines emanating from the Bretton Woods institutions are of more consequence. The brand of developmentalism they advocate is therefore what matters in aid practice. Within the EU parts of the traditional national identity and related interests such as state autonomy may be devaluated, while aspects dealing with economic well-being and mental dimensions (such as 'collective self-esteem' in Wendt's terms) obviously gain in importance. As the Decision-in-principle of 1996 solemnly declares, 'development cooperation is the mark of a civilized nation' [*MFA, 1996: 5*]. Few other donors would feel the need to emphasise this.

What does the analysis put forward here entail for Finnish aid policies? The best possible scenario would be that the developmentalist trends now visibly activated will strengthen further, and donors and recipients will together find forms of genuine co-operation and working solutions to their problems. In the worst scenario the present policies will explode, or perhaps implode, in their internal contradictions, and old or new instrumentalisms will take over. For Finland, and perhaps for some other minor European donors, much will depend on the EU, as Union membership has now become a part of their state identity. But at least as much will depend on how aid effectiveness is seen to develop, since the growing developmentalist case essentially rests on its promise. In this respect much now depends on the argument about the crucial role of good policies *à la* World Bank – probably too much.

REFERENCES

Collier, P. and D. Dollar, 1999, *Aid Allocation and Poverty Reduction*, World Bank Policy Research Working Paper No. 2041, Washington, DC: World Bank.
Collier, P. and D. Dollar, 2000, 'Can the World Cut Poverty in Half? How Policy Reform and Effective Aid Can Meet the International Development Goals', Washington, DC: World Bank, mimeo.
Kehitysyhteistyökomitean mietitintö (Report of the Committee on Development Co-operation), Komiteanmietintö 1978: 11, Helsinki.
Koponen, Juhani, 1999, *Developmentalism vs. instrumentalism: an interpretation of the history of Finnish development aid*, FAD Working Paper 1/1999, Helsinki: University of Helsinki, Institute of Development Studies.

Koponen, Juhani and Päivi Mattila-Wiro, 1996, *Effects or impacts? Synthesis study on evaluations and reviews commissioned by Finnida, 1988 to mid-1995*, Report of Evaluation Study, 1996: 1, Helsinki: MFA.
Koponen, Juhani and Lauri Siitonen, 2001, *Forestry Sector Aid, Sustainability and Participation – Zanzibar and East Usambara Compared*, FAD Working Paper 1/2001, Helsinki: University of Helsinki, Institute of Development Studies.
Kumppani, 2001, No. 5 (Magazine of KEPA, the umbrella organisation of Finnish NGOs).
MFA (n.d.), *Concessional Loan Scheme to Developing Countries* provided by the Ministry for Foreign Affairs of Finland. Leaflet, available through the Internet at http://global.finland.fi.
MFA, 1993, *Finland's Development Co-operation in the 1990s*, Helsinki: Ministry for Foreign Affairs.
MFA, 1996, *Decision-in-principle on Finland's Development Co-operation*, Helsinki: The Government (12 September).
MFA, 1998, *Finland's Policy on Relations with Developing Countries*, Helsinki: The Government (15 October).
MFA, 2000, *Suomen kehitysyhteistyö 2000* [Annual Report on Finland's Development Co-operation submitted to Parliament]. Helsinki: Ministry for Foreign Affairs.
MFA, 2001a, *Operationalization of Development Policy Objectives in Finland's International Development Cooperation*, Government Decision-in-Principle (22 February).
MFA, 2001b, *Suomen kehitysyhteistyö 2001* [Annual Report on Finland's Development Co-operation submitted to Parliament]. Helsinki: Ministry for Foreign Affairs.
MFA, 2004, *Development Policy*, Government Resolution, 5 February 2004.
Moore, Mick et al., 1996, *Ownership in the Finnish Aid Programme*, Report of Evaluation Study, 1996: 3, Helsinki: MFA.
OECD, 1999, *Finland. Development Co-operation Review Series* No. 31, Paris: OECD, DAC.
OECD, 2003, *Finland. DAC Peer Review*, Paris: OECD, DAC.
Osterbaan, Maaike and Raili Kajaste, 1999, *Thematic Evaluation on Environment and Development in Finnish Development Co-operation. Concessional Credits*, Report of Evaluation Study, 1999: 6, Helsinki: MFA.
Siitonen, Lauri, 1996, *Social Science Literature on Finnish Development Aid: A Review and Bibliography*, FAD Working Paper 1/1996, Helsinki: University of Helsinki, Institute of Development Studies. (Also available through the Internet at http://www.valt.helsinki.fi/kmi/fad/finn-oda.htm.)
Siitonen, Lauri (forthcoming), *Small Donors and Aid Regime Norms: A Comparative Study of the Determinants of Aid Policies*.
Telford, John, 2002, *Synthesis Study of Eight Country Programme Evaluations*, Evaluation Report, 2002: 8, Helsinki: MFA.
The Government Programme of Prime Minister Matti Vanhanen's Government on 24 June 2003. http://www.valtioneuvosto.fi/tiedostot/pdf/en/39357.pdf
Wendt, Alexander, 2000, *Social Theory of International Politics*, Cambridge: Cambridge University Press.
World Bank, 1990, *World Development Report 1990*, New York: Oxford University Press for the World Bank.
World Bank, 1998, *Assessing Aid, What Works, What Doesn't, and Why*, New York: Oxford University Press for the World Bank.
World Bank, 2002, *The Role and Effectiveness of Development Assistance. Lessons from World Bank Experience*, Washington, DC: World Bank.

7

French Development Co-operation Policy

JEAN-JACQUES GABAS

French development co-operation policy is part of a long story which cannot be separated from the country's colonial past, or the close relations between the public authorities and private interests and, more generally, the role France wants to play on the international stage. Relations between France and Africa have been built on decolonisation, interwoven public and private interests, and the role France has played in maintaining certain governments in power in accordance with the military co-operation agreements it has signed with the newly independent states.

Since the early 1960s, observers, groups representing civil society, senior civil servants and ministers have written close on 40 more or less official reports (see Annex 7.1) that, in general, analysed critically the effectiveness of development-oriented French co-operation and highlighted both the strengths and the shortcomings in the system. As the twenty-first century makes its debut, the repeated criticism leads us to question recent trends and the nature of French development co-operation policy, analysis of which begins with the 1998 reform, its aid allocation strategy and its main orientations.

I. WHAT ARE THE CHARACTERISTICS OF FRENCH DEVELOPMENT AID?

1. Official Development Assistance (ODA) on the Decline

Between 1994 and 2002 French official development assistance (ODA) dropped by 38 per cent, from €7.17 billion to €5.5 billion, although Table 7.1 shows a slight increase in 2001,[1] 2002 and 2003 (forecast), which may be a sign of a slight reversal of earlier tendencies. But even these recent commitments do not reach the level of the early 1990s. In this declining aid, the share of multilateral aid (not in volume) has gone up fairly steadily (24.4 per cent in 1994, close to 36 per cent in 2002 and 31 per cent in 2003).

TABLE 7.1
FRENCH OFFICIAL DEVELOPMENT ASSISTANCE (€m.)

Net payments	1994	1995	1996	1997	1998	1999	2000	2001	2002	2003
Total ODA	7.166	6.424	5.811	5.611	5.164	5.293	4.454	4.631	5.499	6.071
I. Bilateral aid	5.596	4.891	4.491	4.250	3.763	3.874	3.069	2.653	3.329	4.008
Technical assistance	1.540	1.537	1.551	1.506	1.391	1.341	1.313			
Project aid	1.015	909	774	586	536	419	319			
Incl. Treasury loans/grants	550	423	336	207	133	43	-69			
Incl. FSP (ex- FAC)	189	199	181	156	161	132	133			
Programme aid	532	391	277	127	62	72	-25			
Debt alleviation	1.336	862	756	979	636	790	520	388	1287	1881
TOM*	730	754	714	686	744	753	168	189	192	195
Other	443	437	418	366	395	500	774			
Incl. aid to refugees	86	95	61	51	72	147	160			
Incl. administrative costs	234	240	236	237	244	234	205			
II. Aid through the EU	775	749	659	784	703	750	859	1.165	1.298	1.201
III. Other multilateral aid	795	784	661	577	697	669	526	624	680	667
GDP/GNI (billion euros)	1142.3	1181.8	1212.2	1251.2	1301.4	1344.4	1394.1	1469	1510	1567
ODA as % of GDP/GNI	0.63	0.54	0.48	0.45	0.40	0.39	0.32	0.32	0.36	0.39

Sources: The table is prepared by using data from HCCI, PLF 2003 (Project de Loi de Finances) and CICID.
Note: *TOM = Mayotte et Wallis-et-Futuna.

Why is France withdrawing from development co-operation? There are some highly contrasting factors that explain this downward trend. To begin with, it is essential to reposition France's foreign policy in the full range of actions undertaken by the country. At the beginning of the 5th Republic, General de Gaulle gave high priority to French foreign policy. When François Mitterrand was in office, this was only briefly the case, namely in 1991, two years after the collapse of the Berlin Wall, when there was great concern about ensuring that France made its presence felt in the Central and East European countries (CEEC). In 1993 the Ministry of Foreign Affairs introduced a multi-

---

1. In 2001, France (US$4.2bn.) ranked fifth among donors, after the United States ($11.4bn.), Japan ($9.8bn.), Germany ($5.0bn.) and the UK ($4.6bn.). As a percentage of GNI, France contributed 0.32 per cent of generated wealth. The average for the DAC countries was 0.22 per cent ranging from Denmark (1.03 per cent) to the United States (0.11 per cent) [*OECD, 2003b: Table 13*].

year rationalisation plan for its network abroad and made major budget cuts. From that time up until 2001 the development co-operation budget shrank, forced down even further by short-term economic factors.

This was the case in 2000 when the OECD's Development Assistance Committee (DAC) recommended that, under aid for overseas territories, only funds for Mayotte and Wallis and Futuna should be recorded as ODA. This statistical correction, called for by the other DAC member states for a number of years, caused a significant drop in total French ODA since Polynesia and New Caledonia were the main beneficiaries of French aid.

Furthermore, there was a very big drop in project aid. As Table 7.1 shows, between 1994 and 2000 project aid dropped by 60 per cent, due to smaller allocations to the Priority Solidarity Fund (FSP) – the successor to the *Fonds d'aide et de coopération* (FAC) – managed by the Ministry of Foreign Affairs, and lower allocations for the financial protocols managed by the Ministry of Finance and Economic Affairs (Treasury). This latter decrease was caused by reform of the financial protocols which were replaced by the Emerging Country Reserves, and by far closer attention being given to the choice of projects by the Ministry of Finance. This system restricted the number of aid beneficiary countries and contributed to a geographic concentration of financial flows.

During the 1990s, the aid programme was downsized significantly as a result of the combination of two factors: a decrease in funding when the CFA franc was devalued in 1994, and a reduction in funds allocated to structural adjustment programmes. For example, in 1995 aid for structural adjustment amounted to €331 million but was only €29 million in 2001.

The capacity of the beneficiary country to absorb aid should not be overestimated. It may have limited institutional capacity or be in a state of conflict. This applies especially to French ODA which is concentrated on Africa (in the 1996–2000 period, 41 per cent went to sub-Saharan Africa and 16 per cent to North Africa), a continent with 15 countries at war.

Finally, although it may seem paradoxical, efforts at debt relief declined until 2001. The numerous prerequisites for debt reduction under the HIPC (Highly Indebted Poor Countries) Initiative guaranteed slow implementation. In 2000, for instance, only two countries had reached the completion point. This low rate of payments might seem to contradict France's international commitments. In fact, the opposite was the case. Since 2002 aid has been going up, as may be seen from the accelerated debt cancellations and consolidations connected mainly with the application of the HIPC Initiative.

France has agreed to cancel 100 per cent of its claims on ODA loans for countries that are eligible for the HIPC Initiative (Table 7.2). This additional bilateral effort is based on agreements called Debt Reduction-Development Contracts (C2D or CDD), which are contracts based on a refinancing mechanism. Rather than straightforward debt cancellation, governments continue to repay their debts to France but when the debt has been reimbursed, the equivalent amount is placed in a special account for these states in the Central Bank to fund poverty reduction and sustainable development programmes. Programmes and projects funded under the C2D are country-specific but must fit in with the orientations set out in the Poverty Reduction Strategy Papers (PRSP) and concur with France's strategy for the country in question. Apparently four areas were considered as priorities by the Interministerial Committee for International Co-operation and Development (CICID): basic education and professional training, primary health care and combating major endemic diseases, equipment and infrastructure for local communities, and land planning and natural resources management. The following countries are eligible for the so-called C2D in the priority solidarity zone (ZSP): Mozambique, Mauritania, Cameroon, Guinea, Madagascar, Burundi, Congo, Ivory Coast, the Democratic Republic of Congo, Uganda, Tanzania, Rwanda, São-Tomé and Principe, and Sierra Leone. The only countries that have signed a C2D are Mozambique, Cameroon and Mauritania. For France, which, together with Japan, is their main provider of ODA, the amounts involved are considerable.

TABLE 7.2
BILATERAL DEBT RELIEF FOR HIGHLY INDEBTED POOR COUNTRIES (HIPC) (€m.)

Country	Total debt to be recycled	Amount of the first C2D contract
Ivory Coast	1420	440
Cameroon	1081	283
Congo	343	58
Guinea	187	43
Mozambique	91	20
Mauritania	69	14
Madagascar	68	21

Source: French Development Agency (AFD).

All these reasons explain why ODA has declined for a long period of time, but it was surprising to hear international bodies pleading for solidarity precisely when such a radical decline was taking place, especially since the economic situation provided a context that was conducive to aid-giving, particularly between 1995 and 2000.

This financial disengagement went hand in hand with less creative ideas and proposals on the international scene and thus loss of influence to other donors. Also, and more important, it led to geographic choices for aid allocations that were based on political criteria which seemed more coherent, although the financial realities suggested the opposite.

2. Does France Have a Clear Policy for Deciding where to Spend Its ODA?

During the 1998 reform of French development co-operation, a priority solidarity zone (ZSP) was defined. It was stated that bilateral French development aid should be selective and concentrated, and should include the least developed countries that do not have access to financial markets. Furthermore, in February 2002 the CICID introduced an aid suspension regime for countries in the zone that adopted policies which hindered international co-operation. Two suspension criteria were adopted: countries subject to suspension provided for under the Cotonou Agreement (because of human rights violations, serious corruption, non-respect of democratic principles, violations of the rule of law) and countries on the FATF (Financial Action Task Force on Money Laundering) black list (because of money laundering or due to the fight against terrorism). Furthermore, the ZSP was defined to include 54 states plus the Palestinian Administered Territories, and was based on criteria reflecting their level of development. Since 1998, there is no longer mention of *pays du champ,* which meant the former French colonies, aid for which was managed by the former Ministry of Co-operation. It was recommended that the new definition of the priority zone for co-operation should make a break with the colonial past, as had been strongly urged already in most reports on French development co-operation policy for the past 40 years. During the CICID meeting of 14 February 2002, the ZSP was changed to include two least developed countries (Sudan and Yemen) and to transfer several countries to other regional DOM-TOM co-operation mechanisms. The CICID will be updating the configuration of the Zone regularly.

This new definition of the ZSP represents a very important change, because it includes states that have close cultural and political ties with France but were not previously included in the *pays du champ* of the Ministry of Co-operation, such as Lebanon, the Palestinian Administered Territories, Vietnam, Laos and Cambodia. Other choices seem to reflect a more recent desire for political dialogue and economic solidarity with an eye to development assistance, for example, Cuba, San Domingo, South Africa and Kenya. Even though the aforementioned reasoning applies to cases such as Uganda, Eritrea, Ethiopia, Liberia, Sierra Leone and Tanzania, the overriding reason for the French decision to include them in the ZSP was their level of poverty. In general, the list of ZSP countries is closely aligned to the list of ACP states that are associated with the European Union through the Cotonou Agreement, although there are notable exceptions, but this may be due to shortage of resources.

In view of these clear new political orientations that seek to move away from the system of allocating aid on the basis of networked alliances between the *pays du champ* and the Ministry of Co-operation, the question is whether there have been any changes in geographic orientations. Frankly, for the last ten years the geographic distribution of French aid has not budged, despite changes in the statistical notifications to the Development Assistance Committee of the OECD and the introduction of the ZSP. Up until 1999, 40 per cent of French bilateral aid was concentrated on ten countries: French Polynesia, New Caledonia, Ivory Coast, Morocco, Egypt, Cameroon, Senegal, Madagascar, Algeria and Tunisia. The trend during the period 1995 to 2001 shows a net decline for all regions, and specifically for sub-Saharan Africa. For the final year 2001, New Caledonia and French Polynesia received US$679 million and the countries of sub-Saharan Africa received $944 million.

What has happened to diversification and the level of poverty as criteria for aid allocation? In fact, the least developed countries (LLDCs) are receiving a decreasing share of total aid, which itself is on the decline. The share of French bilateral ODA for the poorest countries is going down, as can be seen from Tables 7.3 and 7.4. In 2001, the main beneficiaries of French ODA and official assistance (OA) were countries or territories most of which were not least developed countries: French Polynesia (US$384m.), New Caledonia ($295m.), Egypt ($201m.), Poland ($183m.), Morocco ($74m.), Mayotte ($119m.), Ivory Coast ($110m.), Senegal ($102m.), Tunisia ($88m.), while Mali ($60m.) and Burkina Faso ($44m.), for example, received far less, although their Human Development Index (HDI) is much weaker. The impression is that aid recipients are not selected on the basis of their needs for human development.

TABLE 7.3
NET DISBURSEMENTS OF ODA AND OA TO MAJOR REGIONS (US$m.)

Region	1995	1998	2001
North of Sahara	931	714	529
South of Sahara	2700	1520	944
Central and South America	377	175	112
Middle East	135	109	71
South Central and Far East Asia	648	351	279
CEECs/NIS	365	435	256
Oceania	897	765	738
including			
New Caledonia	441	335	295
French Polynesia	444	368	384

Source: OECD [2003a].
Note: ODA = Official Development Assistance, OA = Official Assistance – DAC definitions.

Similarly, the ZSP, as a percentage of bilateral ODA, dropped from 52.4 per cent in 1994 to 44.2 per cent in 1999, while the percentage of French development assistance for countries with an intermediate level of income rose from 25.5 per cent in 1979 to 34.4 per cent in 1999. There is a clear contradiction between political rhetoric and actual performance.

TABLE 7.4
PERCENTAGE OF BILATERAL ODA FOR LLDCS AND LICS

	1988–89	1998–99	2000–1
LDCs	32.5	21.4	28
LICs (not LDC with <$760 per capita)	26.8	27.1	14
Total	59.3	48.5	42

Sources: HCCI based on parliamentary reports and PLF 2003.
Note: LICs = low-income countries, LLDCs = least developed countries.

In organising its external policy, France is prioritising the construction of Europe, with, in second place, what we might call the *francophony–Africa–Mediterranean* trio. These three pillars will be the backbone of French development co-operation policy, as shown in its financial choices. Africa is still the top priority. The political authorities proclaim this priority, despite the weaker financial flows and the speech by President François Mitterrand at the Baule Summit in 1990, as reasserted by Prime Minister Edouard Balladur in September 1993 when he referred to the requirements of solidarity by saying, 'if certain countries prefer to stay at a distance from the international community and the rules of good governance, France cannot do anything for them'. This doctrine, indeed, squares with a drop in the available funding. France is trying to extract itself from the image it portrayed in the 1990s, of a *France–Afrique* based on networks of military, political and financial influences stemming from the colonial period, with attendant actions in several countries (in particular, Chad, Gabon, Comoros, Djibouti, Angola) that did not seem beyond reproach [*Glaser and Smith, 1992, 1997*]. Can France erase this image? It is difficult to say 'yes', but apparently the habits and actions of old, like those so strongly castigated, the Foccart-type actions, have disappeared: information is serving to rein in excess, democracy is making inroads (admittedly not all at the same pace), but interference is strongly condemned and French public opinion is very much against this type of influence. Parliament is asking the government more and more questions.

The nature of its connections is changing, but what kind of an equilibrium is France moving towards? The recent Ivoirian crisis is indicative of the very limited leeway France has in striving to alleviate the devastating consequences of internal political tensions and in activating the military co-operation agreements[2] without being guilty of intrusion. To clarify relations, ways and means must be found to build up or at least support new development policies. Jean-Michel Severinot, the Director General of the French Development Agency, recently referred to this point.[3] 'France underestimated poverty and its effects on the continent. ... The 1980s were marked by the post-colonial situation and East–West friction. The French authorities had close personal ties with the African heads of state; in this vein, Jacques Chirac is part of the old generation. Links have become looser. ... The *France-Afrique* mindset and the "briefcase carriers" are no longer phenomena that structure relations between Paris and Africa.' France's share in the funding for sub-Saharan Africa is constantly declining. The sources of funding are becoming more diverse, for example, Japan, and the United States (see Table 7.5) which is assuming an increasingly significant role, even if the recipient countries are different (anglophone countries for the United States and Japan, francophone countries for France). Is France allocating the resources it needs for close co-ordination with other donors?

---

2. Military co-operation is not part of ODA but since 1999 the military assistance budget has been declining. There are about 357 military *coopérants* (military assistants), mostly on assignment in the following countries: Cameroon, Ivory Coast, Djibouti, Morocco, Senegal, Chad and Gabon. The number is falling.
3. In an interview in the weekly *Jeune Afrique, l'Intelligent*, 9 March 2003.

TABLE 7.5
TREND OF FRANCE'S RELATIVE IMPORTANCE IN SUB-SAHARAN AFRICA

	1995		1998		2001	
	US$m.	in %	US$m.	in %	US$m.	in %
Total aid to sub-Saharan Africa	18,420		13,900		13,530	
Total aid of France to sub-Saharan Africa	2,700	14.6	1,520	11.0	944	7.0
Total aid of United States to sub-Saharan Africa	1,050	5.7	713	5.0	1,375	10.2
Total aid of Japan to sub-Saharan Africa	1,352	7.3	948	6.8	850	6.3

Note: computed from DAC statistics (on disbursements basis). Each per cent is the ratio of: Total aid of a donor to sub-Saharan Africa/total aid to sub-Saharan Africa from all sources.

On the one hand, France's Africa policy focuses mainly on the francophone countries and the construction of regional 'spaces'. France has a unique leading role since it is the backbone of the franc zone construct. The West African Economic and Monetary Union (WAEMU) is a monetary zone in which efforts to harmonise business law are largely impelled by France. Does this regional integration make economic and political sense in the present West African context? How can this entity be linked to the Economic Community of West African States (ECOWAS) which includes the anglophone countries, in particular Nigeria and Ghana? On the other hand, how can this North–South regionalisation fit in with the Economic Partnership Agreements (EPAs) that are to be developed between the European Union and the ACP countries via the Cotonou Agreement? These are two vital political challenges that France must face if it wants to redesign its African co-operation.

The Mediterranean is another region where France is playing a role, or intends to do so. Three scenarios have been considered. First, relations with its former colonies (Algeria, Morocco and Tunisia), a highly sensitive scenario. Second, the role France would like to play in solving the Israel–Palestine conflict. Third, building up preferred relations with countries in the south and east Mediterranean through the 1995 Barcelona process.

Outside these priority zones, French participation is very weak. France has a 'strategy' for wielding influence but very little political clout in South and South-East Asia, the Caribbean and even in Latin America. Competitors are clearly on site, for example, the United States in Latin America, Japan in Asia, and the future EU member states which are edging closer to NATO and are receiving major attention from Germany. France wants to maintain itself in its francophony–Africa–Mediterranean solidarity circle, but has to cope with rivalry; for example, the French stand on the peace process in the Middle East is far from unanimously applauded by Israeli politicians, and the US and Britain are still very influential in South Africa and Nigeria, two of the main economic powers in Africa.

Despite the relative growth in aid channelled through multilateral agencies, French influence in international circles is clearly losing momentum. This situation reflects the gradual disengagement process of the 1990s which could be seen during the negotiations of the Cotonou Agreement between the European Union and the ACP countries. The French Presidency of the European Union as of June 2000 was prepared very (too?) late, and did not involve the whole of the administration, or even consult civil society. The same is true for the Bretton Woods international financial institutions. France contributes only marginally to discussions on alternative ideas to the Washington Consensus, although there is strong opposition in France to the structural adjustment programmes. There are several explanations for this. First there is the specifically French way of separating what is called the world of research and the world of political decision-makers. Unlike other donors, France is unable to make the connection between these two worlds. Furthermore, the French representatives in international financial fora come from the Treasury Department, thus bearing the mark of a strong economic culture which approves of the general orientation of these structural adjustment programmes. These representatives are in fact autonomous, and not affected by parliamentary policies, although the 'Government Report to Parliament on the Activities of the International Monetary Fund and the World Bank', submitted to the National Assembly on 5 February 2003, reads: 'France favours increased participation by parliaments and civil society in designing development policies ... and wants to set the example in this field.' This report puts forward the idea that changes need to be made in the international financial institutions, and stresses the need for IFI positions to be consistent with those of the United Nations and those supported by European development policy. It is clear, however, that the majority of multilateral funds (73 per cent during the 1990s) have been oriented towards francophony.

Lastly, the extent to which co-operation policies can be influenced depends on the recipient states. Representatives of French administrations often have limited influence in co-ordination and consultation bodies. Since 1994, France does not seem to have the resources needed to fulfil its ambitions.

II. DOES FRANCE HAVE A DEVELOPMENT CO-OPERATION POLICY?

1. Has French Development Co-operation Policy Become Clearer since the 1998 Reform?

Reform of the French development co-operation mechanisms was started in 1998 when the Ministry of Co-operation was merged with the Ministry of Foreign Affairs, and the French Development Agency (AFD) was given a central role in implementing co-operation projects and defining France's development co-operation policy. The reform was designed to make aid activities clearer, to move French development co-operation away from its colonial country attachments *(pays du champs)* and to avoid dispersing aid funds among too many actors. It had both institutional and political facets. The institutional arrangements were finalised on 28 January 1999 during the first CICID, the main role of which was to co-ordinate ministerial actions. Then on 28 November 1999, the High Council for International Co-operation (HCCI) was established as an advisory body consisting of 60 members (45 members for the second HCCI mandate which started in 2003), namely, members of Parliament, elected local officials, representatives of socio-professional organisations which have developed partnerships in their international co-operation work, and representatives of universities and scientific research organisations devoted to international co-operation and development issues.

Four years later, institutional control of ODA is still fragmented. The Ministry of Economic Affairs, Finance and Industry controls over 40 per cent, with the Ministry of Foreign Affairs and the other technical ministries sharing the remaining 60 per cent, more or less equally. This breakdown of ODA by ministries demonstrates the major role of the Ministry of Finance in the allocation of aid and especially in the allocations to emerging countries, in order to be consistent with the activities of the French companies which are looking for an increasing share of the market. This trend will continue along the same path for some time to come. It is a strong trend.

TABLE 7.6
THE IMPORTANCE OF MAJOR MINISTRIES IN THE ADMINISTRATION OF ODA (%)

	2001	2002	2003
Ministry of Foreign Affairs	35	29	25
Ministry of Finance and AFD	16	32	42
Other ministries	49	39	33

Note: Each per cent is the ratio of total aid managed by a ministry/total aid provided by France.

The institutional reform in fact concerns only a small number of actors. The Ministry of Foreign Affairs (formerly via the ex-Ministry of Co-operation) administers only one-quarter of the total ODA. The breakdown after five years includes three cultures. First there is the diplomatic culture, that is, the Ministry of Foreign Affairs, in particular the ambassadors who are responsible for the expenditure of development co-operation funds in recipient countries where French cultural horizons and concepts of *francophony* are essential: here there has been a major change, since before the reform the two functions of ambassador and chief of French development co-operation mission were carried out by two different people. The ambassador had the diplomacy function and the chief of the French co-operation mission had the responsibility for development co-operation. Now, it is the ambassador with his primary diplomatic function who will implement the projects and programmes financed by France; development will be introduced as a second priority.

Second, there is the culture based on the macroeconomic balance and the extension of commerce, that is, the Ministry of Economic Affairs, Finance and Industry, and, third, the more marginal development culture embodied previously by the Ministry of Co-operation and now reflected by the Directorate General for International Co-operation and Development (DGCID) and the French Development Agency (AFD). Although the reform was undertaken to provide simplicity and greater transparency, this co-existence of objectives and the expectations of the various actors have often led to tensions and lack of clarity for the partners in both the South and in Eastern Europe.

The Presidency of the Republic has continued to play a decisive role under the political *co-habitation*[4] that marked France between 1997 and 2002. With regard to relations between France and Africa, the political minds of the two heads of the executive branch did not always meet, for example, over the Ivory Coast.

---

4. *Co-habitation* in France refers to the fact that the President of the Republic, Jacques Chirac, belonged to the Liberal Party (Union pour la Majorité Présidentielle – UMP) and Prime Minister Lionel Jospin belonged to the Socialist Party (PS).

> *BOX 7.1*
> REPORTS SUBMITTED TO THE PRIME MINISTER DURING THE FIRST HCCI MANDATE 1999–2002
>
> - Relations between NGOs and international institutions – 29.10.2002
> - Democratic governance and international co-operation – 24.09.2002
> - Higher education, research and international co-operation with developing countries – 24.09.2002
> - Co-operation in the health sector with developing countries – 25.06.2002
> - What human resources for what co-operation? – 25.06.2002
> - Co-operation priorities for sub-Saharan Africa and the New Partnership for Africa's Development (NEPAD) – 19.04.2002
> - Implementation of the Cotonou Agrement between the EU and the ACP countries – 26.11.2001
> - Co-operation with regard to professional education with the countries of the priority solidarity zone – 8.10.2001
> - Co-operation in the basic education sector with the countries of sub-Saharan Africa – 8.10.2001
> - World food summit: five years after – 26.09.2001
> - Co-operation and human rights – 10.07.2001
> - Towards co-operation adapted to the needs of LDC countries – 2.04.2001
> - French positions taken in the international financial institutions (IFIs) – 6.12.2000
> - Crises, co-operation and development – 23.11.2000
>
> (All these reports are available online http//:www.hcci.gouv.fr)

The High Council for International Co-operation (HCCI), as its name indicates, serves as an advisory body. Since its inception, the HCCI has written several notes to the Prime Minister to ensure that members of the French administration participate in all its working groups. It is difficult to judge the effectiveness of this type of advisory body (which does not exist in most European countries) with regard to political decisions, but the constant recommendation to stop cutting French ODA was heard only in 2001, and the reaction was very limited, despite the many dossiers and opinions on orientations for development co-operation in various fields that were sent to the government, together with an opinion note to the Prime Minister.[5] The lack of actual participation in public policy formulation and budgetary allocations made this body less effective than it wanted to be. The drop in aid certainly decreased the importance of the HCCI in international discussions, which all the designers of development co-operation policy strongly regretted. They also had the feeling (perhaps erroneously) that they were being less proactive in proposing development policies and strategies. For the second HCCI mandate, which started in January 2003,[6] the three-year work programme is to be partly defined by the Prime Minister. This is a completely new departure and may give the HCCI more of an operational role, making it more interested in implementing recommendations than in political reflection.

The HCCI, by its very existence, reflects the French authorities' decision to be open with civil society and the scientists. It is a strongly political sign. Admittedly, during the past three years the French authorities have tried to create some bridges (links) between decision-makers and researchers, but the attempt has always been timid, and the right way has not yet been found to work in harmony. Very often, the scientific community does not want to work with the administration, the argument being that applied research is not academic research and is not much appreciated in a researcher's career. On the other hand, the administration does not know how to translate the results of the research for its decisions, and also has its own production (coming from the bureaucracy) of policies. But the HCCI should now move much closer to organisations devoted to international solidarity, because, on the international scene, France now stands out as an exception in the field of non-governmental co-operation. The total funding for voluntary associations, co-operation with organisations devoted to international solidarity, and decentralised co-operation is barely more than 0.57 per cent of total French aid for 2003, as against 0.74 per cent in 2001! Non-governmental actors are welcome in word, but in fact, they are excluded from programme implementation.

---

5. See HCCI: www.hcci.gouv.fr
6. The number of HCCI members dropped from 60 to 45, and there were more representatives of decentralised co-operation in the second mandate than in the first one.

BOX 7.2
FRENCH NGOS

The number of French international solidarity associations is unknown, and this is a challenge in debate. Among the hundreds of thousands of NGOs (Association Law 1901), several thousand have a social objective related to co-operation, the development of poor countries and/or international solidarity. Most of these international solidarity associations derive from six types of initiatives :

- The initiative of public authorities: as with the *Comité français pour la campagne mondiale contre la faim*, which became *Comité français pour la solidarité internationale* (CFSI) or the *Association française des volontaires du progrès* (AFVP), set up in 1963 at the initiative of the French Ministry of Co-operation.
- The initiative of private institutions: Catholic relief (*Secours catholique*) established in 1947, by a decison of the French episcopate, the *Comité catholique contre la faim et pour le développement* (CCFD) set up in 1961, by an initiative coming from different Catholic associations, *Agriculteurs français et développement international* (AFDI) created by agricultural professionals (in particular, FNSEA). The CCFD highlights the major role of the Catholic Church.
- The initative of associations which decided to add international activities to their traditional work – particularly the case with youth movements and popular education.
- The initiative of people with the same job: especially the case with all the without-borders organisations: *Médecins sans frontières, Pharmaciens sans frontières, Vétérinaires sans frontières, Reporters sans frontières...* .
- The initative of a group of private people: the case for most of the international solidarity organisations.
- The initiative of immigrants who have established NGOs in order to support development projects in their home countries.

If we compare French NGOs with the other European NGOs, the French situation is characterised by an extreme dispersion and very low budgets. Only one French NGO belongs to the top five European NGOs. Only two NGOs have an annual budget higher than FF300 million: *Médecins sans frontières* (MSF) and *Médecins du monde* (MDM). Five have a budget of between FF100m. and 200m.: the *Agence d'aide à la coopération technique et au développement* (ACTED), the *Association française des volontaires du progrès* (AFVP), the Red Cross (*Croix Rouge française* only for its international activity), *Première urgence* and *Pharmaciens sans frontières* (PSF).

The average level of official finance for all French NGOs amounts to about 40 per cent of their total resources; official international finance, and more specifically European funds, are the most important. This means that private resources give the associations an unequal autonomy for action and proposition.

Source: Drawn from Hatton [2002].

2. Objectives of French Co-operation Policy

French aid policy on sustainable development, according to the official discourse, gives pride of place to reduction of poverty and inequality, and draws on the seven international objectives for development that the DAC member countries adopted in 1996 and on the political orientations spelled out by Prime Minister Lionel Jospin at the World Bank conference in Paris on 22 June 1999 and by the DGCID, a specialised department of the French Ministry of Foreign Affairs. These objectives were reiterated in September 2000 as part of the United Nations' millennium objectives, and the policy was reconfirmed at the CICID meeting in 2002. Let us look at some of the major characteristics.

First, the question of development is situated within the broader context of globalisation and the need for strict governance of the international financial economy. The word 'regulation' is heard over and over again. The role of the IMF, with its focus on its capacity to anticipate financial crises, and the best pace of financial liberalisation for each individual country, need to be reviewed. Next, the economic, social and health situation in each country in the ZSP must be considered objectively in order to adapt development co-operation to specific situations. A prerequisite will be to set up a solid partnership and to select recipients on the basis of performance and the implementation of economic political reforms. Last, development co-operation should be guided by three priorities: sustainable development, which means strengthening institutions at all levels (regional, national and local), basic education and professional training, and primary health care.

Furthermore, global public goods is a theme that is given major importance in the French Development Agency's (AFD) strategic orientation plan. This is a singular approach, essentially initiated by the UNDP, which France defended at the Monterrey Summit on development financing in March 2002. It is an innovative approach to international co-operation which did not interest the other donors because a) they saw the concept as an empty catchword, a sort of catch-all, and b) it is risky for a donor to suggest support for global public goods via an add-on strategy rather than maintaining traditional ODA, even though the latter is constantly decreasing.

Most documents stress the importance of the participation of beneficiaries in designing local development and micro-financing projects. France is unquestionably one of the donors that has acquired the most experience in this field, thanks to the social development projects it set up after the devaluation of the CFA franc and more recently through its participation in the decentralisation process linked to the creation of local investment funds and support for the implementation of political decentralisation in various countries.

The same is true for the good governance approach. France is playing a very active role in using development co-operation to promote a state governed by the rule of law, which entails both improvement of administrative capacities and local development. Here again, France has acquired considerable experience and know-how in supporting democratic institutions and the electoral process, training magistrates, co-operation to ensure security, training for the police and furthering decentralisation processes. At present, this institutional type of co-operation lacks 'clarity'.

One theme that is often highlighted in France is the fight against poverty and inequalities. The latter is actually becoming the focus of special French attention. This inequalities alleviation approach is not the same, at least in words, as the fight against poverty. The writings of Amartya Sen and Joseph Stiglitz are systematically referred to, and not merely in academic terms. They reflect a strong will to give unique content to this objective. Three aspects are brought to the fore: first, capacity-building for stakeholders to empower them during negotiations with other stakeholders, both public and private; second, building up policies that reduce vulnerability to food and climate risks and the risks of armed conflict; and, third, developing accountability in public affairs, and, even more so, developing the capacity of the local populations or their representatives in civil society to direct budget allocations to priority social sectors. This means that French development co-operation wants to distance itself from the exclusively safety-net approach and from projects designed only to reduce the negative effects of macroeconomic policies. Current policies to stem inequalities mainly emphasise the impoverishment process by looking at the causes of the inequalities. To operationalise this approach, French actors need to negotiate further in the beneficiary countries to ensure the participation of the local population in designing the programmes and in ensuring due consideration of what they, themselves, want. This means considering questions of programme design together with civil society, and the society's legitimacy and representativeness. A corollary may be a slower rate of disbursements. The point to remember is that this consideration of the linkage between projects to fight poverty and inequalities and the political level lies at the heart of the French development co-operation mechanism.

Is there a culture for evaluating co-operation projects and policies? This is an important question, almost a nagging one that has been with us for more than 30 years. France does not have a clear-cut answer. Attention could be drawn to a number of characteristics of the evaluation process. Until 1980, evaluations were very rare and only used for projects. Between 1981 and 1986, the former Ministry of Co-operation worked hard to capitalise on its experiences. Many rural development actions (projects and programmes) were assessed (using the effects method – meaning the evaluation of direct and indirect distribution of added value on different actors) and made public, in particular Freud [*1988*]. But with a change of government, the *Direction de l'évaluation et des études économiques* of the Ministry of Co-operation was gradually deprived of the necessary resources. It was not until 1997 that evaluations were undertaken properly again, a shortcoming pointed out repeatedly by France's peers in the DAC. Since 1997, evaluations have focused more on policies and countries, and are based on the effects of the organisation of the agricultural sector on the fight against poverty. The main lines used previously in fighting poverty are reflected in the orientations of the assessment. In methodological terms, the highly economic approach, centred on the use of the effects method, has gradually been replaced by an analysis of both the production sub-sectors and the institutions that the stakeholders are setting up. In more concise terms, the evaluation systematically repositions the question of institution-building *vis-à-vis* market and state. These institutional analyses, which draw on information on co-ordination mechanisms, are being developed but are still insufficient, in view of the immense amount of knowledge that has been accumulated both at the Ministry of Foreign Affairs and the French Development Agency (see Annex 7.2).

The number of people hired for French technical assistance service is decreasing sharply. In 1996 France was second among the DAC countries (after the US), while in 2000 it was fourth, after the US, Japan and Germany. During the last 40 years France has mobilised some tens of thousands of technical assistants and experts, many more than any other bilateral or multilateral donor. These technical assistants were basically supported by the budgets of the Ministry of Foreign Affairs or the Ministry of Co-operation; the rule was to place technical assistants at the disposal of each beneficiary country within the framework of a co-operation agreement. Theoretically, these countries reimbursed France for part of the total cost. At the beginning of the 1980s the number of French technical assistants was more than 20,000; 50 per cent of them in the 'pays du champ' (and half of these in four countries) and 50 per cent in the 'pays hors champ' (90 per cent of these in the Maghreb). The major part of the technical assistance personnel was engaged in the health sector and in training. The number of technical assistants (and junior experts) financed by the Ministry of Foreign Affairs has declined sharply; now there are only about 500. But, if the number of technical assistants and experts has been reduced considerably during that decade, of greater importance is the nature of technical assistance that was revised in the 1998 reform. What used to be long-term assignments, sometimes viewed as personnel substitution, was replaced by short-term technical assistance, expertise and counselling for governments in fields such as governance, support for civil society, economic and agricultural development, and health care systems.

Does France have a European policy? *A priori*, this question seems quite out of place. If we analyse the French contribution to the 9th European Development Fund (EDF) during the period 2000–2005, it is not negligible: €3.35bn., amounting to 24.3 per cent of the global EDF envelope, substantially higher than the 17 per cent contribution to the Community budget. On the other hand, France insists firmly on its preference for four major political orientations. First, enlargement of the policy dialogue with the introduction of the fundamental near essential elements (human rights, democracy etc.): good governance and the fight against corruption. Second, the design of development strategies on a national level based on consultation with civil society. Third, France insists on the establishment of performance criteria for better management of aid, rejecting the principle of the automatic allocation of aid. Lastly, France considers that the economic agreements (*accords de partenariats économiques réciproques*) planned for 2008 must be signed because of their compatibility with the main guidelines of the World Trade Organisation. It is certain that France has played an important role during the preparation of the new Cotonou Agreement, as at the first Lomé Convention in 1975 with the innovations of Stabex and Sysmin, and more generally on the question of how to tackle the problem of instability of prices and exports. The same implications were observed at the beginning of the 1980s when France introduced the concept of food strategies and new thinking on food aid (under the influence of Edgard Pisani, the former European Commissioner).

But, today, does France support these strategic orientations? Where is the policy consistency? In France an important debate is emerging between advocates of the opening up of commercial relations in 2008 and those who estimate that this liberalisation will generate a competition for which the ACP countries are not all prepared, consequently reducing all the efforts made by the EU in its development co-operation policy. It is the same with the Common Agricultural Policy (CAP), the negative effects of which on agriculture in the South are well known to everyone in France; on this latter topic, an inflecting French position does not seem discernible. On the other hand, as pointed by Y. Dauge in a parliamentary report [*Dauge, 1999*], relations with countries like Chad, the Central African Republic or Togo, to quote only these examples, are in total contradiction with major principles of the Cotonou Agreement: these countries are really wretched partners.

How is it possible to analyse the French attitude *vis-à-vis* the United Nations system? A recent evaluation report [*Ministry of Foreign Affairs, 2000*] came to the conclusion that the UN system was viewed by the French authorities with ambivalence. On the one hand, they recognise a real legitimacy but, on the other hand, they reproach the UN for its limited effectiveness. The UNDP inputs into the intellectual debates on economic development and development co-operation are much appreciated but are not seen as practicable. The UN is considered too heavy and too bureaucratic. And, from a more general point of view, it is perceived as having a diplomatic function and only a smaller role in the development field.

III. CONCLUSION

There are a number of objectives in French development co-operation policy: combating poverty and inequalities but also contributing to restoring the macroeconomic equilibrium; promoting the commercial integration of developing countries, particularly those in the prioriy solidarity zone (ZSP); financing global public goods, especially the fight against AIDS; developing influence-generating co-operation (*coopération d'influence*) with concern for expanding French cultural horizons by facilitating *francophony*; and ensuring involvement in matters of sustainable development. All these priorities have been reasserted during the last few years. There are two important issues at stake: how all these paradigms can be interconnected, and how these objectives can be realised in beneficiary countries and international bodies.

Globally, there seems to be a noticeable, at least oral, shift from development plans – usually designed outside the beneficiary countries as working diagrams for various strategies, each with its own development co-operation policies (sectoral development, balanced development, fight against poverty, etc.) – to an attitude characterised by far greater modesty (*retenue*), analysis and support for the wishes of the local actors. Far more attention is being given to the political side of development than to the technical, mechanical or even technocratic side. The real difficulty lies in translating this approach, which at least in political discourse characterises the new orientation of the French development co-operation policy, into political action. This leads to three questions. The first concerns the feasibility of this model of democracy. Can it serve to legitimate the governments' political orientations and their development choices? The second concerns the real level of priority to be given to these objectives within a broader framework that covers geopolitics, geostrategy and global security (in particular the fight against terrorism). The third concerns the feasibility of the objectives in the light of the international financial, commercial and monetary macrosystems. This leads to a much more fundamental question focusing, *inter alia,* on development financing, the status of aid, the coherence of policies and the construction of national financial systems.

REFERENCES

Billets d'Afrique, monthly journal of the Association Survie, Paris.

Bocquet, D., 1998, *Quelle efficacité économique pour Lomé? Redonner du sens au partenariat entre l'UE et les pays ACP* (Which efficiency for the Lomé convention? Making sense of the partnership between the EU and the ACP countries), collection Les rapports officiels, Paris: La documentation française.

Charasse, M.,1999, *Annexe No. 2 Affaires étrangères: coopération*, rapport général no. 89, Sénat, 25 novembre, Paris.

CICID, 2000, *La coopération française au développement: une refondation* (French co-operation: a rebuilding), Paris: La documentation française.

CRID, 2000, *La réalité de la pauvreté. Politiques de coopération 1999/2000* (The reality of poverty. Co-operation policies 1999/2000), Paris (octobre): Les Cahiers de la solidarité, CRID.

Dauge, J., C. Lefort and M. Terrot, 1999, *La réforme de la co-opération vue de Bangui et N'Djamena: l'urgence d'une explication* (The reform of French co-operation seen from Bangui and N'Djamena: the urgency of an explanation), Paris: Les documents d'information, Assemblée nationale, report no. 1701.

Dauge, Y., 1999, *Le nouveau partenariat UE–ACP: changer la méthode* (The new partnership EU–ACP: changing the method), Paris: rapport d'information no. 1776, Assemblée nationale.

Freud, C.,1988, *L'évaluation de la coopération française* (Evaluation of French co-operation), Paris: Karthala.

Glaser, A. and S. Smith, 1992, 1997, *Les Messieurs Afrique*, Vol. 1 and 2, Paris: Calmann Levy.

Hatton, J.M., 2002, *Les organisations de solidarité internationale* (French international solidarity organisations), Paris: Commission coopération développement.

HCCI, 2000a, Avis au Premier Ministre en vue des débats parlementaires sur la coopération et sur la présidence de l'Union européenne adopté le 18 avril 2000 (Recommendation to the Prime Minister concerning parliamentary debates on co-operation and the French presidency of the European Union), Paris: HCCI.

HCCI, 2000b, *Avis au Premier Ministre sur crises, coopération et développement* (Advice to the Prime Minister on crises, co-operation and development), Paris (novembre): HCCI.

Ministry of Foreign Affairs, 2000, *Evaluation des relations avec les partenaires multilatéraux: cofinancements du ministère de la coopération (1990–1997)* (Evaluation of French relations with multilateral organisations 1990–1997), DGCID collection évaluations no. 43, Paris: Ministry of Foreign Affairs.

Observatoire français de la Coopération internationale (OFCI), *Annual Reports* since 1995 and the last one, *Rapport 2002–2003*, Paris: Karthala (2003).

OECD, 2003a, *Geographical Distribution of Financial Flows to Aid Recipients (1997–2001)*, Paris: OECD.

OECD, 2003b, *The DAC Journal Development Co-operation, 2002 Report*, Paris: OECD DAC.

Tavernier, Y., 1998, *La politique de coopération française au développement* (French development policy co-operation), Paris: La documentation française.

Tavernier, Y., 2000, *Fonds monétaire international et Banque mondiale: vers une nuit du 4 août?* (The IMF and the World Bank: towards a 4 August night?), Paris (decembre): Les documents d'information de l'Assemblée nationale (DIAN) no. 2801.

ANNEX 7.1

LIST OF REPORTS ON FRENCH DEVELOPMENT CO-OPERATION SINCE 1960

Pierre Abelin, 1960, *Les relations entre la France et les pays en voie de développement* (published by Documentation française).

Jean Mersch, 1960, *Les relations entre la France et les pays en voie de développement.*

Jean-Marcel Jeanneney, 1963, *La politique de coopération entre les pays en voie de développement et la France* (published by Documentation française).

Pierre Abelin, 1975, *Rapport sur la politique française de coopération* (published by Documentation française).

Jacques Rigaud, 1979, *Les relations culturelles extérieures.*

Charles Magaud, 1981, *Les orientations de la politique française à l'égard des pays les moins avancés.*

Jacques Berque, 1982, *Recherche et coopération avec le Tiers Monde.*

Ignacy Sachs et Christian Comeliau, 1982, *L'impasse Nord–Sud: quelles issues?*

Yves Berthelot et Jacques de Bandt, 1982, *Impact des relations avec le Tiers Monde sur l'économie française* (published by Documentation française).

Alain Vivien, 1982, *Rapport sur le personnel d'assistance technique et de coopération.*

Claude Freud, 1986, *Une évaluation de l'efficacité de l'aide en longue période* (published by Karthala).

Jerome Cazes, 1987, *L'aide française au Tiers Monde.*

Jean Thill, 1989, *La coopération française et les entreprises en Afrique Sud Saharienne.*

Denis Samuel Lajeunesse, 1989, *Les orientations à moyen terme de notre politique d'aide au développement.*

Stéphane Hessel, 1990, *Les relations de la France avec les pays en voie de développement.*

Alain Vivien, 1990, *La rénovation de la coopération française.*

Bernard Husson, 1990, *Recherche pour la diversification et la contractualisation des rapports entre les ONG et les institutions publiques.*

Serge Michailoff, 1992, *Rapport du groupe de prospective sur la coopération* (published by Karthala).

Pierre Thenevin, 1994, *L'aide française durant la période 1980-90.*

Jean-Paul Fuchs, 1994, *Pour une politique de développement efficace, maitrisée et transparente.*

M. Marchand, 1996, *Une urgence, l'afroréalisme: pour une nouvelle politique de l'entreprise en Afrique subsaharienne* (published by Documentation française).

Jean René Bernard et Anne Paugam, 1996, *L'évolution de l'efficacité de l'APD.*

Bernard Prague, 1996, *Pour une meilleure articulation des relations bilatérales et multilatérales de la France.*

M. Bukspan, 1996, *L'accès des entreprises françaises aux financements multilatéraux.*

Robert Cazenave, 1997, *Les contributions volontaires de la France à l'ONU.*

Sami Nair, 1997, *Rapport de bilan et d'orientation sur la politique de codéveloppement liée aux flux migratoires.*

Jacques Attali, 1998, *Pour un modèle européen d'enseignement supérieur.*

Dominique Bocquet, 1998, *Quelle efficacité économique pour Lomé?* (published by Documentation française).

François Perigot, 1998, *Les relations entre l'Union européenne et les pays méditerranéens: quelle rôle pour la France?*

Patrick Bloche, 1998, *La présence internationale de la France et la place tenue dans la société de l'information.*

Yves Tavernier, 1998, *La coopération française au développement* (published by Documentation française).

Yves Dauge, 1999, *Le nouveau partenariat pour le développement des pays ACP* (published by Assemblée nationale).

Catherine Lalumière et Jean-Pierre Landau, 1999, *Rapport sur les négociations commerciales multilatérales.*

Jean Nemo, 1999, *Recherche et pays en développement.*

Jean Nemo, 1999, *Evolution des dispositifs d'assistance technique.*

Ministère des finances, 1999, *Rapport au Parlement sur les activités du FMI et de la Banque mondiale.*

Jeanine Cayet, 1999, *Prospective pour un développement durable: quelle politique de coopération avec les pays de l'UEMOA?*

These reports are available at the Ministry of Foreign Affairs or published by Karthala or La Documentation française as quoted.

ANNEX 7.2
EX-POST EVALUATIONS AND STUDIES FINANCED BY THE
MINISTRY OF FOREIGN AFFAIRS – DGCID – DURING THE PERIOD
2000–2003

Evaluations

Évaluation de 'Ville.Vie.Vacances-Solidarité internationale' et de 'Jeunesse-Solidarité internationale' (Town, life, holidays-international solidarity), 2003.

Evaluation de l'aide française dans le secteur pharmaceutique (Evaluation of French aid in the pharmaceutical sector), 2003.

Évaluation de la formation des personnels de santé en Afrique subsaharienne et à Madagascar de 1987 à 1998 (Evaluation of the staff training in the health sector in sub-Saharan Africa from 1987 to 1998), 2003.

Évaluation des actions de l'aide française dans le secteur agricole et l'environnement à Madagascar (Evaluation of French aid in the agricultural and environmental sectors in Madagascar), 2003.

Évaluation rétrospective des programmes d'actions intégrées (PAI) dans le domaine de la recherche avec le Maroc (Retrospective evaluation of integrated programmes in the field of research with Morocco), 2003.

Evaluation des filières francophones implantées dans les pays d'Europe centrale et orientale et dans les nouveaux Etats indépendants (Evaluation of the francophone channels in ECEC and in NIS), 2003.

Pôle régional de recherche appliquée au développement des savanes d'Afrique centrale (PRASAC) (Regional applied research centre for the development of savannas in Central Africa (PRASAC), 2003.

L'aide française à Djibouti (French aid to Djibouti), 2002.

Formation des personnels de santé en Afrique sub-saharienne et à Madagascar (Health staff training in sub-Saharan Africa and Madagascar), 2002.

Programmes prioritaires Palestine et Vietnam (Priority programmes in Palestine and Vietnam), 2002.

Programmes de lutte contre le VIH/SIDA (Programmes against HIV /AIDS), 2002.

Programme développement local et migration au Mali et au Sénégal (1991–1998) (Local development programme and migration in Mali and Senegal (1991–1998)), 2001.

Projet Santé Abidjan – Évaluation des formations sanitaires urbaines à base communautaire (Health project in Abidjan. Evaluation of a urban health training on communautarian basis), 2001.

Projet ARCHES (ARCHES project), 2001.

Secteur santé au Cambodge (The health sector in Cambodia), 2001.

Programme CAMPUS – Afrique et Madagascar (1986–1998) (CAMPUS programme – Africa and Madagascar (1986–1998), 2000.

Politique française d'aide dans le secteur Jeunesse et Sports dans les pays de l'ancien champ (1987–1997), (French aid policy in the youth and sports sectors in the former pays du champ (1987–1997), 2000.

Aide française dans le secteur hospitalier – Afrique subsaharienne et Madagascar (1987–1996) (French aid in the hospital sector – sub-Saharan Africa and Madagascar (1987–1996), 2000.

Studies

Modes d'organisation des filières et lutte contre la pauvreté – Les cas du coton et du cacao (Main forms of organisation in the cocoa and cotton sectors and the impact of these forms on poverty), 2003.

L'état des sciences en Afrique (Inventory of sciences in Africa), 2002.
Analyse comparative des processus d'intégration économique régionale (Comparative analysis of regional integration processes), 2002.
Performances commerciales, compétitivité et diversification des économies subsahariennes (Trade performances, competitiveness and diversification of sub-Saharan countries), 2001.
Étude comparative des dispositifs d'analyse économique en Afrique (African capacity in economic policy advice : actors and system), 2001.
La mobilisation des ressources locales au niveau des municipalités d'Afrique subsaharienne (Mobilisation of local resources at the municipalities levels in sub-Saharan Africa – English version), 2001.
Actes du séminaire sur la prospective africaine (Proceedings of the seminar on African prospects), 2001.
Filières agroalimentaires en Afrique – Comment rendre le marché plus efficace? (Agricultural sectors in Africa – How to improve market efficency? – English version), 2001.
Mise en oeuvre des accords multilatéraux sur l'environnement: quels instruments économiques pour les pays de la ZSP? (Implementation of multilateral agreements on the environment: which economic tools for the ZSP? – English version), 2001.
Droits fonciers délégués en Afrique de l'ouest: reconnaître et sécuriser les procédures (ground laws in West Africa: how to secure the procedures), 2001.
Le risque maladie dans les assurances sociales: bilan et perspectives dans les PVD (Sickness risk in social insurance – English version), 2001.
Formation professionnelle et investissement (Vocational training and investment – English version), 2001.
Compétitivité des productions animales en Afrique subsaharienne et à Madagascar (Animal production competitiveness in sub-Saharan Africa and Madagascar), 2001.
Les apprentissages en milieu urbain (Training in urban areas), 2001.
Formation professionnelle dans le secteur informel en Afrique (Vocational training in the informal sector in Africa), 2001.
Le rôle des pays prescripteurs sur le marché et dans le monde de l'art contemporain (The role of stipulate countries on the contemporary art market), 2001.
Développement: 12 thèmes en débat (Development: 12 themes in debate), 2000.
Bilan de la prospective africaine – volume 1 (Review of the African prospects), 2000.
Lutte contre la pauvreté et les inégalités: synthèse de l'étude bilan sur les actions de la coopération française (Fight against poverty and inequalities: review of French co-operation), 2000.
Problématique de trois systèmes irrigués en Afrique (Problematic of three irrigated systems in Africa), 2000.
Prospective de l'agriculture irriguée en Afrique subsaharienne (Prospects of irrigated agriculture in sub-Saharan Africa), 2000.
L'analyse économique des filières agricoles en Afrique subsaharienne (Economic analysis of agricultural sectors in sub-Saharan Africa), 2000.

The evaluations and studies are available online http://www. diplomatie.gouv.fr. Some of them are translated into English, this being noted in the list. Each of them has a summary in English.

8

Germany's Development Co-operation Policy since the Early 1990s: Increased Conceptual Ambitions in Times of Severe Financial Constraint

GUIDO ASHOFF

Germany is one of the major bilateral donors, and its policy of development co-operation has found itself in an increasingly paradoxical situation since the early 1990s. While its conceptual approach has clearly become much more comprehensive and ambitious, its net disbursements of official development assistance (ODA) have experienced an unprecedented decline in both real terms and as a share of gross national income (GNI).[1] In spite or because of the resource constraints, the past decade and especially the last few years have witnessed a number of reforms. Yet German development co-operation still faces considerable challenges if its new objectives are to be achieved.[2]

I. MOTIVES (JUSTIFICATIONS) AND OVERALL OBJECTIVES OF GERMAN AID POLICY

1. The Basic Question: Development Co-operation and German Interests

German development co-operation reflects a complex mix of motives that has changed over time.[3] The German Federal government has

---

1. Between 1990/1 and 2000/2001, Germany's net ODA disbursements declined in real terms (at 2000 prices and exchange rates) from $6,085m. to 5,050m. and as a share of GNI from 0.40 to 0.27 per cent. In 1980/1, the corresponding figures had been $5,064m. and 0.45 per cent [*OECD, 2003: 237*]. See also Table 8.1.
2. This chapter follows the structure proposed by the convenors of the present volume in order to ensure comparability with the other chapters. As a consequence, however, several inter-related aspects have to be dealt with in different sections to comply with the overall structure.
3. The following sections can give only a very rough idea of the mix of motives. For more detailed overviews see e.g. Bodemer [*1974*], Hofmeier and Schultz [*1984*], Claus and Lembke [*1992*] and Nuscheler [*1995*].

always regarded development co-operation as part of its overall foreign relations. At the same time, development policy, like every other government policy, is subject to the constitutional imperative 'to benefit the German people and avert danger from them' (wording of the official oath every minister has to swear on taking office), and is therefore obliged to serve German interests. For development policy it is the definition of German interests that matters. They can relate either to overriding goals such as securing peace or our common future, or to specific foreign policy, export, employment or other domestic interests. The weight and influence of such interests have changed in the course of time, depending on a number of factors such as the domestic and international political and economic situation, the political 'colours' of the governing coalitions, the public debate on development co-operation, and, last but not least, the role played by the development co-operation ministry *vis-à-vis* other government departments.

2. Motives Underlying German Aid Policy

(i) Humanitarian (altruistic) motives. Part of the justification of German aid has traditionally originated from humanitarian or altruistic motives together with the acknowledgement of responsibility that Germany has to take as one of the richest nations in the world and a country which itself benefited considerably from foreign aid under the Marshall Plan during its post-war reconstruction.[4] The humanitarian strand of aid justification is reflected not only in the declared goal of German aid,[5] but also, for instance, in the fact that among the partners of German development co-operation there continue to be quite a number of countries, particularly least developed ones in sub-Saharan Africa (see Table 8.3 and Box 8.5), that are of no major foreign policy or economic interest to Germany. Another example is that over time the Federal Ministry for Economic Co-operation and Development (BMZ) has channelled a rising proportion of its aid disbursements through NGOs (including the churches).[6]

---

4. In the late 1950s and early 1960s, it was also pressure from the United States that 'reminded' the German government of the need for burden-sharing and growing responsibility in the common aid efforts intended to serve overall Western interests in the Cold War period [*Bodemer, 1974: 33–44*].
5. See, for example, the still valid Basic Principles of the Federal government's Development Policy of 1986, according to which 'the aim of German development policy is to improve the economic and social situation of the people in developing countries' [*BMZ, 1986: 20*].
6. The share of the BMZ's aid disbursements channelled through NGOs (percentages related to the churches in parentheses) was 5.9 (3.4) per cent in 1970, 6.2 (2.7)

(ii) Foreign policy interests. In the 1950s and 1960s, when Germany was striving for international recognition,[7] the foreign policy interest in image-building and making friends worldwide prompted the government to extend development co-operation to almost every developing country, thus applying what critics have labelled a 'watering-can approach'. In addition, the Hallstein Doctrine prescribed the termination of diplomatic relations and development co-operation with countries that officially recognised the former East Germany. As a consequence, aid to several countries was stopped while others were prevented from recognising East Germany. In the 1970s, this massive influence on aid policy decreased, but foreign policy continued to leave its traces on development co-operation. Examples are the country allocation of aid (large amounts of aid per capita provided to Israel, for example, because of the legacy of the Third Reich, to Turkey and Egypt for geopolitical reasons or to the Palestinian Administered Areas as a contribution to solving the Middle East conflict), the continuation of development co-operation with China in the late 1980s and early 1990s despite the latter's deteriorating human rights record [*Ashoff, 1999a: 135–6*], and the still persisting divergence of opinion between the Foreign Office and the BMZ on the adequate representation of the German aid administration in recipient countries (see section VI below).

(iii) Foreign trade and other economic interests. From the very beginning, development aid was also regarded as a suitable instrument for supporting Germany's efforts to conquer export markets. According to an early argument in favour of aid, 'today's development co-operation partners are Germany's customers of the future'. The export promotion interest was reflected in the sectoral allocation of aid (high shares of transport, energy and industry in German bilateral aid up to the early 1990s; see Table 8.2) and the relatively high share of aid tied to German supplies until the mid-1990s when tying became more difficult as a result of the Helsinki Consensus of 1992 among OECD

---

per cent in 1980, 9.2 (3.5) per cent in 1990, 10.7 (3.8) per cent in 2000 and 11.2 (3.8) in 2001 [*BMZ, 2002: 375; www.bmz.de*].
7. Germany had regained its full sovereignty only in 1955 and was not yet a member of the United Nations at that time (it joined the UN – together with former East Germany – in 1973).

members on tied aid.[8] Securing domestic jobs became another argument in favour of tied aid, particularly in the 1980s, when the BMZ, against the background of rising domestic unemployment, required project proposals to be assessed not only in development policy terms but also in terms of possible job creation in Germany.[9] Although the scope for directly linking export and labour market interests with ODA has diminished, economic self-interest as a motive of aid policy has not disappeared. For example, in the last few years Germany has adopted a pro-untying position in OECD negotiations on further untying of aid with a view to getting other major donors to open up further their restricted aid procurement to international competition and thus to possible contracts for German suppliers [*Ashoff, 2000a: 59*].

(iv) Overriding political and developmental goals. The ultimate justification of German development co-operation has traditionally been to form part of the government's overall policy to serve the cause of security, international peace and stability.[10] Until 1989 the idea of peace was inseparably linked with the East–West conflict. Not surprisingly, the Basic Principles of the Federal government's Development Policy of 1986 [*BMZ, 1986: 21–2*] regarded 'development co-operation with Third World countries as an instrument for promoting a policy of genuine non-alignment (...); consequently the Federal Republic is opposed to any attempt to carry the East–West confrontation over into the Third World'.

After the end of the Cold War, the term 'security' emerged again as a key word in speeches by the BMZ Ministers and their deputies (for example, 'more security by development', 'securing our future' or 'development co-operation for global human security'). The reason is obvious: the disappearance of the traditional security of the Cold War

---

8. The share of tied aid in German bilateral aid (commitments, excluding technical co-operation and administrative costs) declined from 52.6 per cent in 1990 to 39.7 per cent in 1995 and 15.4 per cent in 2001 [*OECD, 1992: A44; 1998a: A50; 2003: 273*].

9. In the 1980s, with a view both to mobilising additional resources mainly for large infrastructure and industrial projects and to securing German exports and jobs, the German authorities created a 'mixed financing' scheme under which the Kreditanstalt für Wiederaufbau (KfW) combined funds from the aid budget with its own funds raised on capital markets and extended the blend as ODA loans. These loans were tied to German supplies because the market component was guaranteed by the German government through the Hermes credit insurance. Due to the tightened 'Helsinki disciplines', the volume of mixed loans dropped markedly in the 1990s [*OECD, 1998b: 44*].

10. In 1966, the first Development Co-operation Minister Walter Scheel declared [*1966: 5–6*]: 'Development policy is security policy', intended to contain the influence of communism in the Third World. In the 1970s and 1980s, development policy was considered to be 'part of the government's overall policy which serves the cause of international peace and stability' [*BMZ, 1986: 21*].

based on nuclear deterrence gave rise to new threats in the form of regional and ethnic conflicts to which other transnational threats such as environmental degradation, population growth, poverty and disease, violations of human rights and drug trafficking had to be added. Hence a former BMZ Under-Secretary of State concluded [*Repnik, 1993: 788*]: 'Our security is more closely linked with the future of the developing countries in political, economic, ecological and social terms than ever before.'[11]

Against this background, and in view of globalisation, the BMZ has increasingly conceived of development co-operation as part of a 'global structural policy'. The essentially new feature in the justification of development co-operation is that it is no longer confined to improving living conditions in the partner countries in the South and East, but has assumed the additional mandate to contribute to safeguarding our global future by helping to shape global framework conditions in such areas as the environment, trade, the international financial architecture and the prevention and settlement of conflicts. The terrorist attacks of 11 September 2001 prompted an intensive debate on the role of development policy as part of a comprehensive peace and security approach in view of the growing terrorism and privatised violence.[12]

3. Increased Developmental Profile of German Aid Policy

The evolution of German development co-operation throughout the past four or five decades has witnessed an increasing influence of the BMZ on the conceptual framework, policies, programmes and country allocations of German aid. Of course, the influence of other government departments, particularly the Foreign Office, the Ministry of Economic Affairs and the Ministry of Finance, has not disappeared, as may be seen, for instance, from the severe constraints that have affected the aid budget in the past decade or the BMZ's ongoing efforts towards enhanced policy coherence (for both aspects, see below). It is fair to say, however, that development co-operation nowadays has a profile of its own at both the conceptual and implementation levels to a much greater extent than in previous decades.

---

11. However catchy the term 'security' is, it should be noted that it is open to both an offensive and a defensive interpretation. The former refers to 'global human security', that is, oriented towards sustainable development and survival in the one world and therefore also requires changes in the North. In contrast, the defensive understanding of 'security' considers development co-operation as part of an overall policy whose aim is to prevent the conflicts and crises in the South and East from spilling over to the North.

12. For a conceptual answer of the BMZ to the new threats see Wieczorek-Zeul [*2002*].

The reasons are manifold: the increased developmental professionalism of the aid administration, the continued existence of the BMZ as a separate development co-operation ministry with full Cabinet rank,[13] the changed international development agenda (for example, the consensus among donors and with partner countries on the basic development goals), the greater discipline through donor co-ordination (for example, concerning tied aid), the critical support for the cause of aid on the part of NGOs, and, last but not least, the leadership by the BMZ and development politicians in the governing coalitions.[14]

Compared with other donors, when it comes to the relationship between aid policy and specific self-interests, Germany stands midfield. It has never subordinated its development co-operation policy to geopolitical, foreign policy or security interests with the blatancy shown by the United States, nor as, for instance, Japan had done in the 1980s and early 1990s, to foreign trade and investment interests.[15] On the other hand, Germany has never sought a developmental profile as distinct as various 'like-minded countries', for example, some Scandinavian countries and the Netherlands, have achieved (for instance, by convincing their parliaments and taxpayers to mobilise ODA volumes well above the internationally agreed target of 0.7 per cent of GNP).

II. STRATEGIES, MAJOR PRINCIPLES AND GUIDELINES

The conceptual framework of German development co-operation has been modified or reoriented in several ways since the early 1990s. Especially in the last few years, there were a number of major policy

---

13. The BMZ was founded in 1961 (several years after German development assistance had started) because of co-ordination problems among the government departments managing the various components of German aid at that time, and because of 'coalition arithmetic' (a Cabinet post had to be found for the leader of the smaller partner in the new governing coalition). In the course of time, the BMZ's existence was repeatedly questioned, for example, during the election campaigns at the federal level in 1994, 1998 and 2002 when representatives of the parliamentary opposition called for merging the BMZ with the Foreign Office.

14. To give just two examples: In 1991 the BMZ, led by Development Co-operation Minister Carl-Dieter Spranger, established five political criteria for decisions on aid allocation (see below) against some reluctance from the Foreign Office. In the coalition agreement of the first 'red–green' coalition, which took office in 1998, the development policy wings of the Greens and the Social Democratic Party succeeded in strengthening the BMZ's competences in inter-ministerial co-ordination (see the section on policy coherence below).

15. For an overview of the criticism levelled against Japan's aid policy up to the early 1990s see e.g. Förster [*1994: 9–11*] and Nuscheler [*1996: 134–5*].

initiatives, two of which will be dealt with more specifically (for further initiatives and decisions see section V): the Programme of Action 2015 for Poverty Reduction approved by the Federal Cabinet in 2001 and the efforts to enhance coherence between development co-operation and other policies affecting developing countries.

1. Changes in the Conceptual Framework since the Early 1990s

The declared goal of German development co-operation – to support economic and social development in the recipient countries by help for self-help, and thereby to contribute to improving living conditions – has basically not changed since it began (see footnote 5 above). The concepts and policies for pursuing this goal have, of course, changed repeatedly in the past because of both growing experience and new challenges.[16] Since the early 1990s, there have been major changes in several respects.

(i) Political conditionality and 'positive measures'. The first major change concerned the shift from the economic conditionality of the 1980s to political conditionality and 'positive measures' to support human rights, participation and good governance. In October 1991, the BMZ established five criteria for decisions with regard to the volume, instruments and areas of co-operation with individual countries: (i) respect for human rights, (ii) involvement of the population in the political decision-making process, (iii) guarantee of the rule of law, (iv) creation of a market-friendly economic system, and (v) the development commitment of government activities. While the last two criteria are closely related to the idea of structural adjustment, the first three are essentially political. Development co-operation has thus admittedly become more political [*Waller, 1994*]. Though the importance not only of economic but also of political framework conditions which are conducive to development had already been recognised, only the collapse of socialism and the wave of democratisation attempts in the South and East have given development co-operation the chance to become more independent of geopolitical considerations

---

16. German development co-operation has gone more or less all the way with the changes in international development concepts over the years. Expressed in very simple terms, these saw a strong emphasis on economic infrastructure and education in the 1960s; on satisfying basic needs in the 1970s; on structural adjustment, policy dialogue and private sector development in the 1980s. The change in the concepts, which the German aid administration regarded as sometimes too rapid, did not lead every time to fundamental changes in developmental practice, but rather to gradual adjustments [*Ashoff, 1996: 30*].

and enlarged the scope for linking development assistance with the political situation in the partner countries.

(ii) Crisis- and conflict-oriented development co-operation. Because of the increasing number of countries which find themselves in violent conflicts, German development co-operation enlarged the scope of its action in two steps. Initially it was recognised that development assistance had to go beyond immediate humanitarian and emergency aid and pave the way for the resumption of 'classical' development assistance once conditions had normalised. Such 'crisis-oriented development co-operation' focuses on issues such as the quick rehabilitation of the infrastructure, the demobilisation and reintegration of former combatants, the supply of urgently needed goods to restart agricultural production, and educational and training programmes for refugees to give them a longer-term perspective. Since the mid-1990s, this approach has been enlarged to cover 'crisis prevention and conflict settlement' [see *Klingebiel, 2001*]. Efforts in this respect include the development of an indicator model to identify the tendency for conflicts to occur in partner countries, the integration of conflict analysis in project planning and management, evaluation of the role of German development co-operation projects in conflict situations, the creation of a Civil Peace Service [see *Wilhelm, 2002*], and the BMZ taking its seat in the Federal Security Council which is responsible for basic decisions on German arms exports.

(iii) Poverty alleviation. Poverty orientation was given more prominence in the justification and conceptual orientation of German development co-operation, as can be illustrated by two milestones. In October 1992, the BMZ presented its concept of the 'Main elements of poverty alleviation' [*BMZ, 1992*], which defined three approaches: (i) structural reforms aiming at eliminating the causes and not merely the manifestations of poverty, (ii) direct poverty alleviation in favour of poor target groups, and (iii) indirect poverty alleviation (projects without an identifiable group of beneficiaries, but incorporated into poverty-oriented sectoral policies). The second milestone is the Programme of Action 2015 for Poverty Reduction approved by the Federal government in April 2001 (see below).

These are but three changes to which more need to be added, for example, greater concern for the protection of the environment and the sustainable use of natural resources (to be promoted essentially by capacity-building projects) and increasing efforts to support private-sector development in the partner countries (by seeking, *inter alia,* the co-operation of German private firms through 'public–private partnerships').[17]

The conceptual framework has thus become clearly much more comprehensive. In its Eleventh Report to Parliament on Development Co-operation of May 2001, the Federal government characterised its development co-operation by a holistic approach involving four dimensions [*BMZ, 2001a: 62*]: (i) social justice (framework conditions that help reduce poverty and social inequality), (ii) economic performance (poverty-reducing growth and economic co-operation), (iii) political stability (peace, democracy, human rights and gender equality), and (iv) ecological balance (preserving natural resources as the foundations of life).

2. Programme of Action 2015 for Poverty Reduction

In response to the internationally agreed target of halving by 2015 the proportion of people living in extreme poverty[18] (people who have less than one dollar per day),[19] the German government adopted in April 2001 its Programme of Action 2015 for Poverty Reduction [*BMZ, 2001b*]. Cabinet approval of the Programme constitutes a milestone in the conceptual framework of German development co-operation in two respects:

– Although poverty reduction had already been defined as the overall task of German development assistance more than a decade previously, the BMZ had for several years been reluctant to accept outcome targets such as the seven International Development Goals (IDGs), the predecessors of the Millennium Development Goals (MDGs),[20] of which halving the proportion of people

---

17. For a balance of the achievements of the public–private partnership programme since 1999 see Rabe [*2002*].
18. This target was promulgated by the UN General Assembly Special Session on Copenhagen+5 (held in Geneva in 2000) and reaffirmed by the heads of state and government at the UN Millennium Summit (September 2000) on the basis of the World Social Summit in Copenhagen in 1995.
19. People in extreme poverty are mainly living in developing countries, including Eastern Europe and Central Asia. A total of 29 per cent of the population in those countries were extremely poor in the reference year, 1990; this is the proportion to be halved. In terms of numbers, 1.3 billion people were living in extreme poverty [*BMZ, 2001b: 3*].
20. In the document *Shaping the 21st Century: The Contribution of Development Co-operation* ('S 21') of 1996, the members of the OECD's Development Assistance Committee (DAC) agreed to contribute to the achievement of seven international development goals adopted at previous UN conferences. The IDGs were reaffirmed in the UN Millennium Declaration of 2000 and reformulated as 'millennium development goals' in the 'Road map to the implementation of the United Nations

living in extreme poverty by 2015 is the first and most important one, as an explicit frame of reference for decisions on its aid policy.[21] Against this background, the preparation and adoption of the Programme of Action 2015 indicate a remarkable change.

- The Programme of Action 2015 is not just a BMZ document but was approved by the Cabinet. It not only sees poverty reduction as a task for development co-operation but also calls for a coherent approach by the environmental, agricultural, trade, economic, finance, science and technology policies [*BMZ, 2001b: 7*]. The BMZ now has an important chance to urge other government departments to share responsibility for development policy.

The Programme defines ten priority areas for action (see Box 8.1) and within each area indicates specific actions to be taken (in total, the Programme comprises 75 actions). The individual measures for implementation relate to three levels of action: international structures, structures in partner countries, and structures in Germany, Europe and other industrialised countries.

BOX 8.1
PRIORITY AREAS FOR ACTION TO BE TAKEN BY THE GERMAN GOVERNMENT IN IMPLEMENTING ITS PROGRAMME OF ACTION 2015 FOR POVERTY REDUCTION
Boosting the economy and enhancing the active participation of the poor Realising the right to food and implementing agrarian reform Creating fair trade opportunities for the developing countries Reducing debt – financing development Guaranteeing basic social services – strengthening social protection Ensuring access to vital resources – fostering an intact environment Realising human rights – respecting core labour standards Fostering gender equality Ensuring the participation of the poor – strengthening good governance Resolving conflict peacefully – fostering human security and disarmament
Source: BMZ [*2001b*].

Millennium Declaration' prepared by the UN Secretariat in 2001. For further information see OECD [*2002a: 125–31*].

21. For instance, during the deliberations in the DAC on 'S 21', Germany was not a supporter of these targets on the grounds that partner countries would be mainly responsible for their achievement and development co-operation would take too much blame for any failures. While finally agreeing to the adoption of the document in the DAC, the BMZ did not give it much prominence in the next few years [*Ashoff, 2000b: 4*].

The Programme does not include prioritisation of countries, sectors and instruments and is yet to be operationalised. A detailed implementation strategy is currently being developed and has to involve the other relevant government departments. Implementing the Programme will require a more precise definition of priorities as regards countries, sectors and instruments as well as further co-ordination with other donors and with the partner countries especially in the framework of poverty reduction strategy papers (PRSPs) [*OECD, 2001: II–16*].

3. Efforts to Enhance Policy Coherence

The issue of policy coherence has played a growing role in the development policy debate in Germany since the early 1990s for a number of reasons:[22]

- First, there were several cases of manifest incoherence between development co-operation and other policies (such as trade, agricultural, fisheries and arms export policies as well as fiscal policy as far as the treatment of corruption was concerned).[23]
- The Treaties of Maastricht (1992) and Amsterdam (1997) first enshrined in law the requirement of coherence between development co-operation and other policies. Even though the relevant article[24] formally applies only to the Community and not to the member states (which are, however, required by Article 10 to act in the Community's best interest) and is worded in rather guarded terms, it represents an important point of reference.
- The real decline of German ODA increased the pressure to improve the efficiency and effectiveness of development co-operation and, to this end, to ensure greater policy coherence.
- As mentioned before, the Programme of Action 2015 explicitly calls for a coherent approach of quite different policies with a view to effectively contributing to halving extreme poverty by 2015.

The various cases of policy incoherence were criticised not only by development co-operation institutions but also by NGOs and the churches and by the development policy committees of the Christian

---

22. This section largely draws on Ashoff [*2002*].
23. For a detailed discussion of these cases of policy incoherence see Ashoff [*1999a*].
24. Article 178 of the Treaty of Amsterdam: 'The Community shall take account of the objectives [of its development policy] in the policies that it implements which are likely to affect the developing countries.'

Democratic Union (CDU), the Social Democratic Party (SPD) and the Greens, especially before the federal parliamentary elections in 1994 and 1998, and were seen as a reason for proposing various improvements in coherence (see Box 8.2). Regarding this process, two things are noteworthy:

– Development policy initiatives, NGOs and the churches made a wider public aware of the incoherence in the case of the EU's subsidised beef exports and fisheries agreements, fiscal policy (corruption) and arms exports by holding hearings and conducting information campaigns (some at European level) and so exerted political pressure, which helped to ensure that adjustments were made (for example, reduction of beef export subsidies in the mid-1990s).
– The proposals for greater coherence referred to in Box 8.2 all primarily addressed the organisation of decision-making on policy (areas of responsibility). Important though this aspect is, it should not be given absolute primacy over the other causes of incoherence, as this may arouse unjustified expectations.[25] What is ultimately decisive (besides solutions to the problems of analysis and information) is the political will for greater coherence that results from the play of political forces.

After the present 'red–green' coalition took office for the first time in late 1998, several remarkable, mostly institutional, steps (in addition to the adoption of the Programme of Action 2015) were taken to achieve greater policy coherence (see Box 8.3). Against this background, efforts to further improve coherence should start by exploiting to the full the BMZ's extended opportunities for taking action. Whether the BMZ actually requires further areas of responsibility to improve coherence should be carefully considered as and when necessary.

---

25. The main causes of policy incoherence are: (i) divergent political interests, complicated by (ii) different areas of responsibility at national and EU level, (iii) partner countries' failure to take countermeasures (e.g. by protecting their agriculture against subsidised EU farm exports with countervailing duties), (iv) deficiencies in the organisation of decision-making on policy, (v) information deficits, (vi) complexity of the development process. For further information see Ashoff [*1999a: 131–4*].

> *BOX 8.2*
> EARLIER PROPOSALS FOR GREATER POLICY COHERENCE IN GERMANY
> - Strengthening development policy through a separate law, as in several other donor countries
> - Creating a 'development cabinet' chaired by the Chancellor and involving the relevant government departments, to ensure greater coherence among the various policies to the benefit of development
> - Strengthening the BMZ's co-ordinating and monitoring role within the Federal government (e.g. by means of guidelines on sustainable development to be used by the BMZ to monitor other policies)
> - Increasing the BMZ's areas of responsibility:
>   - *Variant I:* Transferring to the BMZ other departments' responsibilities relating to development co-operation (e.g. humanitarian aid, currently a Foreign Office responsibility)
>   - *Variant II:* Upgrading the BMZ to a 'Ministry for International Co-operation and Sustainable Development', embracing pivotal aspects of international structural policy (e.g. debt and raw materials policies, Rio follow-up process and international agricultural and technology policies)
> - Integrating the BMZ into the Foreign Office to give development co-operation more weight in foreign policy and better representation in partner countries
>
> Source: Ashoff [2002: 3].

In general, however, it is true to say that development policy is not just a matter for the BMZ, but a cross-sectoral concern of many policies. The BMZ should not therefore try to take on the tasks of other policies and so overstretch itself, but urge and help other government departments to share responsibility for development policy.[26] In 2001, the BMZ started talks with other federal ministries on issues of policy incoherence. Achieving more coherence will require the BMZ:

– to increase its analytical and bargaining capacity in order to identify cases of incoherence very early and to develop a well-founded strategy for negotiations with other ministries;
– to pursue an intensive coherence-related dialogue with other government departments and at EU level;
– to mobilise additional political support (within the German and European Parliaments, by NGOs and research institutes); experience shows that skilled analyses, public relations work and political initiatives by parliamentarians, NGOs and the media may be important allies of development policy in efforts to achieve greater coherence.

---

26. In addition, it is not very realistic to expect other government departments to transfer major responsibilities to the BMZ. This was clearly confirmed by the coalition agreement signed by the 'red–green' federal government after its re-election in late 2002. In contrast to the coalition agreement of 1998, the new one, while recognising the international responsibility of many German policies, did not transfer any new responsibilities to the BMZ. After some disputes between the BMZ and the Foreign Office, the agreement stated: 'Development policy is an independent element of German foreign policy.'

> *BOX 8.3*
>
> CHANGES IN RECENT YEARS TO IMPROVE POLICY COHERENCE IN GERMANY
>
> – In July 2000, as part of the Joint Standing Orders (GGO) of the federal ministries, the BMZ was granted an extended right to examine legislation planned by other ministries for its impact on development. While under the old GGO the BMZ did not receive other departments' proposals for legislation until late in the process, it is now involved at an early stage and itself considers whether development policy interests are affected.
> – Transfer to the BMZ of further responsibilities relevant to development co-operation (e.g. EU development co-operation and the TRANSFORM Programme for co-operation with Central and Eastern European countries).
> – Inclusion of the BMZ in the Federal Security Council, which is responsible, among other things, for principles and decisions relating to arms export policy.
> – Development policy interests taken into account in the amended Political Principles for Arms Exports and the Guidelines for Granting Export Guarantees ('Hermes Guarantees').
> – Creation of the Task Force 2015 within the BMZ, which, among others, is responsible for co-ordinating and monitoring the implementation of the 75 actions to be taken by the various government departments, and appointment of contact staff for the Programme of Action in each department.
>
> *Source*: Ashoff [2002: 2–4].

III. VOLUME TARGETS

In terms of absolute ODA volumes, Germany is one of the largest bilateral donors, ranking third after the United States and Japan in 2001 and 2000 and fourth after Japan, the US and France from 1997 to 1999. But in terms of its ODA/GNI ratio of 0.27 per cent in 2001, it ranked only thirteenth among the 22 members of the OECD's Development Assistance Committee [*OECD, 2003: 228–9*].

In contrast to its conceptual shift to more ambitious goals in the 1990s, Germany's ODA/GNI ratio, which had reached its peak in 1982 at 0.48 per cent, steadily declined during the past decade from 0.42 per cent in 1990 to 0.27 per cent in 2001 (see Table 8.1). While it is true that Germany has not been the only donor showing declining ODA/GNI ratios, the drop in its ratio was disproportionately large, however, when measured against the unweighted DAC average (1990/91: 0.47 per cent; 2001: 0.40 per cent [*OECD, 2003: 229*]). German ODA had become a victim of budget constraints, due to several factors:

TABLE 8.1
LONG-TERM QUANTITATIVE TRENDS IN GERMAN DEVELOPMENT CO-OPERATION, 1960–2001
(€m.)

	1960	1970	1980	1990	1995	2000	2001
Net ODA[a] disbursements	480.0	1,126.3	3,311.2	5,221.9	5,515.4	5,458.1	5,571.3
% of GNP/GNI[b]	(0.31)	(0.33)	(0.44)	(0.42)	(0.31)	(0.27)	(0.27)
bilateral net ODA	353.2	872.2	2,157.1	3,700.9	3,529.5	2,915.3	3,186.1
% of total net ODA[c]	(73.6)	(77.4)	(65.1)	(70.9)	(64.0)	(53.4)	(57.2)
multilateral ODA (incl. EU)	126.8	254.1	1,154.0	1,521.0	1,985.9	2,542.9	2,385.2
% of total net ODA[c]	(26.4)	(22.6)	(34.9)	(29.1)	(36.0)	(46.6)	(42.8)
contributions to the EU[d]	-	126.6	440.8	638.6	1,145.8	1,342.0	1,275.3
other multilateral ODA	-	127.4	713.2	882.5	840.1	1,200.9	1,109.9
Gross ODA[a] disbursements	-	1,303.6	4,430.7	6,991.7	6,512.5	6,299.1	6,547.1
bilateral share (%)[c]	-	(80.8)	(74.0)	(78.2)	(69.5)	(59.6)	(63.6)
multilateral share (incl. EU) (%)[c]	-	(19.2)	(26.0)	(21.8)	(30.5)	(40.4)	(36.4)
BMZ budget[e] (gross disbursements)	-	1,031.8	2,761.9	4,067.1	4,116.8	3,675.2	3,789.8
% of total gross ODA	-	(79.1)	(62.3)	(58.2)	(63.2)	(58.3)	(57.9)
% of federal budget	-	(2.3)	(2.5)	(1.8)	(1.7)	(1.5)	(1.6)
Net official aid (OA)[f] disbursements	-	-	-	3,185.6[h]	3,309.4	702.4	766.7
% of GNI[b]	-	-	-	(0.21)[h]	(0.18)	(0.03)	(0.04)
Private development assistance[g]	-	145.5	390.6	625.2	814.9	917.7	902.7
% of GNP/GNI[b]	-	(0.042)	(0.052)	(0.050)	(0.045)	(0.045)	(0.044)
compared to total net ODA (%)	-	(12.9)	(11.8)	(12.0)	(14.8)	(16.8)	(16.2)

Notes:
[a] Grants and concessional loans to countries and territories in part I of the DAC list of aid recipients (developing countries).
[b] GNP = gross national product (until 1990); GNI = gross national income (since 1991).
[c] See footnote 29.
[d] Including ODA contributions to the European Development Fund, the European Investment Bank and the EU budget.
[e] Disbursements from the BMZ budget are for the most part, but not totally, ODA (a small share is OA).
[f] Grants and concessional loans to countries and territories in part II of the DAC list of aid recipients, i.e. countries in transition which comprise (i) more advanced Central and East European countries and new independent states of the former Soviet Union, (ii) more advanced developing countries.
[g] Aid from non-governmental organisations (e.g. the churches and other NGOs) from own funds and donation campaigns to countries and territories in part I of the DAC list of aid recipients (developing countries).
[h] 1991.

Source: www.bmz.de; OECD [*1974*: 222; *1984*: 253; *1995*: C 10; *1999*: A 22; *2003*: 249].

– Germany's reunification entailing huge transfers to the new federal states in East Germany (in the mid-1990s about DM150 billion or roughly €75 billion a year, that is, more than total ODA from all donors to all developing countries);

- its considerable official aid (OA) to the Central and East European countries and the new independent states of the former Soviet Union (see Table 8.1), which placed Germany, together with the US, at the top of the list of donors providing OA [*OECD, 1999: A 92*];
- its commitment to respecting the macroeconomic stability criteria related to the introduction of the Euro, which required increasing budget discipline in order to prevent public budget deficits from exceeding the ceiling of 3 per cent of GDP.

On the other side, however, the diminished political weight of development co-operation in the federal budget should not be overlooked. The share of the BMZ's budget, which covers most of German ODA,[27] in the federal budget dropped from 1.8 per cent in 1990 to 1.5 per cent in 2000 and 1.6 in 2001 (see Table 8.1; its peak was 2.6 per cent in the mid-1980s). Until recently, there was not much hope that the past trend would change substantially in the next few years because of the government's overall objective of balancing the federal budget by 2006 (the deficit in the autumn of 2003, when the final version of this chapter was written, was about 3.8 per cent of GDP).

At the Financing for Development Conference held in Monterrey in March 2002, however, Germany committed itself to raising its ODA/GNI ratio to 0.33 per cent by 2006 as part of the donor community's effort to achieve the international development goals (IDGs). This commitment is substantial in both quantitative and political terms:

- According to estimates by the DAC Secretariat, it means, compared with German ODA in 2000 ($5,030m.), a real increase of $2,087m. or 41 per cent calculated at 2000 prices and exchange rates [*OECD, 2002b*].
- After its election in 1998, the 'red–green' federal government had declared a clear commitment to development co-operation and to reversing the previous downward trend of the ODA/GNI ratio, particularly because of the great importance traditionally attached to Third World issues by the Greens. However, despite efforts by the Development Co-operation Minister to increase the BMZ budget, the latter was not exempted from budget cuts in 1999 and 2000 and has not yet recovered its 1998 level. The Monterrey commitment (confirmed in the coalition agreement of October

---

27. See note e in Table 8.1.

2002), if implemented, can help restore the credibility of aid policy which has partly been lost in the past few years.

IV. MAIN CHANNELS AND FORMS OF AID

1. Main Actors in Bilateral and Multilateral Aid

The actors concerned with German ODA are very diverse, both with respect to the financing (provision of funds) as well as the implementation of projects and programmes. German bilateral and multilateral ODA reported to the DAC embraces:

- BMZ development assistance, accounting for about 58 per cent of total (gross) ODA in 2000/2001 (see Table 8.1), making it the most important actor;
- development assistance of other government departments such as the Foreign Office (responsible for humanitarian and emergency aid, materials and equipment assistance, cultural co-operation, assistance to promote democracy and contributions to several UN organisations) and the Ministries of Education and Research, Economics and Technology, Agriculture, Environment, and Health;
- development assistance of the federal states (Bundesländer): (i) imputed costs of providing tuition to students from developing countries studying in Germany, (ii) development co-operation projects and programmes;
- contributions to the EU's development co-operation budget (calculated *ex post* on the basis of Germany's share in the overall EU budget),[28] debt-service relief, and part of the expenditure on political refugees from developing and transition countries in Germany.

The involvement of several ministries in ODA is not a unique feature of German development co-operation but may be observed in other donor countries (for example, France, the US and Japan). A peculiarity of German ODA, however, is the developmental commitment of the federal states. They are involved primarily in education, but also run their own development co-operation projects and programmes. German municipalities also engage in developmental activities in the context of partnerships and local government co-operation.

---

28. Germany's contributions to the European Development Fund, which is not part of the EU budget, are financed from the BMZ budget, however.

2. Bilateral aid

(i) Main components, forms and implementing organisations. Over the past ten years, bilateral aid accounted on average for 68 per cent of Germany's total gross ODA disbursements and for 62 per cent of total net ODA disbursements.[29] In 2000/2001, 62 per cent of gross bilateral ODA[30] disbursements were financed from the BMZ budget [*BMZ, 2002: 359; OECD, 2003: 249*]. Its main components are financial co-operation (40 per cent of BMZ's bilateral disbursements in 2001/2002), that is, ODA loans and grants managed by the Bank for Reconstruction and Development (KfW), and technical co-operation (32 per cent of BMZ's bilateral disbursements in the same period), about two-thirds of which were carried out by the German Agency for Technical Co-operation (GTZ) [*BMZ, 2002: 359*]. The GTZ provides advice on, *inter alia,* economic reform, financial systems, privatisation, decentralisation, legal and administrative reforms, the environment, and strengthening civil society institutions and the private sector [*OECD, 2001: II–25*]. In addition to financial and technical co-operation, the BMZ supports projects and programmes implemented by NGOs. In 2000/2001, it channelled almost 17 per cent of its bilateral budget through NGOs [*www.bmz.de; BMZ, 2002: 375*]. Box 8.4 gives an overview of the various forms of German bilateral ODA and the main organisations implementing them.

(ii) Allocation by major purposes. Table 8.2 breaks down bilateral ODA by main purposes and shows a decrease in the past decade in the shares of economic infrastructure and services (particularly transport and energy) as well as the production sectors (especially industry),[31] whereas the shares of social infrastructure and services (particularly water supply and sanitation) multi-sector aid as well as emergency assistance increased. A few comments are necessary for clarification:

---

29. Most statistics on bilateral and multilateral ODA are based on net disbursements, that is, gross disbursements less repayments on earlier ODA loans. This is correct when the focus is on aid flows. Gross disbursements, however, give a more accurate picture of the government's decision on what proportion of the aid budget in a given year is to be allocated for bilateral and multilateral purposes since in Germany repayments of former ODA loans flow back to the general budget instead of being available to the aid budget. The above-mentioned shares refer to the period 1991–2001 and were calculated on the basis of OECD [*1996: A23; 2000: 189; 2003: 249*].
30. In 2000/2001, the grant share of bilateral ODA was 82.7 per cent, the fifth lowest of all DAC members. The grant element of ODA loans was 65.3 per cent and below the DAC average of 71.4 per cent [*OECD, 2003: 270*].
31. The sharp decline in the share of industry, mining and construction largely reflects a shift in the KfW loan portfolio towards social infrastructure.

BOX 8.4

MAIN FORMS AND IMPLEMENTING ORGANISATIONS OF GERMAN BILATERAL ODA

Form of co-operation	*Main implementing organisations*
1. Financial co-operation – Forms: financing for investment projects and programmes; import financing; structural adjustment aid – Conditions: grants (for LDCs) or favourable loans (depending on development level: interest rates of 0.75 or 2 per cent and 40- or 30-year maturities with 10-year grace periods)	Kreditanstalt für Wiederaufbau (KfW) (Bank for Reconstruction and Development)
2. Technical co-operation – Forms: provision of experts, material and equipment assistance and training – Conditions: grants	Deutsche Gesellschaft für Technische Zusammenarbeit (GTZ) (German Agency for Technical Co-operation)
3. Manpower co-operation (a) Advanced training programmes for skilled employees, executives and senior civil servants from developing countries	InWEnt: merger of former Carl-Duisberg-Gesellschaft (CDG) (Carl Duisberg Society) and Deutsche Stiftung für internationale Entwicklung (DSE) (German Foundation for International Development)
(b) Co-operation at university level (scholarships for graduates from developing countries; secondment of teaching staff to universities in developing countries)	Deutscher Akademischer Austauschdienst (DAAD) (German Academic Exchange Service); Alexander von Humboldt Stiftung (AvH) (Alexander von Humboldt Foundation); Deutsche Forschungsgemeinschaft (DFG) (German Research Foundation)
(c) Assistance to returnees to set up own businesses or find employment in their home countries	Zentralstelle für Arbeitsvermittlung der Bundesanstalt für Arbeit (ZAV) (Central Placement Office of the Federal Employment Office); Deutsche Investitions- und Entwicklungsgesellschaft (DEG) (German Investment and Development Corporation)*
(d) Secondment and placement of experts in developing countries	GTZ; Centrum für internationale Migration und Entwicklung (CIM) (Centre for International Migration and Development)
(e) Volunteer services	Deutscher Entwicklungsdienst (DED) (German Development Service); Arbeitsgemeinschaft für Entwicklungshilfe (AGEH) (Personnel Agency of the German Catholics for International Co-operation); Dienste in Übersee (DÜ) (Service Overseas), now part of the Evangelischer Entwicklungsdienst (EED) (Church Development Service; Protestant); Weltfriedensdienst; Eirene and some others
4. Economic co-operation – Objective: promotion of private-sector development in partner countries – Forms: promotion of co-operation between firms and business associations in Germany and partner countries	DEG*; KfW; GTZ; Senior Expert Service (SES); Stiftung für wirtschaftliche Entwicklung und berufliche Qualifizierung (SEQUA) (Foundation for Economic Development and Vocational Training); Sparkassenstiftung für internationale Kooperation (SIK) (Savings Banks Foundation for International Co-operation)

5. Private development co-operation by NGOs funded by the BMZ (see also 3.e)	
(a) Political foundations	Friedrich Ebert Stiftung (FES) (Social Democrat) Friedrich Naumann Stiftung (FNS) (Liberal) Hanns Seidel Stiftung (HSS) (Christian-Social) Heinrich Böll Stiftung (HBS) (Green) Konrad Adenauer Stiftung (KAS) (Christian-Democrat) Rosa Luxemburg Stiftung (RLS) (Democratic-Socialist)
(b) Churches – Protestant	Evangelische Zentralstelle für Entwicklungshilfe (EZE) (Protestant Association for Co-operation in Development), now part of the Evangelischer Entwicklungsdienst (EED) (Church Development Service)
– Catholic	Katholische Zentralstelle für Entwicklungshilfe (KZE) (Catholic Central Agency for Development)
(c) Others	Deutsche Welthungerhilfe (German Agro Action); Terre des Hommes; and many others
6. Others forms (e.g. food aid; humanitarian and emergency aid; debt relief)	Various
Source: Compiled from BMZ [2002].	
Note: *In 2001, the DEG became a subsidary of the KfW.	

- The high share of education, which is well above the DAC average, involves the imputed costs of tuition provided to developing country students studying in Germany; this item accounted for 13 per cent of bilateral net ODA in 1999/2000 [*BMZ, 2002: 354, 360*].
- The upward trend in the share of 'multi-sector aid' reflects the growing importance attached to programme aid[32] frequently involving different sectors and intended as a means to increase the impact of German aid, and the rising engagement in joint financing with other donors such as sector-investment programmes and sector-wide approaches.
- The decreasing share of 'commodity and programme aid' (see footnote 31) is mainly due to the decline in commodity aid, whose peak at the beginning of the 1990s resulted from a large volume of assistance to countries affected by the Gulf War, whereas the share of food aid has been fairly stable (between 2

---

32. This type of multi-sectoral programme aid should be distinguished from the 'commodity and programme aid' mentioned as a separate item in Table 8.2. According to the DAC's directives for the reporting on the purpose of aid, 'commodity and programme aid' includes 'structural adjustment assistance with World Bank/IMF', 'developmental food aid/food security assistance' and 'other general programme and commodity assistance' (balance-of-payments support, budget support and import support).

and 3 per cent in most years). On the other hand, the low share of 'commodity and programme aid' also reflects Germany's reluctance concerning budget support because of scepticism, shared with some other donors, about recipients' financial and auditing capacities.

TABLE 8.2
BILATERAL GERMAN ODA BY MAIN PURPOSES IN 1985/86, 1990/91, 1995/96 AND 2000/2001 (% OF TOTAL BILATERAL ALLOCABLE ODA COMMITMENTS)

	1985/86	1990/91	1995/96	2000/01	Total DAC 2001
Social infrastructure and services	35.9	24.8	34.4	42.2	32.4
Education	19.9	13.5	16.6	17.4	8.6
Health	1.8	1.2	2.2	1.9	4.1
Population programmes	0.4	0.2	1.4	1.6	2.7
Water supply and sanitation	7.6	4.2	6.5	11.3	4.8
Government and civil society	4.0	2.6	2.8	5.4	6.8
Other social infrastructure and services	2.3	2.9	4.9	4.6	5.4
Economic infrastructure and services	25.5	24.5	23.7	16.9	14.9
Transport and storage	8.7	10.8	15.1	5.4	9.2*
Communications	1.6	1.8	0.7	0.5	
Energy	9.1	10.4	5.1	5.4	3,7
Banking and financial services	4.4	1.3	2.2	3.1	2.0**
Business and other services	1.7	0.2	0.6	2.5	
Production sectors	16.3	11.6	10.8	5.5	8.8
Agriculture, forestry and fishing	10.2	6.7	7.4	4.4	6.7
Industry, mining and construction	6.0	4.3	1.6	0.8	1.4
Trade and tourism	0.1	0.6	1.0	0.3	0.6
Other	-	-	0.8	-	-
Multi-sector aid	0.1	2.7	3.9	13.2	7.2
Commodity and programme aid	3.2	9.0	4.0	0.6	6.8
Action relating to debt	8.7	10.8	8.5	4.3	9.8
Emergency assistance	0.7	4.1	5.0	6.5	7.1
Administrative costs of donors	1.5	3.3	3.8	7.0	6.5
Unspecified	8.1	9.2	5.9	3.8	6.6

Source: OECD [*1998b: 60; 2003: 269*]; www.bmz.de.
Notes: * includes 'communications'; ** includes 'business and other services'.

TABLE 8.3
GEOGRAPHICAL DISTRIBUTION OF GERMAN ODA BY MAIN DEVELOPING REGIONS AND INCOME GROUPS IN 1995/96 AND 2000/2001 (% OF GROSS DISBURSEMENTS OF BILATERAL ALLOCABLE ODA)

	Germany		Total DAC members	
	1995/96	2000/2001	1995/96	2000/2001
Main developing regions				
Europe	9.9	12.3	4.2	6.8
North of Sahara	8.7	6.5	7.7	7.0
South of Sahara	24.1	23.8	28.3	27.0
Latin America and Caribbean	16.2	14.6	13.2	14.5
Middle East	7.8	9.3	7.6	4.3
South and Central Asia	11.9	13.1	12.3	15.2
Far East Asia	21.2	20.3	22.2	22.9
Oceania	0.3	0.2	4.5	2.2
Income groups				
Least developed countries	20.6	21.8	23.6	26.5
Other low-income countries	37.8	32.0	29.1	32.8
Lower-middle-income countries	32.1	36.5	33.4	35.0
Upper-middle-income countries	6.1	9.7	6.3	5.6
High-income countries	3.4	0.0	7.6	0.1

Source: OECD [*1998a: A 59, A 66; 2003: 295, 305*].

(iii) Geographical distribution. Table 8.3 breaks down bilateral ODA by main recipient regions and income groups and shows that the regional distribution corresponds roughly with the DAC average, except that the shares of Europe and the Middle East are higher in German ODA than in total DAC ODA whereas the share of sub-Saharan Africa is somewhat lower. The share of the least developed countries in German ODA is lower than in total DAC ODA whereas upper-middle-income countries attract a higher proportion of German ODA. In 2000/2001, the major recipient countries of German bilateral ODA were China, India, Indonesia, Turkey and Egypt, compared with Turkey, India, Israel, Egypt and Kenya in 1990/91 [*OECD, 2003: 295*].

3. Multilateral Aid

Germany has always considered its development co-operation as primarily a bilateral policy – just like most other donor countries.[33] This

---

33. In 2000, for instance, there were only two DAC member countries whose multilateral ODA exceeded 50 per cent of their total ODA: Italy and Greece [*OECD, 2002: 214–15*]. Incidentally, donors with small ODA volumes do not channel a significantly higher share of their ODA through multilateral organisations than large donors do (the 2000 data show no correlation between DAC members' total ODA

is due not only to the various domestic interests linked to German aid policy (see above), but also to the Federal government's claim, as one of the major donors, to operate its own development co-operation policy, for the implementation of which it has specialised and experienced organisations and a comprehensive set of instruments (see Box 8.4). In addition, it has repeatedly criticised the lack of efficiency of some parts of the development co-operation of the EU and the UN.

On the other hand, from the beginning there has never been any doubt about Germany's willingness to support multilateral co-operation.[34] In fact, the very start of German aid consisted of a financial contribution to a technical assistance programme of the UN in 1952. In the 1960s, Germany channelled on average 15.7 per cent of its net ODA disbursements through multilateral programmes. In October 1970, when the UN adopted the 'international strategy for the second development decade', the German government declared its intention to spend at least 20 per cent of its ODA on multilateral programmes [*BMZ, 1971: 56–60*].

In the following years, the multilateral share of German net ODA grew so fast (from 22.6 per cent in 1970 to 29.3 per cent in 1974) that in 1975 the Federal government announced its intention not to increase the multilateral share beyond 30 per cent [*BMZ, 1975: 38*]. Nevertheless, the multilateral share continued to rise, reaching 39.8 per cent in 1977. Also, in the long run, it went up from 31.7 per cent on average in the 1970s to 32.2 per cent in the 1980s and 35.6 per cent in the 1990s[35] – despite sometimes considerable annual fluctuations due primarily to the cycle of replenishment and changes in exchange rates rather than to specific political decisions. The increase of the multilateral share in the last few years (up to 46.6 per cent in 2000) was mainly due to the decline of bilateral ODA (see Table 8.1). The experience that, contrary to the government's announcement in 1975, it was not so easy to control the multilateral share of total

---

volumes and the corresponding multilateral shares: the Spearman rank correlation coefficient was -0.0378).
34. In this context, the term 'multilateral co-operation' refers to German support of ODA activities of the EU, the UN system, the World Bank and the regional development banks.
35. The average shares of multilateral ODA in total German ODA in the 1960s, 1970s, 1980s and 1990s were calculated on the basis of statistics published in BMZ [*1993: 153; 1995: 162; 2002: 354*]. It has to be recalled that these shares are based on net ODA disbursements and therefore overstate the importance of multilateral ODA as explained in footnote 29. For the difference between the multilateral shares of total net and gross ODA in selected years see Table 8.1.

German ODA[36] prompted the Budget Committee of the German Parliament to request a ceiling of 30 per cent for multilateral contributions to be introduced in the BMZ's budget. This limit has largely, although not entirely, been respected.[37]

After Japan and the US, Germany is the third largest contributor to multilateral co-operation and by far the most important contributor to the development co-operation of the EU [*OECD, 2003: 260–1*]. From 1996 to 2001, 54 per cent of Germany's multilateral ODA went to the EU, 21 per cent to the World Bank group, 15 per cent to the UN, 5 per cent to the regional development banks and 5 per cent to others [*www.bmz.de*].

Germany has from the beginning supported the development co-operation of the EU, although the latter had its roots primarily in the colonial legacy of France and (at a later stage) Great Britain. Germany accepted the strong influence of French and British interests on the EU's aid policy (for example, regarding the preferential treatment given to the ACP countries) because of the overriding German commitment to European unification. This explains why Germany, despite being the main financial contributor to the EU development co-operation, had not played a leading role in its design for a long time. In the 1990s, it increasingly advocated better co-ordination of development co-operation within the EU (although not a stronger Europeanisation of the member countries' individual aid programmes) and took a more demanding stance regarding the efficiency and effectiveness of EU aid.

Regarding the UN, the World Bank and the regional development banks, Germany has been an active member in, and financial contributor to, most institutions right from their inception. Scepticism with regard to the efficiency and effectiveness of some parts of multilateral aid prompted the Federal government to press for reforms and also led to temporary cuts in financial allocations (for example, to UNDP in 1998). On the other hand, Germany abstained from spectacular steps, such as the withdrawal of the US and Britain from UNESCO.

---

36. There are several reasons for this. First, the total ODA volume is only known *ex post* because of the many components of German ODA reported to the DAC (including contributions to the EU's aid budget, which, as mentioned above, are calculated on the basis of Germany's share in the overall EU budget). Second, the financing and management of German multilateral assistance is shared between the BMZ, the Foreign Office and some other federal ministries. Third, the cycle of replenishment of several multilateral funds and exchange-rate fluctuations influence the volume of multilateral ODA and its share of total ODA.

37. In the 1990s, the multilateral share of the BMZ budget fluctuated between 28.9 per cent (1996) and 34.8 per cent (1994) [*bmz.www.de*].

V. RECENT CHANGES AND NEW 'FOOTPRINTS' IN BILATERAL AND MULTILATERAL ODA

1. Bilateral Development Co-operation

Persistent budget constraints since the early 1990s caused the BMZ to take several major decisions with a view to improving the efficiency, effectiveness and impact of bilateral aid:

(i) Introduction of country assistance strategy papers ('country concepts'). In order to streamline the planning, implementation and monitoring of projects and programmes, the BMZ decided in 1991 to base its development co-operation with about 40 partner countries on so-called 'country concepts'. These are intended to define a limited number of priority areas on which bilateral ODA with the partner countries is to be concentrated, and to work out a coherent strategy for the co-operation in each area. According to the BMZ, the country concepts constitute a key instrument for managing German bilateral aid. Following an evaluation of nine such country concepts in 1996/97, it was decided to complement the country concepts by specific strategy papers for the priority areas they define (for further details see Ashoff [*1999b*]).

(ii) Reduction of the number of partner countries. In 1999 the Federal Minister of Finance presented a medium-term budget plan according to which the BMZ's budget was to be reduced by more than 13 per cent over the next four-year period. As a consequence, the BMZ decided in 2000 to reduce the number of recipient countries gradually to 37 'priority' and 33 'partner' countries. To understand this decision, it has to be recalled that, in line with the traditional worldwide approach of German development co-operation, the two main implementing agencies were still engaged in 2001 in 113 countries (GTZ) and 102 countries (KfW) respectively.[38] Box 8.5 gives the list of 'priority' and 'partner' countries, which was recently updated and now contains 40 'priority' and 35 'partner' countries. Implementing the decision to reduce the number of countries for bilateral ODA will, of course, take several years, since ongoing projects and programmes will not be abruptly stopped.

---

38. Only developing countries according to part I of the DAC list of aid recipients. Altogether, the GTZ and KfW were engaged in 131 and 106 countries respectively [*GTZ, 2002: 4*] (additional information provided by the KfW).

BOX 8.5
'PRIORITY' AND 'PARTNER' COUNTRIES OF GERMAN BILATERAL DEVELOPMENT CO-OPERATION (AS OF 2002)

Region	'Priority' countries (40) (three focal areas for co-operation)	'Partner' countries (35) (one focal area for co-operation)
Central and Eastern Europe + new independent states of former Soviet Union	Albania, Bosnia and Herzegovina, Georgia, Macedonia	Armenia, Azerbaijan, Kazakhstan, Kyrgyz Republic, Tajikistan, Uzbekistan
North Africa and Middle East	Egypt, Morocco, Palestinian Administered Areas, Turkey, Yemen	Algeria, Jordan, Mauritania, Syria, Tunisia
Sub-Saharan Africa	Benin, Burkina Faso, Cameroon, Ethiopia, Ghana, Kenya, Malawi, Mali, Mozambique, Namibia, Rwanda, Senegal, South Africa, Tanzania, Uganda, Zambia	Burundi, Chad, Eritrea, Guinea, Ivory Coast, Lesotho, Madagascar, Nigeria, Niger
Asia	Afghanistan, Bangladesh, Cambodia, China, India, Indonesia, Nepal, Pakistan, Philippines, Vietnam	East Timor, Laos, Mongolia, Sri Lanka, Thailand
Latin America	Bolivia, El Salvador, Honduras, Nicaragua, Peru	Brazil, Chile, Columbia, Costa Rica, Cuba, Dominican Republic, Ecuador, Guatemala, Mexico, Paraguay

Source: BMZ [2002: 168].

(iii) Reduction of the number of focal areas in the co-operation with individual countries. Parallel to reducing the number of countries, the BMZ decided to concentrate bilateral ODA on three focal areas in the 'priority countries' and one focal area in the 'partner' countries.

(iv) Strengthening the outcome orientation of bilateral ODA. The planning approach of German bilateral ODA (whether at country, sectoral, programme or project level) was until recently predominantly input-oriented in the sense that resource allocation was given much more attention than the intended development impact in terms of measurable outcomes. The BMZ is trying to introduce outcome considerations into the planning and decision-making process, which to some extent requires a change of 'culture' in the German aid system.

There are now several major challenges to bilateral German ODA: (i) to implement the aforementioned decisions, (ii) to adjust to new approaches such as the Comprehensive Development Framework (CDF) and the Poverty Reduction Strategy Papers (PRSPs), (iii) to take greater care of the principles of partnership and ownership in aid delivery, and (iv) to harmonise administrative procedures together with other donors.

2. Multilateral Development Co-operation

Considering development co-operation as an important part of global structural policy, the 'red–green' coalition government, on coming to office in late 1998, took a number of initiatives at the multilateral level, as follows:

- the HIPC initiative presented at the G-8 summit at Cologne (in June 1999) and approved in September 1999 at the annual meeting of the IMF and World Bank;
- the Stability Pact for South-east Europe agreed in June 1999 by more than 40 states, international financial institutions and multilateral organisations on Germany's initiative;
- concrete measures to control small arms decided by the EU on the initiative of Germany;
- efforts to improve the efficiency and effectiveness of the EU's development co-operation (several initiatives during the German presidency of the EU in the first half of 1999, for example, concerning country assistance strategy papers to be prepared for EU development co-operation).

The Federal government has re-emphasised the role of the UN (especially after the events of 11 September 2001) and increased its ODA contributions to the UN organisations (from €229m. in 1997 to €515m. in 2001) [*www.bmz.de.*].

VI. INSTITUTIONAL SET-UP, PROBLEMS, CHANGES AND CHALLENGES

The institutional set-up of German development co-operation is characterised by strengths and weaknesses. While some changes have occurred in the past few years, bilateral co-operation still faces a number of challenges.

1. Strengths

(i) Existence of a separate development co-operation ministry (BMZ). Unlike most other donor countries, Germany has a fully-fledged development co-operation ministry.[39] Development co-operation

---

39. It must be recalled that the BMZ, according to the latest figures, is responsible for only less than 60 per cent of German ODA (see Table 8.1).

therefore enjoys full Cabinet rank and is in a privileged position to bring its voice and experience to bear in the government's discussions and decision-making and to foster policy coherence. In domestic politics, however, this view is not undisputed. Critics advocate integrating the BMZ into the Foreign Office with a view to increasing the political weight and influence of development co-operation, not least because of the BMZ's weak representation in partner countries (see below) [*Köhler, 2002*] (for more detailed background information see Ashoff [*1999a*]).

(ii) Implementation of financial and technical co-operation by two major semi-autonomous and experienced organisations. KfW and GTZ are the two main pillars of German bilateral development co-operation. Most other donors, where implementation is incumbent upon development co-operation authorities or agencies, have such organisations only to a limited extent or not at all. The KfW and GTZ can work more flexibly than a co-operation authority. They have accumulated a great deal of regional and sectoral expertise, and in this respect enjoy international repute. In many cases, they have more expertise than the BMZ and have played an important role in developing new sectoral concepts and methods of project planning, monitoring and evaluation.

(iii) Institutional pluralism. The German aid delivery system involves a large array of parastatal organisations, NGOs and private sector organisations (see Box 8.4). The advantage of this structure is that the Federal government can count on numerous experienced organisations when it comes to the implementation of the different ODA programmes and projects. In addition, the complex structure delivers pluralism, offers possibilities for specialisation, and helps increase the public's acceptance of development co-operation.

(iv) Political foundations as a German particularity. There are six political foundations in Germany (see Box 8.4), each of which is affiliated to one of the main political parties. The foundations, whose main area of activities is political education in Germany, also conduct projects in developing countries financed from BMZ funds and primarily intended to support democratic development and respect for human rights. In doing so, they co-operate with many groups in the partner countries, such as political parties, trade unions, employers' associations, co-operatives, self-help groups and the mass media.

2. Weaknesses

(i) Inadequate division of labour between BMZ and the implementing organisations (despite recent improvements). The BMZ's responsibilities are not confined to the classic ministerial functions.[40] According to the Federal government's guidelines on financial and technical co-operation of 1996 [BMZ, 1996: 7], the BMZ's competence ranges from decisions on principles to planning, monitoring and controlling every individual project from the development policy perspective (whereas the specialist-technical planning, management and monitoring of the projects is left to the implementing agencies). In practice, this has often led to an overlapping of work between the BMZ, on the one hand, and KfW and GTZ, on the other, and has cramped the BMZ's capacity for conceptual work. Although the BMZ has taken several steps to withdraw from the project level,[41] the point has not yet been reached where the decisions on the country concepts and strategy papers for the priority areas of co-operation are taken by the BMZ in consultation with the partner countries while decisions on individual projects to implement the strategies are left to the executing agencies.

(ii) Institutional separation of financial and technical co-operation. This separation is unknown to most other donors and entails problems of co-ordination. Although several efforts have been made to improve co-ordination, it does not yet appear sufficient in view of the trend to more comprehensive forms of programme aid (such as sector investment programmes, sector-wide approaches or participation in poverty reduction strategies) and the need for intensive policy dialogue with the partner countries and co-ordination with other donors.

(iii) Complexity of the aid system. The disadvantage of the institutional pluralism of Germany's aid system is its complexity, which not only entails problems of co-ordination but also puts an additional burden on the partner countries' administrative capacity. In the past, the Federal government tried to mitigate this problem in three ways: (i)

---

40. Definition of the principles and guidelines of development co-operation; decisions on the basic allocation of the aid budget according to partner countries and priority areas; representation of development co-operation within the Cabinet, in Parliament and in major international organisations; co-ordination with other donors at the political level.

41. One example is the BMZ's decision, taken in 1997 following a recommendation by the DAC, to transfer the responsibility for the evaluation of individual projects to KfW and GTZ while the BMZ concentrates on thematic and cross-sectoral evaluations.

by assigning a precise role to each organisation in order to achieve a clear division of labour; (ii) by enhancing co-ordination between organisations operating in the same areas (for example, between KfW and GTZ or between GTZ and the political foundations), and (iii) by providing information to the partner countries on the German system. Nevertheless, some co-ordination problems persist.

(iv) Insufficient representation in developing countries. A traditional weakness criticised repeatedly by the DAC is the lack of strong representation with decision-making powers in the partner countries. In the donor community, there is a broad consensus that development co-operation cannot be run efficiently and effectively from headquarters, but needs to be decentralised if it is to meet the increasing requirements for dialogue with the partner countries and for co-ordination with other donors. Although progress has been made over time in terms of an increasing number of field offices, the overseas representation of German development co-operation cannot yet match the scale of a number of other donors and the degree of autonomy enjoyed by their country offices. Opposition from the Foreign Office, in fear of the establishment of parallel embassies, and the fact that decision-making is still centralised to a considerable extent in the BMZ have long prevented any substantial change involving the transfer of decisions to the field level.[42]

3. Recent Changes

In the past few years, efforts have been made to simplify the planning and monitoring of projects, to delegate a greater share of project-related activities (including project evaluation) from the BMZ to the implementing agencies, and to increase the number of GTZ and KfW offices (but not of BMZ professionals seconded to embassies) in partner countries. Several parastatal organisations were merged or integrated (see also Box 8.4): CDG and DSE, which implement advanced training programmes, were merged into InWEnt; in 2001, DEG became a subsidiary of KfW, and in 2003, the Deutsche Ausgleichsbank (DtA), which promotes business start-ups of returnees in their home countries, was merged into KfW. There are currently efforts to improve the co-ordination between KfW and GTZ further, particu-

---

42. The present situation may be described as follows: there are 25 professionals from the BMZ posted as development co-operation advisers in German embassies and permanent missions [*BMZ, 2002: 49*]; GTZ currently maintains 63 offices abroad [*GTZ, 2002: 4*]; KfW has 24 offices in developing countries (outside the TRANSFORM programme) [*KfW, 2001*]. The main point, however, is that decision-making still takes place to a high degree in the BMZ and the headquarters of GTZ and KfW.

larly in the context of new forms of programme financing; the institutional separation of financial and technical co-operation, however, remains basically unchanged.

VII. DEVELOPMENT CO-OPERATION AMONG POLITICAL PARTIES, PARLIAMENT AND THE PUBLIC

Development co-operation policy enjoys a positive consensus among the main political parties, Parliament and the public, despite many differences (which are also partly a matter of style for political 'show business' reasons) and some scepticism about the practice of co-operation. This is a great developmental asset, even if development co-operation has naturally far less importance in political and public minds than domestic and economic topics.[43]

Development co-operation is on the agenda of all parties represented in the Federal Parliament (Bundestag): not only for their manifestos and election campaigns, but also for parliamentary debates, question times and, not least, in consultations and decision-taking on the BMZ's budget. Development policy is debated intensively in the Committee for Economic Co-operation and Development, whose influence on the practice of co-operation is less than that of the Budget Committee, however.

There is an active development-oriented civil society (churches, NGOs, political foundations, Third World groups), and the public response to the churches' annual fund-raising campaigns or to special campaigns following disasters has in the long run remained undiminished and in the last few years has even increased. In the period from 1970 to 1990, private development assistance from own funds and donations[44] accounted for an amazingly constant equivalent of about 12 per cent of ODA; in the second half of the 1990s, this rose to 16 per cent in 2001 (see Table 8.1). The aid/GNI ratio of German NGOs remained stable over the past decades (at about 0.05 per cent; see Table 8.1) and was clearly higher than the DAC average of 0.03 per cent [*OECD, 2002a: 207*].

According to opinion polls, the majority of Germans (in both the old and the new federal states) tacitly endorse development assistance, even in hard economic times, and this is reflected in their

---

43. This section draws largely on Ashoff [*1996: 31 ff.*]. For further information see also Wiemann [*1997*].
44. The private development assistance by the churches and other NGOs must not be confused with the subsidies these organisations receive from the government for aid purposes.

readiness to make donations. On the other hand, the polls show that aid to developing countries is not of high explicit interest to the public: in an opinion poll commissioned by the BMZ in 2000, only 11 per cent of respondents expressed strong conviction on the importance of development co-operation and 27 per cent showed very little or no interest in the topic [*OECD, 2001: II–19*]. The results of previous polls were mixed. They showed:

- an increasing awareness of the interdependence of developments in the North, South and East;
- overestimation of not only the quantitative significance but also the possible impacts of development co-operation, which is probably partly due to its sometimes overambitious rhetoric (creating an 'omnipotence trap');
- low awareness of government development assistance compared with the work of NGOs, which seems to enjoy a better reputation; and
- scepticism about the practice of official development co-operation, based on reports of failures, doubts about the 'feasibility' of aid and the credibility of self-evaluations carried out by implementing agencies, and suspicion that ODA is only an 'alibi' to camouflage the pursuit of economic self-interest (for example, promotion of German exports).

In the last few years, the BMZ has steadily increased the funds for public information to €4.1 million or 0.1 per cent of the BMZ budget in 2002 [*BMZ, 2002: 334, 359*]. However, this still falls short of UNDP's recommendation for donors to allocate 2 per cent of their development budget for this purpose [*OECD, 2001:II–19*].

VIII. CONCLUDING REMARKS

A few years ago, the editor of a volume including short presentations of 33 donor countries characterised Germany's development co-operation as 'reform-minded but politically weak' [*Thiel, 1996: 5*]. Both of these labels are correct and yet need some qualification. There is no doubt that in Germany (just as in most donor countries) development co-operation has far less importance in political and public minds than domestic and economic topics. This has been particularly true since the early 1990s when Germany first had to manage the process of reunification and now is facing the unprecedented challenge of reforming major pillars of its post-war growth and welfare model. The sharp decline in Germany's ODA/ GNI ratio occurring during the 1990s is a clear sign of the reduced importance attached to development assistance.

On the other hand, German aid policy, which originated from and was shaped by a mix of quite different motives throughout the previous decades, has increasingly developed a profile of its own because of growing developmental professionalism of the aid administration and a strong reform drive resulting particularly from the challenges of globalisation and the need for aid policy to redefine its role in Germany's response to these challenges. The conceptual shift towards a policy that, in addition to improving living conditions in the partner countries, contributes to safeguarding our global future has given development co-operation a chance to get more political weight. In contrast to most other ministries, the BMZ has not only a sizeable budget that may be used for global action but also relevant expertise (accumulated also in its executing agencies) that is frequently asked for by other government departments and now puts the BMZ in a better position to advocate more policy coherence.

For its greater political weight to be effectively used, however, German development co-operation needs strong commitment in several respects. Raising the ODA/GNI ratio to 0.33 per cent by 2006, as announced by Chancellor Gerhard Schröder at the Monterrey summit in 2002, will be a litmus test and may well become a goal very difficult to achieve in the face of the current budget constraints. Developing a detailed strategy for implementing the ambitious Programme of Action 2015 for Poverty Reduction will be another major challenge to which further reform of the institutional set-up has to be added.

Compared with other donors, German aid policy may, roughly speaking, be classified as being somewhere between that of 'progressives' showing high financial commitment and a distinct developmental profile and often setting the international development agenda, on the one hand, and 'less progressives', on the other. German aid policy resembles a tanker. The size of its aid programme in absolute terms gives it a considerable weight in the convoy of donors and, for example, in efforts to improve donor co-ordination. Like a tanker, German development co-operation has changed its course over time somewhat more slowly than some other reform-minded countries (not because of its sheer size but because of the influence of other interests sometimes including institutional self-interests within the aid administration), but once a change of course has occurred, Germany has followed this course fairly steadily and reliably – again like tankers in real life.

REFERENCES

Ashoff, Guido, 1996, 'The Development Policy of the Federal Republic of Germany', *D + C (Development and Co-operation)*, No. 5, Berlin: German Foundation for International Development.
Ashoff, Guido, 1999a, 'The Coherence of Policies Towards Developing Countries: The Case of Germany', in Forster and Stokke (eds).
Ashoff, Guido, 1999b, 'Country Assistance Strategies as a Management and Evaluation Instrument for Donors: Some Conclusions Drawn from German Experience', *Evaluation and Effectiveness*, No. 2, Paris: OECD, DAC.
Ashoff, Guido, 2000a, *Der Entwicklungshilfeausschuss der OECD und die deutsche Entwicklungszusammenarbeit – Ein Verhältnis auf dem Prüfstand*, Bonn: Weltforum Verlag.
Ashoff, Guido, 2000b, *The OECD's Development Assistance Committee and German Development Co-operation: A Relationship under Scrutiny*, GDI Briefing Paper, No. 1/2000, Berlin: German Development Institute.
Ashoff, Guido, 2002, *Improving Coherence between Development Policy and Other Policies – The Case of Germany*, GDI Briefing Paper, No. 1/2002, Berlin: German Development Institute.
BMZ (Bundesministerium für wirtschaftliche Zusammenarbeit und Entwicklung, Federal Ministry for Economic Co-operation and Development), 1971, *Die entwicklungspolitische Konzeption der Bundesrepublik Deutschland und die Internationale Strategie für die Zweite Entwicklungsdekade*, Bonn: BMZ.
BMZ, 1975, *Zweiter Bericht zur Entwicklungspolitik der Bundesregierung* (Anlage 2: '25 Thesen zur Politik der Zusammenarbeit mit Entwicklungsländern, verabschiedet vom Bundeskabinett in seiner Sondersitzung am 9. Juni 1975'), Bonn: BMZ.
BMZ, 1986, *The Basic Principles of the Federal Government's Development Policy*, Bonn: BMZ (original German version: 1986, English translation: 1989).
BMZ, 1992, 'Hauptelemente der Armutsbekämpfung', *BMZ aktuell*, Bonn: BMZ.
BMZ, 1993, 'Neunter Bericht zur Entwicklungspolitik der Bundesregierung', *Entwicklungspolitik Materialien*, Bonn: BMZ.
BMZ, 1995, 'Zehnter Bericht zur Entwicklungspolitik der Bundesregierung', *Entwicklungspolitik Materialien*, Bonn: BMZ.
BMZ, 1996, 'Leitlinien für die bilaterale Finanzielle und Technische Zusammenarbeit mit Entwicklungsländern vom 23. Februar 1984 (aktualisierte Fassung 1/96)', *BMZ aktuell*, No. 061, Bonn: BMZ.
BMZ, 2001a, 'Elfter Bericht zur Entwicklungspolitik der Bundesregierung', *BMZ Materialien*, No. 111, Bonn: BMZ.
BMZ, 2001b, 'Poverty Reduction – a Global Responsibility. Program of Action 2015. The German Government's Contribution towards Halving Extreme Poverty Worldwide', *BMZ Materialien*, No. 108, Bonn: BMZ.
BMZ, 2002, *Medienhandbuch Entwicklungspolitik 2002*, Berlin.
BMZ, www.bmz.de.
Bodemer, Klaus, 1974, *Entwicklungshilfe – Politik für wen? Ideologie und Vergabepraxis der deutschen Entwicklungshilfe in der ersten Dekade*, Munich: Weltforum Verlag.
Claus, Burghard and Hans H. Lembke, 1992, 'The Development Co-operation Policy of the Federal Republic of Germany', in Yamazawa and Hirata (eds).
Cox, Aidan *et al.* (eds), 1997, *How European Aid Works*, London: Overseas Development Institute.
Förster, Andreas, 1994, *Japans Zusammenarbeit mit der Dritten Welt zwischen Entwicklungsorientierung und außenwirtschaftlichen Prioritäten*, Berlin: Deutsches Institut für Entwicklungspolitik.
Forster, Jacques and Olav Stokke (eds), 1999, *Policy Coherence in Development Co-operation*, London: Frank Cass.

GTZ (Deutsche Gesellschaft für Technische Zusammenarbeit, German Agency for Technical Co-operation), 2002, *Jahresbericht 2001,* Eschborn: GTZ.

Hofmeier, Rolf and Siegfried Schultz, 1984, 'German Aid: Policy and Performance', in Stokke (ed.).

KfW (Kreditanstalt für Wiederaufbau = Bank for Reconstruction and Development), 2001, Jahresbericht über die Zusammenarbeit mit Entwicklungsländern 2001, Frankfurt: KfW.

Klingebiel, Stephan, 2001, *Approaches to Crisis-preventing and Conflict-sensitive Development Co-operation,* GDI Briefing Paper, No. 4/2001, Bonn: German Development Institute.

Köhler, Volkmar, 2002, 'Zur Zukunft des Auswärtigen Dienstes aus der Sicht der Entwicklungspolitik', *E + Z (Entwicklung und Zusammenarbeit),* Vol. 43, No. 6, Bonn: Deutsche Stiftung für internationale Entwicklung.

Nuscheler, Franz, 1995, *Lern- und Arbeitsbuch Entwicklungspolitik,* 5th edition, Bonn: Bundeszentrale für politische Bildung.

Nuscheler, Franz, 1996, 'Japan – Weltmeister in Entwicklungshilfe', in Thiel (ed.).

OECD, 1974, *Efforts and Policies of the Members of the Development Assistance Committee,* 1974 Report, Paris: OECD, DAC.

OECD, 1984, *Efforts and Policies of the Members of the Development Assistance Committee,* 1984 Report, Paris: OECD, DAC.

OECD, 1992, *Efforts and Policies of the Members of the Development Assistance Committee,* 1992 Report, Paris: OECD, DAC.

OECD, 1995, *Efforts and Policies of the Members of the Development Assistance Committee,* 1994 Report, Paris: OECD, DAC.

OECD, 1996, *Efforts and Policies of the Members of the Development Assistance Committee,* 1995 Report, Paris: OECD, DAC.

OECD, 1998a, *Efforts and Policies of the Members of the Development Assistance Committee,* 1997 Report, Paris: OECD, DAC.

OECD, 1998b, 'Germany', *Development Co-operation Review Series,* No. 29, Paris: OECD, DAC.

OECD, 1999, *Efforts and Policies of the Members of the Development Assistance Committee,* 1998 Report, Paris: OECD, DAC.

OECD, 2000, *Efforts and Policies of the Members of the Development Assistance Committee,* 1999 Report, Paris: OECD.

OECD, 2001, 'Review of the Development Co-operation Policies and Efforts of Germany', *DAC Journal,* Vol. 2, No. 4, Paris: OECD, DAC.

OECD, 2002a,'Efforts and Policies of the Members of the Development Assistance Committee, 2001 Report', *DAC Journal,* Vol. 3, No. 1, Paris: OECD, DAC.

OECD, 2002b, ODA Prospects after Monterrey, Note by the Secretariat, DCD/DAC (2002) 8, 19 April.

OECD, 2003, 'Efforts and Policies of the Members of the Development Assistance Committee, 2002 Report', *DAC Journal,* Vol. 4, No. 1, Paris: OECD, DAC.

Rabe, Hans-Joachim, 2002, 'Public Private Partnerships', *D + C (Development and Co-operation),* No. 4, Berlin: German Foundation for International Development.

Repnik, H.P., 1993, 'Entwicklungszusammenarbeit für umfassende menschliche Sicherheit', *Bulletin,* No. 76, 22 September, Bonn: Presse- und Informationsamt der Bundesregierung.

Scheel, Walter, 1966, *Neue Wege deutscher Entwicklungspolitik. Durch Entwicklungshilfe zu Sicherheit und wirtschaftlichem Fortschritt,* Bonn: BMZ.

Stokke, Olav (ed.), 1984, *European Development Assistance,* Vol. 1, *Policies and Performance,* Tilburg: EADI.

Thiel, Reinold (ed.), 1996, *Entwicklungspolitiken – 33 Geberprofile,* Hamburg: Deutsches Übersee-Institut.

Waller, Peter P., 1994, 'Aid and Conditionality', *D + C (Development and Co-operation),* No. 1, Berlin: German Foundation for International Development.

Wieczorek-Zeul, Heidemarie, 2002, 'Development Policy after September 11. Towards a Comprehensive Peace and Security Approach', *D + C (Development and Co-operation)*, No. 2, Berlin: German Foundation for International Development.

Wiemann, Jürgen, 1997, 'German Development Aid', in Cox *et al.* (eds).

Wilhelm, Jürgen, 2002, 'Civilising Conflicts. Some Reflections on the Civil Peace Service', *D + C (Development and Co-operation)*, No. 6, Berlin: German Foundation for International Development.

Yamazawa, Ippei and Akira Hirata (eds), 1992, *Development and Co-operation Policies of Japan, United States and Europe,* Tokyo: Institute of Developing Economies.

9

The Foreign Aid Policy of Ireland

HELEN O'NEILL

I. INTRODUCTION

The Irish government established its official aid programme in 1974 when the country entered the European Economic Community, now the European Union (EU). Total expenditure on official development assistance (ODA) that year, both bilateral and multilateral, was €1.9m. or 0.05 per cent of Gross National Income (GNI). In the intervening years, the ODA budget has grown significantly, reaching 0.4 per cent of GNI in 2003. In 2000, the Taoiseach (prime minister) made a solemn commitment at the UN Millennium Summit to reach the UN target of 0.7 per cent by the end of 2007 and to meet an interim target of 0.45 per cent by the end of 2002. This commitment was reiterated by the Taoiseach during his address to the UN Conference on Sustainable Development in Johannesburg in August 2002 and, most recently, in his address to the UN General Assembly in September 2003.

Since the official aid programme was established, the world economy and polity have both changed dramatically and the global context within which aid programmes operate today is very different from what it was in 1974. The evolution of Irish aid policy – and wider foreign policy – reflects these changes. More importantly, it reflects domestic economic and social developments, brought about by the transformation of the country from one of the poorest and most conservative in the EU to an 'average' north European country in terms of both income and social attitudes.

II. OVERALL OBJECTIVES OF IRISH AID POLICY

From its beginnings in the 1970s – and not surprisingly, given that missionaries had pioneered Ireland's involvement in developing countries – the Irish aid programme has been imbued with a strong humanitarian motivation. Early ministerial speeches and departmental documents stressed a 'moral obligation' to help 'poor countries and poor people' and to 'promote the development of developing countries'. Interestingly, promotion of human rights – a huge issue at the global level today – was included in the aims of the Irish aid programme as early as 1979.[1] Poverty reduction, satisfaction of basic needs, an equitable internal distribution of the benefits of economic development, and promotion of self-reliance were repeatedly cited as the main aims of the programme from its earliest days. Indeed, official Irish publications in the early 1960s, even those addressing domestic development problems, had acknowledged that aid to developing countries was an 'obligation' because Ireland was perceived to be 'well-off' relative to such countries. By the beginning of the 1980s, the North–South 'mutuality of interests' argument began to be advanced.[2]

While the overall aims, set out in the 1970s, of promoting development and reducing the poverty of 'poor countries and poor people' have remained at the centre of the Irish aid programme, many new elements such as gender equality and conflict prevention have been added in the meantime. For instance, the 1997 strategy statement of the Department of Foreign Affairs (DFA), *Pursuing Ireland's External Interests,* included gender equality, human rights, sustainable development and a more active participation in multilateral institutions.[3]

---

1. In a foreword to the first report on the bilateral aid programme published by the Department of Foreign Affairs in 1979, the then Minister for Foreign Affairs wrote: 'There is too much to be done, and there are too many people at present suffering from the indignities of hunger, poverty and the deprivation of the most fundamental of human rights waiting for our response' (Michael O'Kennedy, TD, Minister for Foreign Affairs, 'Foreword', *Development Cooperation: Ireland's Bilateral Aid Programme,* Dublin: Department of Foreign Affairs, 1979).
2. While reiterating 'our responsibility to play our part in the process of developing the third world', the motivation for giving aid was described as being based 'not only on charity or altruism but on a realisation that our future is inextricably bound up with the developing world' (Jim O'Keeffe, TD, Minister of State for Development Cooperation, Address to Comhlamh, the returned Development Workers' Association, Cork, 8 October 1981, p.1).
3. The 1997 strategy statement spelled out the objectives as follows: 'To contribute to the development needs of poor countries in partnership with the governments and people of those countries and in line with their priorities; support a process of self-reliant, sustainable, poverty-reducing and equitable sustainable development in all its aspects including material well-being, human rights, fundamental freedoms, gender equality, protection of the environment, support for civil society and processes, as well as mechanisms to prevent, resolve and recover from conflict; ensure rapid and effective response to humanitarian emergencies; maintain coherence in all aspects of Ireland's relations with developing countries; and promote active participation by Ireland in multilateral institutions concerned with development' [*DFA, 1997: 18*].

The 2002 report of the Ireland Aid Review Committee (IARC) reiterates the 'overarching objective' of Ireland's aid policy as the reduction of poverty. 'The Ireland Aid programme should have as its absolute priority the reduction of poverty, inequality and exclusion in developing countries.' It also adds that it should 'aim for sustainable development' and 'prioritise the objectives of gender equality and environmental protection' [*DFA, 2002: 22*].

III. PRINCIPLES GUIDING THE IRISH AID PROGRAMME

From its beginnings, the Irish aid programme has been focused on a small number of low-income programme countries mainly in sub-Saharan Africa. Five such countries were selected in 1974: Lesotho, Zambia, Tanzania, Sudan and India. India was soon dropped because of its size and Sudan was dropped in more recent times. In 1994, two new countries, Ethiopia and Uganda, were added to the list and in 1996 Mozambique brought the total to six. All are now located in sub-Saharan Africa, all are classified as least-developed by the United Nations and the World Bank, and all are ranked among the lowest 36 countries (out of 162) in the UNDP's human development index. With the exception of Lesotho, they are all classified among the heavily indebted poor countries (HIPC) in terms of their debt burdens. Ireland has very few trade or investment links with any of these countries. Their choice reflects the focus of the Irish aid programme on poor people and poverty reduction.

A cursory examination of expenditure over the years might suggest that there has been little change in the distribution of Irish aid by sector. Today, the focus is still on health, education and rural development, but both the drivers and the delivery mechanisms are very different from what they were nearly 30 years ago. In the early days, there appeared to be an underlying assumption – in keeping with modernisation theory – that assistance to promote growth in the productive sectors as well as assistance to develop high-level skills in the services sector would promote poverty reduction. Aid from what was a very small programme was focused on small-scale agriculture, upgrading of the quality of livestock, hospitals and universities, and development of accounting skills in specialised training institutes. Over the years, and especially since the early 1990s, a more direct poverty-reduction strategy has been adopted. Although the focus on health, education and rural development continued, there have been significant shifts *within* these sectors. Within the health sector, the focus is on primary health care; within education, on primary and informal schools, teacher training, and adult literacy. Within rural development, the focus is on food security, rural roads (especially those that can be built with local labour), provision of clean water and sanitary services, and micro-credit (especially for women). Indeed, sensitivity to the gender implications of all aspects of the aid programme is a key principle.

It might be argued today that the current 'overarching' focus on direct poverty reduction may be somewhat overdone. After all, it is surely impossible to reduce poverty in a very poor country if there is no economic growth – especially if increased food output and better agricultural methods are not being promoted. An agriculture desk was established at headquarters during 2003. It is to be hoped that it will also examine ways to stimulate the processing of agricultural products and encourage greater linkages between the agriculture and industry sectors in its programme countries in order to expand employment opportunities and add value to the output sold on both the domestic and export markets.

From its beginnings, Irish aid has been in grant form and remains so today. Promotion of Irish exports or Irish investments has never been a guiding principle. There is no tying of aid to purchase of Irish goods. Of course, Ireland's trade links with its programme countries are minuscule. Today, Ireland is in the forefront of donors demanding full untying of all aid. Because Irish aid has been delivered in the form of grants, it creates no debt. The 'guiding principles' that inform the bilateral aid programme today can be summed up as: a focus on poverty reduction, participation and local ownership, gender sensitivity, and sustainability. The principles also include 'a commitment to peace, human rights and democracy'. The 2002 IARC report restates the partnership principle of the Irish aid programme as follows: 'It should incorporate a high degree of partnership with recipient countries and also with the international donor community and NGOs both at home and abroad' [DFA, 2002: 22].

Principles relating to accountability and involvement in partner country policies have undergone significant change in line with the approaches of all donors. In the 1980s, it could be stated that, having made its choices regarding partner countries no further demands are made in relation to the recipients' planning capacity and follow-up ability. Today, while partnership is at the heart of the guiding principles, transparency and accountability are now demanded. The 2002 IARC report includes among the 'key principles' of the Irish aid programme that 'it should prioritise effectiveness, value for money, transparency and accountability'. Good governance is also considered to be 'of critical importance'. Democracy is described as 'both means and end, a valid moral and political objective in itself and at the same time providing a context in which poverty and inequality can be overcome'. The report concludes that 'Measures to increase fairness, transparency and accountability in the public sector and to eradicate corruption are vital components of the democratisation process' [DFA, 2002: 22–3].

IV. VOLUME TARGETS AND OUTTURNS

In 1970, at the 25th General Assembly of the United Nations, Ireland accepted the ODA target of 0.7 per cent of gross national product (GNP, now gross national income, GNI) 'in principle'. However, no commitment was made to reaching it by any specific date but rather 'as our resources permit'.[4] Thirty years later at the UN Millennium Summit held in September 2000, the Taoiseach Bertie Ahern pledged that Ireland would reach the UN target by the end of 2007 and would reach an interim target of 0.45 per cent by the end of 2002.

In the intervening years, various targets had been set – in terms of percentages of GNI, total nominal amounts, and also various annual rates of increase expressed as percentages of GNI. With the exception of a few nominal targets in the 1990s, what all these targets had in common is that none of them has been reached. For example, in 1974, 'an annual increase in ODA of the order of 0.05 per cent of GNP taking one year with another, over the next five years' was promised.[5] This was not achieved. In 1981, an explicit and unambiguous commitment was made to reach the UN target of 0.7 per cent of GNI by 1990. This was not reached. Indeed, because of the emergence of a crisis in the public finances in the second half of the 1980s, the aid budget was cut severely. In 1993, the incoming government promised to 'substantially increase' aid that year and by 0.05 per cent of GNI each year thereafter so as 'to make steady progress toward achieving' the UN target and specifically to reach a medium-term target of 0.4 per cent of GNI by 1997.

---

4. Statement on behalf of the Irish government to the United Nations during the debate on the Second Development decade at the 25th General Assembly, September 1970.
5. Garrett Fitzgerald, TD, Minister for Foreign Affairs, Address to the Council of Ministers of the EEC, 30 April 1974, p.1. Text published by the Government Information Service, Dublin.

TABLE 9.1
IRELAND'S ODA, SELECTED YEARS 1974–2002 (€m. AND %)

	1974	1978	1984	1988	1994	1998	2002
Total ODA	1.90	10.70	42.20	41.10	95.50	177.30	422.1
Administration[a]	-	0.10	0.30	0.70	1.60	3.10	22.4
Bilateral aid	0.30	3.30	16.60	17.50	48.40	107.20	261.1
Multilateral aid	1.60	7.30	25.30	22.90	44.70	67.00	138.6
Bilateral as % ODA[b]	15.80	31.80	40.00	44.30	52.60	62.20	67.2
ODA as % GNI	0.05	0.13	0.22	0.18	0.24	0.27	0.40

Sources: DFA, *Ireland's Official Development Assistance,* various years and data supplied in July 2003.

Notes:

[a] Changes in accounting procedures, introduced during 2002, resulted in the transfer of the costs of administering the bilateral aid programme out of the bilateral aid budget into the administration subhead. Consequently, it is not possible to compare administration costs in 2002 with administration costs in earlier years.

[b] Bilateral aid figure used in this calculation includes the administration costs of the entire ODA programme.

ODA flows increased strongly and steadily between 1993 and 1997 but because growth of GNI was exceptionally high during the period and nominal ODA flows did not increase at the same pace, the 0.4 per cent target was not met in 1997. This did not deter the incoming government from committing itself to further strong increases in aid and promising to reach 'an ambitious – but attainable – target of 0.45 per cent of GNI by 2002'.[6] However, unprecedented increases in GNI growth between 1997 and 2001 always made the attainment of this target unlikely despite very significant growth in the aid budget (see Table 9.1). Finally, while the 2001 ODA/GNI percentage reached 0.33 per cent, the interim target of 0.45 per cent was not achieved by the end of 2002.

In addition to the Development Cooperation Division (Development Cooperation Ireland)[7] of the Department of Foreign Affairs, a number of other government departments are involved in spending the total ODA budget. These include Finance, Agriculture and Food, Education and Science, Health, Enterprise Trade and Employment, Public Enterprise, and Social, Community and Family Affairs. Around 70 per cent of the ODA budget is accounted for by DCI. An

---

6. Liz O'Donnell, TD, Minister of State at the Department of Foreign Affairs, Address to the National Committee for Development Education, Dublin, 18 July 1997.

7. The original name was Irish Aid. This was changed to Ireland Aid in 1999 and to Development Cooperation Ireland in July 2003. Even though it was not yet operative during the period covered by this chapter, the term Development Cooperation Ireland (DCI) is used thoughout.

agreement reached by the Ministers for Foreign Affairs and Finance specified that, for the years 2001, 2002 and 2003, the latter allocation was to be fixed at €239m., €372m. and €441m. respectively. This three-year agreement underpinned the 2007 target and was designed to help Ireland Aid to plan its programme on the basis of three-year budgetary commitments – rather than moving targets. In order to reach 0.7 per cent of GNI, the government also agreed annual increases in total ODA and ODA/GNI targets through 2007.

No sooner had these annual targets been spelled out, however, than the government decided that slower than expected growth in the economy, brought about by a slow-down at global level, and the deterioration in the public finances warranted major reductions in planned increases in public expenditure across a wide range of government departments during 2002. The first reduction to be announced was in DCI's budget. Clearly, and not for the first time, the 'resources permitting' constraint was to be applied to the aid budget. It was stressed that the cuts represented 'reductions in planned increases' and that total ODA would still grow by over €100 million in 2002. Moreover, the Taoiseach reiterated Ireland's 'absolute commitment' to achieving the UN target by the end of 2007 during his address to the plenary session of the World Summit on Sustainable Development in Johannesburg on 3 September 2002.[8] He reiterated it again, most recently, at the UN 58th General Assembly in September 2003.[9] Nevertheless, the 'reductions in planned increases' meant that the nominal and percent of GNI interim targets (€466m. and 0.45 per cent respectively) could not be achieved during 2002. The 2003 DAC Peer Review recommended that Ireland should consider reintroducing the inter-ministerial agreement on allocations to DCI, ideally with rolling three-year schedules.

Ireland provided 0.16 per cent of GNI to least-developed countries in 2001, thus meeting the UN target made by donors at various United Nations conferences to devote 0.15 per cent of GNI to the poorest countries. A further DCI subsidiary target to reach a bilateral/multilateral distribution of two-to-one by 1997 was reached in 2002.

---

8. Ireland, Department of the Taoiseach (2002), Address by the Taoiseach to the Plenary Session of the World Summit in Johannesburg, South Africa, 3 September.
9. Ireland, Department of the Taoiseach (2003), Statement of the Taoiseach, Bertie Ahern, TD, to the General Debate at the 58th General Assembly of the United Nations, New York, 25 September.

V. COMPONENTS OF THE IRISH AID PROGRAMME

Sustainable poverty reduction has always been the 'overarching' objective of the Irish aid programme, as clearly expressed in policy statements. The 1999 DAC Peer Review noted that Ireland 'pursues a strategy aimed at poverty reduction at several levels' and added that 'Ireland deserves high marks for steering its programme towards poverty reduction' [*OECD/DAC, 1999: 46*]. The more recent DAC Peer Review stated that the programme 'distinguishes itself by its sharp focus on poverty reduction and its commitment to partnership principles' [OECD/DAC, 2003: 11]. As one way of operationalising that objective, the country focus is on the least-developed countries and, within them, on the poorest people and the social sectors. The 2003 DAC Peer Review noted that half of Ireland's ODA was channelled to least-developed countries – 'the largest share in the DAC' [*ibid.*]. It also noted that 'activities take account of their likely contribution to reducing poverty and achieving the Millennium Development Goals' [*ibid.*]. These goals in turn were described as providing the context in which priority sectors are decided. It added: 'A field visit to Tanzania to prepare for this Peer Review found that Ireland was appreciated as a collaborative partner' [*ibid.*].

Before 1974, almost all Irish aid was provided in the form of contributions (mainly mandatory) to multilateral organisations such as the World Bank. As the aid programme grew over the years, the European Union replaced the Bank as the main conduit of multilateral flows. Voluntary contributions to UN agencies and to debt reduction facilities grew significantly. With huge demands today on the multilateral side for assistance to help fight HIV/AIDS and reduce debt to sustainable levels, the challenge for the rapidly expanding Irish aid budget is to manage the bilateral/multilateral balance and the components within each of them.

1. The Bilateral Side of the Aid Programme

DCI delivers bilateral assistance through: its country programmes; co-financing with NGOs; funding volunteers on assignment in developing countries through the Agency for Personal Service Overseas (APSO); emergency and rehabilitation in countries affected by major emergencies; a special programme to promote democratisation and human rights; assistance to programme refugees; a scheme to provide fellowships to students from developing countries; and support for development education in Ireland.

(i) Country programmes. In 2002, expenditure in the six programme countries accounted for 50 per cent of all spending on the bilateral side of the Irish aid programme. Expenditure in 'other countries' rarely exceeds 4 per cent of bilateral assistance. According to the 1979 report on the bilateral programme, the criteria used to select the first programme countries were: relative poverty; actual need of external aid; ability to absorb and make effective use of aid; suitability of social structures and policies to facilitate the flow of assistance to the 'neediest sectors'; historical and cultural links; the possibility of being able to use English as a working language; familiar legal, administrative and other systems; and 'special considerations' such as the dependence of Lesotho on the then-apartheid South Africa [*DFA, 1979: 137*]. The criteria used to choose new programme countries in the 1990s were fairly similar although use of the English language was less emphasised. By 2002, the criteria had changed to reflect the new orthodoxy in development co-operation approaches. The IARC list includes: consideration of the recipient country's human rights record and status, the quality of its governance, its political stability, its administrative capacity and financial accountability arrangements, its commitment to combating corruption, and its economic policies [*DFA, 2002: 75*].

Programmes and projects supported by Development Cooperation Ireland in poor countries are focused on improvements in primary health care and primary education, on provision of rural roads, and on enhancing human rights and fundamental freedoms including gender equality. All donors find it difficult to measure the impact of their aid programmes on poverty reduction in the countries to which they provide various types of development assistance. One way in which the poverty reduction objective of the Irish aid programme can be assessed is by examining, in the first instance, the geographic and sectoral focuses of expenditure within the country programmes. All of DCI's programme countries are located in sub-Saharan Africa, all are classified as 'least developed' by the World Bank and the United Nations and all are ranked among the lowest 36 countries (out of 162) in the human development index of the United Nations Development Programme (UNDP). With the exception of Lesotho, they are all classified among the heavily indebted poor countries (HIPC) in terms of their debt burdens. Ireland has very few trade or investment links with any of its current programme countries: their choice reflects DCI's focus on poor people and poverty reduction.

In common with many other donors, DCI has moved away from supporting 'project islands' over the years and is now delivering its assistance mainly through targeted support to sub-national districts, to specific sectors or even general support to central government budgets. These approaches are described as area-based programmes (ABPs), sector-wide approaches (swaps), and budgetary support. In the main, aid to ABPs and budgetary support have been provided to the social sectors.

In all of DCI's programme countries (and also in its contributions to many multilateral programmes), major assistance is now being

TABLE 9.2
IRELAND'S BILATERAL AID, SELECTED YEARS 1992–2002 (€m.)

	1992	1994	1998	1999[a]	2000	2002
Programme countries (PCs)	8.4	14.4	53.6	56.0	78.7	130.2
Ethiopia	-	1.7	14.4	14.7	20.6	20.9
Lesotho	2.6	4.1	5.9	6.7	7.6	10.2
Mozambique	-	-	6.5	8.1	14.7	26.5
Sudan	0.6	-	-	-	-	-
Tanzania	3.3	3.8	12.8	9.8	15.9	21.1
Uganda	-	0.9	6.5	9.3	12.4	33.1
Zambia	1.9	3.9	7.5	7.4	7.5	18.4
Other countries	0.7	4.1	6.6	5.5	6.2	16.3
Co-financing with NGOs	2.2	5.1	8.1	9.0	11.1	23.2
Co-financing with multilaterals	0.5	1.5	2.7	3.4	3.8	0.5
Agency for Personal Service Overseas	3.3	8.9	13.6	13.5	14.5	17.6
Emergency humanitarian assistance	2.1	7.4	7.6	15.2	11.8	20.0
Rehabilitation	-	1.7	6.3	7.6	13.5	17.3
Programme refugees	-	0.7	1.4	14.0	14.1	-
Democratisation/Human rights	-	0.6	1.3	1.9	1.9	2.8
Development education	0.7	0.9	1.4	1.3	1.7	2.1
Fellowships and training in Ireland	1.2	1.7	1.9	1.9	1.9	2.1
Information on Ireland Aid	-	-	0.1	0.1	0.1	0.2
Programme support	0.2	1.4	2.0	2.5	3.0	-
Areas of new expenditure	-	-	-	-	-	28.0
Contingencies	-	-	-	-	-	0.8
Total bilateral aid[b]	19.3	48.4	106.6	131.9	162.3	261.1

Sources: DFA, *Ireland's Official Development Assistance*, various years and data provided in July 2003.

Notes:
[a] Including carry-over from 1998 of €5.6m.
[b] Excluding administration costs.

provided to tackle the HIV/AIDS pandemic. DCI regards such expenditure not merely as part of a health-improvement programme, but rather as a cross-cutting issue that has implications for all sectors including agricultural output and education (because of the loss of so many young adult farmers and teachers to the disease).

In Ethiopia, where aid has increased very significantly during the past few years, ABPs have been funded in Gurage and Sidama Zones in the Southern Nations and Nationalities Peoples Region and in the Eastern and Southen Zones of Tigrey Region (which includes a regional food security programme). The swaps focus on assistance to the health and education sectors. An HIV/AIDS baseline survey has been carried out and a number of anti-AIDS clubs have been established in high schools. In the opinion of DCI, the anti-malarial programme is regarded as very successful. Assistance has also been provided to watershed management, food security and civil service reform. In Lesotho, assistance is concentrated on poverty reduction programmes including health and education, water supply and rural roads, as well as on 'good governance' programmes designed to reform the public sector. In Mozambique, aid is provided through two regional ABPs, one in Niassa province in the north and the other in Inhambane province in the south. The objective of these programmes is to help the provincial and district administrators to plan and manage their own programmes and to deliver key services in education, health, water and sanitation, roads, social action, agriculture, and HIV/AIDS. The Tanzanian country programme focuses support on four sectors – health, education, agriculture/natural resources, and governance. Aid is channelled through the relevant central government ministries and institutions and through local councils in the districts of Kilombero, Kilosa, Muheza and Ulanga.

In Uganda, where poverty levels (and the incidence of HIV/AIDS) have fallen significantly in recent years, core Irish assistance is channelled through a Poverty Action Fund that covers sectors such as health and education. Assistance for poverty reduction is aligned to Uganda's own poverty reduction action plan and poverty reduction strategy paper, and is tied to policy reforms and benchmarks within the World Bank's poverty reduction support credit. Assistance is provided through ABPs in the districts of Kibaale, Kiboga and Kumi, where the emphasis is now on capacity building and human resource development, as well as on the education, health, and legal sectors under swaps. In Zambia, where socio-economic conditions have been deteriorating over the years (to such an extent that the country has been reclassified from 'lower middle-income' to 'least developed'), the geographic concentration is on Northern Province, while sectoral assistance is focused on health, education, water and sanitation, and urban development. In all of these country programmes, Ireland Aid has co-operated with other donors.

The 'other countries' assisted by DCI include South Africa, the Balkans, Palestinian Administered Areas, Zimbabwe, Ghana, Nigeria, Eritrea and East Timor. Most of them are defined as either least-developed or low-income countries by the UN and the World Bank. The programme in South Africa aims to assist those who had been disadvantaged by apartheid. Support is focused on education and training, health, human rights and democratisation, as well as rural and community development.

In March 2003, the Minister of State, Tom Kitt, TD, announced that East Timor (now Timor Leste) was to become a programme country. Despite the fact that DCI now has seven programme countries, the very small size of Timor Leste ensures that the Irish aid programme there will be a small one (around €11 million over the three-year period 2003–5) and DCI is still seriously considering adding on another programme country – provided the aid budget continues to increase in real terms. Among the types of candidates being considered are a post-conflict country in sub-Saharan Africa and an Asian country in transition from communism that could act as the regional hub for focused poverty reduction and governance programmes in South-East Asia. Ireland has very few trade or investment links with its sub-Saharan African partners. However, choosing a programme country in Asia is perceived within the Department of Foreign Affairs as sitting well with Ireland's growing economic engagement within that continent (while contributing to its wider foreign policy and the growing presence of 'Ireland in the world') even though its development co-operation policy in Asia would focus on poor countries and poverty reduction.

(ii) Co-operation with the NGO sector. DCI spends a significant and growing part of its aid budget through the development NGOs that implement their own projects in developing countries (see Table 9.2). The 2002 IARC report described the experience and insights of these NGOs as 'a unique resource...that should be fully utilised by Ireland Aid in the planning and implementation of its aid programme' [*DFA, 2002: 80*]. Their work complements that of DCI and provides benefits for both the NGOs through financial support, and DCI through their experience and local contacts. NGOs are also able to provide assistance to developing countries where Ireland has no official representation and no official aid programme. Many projects undertaken by

Irish NGOs are of a long-term developmental nature but, increasingly in recent years, many of them are involved in providing emergency humanitarian assistance in territories affected by conflict, as well as in helping communities in these areas to rehabilitate their societies and infrastructures in the post-conflict period.

The 1999 DAC review team suggested that Ireland should be cautious about increasing substantially the amount of aid channelled through Irish NGOs, arguing that they already absorb 'a relatively large share' of the aid budget and, in addition, 'are generously supported by the Irish public and receive funding from other sources, such as the EU and United Nations agencies' [*OECD/DAC, 1999: 18*]. It suggested, instead, that additional funding channelled through NGOs could be directed towards indigenous organisations in developing countries. In contrast, the 2002 IARC report recommended that funding to NGOs should 'increase significantly as the Ireland Aid budget expands' [*DFA, 2002: 6*].

DCI has been co-financing NGO development activities through four separate funding mechanisms. First, the heading 'co-financing etc. with NGOs' (see Table 9.2) includes both the fund for individual projects and block grants provided to the five biggest Irish NGOs. These latter grants were first introduced in 1994 and have allowed recipients freedom to select projects within criteria agreed with DCI. Next, a local funding scheme administered through Irish embassies in developing countries assists micro projects that are undertaken primarily by indigenous NGOs. Finally, there is a separate budget line for emergency humanitarian assistance and post-emergency recovery. Over 44 per cent of DCI's total expenditure on emergency humanitarian assistance in recent years has been spent through NGOs to help them undertake their own emergency relief and recovery programmes. Other headings under which Ireland Aid has been providing funding to NGOs include: democratisation and human rights, personnel on assignments overseas through APSO, and development education through the National Committee for Development Education (NCDE).

The 2002 IARC report recommended that the various mechanisms being used to co-finance NGO activities should be consolidated into one single fund. This appears to be for transparency as much as efficiency reasons. According to the report, it would 'ensure that NGOs and the public have a full picture of the government funding available' [*ibid.: 81*]. It suggested that specific deliverables and monitoring arrangements be agreed with the NGOs and published, and that they give 'due recognition' to DCI by using its logo. While funding channelled to NGOs should be 'significantly increased' as the aid budget expands, it should be poverty-focused and 'subject to the standards for planning, implementation, effectiveness, evaluation and best practice which apply universally in the DCI programme' [*ibid.: 82*]. The IARC report also recommended that an NGO Liaison Unit be established within the Ireland Aid management structure, that a Development Forum be established 'to enhance policy dialogue between NGOs and Ireland Aid policy-makers at both political and official levels' [*ibid.: 84*] and that, in the interests of accountability, legislation on regulation of charities and charitable fund-raising is to be enacted 'as quickly as possible' [*ibid.: 85*].

The IARC report supported the proposal – which was under discussion as its report was being prepared in 2002 – that a new multi-annual programme scheme (MAPS) be set up between Ireland Aid and some of the larger NGOs who might wish to move away from the block grant scheme on to 'a more programmatic model' with predictable budgetary frameworks over a three-year period [*ibid.: 81*]. This scheme came into operation on 1 January 2003. It has been described as DCI's 'third key partnership component' (in addition to its partnership with the programme countries and its partnership with international organisations) in the delivery of the government's ODA programme.[10] DCI's total support for participating NGOs will amount to €117 million over the next three years – provided the total ODA budget continues to be adequate over the period. By April 2003, five NGOs were participating in the MAPS. The NGOs and the amounts provisionally allocated to each are: Concern (€39.5m.), Trócaire (€34.3m.), GOAL (€32m.), Self Help (€6.9m.) and Christian Aid (€4.5m.). The smaller NGOs will continue to receive funding from DCI under the other schemes as before.

As a result of this huge increase in spending through the NGOs, DCI's annual budget for supporting their activities now amounts to around €70m. – over one-quarter of bilateral aid expenditure in 2002 and 2003 and over 15 per cent of total ODA. This means that Ireland spends a higher proportion of its total aid budget through the NGOs than does any other DAC donor.

Clearly, there are both opportunities and risks for both sets of actors in this very significant expansion of funding for NGOs by DCI, especially for the NGOs that are participating in the MAPS. For them, the scheme provides some significant funding predictability over a three-year period. This secures part of their resource base and helps them to build capacity and work out medium-term strategic plans. The

---

10. DFA, Address by the Minister of State, Tom Kitt, TD, at the official launch of the MAPS, 16 April 2003.

risks are associated with increased dependency on government funding that, in the cases of some of them, now exceeds their income from their own private fund-raising activities. If funding from other official sources such as the EU is factored in, their dependence on public resources is so significant that it raises issues relating to the nature of their partnership with the official aid programme. For DCI, the risks attached to spending taxpayers' money for the benefit of people in other jurisdictions is now spread more widely. On the other hand, the audit and evaluation obligations remain unchanged in essence but become more complex in terms of execution. Partnership Monitoring Committees between DCI and each of the participating NGOs are being put in place. As the Minister of State put it in his speech at the launch of the MAPS: 'These will meet regularly and will help to guarantee that there is a good balance maintained between learning and accountability, and that the programmes are achieving the maximum benefit for the beneficiary communities and value for money for the taxpayer' [*ibid.: 81*]. The ABIA recommendation for the enactment of legislation for charities is timely.

DCI has also been providing assistance to the development activities of Irish missionaries in developing countries through co-financing projects that address basic needs and funding of personnel through APSO. However, the number of missionaries has been falling rapidly in recent years and many of them are currently implementing strategies of transition as they hand over their activities gradually to local groups. The IARC report pointed to the fact that the historical roots of Ireland's official aid programme lie in the work of missionaries over many years. It praised the 'profound impact' they have had at local level in developing countries, especially in the areas of primary health care and education. It also acknowledged the important role that missionaries have played in highlighting development issues within Ireland and thus in strengthening public support for development assistance generally. In its recommendations, the IARC report suggested that 'significantly increased resources' be made available to support the development work of the missionaries, through a dedicated budget in the form of a block grant to co-ordinating organisations such as the Irish Missionary Union and the Church Missionary Society which would be administered by APSO [*ibid.: 89*].

(iii) A declining role for Irish volunteers. APSO has been in existence since the Irish bilateral programme was established in 1974. Over the years, it has been the vehicle through which tens of thousands of young volunteers have been able to work in developing countries. Its activities have included placement and funding or co-funding of vol-

unteers (its own and those of NGOs and missionaries) and training of development workers before they go on assignment to developing countries. It also operates the Rapid Response Register for recruitment of personnel for humanitarian missions and election monitoring. APSO's budget grew significantly during the 1990s. However, it carried out a review of its mission and activities in 2000, recognising that the external environment within which it operates has undergone enormous change since 1974.

As developing countries develop their own human resource base, their demand for expatriate technical assistance has been changing. They no longer require or request qualified but inexperienced volunteers from donor countries. Demand has shifted towards more experienced personnel with greater expertise in more specific professional areas. APSO's 'bulk' programmes involving two-year assignments of, for example, Irish teachers have been wound down in favour of single assignments of specialists at community-based level. From well over 1,000 in the early 1990s, there are now only about 100 APSO direct volunteers in the field. The IARC report paid tribute to APSO's valuable contribution over the years and to its continuing work in the area of recruitment and training – 'an area in which APSO still has much to contribute'. It recommended that APSO be integrated into DCI 'in a carefully planned and phased process which will ensure that APSO's experience is built on and its particular role and value safeguarded' [*DFA, 2002: 90*]. This process began in 2003.

2. Multilateral Aid and Links between Aid and Wider Foreign Policy

Ireland joined the UN in 1955 and the World Bank in 1957. Apart from tiny amounts spent through the Overseas Trainee Fund (used mainly to train Zambian army officers and public administrators in the mid-1960s), all of what could be classified as official development assistance up to 1974 was spent through those two international channels. After establishing the bilateral aid programme in 1974, however, the European Commission rapidly replaced the World Bank as the main conduit of Ireland's multilateral ODA payments. As the bilateral programme was built up, the multilateral side itself began to fall as a proportion of total ODA expenditure (see Table 9.3). With the exception of discretionary payments to UN agencies, as well as replenishments to the World Bank's soft-loan affiliate IDA, the amounts payable on the multilateral side are mandatory and assessed by the individual organisations to which Ireland belongs. In the case of the EU, Ireland contributes to the European Development Fund (EDF) which finances the Lomé and Cotonou Conventions between the EU and the

African, Caribbean and Pacific (ACP) countries, and to the development co-operation part of the EU budget which is spent by the Commission on food aid, humanitarian emergency assistance, and aid to non-ACP developing countries. Contributions to the EDF are paid through DCI, while the budget contribution and payments to the World Bank group are channelled through the Department of Finance, which continues to provide Ireland's alternate executive director to the IMF and its senior adviser to the World Bank. Although these appointments cause no friction between the two Departments, and no issue of policy coherence arises as a result, it might be more appropriate to consider appointing an official from DCI to fill the World Bank position.

As early as 1973, development co-operation was being described as one of the 'basic objectives of Irish foreign policy'.[11] The 1996 White Paper on foreign policy acknowledged the 'inextricable links between development co-operation and other pillars of foreign policy, notably human rights, conflict prevention and resolution, and commercial and financial policies' [*Government of Ireland, 1996: 250, para. 9.89*]. The 1997 strategy document also stresses that development co-operation policy is 'an integral part of Irish foreign policy' which works towards the overall goals of 'international peace, security, and a just and stable global economic system' [*DFA, 1997: 18*].

TABLE 9.3
IRELAND'S MULTILATERAL AID, SELECTED YEARS 1992–2002 (€m.)

	1992	1994	1998	1999	2000	2002
European Union	21.8	31.9	43.8	53.6	50.7	66.7
EU Budget (Development Co-operation)	13.0	21.8	33.9	38.1	43.4	57.2
EDF (Lomé and Cotonou Conventions)	8.8	10.1	9.9	15.5	7.3	9.5
World Bank/IMF group	5.6	5.3	6.7	20.8	12.2	10.5
IDA	5.5	5.2	6.2	6.1	9.1	8.4
WB HIPC Trust Fund/IMF HIPC and ESAF/PRGF	-	-	-	13.5	1.7	1.0
Global Environment Facility	-	-	0.5	1.2	1.0	1.1
IFC	0.1	0.1	-	-	-	-
MIGA	-	-	-	-	0.4	-
United Nations	3.4	8.1	16.5	16.3	24.0	56.3
UN agencies	1.0	5.7	10.7	10.5	15.9	42.5
Other UN organisations	2.4	2.4	5.8	5.8	8.1	13.8
Co-financing with multilaterals/other UN organisations	-	-	-	-	-	5.1
Total multilateral aid	30.8	45.3	67.0	90.7	86.9	138.6

Sources: DFA, *Ireland's Official Development Assistance*, various years and data provided in July 2003.

---

11. Garret Fitzgerald, TD, Minister for Foreign Affairs, Parliamentary Debates, Dáil Eireann, Official Report, Vol. 265, 1973, p. 743.

The most recent restatement of the place of aid policy within the context of overall foreign policy is presented in the 2002 IARC report where development policy is again described as forming an 'integral part of Ireland's wider foreign policy' which, like other countries, 'bears the imprint of its history'. The report notes: 'We have first-hand experience of colonisation. The human cost of poverty and underdevelopment is engraved on the national memory. This shared legacy with developing countries has helped to create a strong bond of understanding and empathy' [*DFA, 2002: 1*]. The breadth of support for Ireland when it was elected to a seat on the UN Security Council for the years 2001 to 2002 'testifies to the strength of our relationship with the developing world' and is 'a validation of the diplomacy of successive Governments and also a recognition of the contribution by Irish missionaries and other development workers over many years'. Aid policy is seen as 'rooted in Ireland's foreign policy' and as having 'above all, a moral and humanitarian imperative' [*ibid.*].

(i) Ireland and the EU. The main issues for DCI in relation to the European Union are policy coherence and the so-called EU ori-entation debate. The 2002 IARC report describes policy coherence as a 'key' requirement which 'should be striven for across the full range of Irish government policies' [*DFA, 2002: 26*]. For DCI, the main challenge is to ensure that other government departments – most notably in this context the Departments of Enterprise, Trade and Employment and Agriculture and Food – incorporate a development perspective into their policy approaches.

At the EU level, development co-operation policy is now decided in the General Affairs and External Relations Council which is responsible for all aspects of the EU's external relations. The 133 Committee, which deals with commercial issues including trade, meets almost every week in Brussels. Before each meeting, an ad-hoc inter-departmental '133 co-ordinating committee', which includes representatives from the Departments of Foreign Affairs, Enterprise, Trade and Employment, and Agriculture and Food, meets in Dublin to co-ordinate the Irish position for the Brussels meeting. Other inter-departmental meetings also take place, some regular and others more ad hoc, which enable each department to ventilate its views before preparing a co-ordinated presentation at other international fora such as the Food and Agriculture Organisation (FAO) and the World Trade Organisation (WTO). One example was the preparation of the Irish speech for the Doha ministerial meeting in November 2001 which included a call for increased aid, more support for building trade capacity in developing countries, and access to medicines under the TRIPs agreement, while also emphasising the importance of agriculture to the Irish economy and 'an appropriate balance between trade and non-trade concerns'.[12] Tensions which arise in trying to achieve that 'appropriate balance' reflect those which arise at North–South level during meetings of the WTO. They also reflect the continuing importance of the farm and fisheries lobbies within the Irish economy. The Department of Agriculture and Food is committed to 'safeguarding the European model of agriculture [which] is of fundamental importance, therefore, to Ireland and the EU'. That department's position includes protecting Irish sugar and beef producers, just as the Department of the Marine is always concerned with protecting the interests of the Irish fishing industry in negotiations on EU fishery policy.

Negotiations between the EU and the ACP states on moving from Lomé-style preferences to Cotonou-style contractual agreements involving regional arrangements got under way in September 2002. During the preparatory discussions within the European Council in May 2002, Ireland and the other members of the EU 'like-minded' group were critical of the Commission's opening proposals, describing them as 'ungenerous' and arguing instead for extension of access for 'everything but arms, duty-free and quota-free' to all developing countries (including all ACPs).

The so-called 'orientation debate' within the EU has been focused on improving the effectiveness of EU actions on the global stage. In relation to aid policy, it has to be admitted that, despite accounting for more than half of all global aid flows, the EU (15 member states plus the Commission) cannot yet be regarded as an 'aid leader' since it does not speak with one coherent voice. The need for co-ordination among member states across the three pillars of EU foreign policy is now seen to be acute. In relation to aid policy, the current orientation debate turns in part on whether aid is an 'integral part' or 'an instrument' of the EU's overall foreign policy. Ireland is concerned that the EU's development policy should not be diluted or 'oriented' away from its poverty focus, nor used as an instrument to achieve broader foreign policy objectives. It sees the need for a balance to be achieved between adhering to its agreed strategic approach and its desire to have a greater global role by responding to every international crisis that emerges.

---

12. Address by Tom Kitt, TD, Minister of State for Enterprise, Trade and Employment, to the WTO Ministerial Meeting, Doha, Qatar, 12 November 2001.

(ii) Ireland and the World Bank and IMF. Until relatively recently, debate on Ireland's role within the World Bank and IMF was rather muted. To be sure, Irish NGOs and academics, just like those in other countries, have often criticised their lending activities, especially in relation to structural adjustment loans and facilities whose conditionalities in the 1980s and early 1990s were seen to have negative implications for social spending and for the poor. But there was little comment on Ireland's contributions to these organisations because payments to them, apart from the IDA, were mandatory. However, when the government announced its decision to contribute to the Enhanced Structural Adjustment Facility (ESAF) in 1994, criticisms were voiced by the NGOs – under the umbrella of the Debt and Development Coalition – and the Minister responded by deferring payments. Four years later, a major contribution was made to the ESAF, as part of a wide-ranging debt relief package jointly announced by the Ministers for Finance and Foreign Affairs.

Ireland provides all its aid in grant form so no developing country owes any official debt to it. Nevertheless, the two Ministers decided in 1998 that it would be appropriate for Ireland to become involved in debt-relief programmes for HIPCs whose development programmes are clearly being severely constrained by their debt-servicing obligations. Five of Ireland's six programme countries (Ethiopia, Mozambique, Tanzania, Uganda and Zambia) are involved in the HIPC initiative while all six (including Lesotho) are eligible for the Poverty Reduction and Growth Facility (PRGF).

The two Ministers jointly announced a debt relief package valued at €40 million in September 1998 with a two-pronged approach – part of the contribution to be made through the World Bank/IMF HIPC Trust Fund and the IMF's ESAF (now PRGF), and part to be contributed bilaterally to Mozambique and Tanzania. The elements of the package were: €19m. to be paid to the World Bank and IMF under the HIPC initiative, €9m. to ESAF/PRGF, and €12m. bilaterally to the two programme countries. When announcing the package, the ministers stressed their appreciation of the work of Irish NGOs in highlighting the debt issue nationally and internationally and stated that they would remain open to receiving further inputs from them on debt and development policy. A total of €18.5m. was paid to the World Bank and IMF in 1999, €13.5m. (as shown in Table 9.3) and a further €5m. which was transferred out of Ireland's balance in the IMF's special contingency account (SCA–2). The remainder of the €40m. is being paid in instalments up until 2008.

The figures in Table 9.3 might suggest that Ireland is now winding down its contributions to debt relief. However, this is not the case. Already, Ireland has paid around €6m. of the EU contribution of over €1b. to debt relief, proportionate to its EU budgetary contributions. Further payments are also being made to the HIPC Trust Fund which are being earmarked for Ireland's programme countries (€4.5m. for Ethiopia in 2001 and around €1m. for Zambia in 2002). During 2002, the Departments of Foreign Affairs and Finance unveiled a major new debt relief strategy – although it was not accompanied by any announcement of new resources.

Neither department is satisfied with the impact to date of the enhanced HIPC initiative. In their joint contribution on debt strategy to the International Conference on Financing for Development held in Monterrey, Mexico in March 2002, concerns were expressed under four headings: the adequacy of HIPC relief; debt sustainability (particularly the selection of indicators and assumptions/projections regarding future trends); debt relief for post-conflict countries; and the future funding of HIPC debt relief. There is particular concern in relation to some countries where the likely impact of HIV/AIDS has not been taken adequately into account. Projections underpinning debt sustainability analysis are seen as based on optimistic growth rates in exports and national income (and thus government revenue). Debt service/government revenue is considered to be a more appropriate indicator of debt sustainability than the debt service/export ratio. There are also concerns regarding ways of dealing with shorter term shocks to which HIPCs are particularly vulnerable because of their export concentration, import dependence, low reserves and aid dependence.

The 1998 debt relief package was based on the following principles: emphasis on human development needs in assessing debt sustainability; increased flexibility to ensure speedier implementation and wider coverage; consultation with Irish NGOs; and promotion of greater transparency in the workings of the international financial institutions. Two others were added in 2002: working with other donors and international agencies (such as Debt Relief International) to assist HIPC governments to develop their capacity to undertake debt strategy analysis and debt management; and support for civil society actors in partner countries to facilitate their active participation in dialogue and strengthen qualitative feedback mechanisms. Ireland's view is that total debt cancellation within the HIPC process, funded from additional donor contributions, could make an important contribution to the achievement of the millennium development goals, but such additional funding from major donors is unlikely in the short run. Other possible sources include building additional funding for debt relief into IDA replenishment (which would provide assured funding and represent structured burden-sharing) and seeking increased voluntary contributions. Ireland's position is that IDA commitments to the HIPC process should be funded from some combination of IBRD net income and 'additional' contributions through the replenishment process [*Government of Ireland, 2002: 17–21*].

The World Bank is the only multilateral development bank – apart from the European Bank for Reconstruction and Development – to which Ireland currently belongs. Following acceptance by the government of the IARC recommendation that Ireland should join the African Development Bank and the Asian Development Bank, 'subject to securing of satisfactory conditions of accession' [*DFA, 2002: 58 and 60*], it was announced that legislation to that effect would be prepared. However, no action had been taken by the end of 2003.

(iii) Ireland and the United Nations. As Table 9.3 shows, funding from DCI to the UN agencies has been rising sharply during the past decade. The 1999 DAC review described Ireland's multilateral involvement as a 'lively and active engagement'. It noted the increasingly large number of UN agencies (39 in 1999) to which Ireland makes voluntary contributions and suggested that a more selective and targeted approach should be adopted if these contributions are to be further expanded. It advised that 'Ireland might benefit from becoming less of a marginal player in some of the multilateral agencies it supports' [*OECD/DAC, 1999: 18*]. The 2002 IARC report supported the suggestion of a more targeted approach and further suggested that DCI enter into 'strategic partnerships' with UNDP, UNHCR and UNICEF 'whose policies and priorities fit well with those of Ireland Aid and which have made a tangible contribution to the alleviation of poverty and suffering' [*DFA, 2002: 95*].

This approach is now being pursued. The number of agencies receiving funding has already been reduced from 35 to around 20 and is still falling, although Ireland Aid recently agreed to contribute €3m. over three years to a new Environmental Trust Fund within the UN Environment Programme. Agencies that have been receiving token amounts are being dropped and major funding is to be provided to the three principal recipients UNDP, UNHCR and UNICEF, whose efficiency and effectiveness (including their 'results-based' budgeting approach) are to be kept under surveillance in co-ordination with other main donors (probably in multi-donor evaluations). Over the period 2000 to the end of 2007, voluntary contributions to all UN agencies are to rise from €15.9m. to around €60m. within a rolling programme underpinned by a more active involvement within the agencies concerned. Contributions from DCI will be provided in the form of core funding and will not be earmarked. Greater representation on the boards of the major UN agencies is to be sought in order to strengthen Irish influence. Ireland has been a permanent member of the executive committee of UNHCR since 1998 and is a member of ECOSOC and the Commission on Human Rights for the period 2003 to 2004. It participated in the working group that established the Global Fund to fight AIDS, TB and malaria. In common with other donors, Ireland is putting increasing amounts into the fight against AIDS. In 2002 alone it contributed around €43m. divided between its bilateral programme and Irish and international NGOs, as well as multilaterally through the Global Fund, the International AIDS Vaccination Initiative and research into microbicides.

Ireland is also strengthening its contribution to UN conferences and to preparations for them within the EU where it usually works in co-ordination with other EU 'like-minded' countries such as the Scandinavians and the Netherlands. At the 2002 Monterrey conference on financing for development, Ireland worked within this group to press for large increases in aid. At the conference planning stage, the average ODA/GNI ratio in the EU was 0.33 per cent (total EU ODA divided by total EU GNI). The official EU proposal at the Monterrey conference was that a ratio of 0.33 per cent should be attained by each individual member state by 2006. This would lift the EU average to 0.39 per cent. Ireland proposed that the 0.39 per cent should be an interim target only and that the EU member states should commit themselves to reaching the UN 0.7 per cent target by a specific date.

Two issues lay at the heart of the Monterrey conference. They have continued to be central during the follow-up phase.

First, the 'Monterrey consensus' was based on a deal between developing and industrialised countries under which the former committed themselves to good governance, human rights and sound economic management and donors committed themselves to reaching agreed ODA targets, increasing investment and opening up their markets to exports from developing countries.

The second issue is that of mutual accountability. Are the commitments being lived up to? How can they be tracked? One tracking mechanism will bring together the DAC members at ministerial level and the Economic Commission for Africa. The DAC will track how its members are meeting their commitments while the secretariat of the New Programme for Africa's Development will report on how NEPAD's peer review systems (one on governance and one on economic reform) are working. Reports will be issued identifying weaknesses and suggesting improvements – all in a spirit of constructive dialogue rather than confrontation.

At UN level, the Monterrey consensus calls for a set of meetings to enable all relevant stakeholders to 'stay engaged' on financing for development. In particular, the consensus encourages the annual spring meeting of the Economic and Social Council of the UN (ECOSOC), the World Bank and IMF (a grouping in which the WTO now also participates) 'to address issues of coherence, coordination and cooperation as a follow up to the Conference'.[13] ECOSOC has a number of subsidiary bodies or commissions that report to it and it then reports in turn to the UN General Assembly. One of these bodies is the Commission on Social Development whose main function is to monitor the follow-up to the social summit held in Copenhagen in 1995. Another is the Population Commission which monitors the follow-up to the population summit held in Cairo in 1994. No commission has been set up to monitor the follow-up to Monterrey because it was not a single-issue conference but rather one with three issues – finance, trade and investment. However, it was agreed that the annual spring meeting of ECOSOC, the World Bank and IMF and the WTO would examine issues of policy coherence and the follow-up to Monterrey. In autumn 2003, a high-level ministerial meeting was held during the UNGA and it focused on the follow-up to Monterrey.

At EU level, the Commission has been given a mandate to track implementation of the eight EU commitments made at Monterrey (the most important of which are the commitment to increase the average ODA/GNI ratio to 0.39 per cent by 2006, increased debt relief and a greater voice for developing countries in the World Bank and IMF). Member states of the EU have been circulated with a questionnaire that, when completed, will show the extent to which each is fulfilling its obligations. This monitoring exercise will be conducted each year including during Ireland's presidency in 2004.

Ireland was a non-permanent member of the UN Security Council for the two-year period 2001 to 2002 where it tried to get African issues higher up the agenda. Since it has no strategic or trade interests in the continent, it got involved in issues such as sanctions,

---

13. Report of the International Conference on Financing for Development, Monterrey, Mexico, 18–22 March 2002 (A/CONF.198/11), Ch. 1, Resolution 1, Annex, para. 69 (b).

peacekeeping and conflict resolution. It is considered an 'honest broker' in relation to the main conflicts, all of which are cross-border in terms of their impacts (DRC, Somalia, Liberia, Sierra Leone). Its growing aid budget contributes to its credibility, as does its willingness to follow up on decisions. Following the elections in Sierra Leone, for instance, Ireland contributed €630,000 to the Special Court. It also took a pragmatic approach to its role, operating in the main on its own and not getting involved in issues where the permanent members had already adopted fixed positions.

VI. THE DOMESTIC DIMENSION

1. The Main Actors

Outside the government sector, the development NGOs are the main actors in the development arena in Ireland. The three largest among them (Concern, Trocaire and Goal) exercise influence through their size, their access to ministers and officials in the Departments of Foreign Affairs and Finance, and their use of parliamentary questions and ability to organise write-in campaigns. On the whole, very good relations operate between the NGOs and DCI. There has always been a high level of complementarity, even similarity, in their projects. Moreover, as the NGOs follow DCI in adopting programmatic approaches, the complementarities and similarities in terms of approach and analysis are tending to increase. Many of the consultants and special advisers employed by DCI have backgrounds in the NGO sector. Although the recent sharp rise in the funding provided by DCI to the NGOs is undoubtedly expanding their resource base, it also imposes on them greater evaluation and reporting conditions, and increases their dependence on government. On the other hand, the operative term may be mutual dependence. The reaction of the NGOs to the 2002 budgetary cuts was relatively muted: significantly, their own funding had not been reduced.

Because, unlike the big or medium-sized countries, Ireland does not pursue global strategic interests, there are few international political issues that give rise to tensions between the NGOs and the Irish government. Any tensions that do arise tend to be in the economic arena (for example, in relation to the impact of the EU's Common Agricultural Policy on developing countries) or, in the early days of the ESAF, in relation to positions adopted within the Bretton Woods institutions.

There are weak institutional links between DCI and the universities, although individual academics are involved in advisory committees and as consultants to the aid programme, and DCI funds a growing number of Study Fellows from the programme countries enrolled on (mainly postgraduate) courses in the universities and other third-level institutions. Funding for research, through the National Committee for Development Education, had been miniscule, but it has increased quite significantly since the establishment of a dedicated research fund within the new Advisory Board of DCI. The level of interest shown by the private sector in the aid programme has been low – probably because of the lack of significant trade and investment links between Ireland and its programme countries which are all among the least-developed countries in Africa. To date, the Irish aid programme has been providing assistance to the private sector in its programme countries only through funding provided to the International Labour Organisation and the African Management Services Company. The Minister of State now wants to get the Irish private sector directly involved in the aid programme. In April 2004, DCI set up a Private Sector Forum which is designed to act as 'a facilator and educator' for Irish entrepreneurs in order to encourage them to get involved in DCI's programme countries and to help promote their growth and international trade. However, in launching the Forum, the Minister of State emphasised that no ODA funding would be made available for the new initiative. 'As in all business ventures,' he stated, 'the risk will remain with the investor and trader....I am appealing overtly to your commercial self-interest and unapologetically I am appealing also to your altruistic sense, what is now being called corporate social responsibility.'[14]

Only a relatively small number of politicians tend to be vocal in the Dáil (lower house) and Seanad (upper house) on the aid programme. On the other hand, those who are outspoken tend to be high-profile politicians who are also members of the Development Cooperation sub-Committee of the Joint Oireachtas (Parliament) Committee on Foreign Affairs. There is a widespread perception among politicians that 'there are no votes in development co-operation'.

2. Public Understanding and Attitudes to the Official Aid Programme
A Eurobarometer survey undertaken in 1997 reported that 95 per cent of respondents in Ireland (compared with the European average of 87 per cent) were in favour of development assistance. The 2002 IARC

---

14. Speech by Mr Tom Kitt, TD, Minister of State for Development and Human Rights at the launch of The Private Sector Forum, Dublin, 19 April 2004.

report states that there is 'broad public support' for aid today, citing the 'extraordinary levels of private generosity' shown by Irish people in responding to humanitarian emergencies and in contributions to development NGOs. The IARC report also claimed that 'there is also firm support for the government's official programme of development co-operation' despite the fact that it states that 'relatively little is known about the breadth and scale of the Irish aid programme [*DFA, 2002:102*]. The general public has always tended to associate development assistance with the work of NGOs, especially during humanitarian emergencies, but has a low level of awareness of the extent to which NGOs and missionaries receive official funding.

The IARC report recommended that increased resources, human and budgetary, should be directed towards an information strategy. It also acknowledged the 'crucial role' that development education has to play 'in enlarging public understanding of development issues, both global and local' and in challenging attitudes which perpetuate poverty and injustice and empowering people 'to act to bring about more equal development'. It stated that an effective development education system 'stimulates greater public interest in these issues and contributes to a greater understanding of the underlying causes of poverty and underdevelopment', adding that a spin-off from this process 'could be greater public awareness of, and support for, the government's official aid programme' [*ibid.: 104*].

Development education may be defined as the process of raising awareness and deepening understanding about development issues, and promoting active and informed responses among all sectors of a population. It has been an area of lively activity in Ireland for many years with involvement by a wide variety of actors within the formal (primary, secondary and tertiary) education sector and the development NGOs. The official Irish aid programme has been supporting their efforts for nearly 20 years through a succession of agencies: Higher Education for Development Cooperation from 1986, the National Development Education Grants Committee from 1990 and the National Committee for Development Education since 1995. The IARC report recommended that NCDE should be absorbed into a dedicated Development Education Unit within DCI, that it should have structured relationships with the Department of Education and Science and with the National Council for Curriculum Assessment, that it should be given a larger budget, conduct attitudinal research, and produce a strategy plan [*ibid.: 105*]. These recommendations were effected during 2003.

A national survey of attitudes to development issues was carried out in October to November 2002 [*DFA, 2003*]. It was the first such national attitude survey of adults in Ireland since the 1985 and 1990 surveys conducted by the Advisory Council on Development Cooperation (ACDC). The survey began by asking people for their 'spontaneous impressions' of 'developing countries'. Not surprisingly, the four most cited impressions were: Africa; poverty/poor countries; famine/hunger, and Third World. There was no further probing of respondents' attitudes following this opening question. One does not have to be an expert on attitude surveys to know that if a follow-up question had asked for spontaneous impressions of a specific developing country (for example, Brazil or Cameroon) there would have been a high likelihood of eliciting a very different type of response (for example, football or sunshine).

The survey then proceeded to ask respondents about various aspects of 'overseas development aid'. It found that a majority of Irish people are in favour of helping developing countries, perceive a need for such assistance and feel they have a responsibility to do so (as in 1985 and 1990). However, most respondents 'did not seem to have a clear idea of what overseas development aid [*sic*] comprises' [*DFA, 2003:25*]. The majority were of the opinion that 'giving money' was the main way in which Ireland helped developing countries. There was a much higher recognition of the role of NGO aid than ODA from the government (as had been the case in 1985 and 1990). Again, this is not surprising, given the vagueness of the term 'overseas development aid'. Almost half of all respondents had 'absolutely no idea how much financial assistance was provided by the Irish government in 2001' (but surprisingly, when informed, felt it was 'about right').

Well over half of the respondents had never heard of DCI – but almost all opined that it should be better known. Interestingly, there was a basic contradiction between perceptions of what was 'very important' in causing poverty (disease, lack of health care, war and conflicts) in developing countries and ways in which it might be alleviated (sending out skilled people to train people, supporting self-help programmes, providing emergency relief). This lack of consistency is in contrast to the opinions expressed in the 1985 and 1990 ACDC surveys where 'lack of education' was perceived as the main cause of poverty and 'sending out skilled people to train and educate people in the third world' was seen as the 'most helpful' way of helping poor people [*ACDC, 1985 and 1990*].

The new Development Education Unit within DCI produced a strategy plan for development education in 2003. It sets out the following six objectives for development education in Ireland:

- to integrate a development education perspective in relevant education policies
- to integrate and support the delivery of development education in selected areas in the formal and non-formal education sectors
- to provide support to civil society organisations to increase public understanding of development issues
- to facilitate capacity-building within the sector
- to promote more effective use of communications to increase public understanding of development issues
- to identify and maximise educational opportunities for public engagement with DCI's programme.

In terms of approach, DCI will work through partnerships with identified groups – such as youth, third-level education, NGO and community groups – within the broad-based and heterogeneous 'development education sector'. DCI does not intend to 'deliver' development education itself. It will be the responsibility of actual or potential groups to organise themselves, come up with a plan, and then engage in a partnership with the Development Education Unit within DCI. It will provide resources through a range of funding mechanisms that will include partnership funds, a civil society development education fund, a capacity-building fund and a media challenge fund. The operational details of the new strategy including the ways in which the partnership arrangements will work still need to be spelled out. According to those already engaged in development education, this will necessitate a deeper level of consultation than what took place before the plan was published. However, what is already clear is that, in contrast to the NCDE model, the new strategy signals an approach by DCI that goes beyond funding to include a partnership between it and those who directly deliver development education programmes.

The IARC report had also called for increased funding for development education. This is essential if the new strategy is to be successfully delivered. An examination of Table 9.2 shows that funding has remained fairly static in recent years. Spending has not kept pace with the growth of the overall ODA programme. Indeed, in 2002, the budget was not fully spent. Moreover, it would appear that spending on development education by DCI still lags behind that of the major NGOs. Data published in 1995 showed that expenditure on development education by DCI in 1994 was less than half that spent by just one major NGO that year [*O'Neill, 1995: 195*]. Data obtained in July 2003 on expenditure by four of the biggest NGO spenders on development education in 2002 (around €2.7m.) shows that it still exceeds both the budget (€2.6m.) and the outturn (€2.1m.) of DCI expenditure under this heading in 2002.

VII. THE INSTITUTIONAL STRUCTURE

Until now, the Irish aid programme has been implemented with a relatively small budget. Total administration costs, including both headquarters and the field, amounted to 5.3 per cent of total ODA in 2002 (see Table 9.1). Staffing levels have not kept pace with the increase in the size of the programme. The 1999 DAC review had recommended that 'staffing, skill mixes and career perspectives must be reinforced' [*OECD/DAC, 1999: 22*]. The 2002 IARC report referred to a '1999 external report' which concluded that 'fundamental changes are needed to the management of the programme' [*DFA, 2002: 108*]. The 1999 DAC Peer Review described 'an impressive process through which Irish Aid has managed to accomplish such volume increases, while simultaneously strengthening the quality and operating professionalism of the programme'[*OECD/DAC, 1999: 15*]. However, while noting that staff levels had increased from 84 in 1998 to 143 in 2003, the more recent Peer Review described the current situation as 'barely adequate' in terms of staff numbers and skills mix [*OECD/DAC, 2003: 60*].

DCI is located within the Development Cooperation Division of the Department of Foreign Affairs. Alternative administrative arrangements, such as the creation of an autonomous aid agency, had been debated over the years but have always been rejected by government. The 1999 DAC review concluded that reinforcing existing organisational structures is preferable in the short term although it added the rider that 'in the longer-term, establishing an independent implementing agency is an appealing option from an operational point of view' [*OECD/DAC, 1999:22*]. The 2002 IARC report (all recommendations of which have been accepted by the government) considered a number of administrative options but concluded that 'a dedicated directorate of the Department of Foreign Affairs offers the best potential for management of the expanding programme' despite 'current shortcomings' such as 'serious understaffing and a lack of managerial flexibility' [*DFA, 2002: 116*]. However, the issue of organisational structures may not have been settled for the long term. The 2003 DAC Peer Review suggested that Ireland should keep its institutional framework 'under review' in order 'to ensure it has, in a medium term perspective, the operational flexibility needed to manage its ODA and the capacity to adapt to rapid change, especially if budget growth projections to 2007 are realised' *[OECD/DAC, 2003: 16]*.

As noted above, headquarters staff has been increased. Managerial improvements are being introduced. The entire process whereby projects are assessed, approved, monitored, reported and evaluated is becoming more 'joined-up'. The relationship between 'desk' personnel at headquarters and field officers is altering as the former assume a more pro-active role. Key risks that might affect the programme, whether strategic, operational, financial or reputational, have been identified, and the implementation of a system of active risk management is under consideration.

The Irish Aid Advisory Committee has been replaced by the new Advisory Board whose tasks are described as including: general oversight and the provision of advice to the Minister and to senior DCI management on the strategic direction of the programme and on the implementation of policy; organisation of the Development Forum; reviewing evaluations of programmes and projects; supervising general reviews of the programme at three-yearly intervals; and commissioning research relevant to the development cooperation policy agenda. According to the 2003 DAC Peer Review, this wide range of responsibilities 'creates a challenge in terms of ensuring that the board maintains a clear strategic view while performing both executive and advisory functions' *[OECD/DAC, 2003: 16]*.

VIII. SOME CONCLUDING COMMENTS

Ireland established its bilateral ODA programme in 1974 a year after it entered the European Economic Community. In the period up until 2003, the programme has grown in volume terms from just under €2 m. to over €450m. As a percentage of gross national income, it has grown from 0.05 to 0.4 in 2002, putting it in seventh place in the DAC donor list. Although it has failed to meet all the targets set for its growth over the years, that of reaching the UN target of 0.7 per cent of GNI is supported at the highest level of government and among the general public.

From its beginnings in the 1970s – and not surprisingly, given that missionaries had pioneered Ireland's involvement in developing countries – the Irish aid programme has been imbued with a strong humanitarian motivation and an overarching focus on poverty reduction. This focus is reflected in its choice of programme countries, all of which are among the poorest in sub-Saharan Africa, and in its sectoral assistance which is directed towards primary education, primary health care, subsistence agriculture and rural infrastructure. In common with many other donors, it has moved away from supporting 'project islands' over the years and is now delivering its assistance mainly through targeted support to sub-national districts, to specific sectors and, in co-operation with other donors, to general support to central government budgets. During the past few years, it has been providing significant amounts to prevention and alleviation of HIV/AIDS both at bilateral and multilateral levels. The number of cross-cutting issues that it is attempting to 'mainstream' now number four: gender, governance, environment and HIV/AIDS. Although Ireland provides all its ODA in grant form, it has begun to make significant contributions to debt relief, bilaterally to some of its programme countries and multilaterally to the World Bank/IMF HIPC scheme. With respect to policy coherence, tensions can arise on the trade front. Ireland is progressive in general terms in trade discussions within the EU and the WTO, being strongly supportive, for example, of the 'everything-but-arms' initiative. However, it remains cautious with respect to reform of the EU's common agricultural policy and, under the influence of the Department of Agriculture and Food, tends to be more responsive to the demands of its own agriculture lobby than to the demands of developing countries to open up EU markets in meat and dairy products.

Development Cooperation Ireland has always worked in close co-operation with Irish NGOs and now spends a higher proportion of its ODA budget through them than does any other donor. Ireland also co-operates with other donors, especially other 'like-minded' EU member states, both in its programme countries and multilaterally in its attempts to promote progressive positions with regard to aid volumes and debt reduction. The 1999 DAC review of Irish aid noted that Ireland has had a 'very lively engagement' with international organisations. The 2003 DAC review described the Irish aid programme as one 'that distinguishes itself by its sharp focus on poverty reduction and its commitment to partnership principles' *[OECD/DAC, 2003: 11]*.

REFERENCES

DF, 2001, Statement by Minister for Finance, Charlie McCreevy, TD, Governor of the World Bank and the International Monetary Fund for Ireland, at the joint annual meeting, September 2000, *Annual Report: Ireland's Participation in the World Bank and the International Monetary Fund 2000*, pp. 27–8.

DFA, various years, *Ireland's Official Development Assistance*, Dublin.

DFA, 1979, *Development Cooperation: Ireland's Bilateral Aid Programme*, Dublin.

DFA, 1993, *Irish Aid: A Strategy Plan*, Dublin.

DFA, 1996, *Challenges and Opportunities Abroad: White Paper on Foreign Policy*, Dublin.

DFA, 1997, *Pursuing Ireland's External Interests: Strategy Statement of the Department of Foreign Affairs*, Dublin.

DFA, 2001, National Statement by Ms Liz O'Donnell, TD, Minister of State at the Department of Foreign Affairs, at the Third United Nations Conference on the Least Developed Countries, Brussels, 14–20 May 2001.

DFA, 2002, *Report of the Ireland Aid Review Committee*, Dublin.

DFA, 2003, *Attitudes Towards Development Cooperation in Ireland: The Report of a National Survey of Irish Adults* (conducted by MRBI in 2002 and edited by John Weafer), Dublin: Department of Foreign Affairs (Ireland Aid).

Government of Ireland, 1996, *Challenges and Opportunities Abroad: White Paper on Foreign Policy*, Dublin.

Government of Ireland, 2002, 'Ireland Aid and Department of Finance Debt Strategy', International Conference on Financing for Development, Monterrey, Mexico, 18–22 March.

OECD/DAC, 1999, *Ireland*, Development Cooperation Review Series 35, Paris: OECD.

OECD/DAC, 2003, *Ireland*, Development Cooperation Review, Paris: OECD.

O'Neill, Helen, 1995, 'Ireland's Foreign Aid in 1994', *Irish Studies in International Affairs*, Vol. 6, pp. 185–98, Dublin: Royal Irish Academy.

10

Trends in the Debate on Italian Aid

JOSÉ LUIS RHI-SAUSI AND MARCO ZUPI

1. MOTIVES AND OVERALL OBJECTIVES OF AID POLICY

The evolution of Italian aid has been marked by a series of ups and downs. Cycles of increased resources have been followed by drastic reductions. This reflects an essential characteristic of Italian development co-operation policy: it is not a structural component of Italian foreign policy, but a circumstantial component dependent on diverse factors.

The initial phase of Italian aid in the late 1970s and early 1980s was marked by humanitarian motivations linked to famine in East Africa, and the Horn of Africa in particular. This led to discussion of a series of legislative initiatives on co-operation with developing countries, with the approval by Parliament of Law No. 49 of 1987, which is still in force.

The 'African' cycle, hinged on aid to the world's poorest countries, gave way, towards the end of the 1980s, to a new motivation for official development assistance (ODA), based this time on economic criteria. Development co-operation was used as an instrument for commercial penetration by Italian firms. This cycle involved emerging markets of particular importance for Italian economic interests, such as Argentina and China. A determining factor for this policy reorientation was the decrease in demand for Italian goods and services on the part of developing countries, especially in the aftermath of the debt crisis. ODA came to represent an important tool for the internationalisation of certain sectors of the Italian economy such as construction and telecommunications.

This brief phase, in which Italian ODA reached historic heights (0.42 per cent of GNP and ranking as sixth donor country), ended badly in the early 1990s. The judicial process against generalised corruption, known as *Tangentopoli* or *Clean Hands*, also involved development co-operation, leaving it discredited in the eyes of the public.

Furthermore, the rules set by the international community limited the use of ODA as an economic instrument by donor countries. Finally, ODA began to reflect the effects of domestic policies aimed at reorganisation of the budget.

These first two phases of Italian development co-operation were promoted and supported by the whole range of political forces. As the data on the budget show, Parliament increased the allocations for ODA throughout the period, in accordance with the government's proposals. To be more precise, one can observe that it was mainly the Christian Democrats who supported the 'African' phase, while the 'economic' phase was conducted by the Socialist Party.

The second half of the 1990s saw the onset of a new cycle of drastic reduction in aid to developing countries, which is still the hallmark of Italian co-operation. Co-operation policy's persistent loss of credibility, restrictions on the budget and Italy's new position within the post-Cold War international order created the conditions for the emergence of a new motivation for ODA: diplomatic interests. The reappropriation of co-operation policy by the Ministry of Foreign Affairs has led to its subordination to the interests and trends of Italian foreign policy. Three motivations for Italian development co-operation can be discerned. On the one hand, there is its reorientation towards the objectives of international co-operation, in particular the emphasis on poverty reduction as the main purpose of aid, and the adaptation of Italian policy to the recommendations agreed by the Development Assistance Committee of the OECD. On the other hand, ODA has been used as a negotiating instrument for the achievement of certain strategic objectives of Italian foreign policy, such as the search for consensus among developing countries regarding a proposal for reform of the UN Security Council which does not exclude or penalise Italy. Finally, Italian development co-operation seems to have found, for the first time, a geopolitical motivation in areas of geographical proximity (the Balkans and the Mediterranean Basin).

In the present cycle, therefore, different motivations for Italian ODA can be discerned alongside a greater fragmentation of international aid policy. One factor which has affected – and may increasingly affect – its declining trend is the renewed attention on the part of civil society organisations to North–South relations. The World Social Forum of Porto Alegre and the protests against the G-8 meeting in Genoa in 2001 have triggered the mobilisation of civil society on globalisation issues. These movements have emphasised two themes: the maintainance of peace and the cancellation of the poor countries' foreign debt. ODA does not, in itself, appear to be a fundamental preoccupation of Italian civil society. There is no significant movement in favour of the 0.7 per cent of GDP target or any other objective aimed at an increase in aid. Moreover, the agents promoting these movements are not the traditional development NGOs. What is in process is a social phenomenon that does not directly concern ODA, but is certainly conditioning its evolution.

The widespread mobilisation in favour of the cancellation of poor country debt (Jubilee 2000) has led to the approval of specific legislation (Law No. 209 of 2000). The policy stemming from it is more far-reaching and less constrained than the existing policies promoted by the IMF and the World Bank, particularly as regards the Heavily Indebted Poor Countries (HIPC) Initiative. In terms of the financial resources destined for ODA, this has translated into an increase in the total amount of ODA, since debt cancellation and reduction are classified as part of ODA rather than as a new budget item. In other words, bilateral debt was not being repaid before it was cancelled, and did not therefore free up resources for new uses.

The debt issue has received strong political and social support. Mobilisation by the Catholic Church, media personalities and a number of organisations probably reflected serious unease about a co-operation policy deemed lacking in quality, efficacy and resources. The political parties simply followed the lead taken by civil society and the Church. The centre–left government which came to power in 1996 responded to this social demand and transformed it into government policy. The importance of social pressure is particularly evident when one notes that among the conditions imposed by Italy to guarantee debt relief is an emphasis on peace, an issue of major importance to Italian society. On the other hand, society's sensitiveness to the issue (crisis prevention, humanitarian aid, post-conflict reconstruction and peace enforcement) has pushed the traditional vocation of Italian ODA further to privilege humanitarian and emergency aid. More than 1.5 million Italians took to the streets in dozens of cities to protest against possible US military action against Iraq. Opinion polls supported that view, with a survey from *Opinioni* showing that more than two out of every three Italians were opposed to any armed conflict over Iraq, and nearly four out of five were opposed to Italian participation in such action, except as part of a UN-sponsored force.

Finally, the widening of the motivations and objectives of Italian development co-operation has also led to a plurality of agents, both as regards aid management and as executive agents of international co-operation. The most significant element in this new direction of Italian aid is the increase – in importance and in the volume of resources mobilised – of decentralised co-operation, as subnational governments (regions, provinces and municipalities) increase their role in Italian co-operation.

II. STRATEGIES, MAJOR PRINCIPLES AND GUIDELINES

The first legislation to deal with aid was Law No. 1222 of 1 December 1971, which established a 'Service for Technical Co-operation with Developing Countries' within the Directorate-General for Cultural, Scientific and Technical Relationships of the Ministry of Foreign Affairs (MFA). But it was Law No. 38 of 9 February 1979 that created a specific structure for the management of bilateral co-operation and aid administration, the Development Co-operation Department (*Dipartimento per la cooperazione*, DIPCO), an autonomous department within the MFA. Law No. 73 of 8 March 1985 added another autonomous structure, the Italian Aid Fund (*Fondo Aiuti Italiani*, FAI) to deal with humanitarian and emergency aid to African countries affected by drought in the Sahel.

On 26 February 1987 Law No. 49, aimed at increasing the funds allocated to aid, reorganised all the existing provisions by setting up an entirely new system and creating a new structure, the Directorate-General for Development Co-operation (*Direzione Generale per la Cooperazione allo Sviluppo*, DGCS). Development co-operation remained a part of the MFA, but it became a clearly separate sector, with a specific institutional branch – the DGCS – entirely devoted to managing the whole of development co-operation policy. Country programming and the management of projects were handed over to the MFA, with Parliament reserving a controlling role without authority to verify or determine the course of action. The DGCS/MFA only has to submit an annual report on its activities to Parliament. The Inter-ministerial Committee for Development Co-operation (*Comitato Interministeriale per la Cooperazione allo Sviluppo*, CICS, ex Art. 3 of Law No. 49 of 1987), set up to define the general objectives and political guidelines of aid policy, was replaced by the Inter-ministerial Committee for Economic Planning (*Comitato Interministeriale per la Programmazione Economica*, CIPE) by Law No. 537 of 1993.

In terms of general principles, Law No. 49 of 26 February 1987, Art. 1 establishes:

> Development co-operation is an integral part of Italy's foreign policy and pursues the ideals of solidarity among peoples, seeking the fulfilment of fundamental human rights, in

accordance with the principles sanctioned by the United Nations and EEC-ACP conventions. This policy is aimed at satisfying basic needs and, primarily, at safeguarding human life; hence, encouraging food self-sufficiency, the enhancement of human resources, environmental conservation, the implementation and consolidation of self-reliant development processes together with the economic, social and cultural growth of developing countries. Development co-operation must have the purpose of furthering the condition of women and children and support the promotion of women.

At the sector level, the planned approach was most recently reiterated by the then Foreign Minister Renato Ruggiero, when he published the Ministry's forecast for ODA for 2002. The document laid down the following eight guidelines:

(i) poverty reduction to be the main objective of development co-operation;
(ii) Italy to contribute, via the adoption of the Genoa Plan for Africa, to the New African Initiative, which centres on democracy and good governance, conflict prevention, food security, education and health, international trade and the promotion of private investment;
(iii) support for the private/public partnership initiative for the global AIDS, malaria and tuberculosis Fund;
(iv) support for education, in particular basic primary schooling;
(v) cancellation of developing countries' external debt;
(vi) creation in developing countries of a climate conducive to foreign investment;
(vii) help for developing countries so that they can participate in international trade;
(viii) promotion of the interests of women, children and the handicapped.

These guidelines reaffirm the principles Italy has subscribed to at international level in the past, the core aim of which is the reduction of poverty.[1] Reaffirming the centrality of Africa and the emphasis

---

1. In the Inter-ministerial Committee for Economic Planning's 1995 framework, poverty reduction was treated as one among many different goals. In 1999, the DGCS addressed this issue by developing specific guidelines on poverty reduction, based on the work undertaken by the DAC Informal Network on Poverty Reduction.

given to education and basic health care are an important corollary to this.

Another important element to be considered is the relevance of the 'food aid' sector. Following the World Food Summit in Rome, in November 1996, Italy committed itself to making food security a first priority in its aid policy. At the end of 1997, Parliament ratified the London Convention,[2] the first consequence being that funding lines to food aid were reopened; in 1998 Italy allocated US$110 million for food aid, most of it for sub-Saharan Africa.

In terms of strategic approach, the Italian guidelines clearly reflect the importance of the DAC guidelines. Their emphasis on gender issues (November 1998), children (November 1998), disabled persons and community-based rehabilitation (July 1999), and, above all, poverty reduction (October 1999) are directly based on the work undertaken by the DAC.[3] In terms of numerical data, however, it is evident that the smallness of the overall resources and the predominance of activities to do with external debt reduction have slowed down the implementation of the understandings which Italy had been enunciating since the mid-1990s as well as in Law No. 49 of 1987.

Again, decentralised co-operation, namely projects financed at local government level by regions and municipalities, seems to be a particular Italian component in broader strategic terms as well.[4] Up until 1993 local authorities looked to the DGCS for financing and coordination, but since the mid-1990s the reduction of the national aid budget and the possibility of allocating up to 0.8 per cent of municipal budgets for development co-operation activities as well as regional financial resources have provided the financial means to support a major role for these new local actors. This new role also implies stronger support for the principles of local development and territorial partnership, which are the direct inspiration of decentralised co-operation. In March 2000, the DGCS issued its guidelines on decentralised co-operation, in May 2001 it signed an agreement with the National Association of Italian Municipalities (*Associazione*

---

2. Food Aid Convention signed by Heads of State and Government who attended the World Food Summit in Rome (November 1996), renewed in 1999.
3. The results of the DAC Informal Network on Poverty Reduction directly influenced the Italian guidelines on poverty reduction.
4. With regard to the co-operation carried out by Regions and Autonomous Provinces, overall ODA disbursement, in 2002, amounted to €34.073m., with commitments for 2003 of €34.974m. (data from Zupi and Mazzali [*2003*], the chapter 'La finanza italiana locale per la cooperazione allo sviluppo' – 'Italian Local Financing for Development'). Data concerning the municipalities' development co-operation are not available, since there are no official statistics on their activities so far. The only available data are to be found in a CeSPI Working Paper [*Stocchiero, 2001*].

Nazionale dei Comuni Italiani, ANCI) and in November 2002 it signed a similar agreement with the Union of Italian Provinces (*Unione delle Province d'Italia*, UPI).

III. VOLUME AND MAIN CHANNELS OF ITALIAN AID

Italian development co-operation consists of grants and soft loans,[5] allocations for which are established on a three-year basis by law. Each year a ministerial budget is attached to the financial estimates, containing the proposals and rationale for the allocation of financial resources, the choice of priorities and sectors, and the measures to be employed.

The size of co-operation policy seems to coincide with the international interests of Italian firms. When the entrepreneurial sector becomes particularly interested in the instrument of development co-operation, and the latter becomes a hinge between the objectives of international solidarity and those of economic and territorial internationalisation, the resources for bilateral ODA tend to increase. What can be observed today, therefore, is a scarcity of resources, together with a conspicuous multilateralisation of policy, functional to maintaining an important role within international organisations[6] (notwithstanding the decrease in the total resources allocated). In the face of the rigidity of commitments on the multilateral front, the brunt of Italian disengagement falls on the bilateral channel.

In absolute values, the 1980s were clearly the ascending phase in the cycle of Italian aid flows. Initially these were directed mainly to sub-Saharan Africa, with spillover effects on other regions in the second half of the decade. The 1990s, in contrast, represented the descending phase of the cycle, with a particularly pronounced effect on sub-Saharan Africa in the early years and a delayed spillover effect. The most emblematic case in this sense is that of Mediterranean Africa, 'discovered' by Italian ODA in the mid-1980s and with substantial increases up to the mid-1990s, only to suffer from the general slump in resources thereafter.

---

5. Italy has two different funds for aid: one for grants (*Fondo di cooperazione*), which is managed directly by the DGCS and requires annual replenishment, and another for soft loans (*Fondo rotativo*), managed by *Mediocredito Centrale*, a bank created by the Italian Treasury, which serves as a revolving fund so that repayment may be used to fund new loans without requiring annual replenishment.
6. The United Nations, the international financial institutions and, above all, the European Union which, as indicated in Table 10.2, has received around half the resources allocated to the multilateral channel.

TABLE 10.1
NET DISBURSEMENTS OF ODA (US$m.) AND ANNUAL VARIATIONS (%), 1990–2003

Year	Disbursements	ODA % GNI	Annual variations
1990	3395	0.31	-6.0
1991	3347	0.30	-1.4
1992	4122	0.34	23.1
1993	3043	0.31	-26.2
1994	2705	0.27	-11.1
1995	1623	0.15	-40.0
1996	2416	0.20	48.9
1997	1266	0.11	-47.6
1998	2278	0.20	80.0
1999	1806	0.15	-20.7
2000	1376	0.13	-23.8
2001	1627	0.15	18.2
2002	2332	0.20	43.3
2003	2393	0.16	2.6

Source: Elaboration on OECD [*2003*] data (IDS online).

In the 1990s Italian ODA amounted to around 0.15 per cent of domestic GNP, as shown in Table 10.1. Occasional improvements (such as the 1998 figure, reflecting the Treasury's obligatory periodic replenishment of multilateral funds following a delay in Parliamentary approval, the 2002 increase, mainly explained by the application of Law 209 of 2000 on bilateral debt relief and by emergency initiatives in the bilateral channel, the 2001 figure, mainly reflecting the increase in programme and project aid in the bilateral channel, and the US$50m. contribution to the FAO Trust Fund for Food Security in the multi-bilateral one and the World Bank replenishment in the multilateral one) could not be considered the beginning of a new positive trend. From the mid-1990s onwards there have been only modest degrees of development co-operation, primarily as part of multilateral programmes (the obligatory payments to EU programmes constitute almost half of Italy's ODA).

In fact, the second major element to be stressed is the current predilection for the multilateral channel, which absorbs 65 to 75 per cent of total Italian ODA[7] (see Table 10.2). This policy of providing such a

---

7. The 2002 data go against this trend, showing 'only' 57.8 per cent of total aid disbursed through the multilateral channel. But this percentage decrease does not reflect a reduction of multilateral aid, which in absolute terms increased from $1,184.75m. in

high proportion of ODA through the multilateral channel is not explicitly set out in the 1995 CIPE Guidelines governing Italian development co-operation, although *Prospective 2000*[8] states that 60 per cent will be implemented in co-operation with international organisations. In 2000 Italian resources underwent a serious reduction (US$1,376.26m.), and co-operation via the bilateral channel correspondingly felt the consequences (falling to $376.8m., or 27.4 per cent of total ODA). The obligatory contributions to multilateral organisations – and particularly to the EU[9] – mean that the level of multilateral aid becomes very high when the total amount is relatively low. Aside from this technical reason, there is a prevailing political interest in maintaining a strong presence in the international organisations (the EU, the UN agencies and the development banks) through the use of both obligatory and voluntary contributions, as well as through the multi-bilateral channel, which is technically part of bilateral aid but is a way of negotiating national interests within multilateral contexts. Along with the obligatory contributions to the various IFIs, Italy also participates in the funding of various trust funds in order to enforce its presence within a wide range of international organisations. As already mentioned, in 2001 the biggest disbursement in the multi-bilateral channel was the contribution given to the FAO Trust Fund for Food Security: US$50m. Two other important UN trust funds are the ILO's Universitas (a programme for Innovation, Education and Training for Decent Work and Human Development) and UNDP's Anti-Poverty Partnership Initiatives (APPI). As regards the IFIs, Italy participates in many Consultant Trust Funds: eight with the World Bank, managed by the Italian Ministry of Economics and Finance and two managed by the Treasury, and also with other development banks: the Inter-American Development Bank, the Asian Development Bank, the African Development Bank and the European Bank for Reconstruction and Development. Finally, Italy participates in a number of multilateral special funds, the most relevant being the Global Fund to Fight AIDS, Tuberculosis and Malaria, to which it has committed US$200m. ($100m. for 2001).

2001 to 1,336.92m. in 2002, and is explained by the strong increase in bilateral aid, resulting from debt reduction and emergency activities.

8. A prospective and programmatic report on development co-operation activities for the year 2000, *Prospective 2000* is the annual report on development co-operation policy. In this report, the Ministry of Foreign Affairs established poverty alleviation as the main priority of the Italian aid programme.

9. Italy contributes to EU aid activity with obligatory contributions both to the EU budget and to the EDF, amounting to about 13 per cent. In 2001 Italian voluntary contributions to the EU amounted to about €13.4m. (€12.4m. for a rehabilitation programme in Somalia and €1m. for the programme 'Junior Experts').

In fact, the latest news from the government shows a tendency towards disengagement also on the multilateral front (in terms of commitments). The Draft of the Italian 2004 Finance Bill forecasts a 20 per cent decrease in the Italian contributions to UN agencies (along with a decrease in bilateral aid), leading UN Secretary-General Kofi Annan to address a letter to Prime Minister Silvio Berlusconi (23 June 2003) expressing his concern about these cuts, and calling for a rethink. So far, no answer has come from the Italian side.

In terms of actual disbursements, Italian aid has encountered difficulties in completing the finalisation of its planned initiatives; in the case of soft loans, for example, payments are well below budget capabilities. The important tool of credit for financing joint ventures, additional to ODA resources (pursuant to Art. 7 of Law No. 49 of 1987) has been largely underused: only 648 million lire, about US$3.25m., was spent on two projects in 2001, while 104 billion lire (about $53m.) were available for this purpose at *Mediocredito Centrale* at the beginning of the year and 126 billion (about $68m.) at the end of the year.

IV. GEOGRAPHICAL DISTRIBUTION OF BILATERAL AID

In the light of the significant reduction in the resources allocated to bilateral aid, it is useful to analyse (in terms of the total, partial or minimal correspondence of disbursements to the amounts programmed) the geographical destination of Italian bilateral aid by country or region. In particular, it is possible to assess the weight of countries belonging to the various geographical regions that receive aid. The reference period is still 1980 to 2001, since in the previous period (1967–79) only limited resources were allocated to ODA: a total of US$1,390.81m. net, the equivalent of what was allocated annually during the second half of the 1980s.

The shifts from the solidarity motivation to help the poorest (African countries) to the economic interest (China and Latin America) and then to the political diplomacy interests (Middle East and North Africa (MENA countries) and the Balkans) are clearly shown in the data. In percentage terms, the 1980s started with 78.7 per cent of aid going to sub-Sahara Africa (compared with 2.9 per cent for the Middle East and a deficit in negative terms for the European countries). Engagement in areas seriously affected by desertification began in 1982 when Parliament approved the 'Italian initiative for the Sahel': interventions during this period were characterised by the adoption of an integrated approach to rural development, including

TABLE 10.2
NET PAYMENTS OF ODA 1997–2000 (US$m.)

	1997	1998	1999	2000	2001	2002
Total ODA	1,265.55	2,355.55	1,805.72	1,376.26	1,626.95	2,332.13
(as % annual variation)		*(80)*	*(-20.7)*	*(-23.8)*	*(18.2)*	(43.3)
Of which: to sub-Saharan African countries	278.63	443.88	246.50	261.09	190.56	n.a.
(as % of total aid)	*(22)*	*(19.5)*	*(13.7)*	*(19)*	*(11.7)*	
Bilateral ODA	453.72	697.45	450.72	376.80	442.19	1,006,56
(as % of total aid)	*(35.9)*	*(30.6)*	*(25)*	*(27.3)*	*(27.3)*	*(43.2)*
Bilateral grants	360.78	624.33	550.76	524.81	545.95	1084.93
(as % of total aid)	*(28.5)*	*(27.4)*	*(30.5)*	*(38.1)*	*(33.6)*	*(46.5)*
Of which: project and programme aid	168.34	152.66	162.44	106.57	176.54	134.62
(as % of total aid)	*(13.3)*	*(6.5)*	*(9)*	*(7.7)*	*(14)*	*(5.8)*
technical co-operation	57.79	40.42	53.19	27.16	92.1	101.58
(as % of total aid)	*(4.6)*	*(1.8)*	*(2.9)*	*(2.0)*	*(4.7)*	*(4.4)*
developmental food aid	16.11	39.43	43.78	31.95	75.63	42.47
(as % of total aid)	*(1.3)*	*(1.7)*	*(2.4)*	*(2.3)*	*(4.6)*	*(1.8)*
emergency and distress relief	50.25	21.25	102.97	72.36	64.79	83.67
(as % of total aid)	*(4.0)*	*(0.9)*	*(5.7)*	*(5.3)*	*(4)*	*(3.6)*
debt relief (forgiveness)	19.82	269.09	101.92	201.47	10.33	619.52
(as % of total aid)	*(1.6)*	*(11.4)*	*(5.6)*	*(14.6)*	*(0.6)*	(26.6)
Extended soft loans	240.65	241.5	82.18	74.66	82.94	
(as % of total aid)	*(19)*	*(10.2)*	*(4.5)*	*(5.4)*	*(5.1)*	
Net soft loans	92.95	73.12	-100.04	-148.01	-103.76	-108.59
(as % of total aid)	*(7.3)*	*(3.2)*	-	-	-	-
Of which: debt relief (rescheduling)	130.27	115.17	3.41	15.04	3.91	32.08
(as % of total aid)	*(10.3)*	*(4.9)*	*(0.2)*	*(1.1)*	*(0.2)*	*(1.4)*
Multilateral ODA	811.83	1,580.86	1,355.00	999.46	1184,75	1325,57
(% of total)	*(64.1)*	*(69.4)*	*(75)*	*(72.6)*	*(72.8)*	*(56.8)*
Of which: UN system	163.60	171.62	151.22	202.36	182.60	203.44
(as % of total aid)	*(12.9)*	*(7.5)*	*(8.4)*	*(14.7)*	*(11.2)*	*(8.7)*
European Commission	613.76	706.80	679.00	637.62	618.95	762.08
(as % of total aid)	*(48.5)*	*(31)*	*(37.6)*	*(46.3)*	*(38)*	*(32.7)*
World Bank Group	17.43	498.78	302.83	18.70	250.50	130.27
(as % of total aid)	*(1.4)*	*(21.9)*	*(16.8)*	*(1.4)*	*(15.4)*	*(5.6)*
Regional development banks	1.35	182.56	146.65	75.89	75.74	46.22
(as % of total aid)	*(0.1)*	*(8)*	*(8.1)*	*(5.5)*	*(4.7)*	*(2)*

Source: Elaboration on OECD [*2003*] data (IDC online).
Note: * Since no data for 2002 are available so far, the volume of debt relief for that year may be taken from Table 10.9.

emergency interventions. The decade closed with 54.5 per cent in 1989 going to sub-Sahara Africa (as against 1.4 per cent for the Middle East and 0.5 per cent for European countries). During that time, the importance of Latin America and the Caribbean had increased: from 10.5 per cent in 1980, the figure more than doubled, reaching 22.7 per cent in 1989 – a decade, therefore, of increasing resources for development co-operation, and a growing 'ODA Latin-Americanisation'. During the 1990s, however, two 'neighbourhood' areas grew in importance and were consolidated: the MENA region and the Balkans, reflecting the importance of geopolitics and security issues along with a particular attention to countries affected by recent conflicts (Kosovo, Albania, the Palestinian Territories).

With regard to the Balkans, after the high ODA volumes in the first half of the 1990s,[10] the region continues to absorb huge (although decreasing) ODA resources: from US$140.83m. (1999) to $99.37m. (2000) to $70.83m. (2001). To take the three main beneficiaries (Albania, Bosnia-Herzegovina and Serbia/Montenegro), one can note that Albania is the only country to receive soft loans. In 2001 Serbia/Montenegro was the prime beneficiary, receiving US$32.09m. in grants, followed by Albania ($21.66m., $11.02m. soft loans) and Bosnia-Herzegovina ($5.01m.).

TABLE 10.3
REGIONS WHICH BENEFITED MOST FROM AID 1980–2001 (% OF TOTAL GROSS DISBURSEMENTS)

	1980–81	1990–91	1995–96	2000–1
Sub-Saharan Africa	55.6	45.1	48.3	46.7
Europe	16.8	12.1	13.2	17.7
North of Sahara	9.5	13.1		7.8
Middle East	2.4	2.1	18.2*	6.1
Sub-total	84.3	72.4	79.7	78.3

Source: OECD [2003].
Note: * The data refer to all MENA countries (North of Sahara plus Middle East).

10. Italian ODA to the Balkans region follows the path of recent history. The peak was reached in 1991 (US$371.18m.) and the following two years (227.36 and 207.1m.) in line with the war; 1996 (after the Dayton Agreement) represents another peak ($162.56m., of which $67.4m. went to Bosnia-Herzegovina. Serbia/Montenegro started to be a main beneficiary after 1999, the year of the Kosovo crisis when grants of $66.29m. and $36.82m. – 72 per cent of total Italian ODA to Europe – were given to Albania and Serbia/Montenegro. As regards Bosnia-Herzegovina, after the post-war escalation (from $18.88m. to $44.22m. to $67.4m. between 1994 and 1996), the data show a fall in Italian aid: excluding 2000 (when BH received $32.97m., almost one-third of the regional total), the annual average in 1997 to 2001 was $8.85m., about 11 per cent of total Italian ODA to Europe.

TABLE 10.4
ODA DISBURSEMENTS IN EUROPE, 1990–2001 (US$m.)

Year	Total ODA	Total grants	Net loans	Loans rec. (Repayments)	Loans ext.
1990	76.19	42.36	33.83	-10.69	44.52
1991	371.18	313.8	57.38	-8.63	66.01
1992	227.36	144.68	82.68	-9.94	92.62
1993	207.1	189.64	17.46	-11.29	28.75
1994	86.85	72.67	14.18	-12.65	26.83
1995	63.54	65.50	-1.96	-10.05	8.09
1996	162.56	165.97	-3.40	-15.48	12.08
1997	52.22	53.19	-0.99	-14.45	13.46
1998	51.81	54.06	-2.24	-12.08	9.84
1999	140.83	143.37	-2.54	-10.12	7.58
2000	99.37	103.29	-3.92	-8.43	4.51
2001	70.83	67.02	3.82	-7.17	10.99

Source: Elaboration on OECD [*2003*] data (IDS online).

TABLE 10.5
ODA DISBURSEMENTS NORTH OF SAHARA,1999–2001 (US$m.)

	1999	*2000*	*2001*
ODA Total net	-2.78	-19.60	-1.66
ODA Loans total net	-30.68	-38.26	-16.38
ODA Grants, total	27.90	18.66	14.72
ODA Loans extended	15.66	19.44	28.81
ODA Repayments of loans (Repayments)	-46.35	-57.70	-45.17

Source: Elaboration on OECD [*2003*] data (IDS online).

TABLE 10.6
ODA DISBURSEMENTS IN THE MIDDLE EAST, 1999–2001 (US$m.)

	1999	*2000*	*2001*
ODA Total net	30.97	25.9	34.47
ODA Loans total net	3.41	4.48	11.43
ODA Grants, total	27.56	21.42	23.04
ODA Loans extended	4.49	6.56	13.06
ODA Repayments of loans (Repayments)	-1.07	-2.09	-1.63
Reorganised debt		1.68	

Source: Elaboration on OECD [*2003*] data (IDS online database).

With regard to the Middle East and North Africa (MENA) countries, Tables 10.5 and 10.6 reveal the profound difference between the two subregions: in 1999 to 2001 Middle Eastern countries received mainly grants, while North African countries received mainly soft loans, this making the regional total net ODA negative. Within the Middle East, while the Palestinian Administered Areas are a traditional beneficiary of Italian ODA (receiving, in the period studied, between half and two-thirds of total regional grants), Jordan and Syria are two emerging beneficiary countries: together, they received almost all the new soft loans to the area.[11] The remaining countries (with the exception of Lebanon) received grants of less than US$1 to 1.5m.

North African countries are of particular interest to Italy. Since they are mainly middle-income countries, as already noted, they mostly receive soft loans rather than grants.[12] This means that Italian aid offers North African countries an essential development tool, because it offers access to credit *per se* (otherwise difficult to find in countries not fully integrated into international finance markets) and on particularly good repayment terms. For Italy too, credit to North African countries seems to work rather well, given that these countries repay their loans (presumably because low interest rates are applied). Italy has also signed agreements, in particular with Morocco and Egypt, for converting old debts from previous aid into social and environmental programmes (debt swaps). The loan repayments are invested in new programmes aimed mainly at creating job opportunities, especially for small and medium-sized firms. This mechanism is one of the most interesting innovations in the field of Italian bilateral aid. Indeed, it has been extended to Peru and Ecuador. Negotiations

---

11. Syria received soft loans for $4.49m. (1999), $4.04m. (2000) and $8.11m. (2001), with very few grants (under $1 million yearly); Jordan received soft loans for $1.68m. in 2000 and three times this amount ($4.63m.) in 2001 with a decreasing amount of grants ($2.26m. in 1999, $2.47m. in 2000, $1.25m. in 2001).

12. In 1999 to 2001 Italian total net ODA to North African countries was negative. Algeria, Lebanon, Libya and Mauritania received (almost) only grants, mainly small amounts. Tunisia, Morocco and Egypt are the main beneficiaries in the region: in 2001 Tunisia received a net total of $6.46m. composed of $2.6m. grants and $3.86m. soft loans (the result of $25.25m. extended loans minus $21.39m. repaid); Morocco in the same year showed a net total of $0.73m., resulting from a $2.84m. loan and the negative value of its total net loans (-$2.11m.). Egypt, which in 2001 had a negative net total (-$7.56m.), in 1999 to 2001 saw a strong reduction in grants (from $20.98m. to $5.38m.) along with a decrease in extended loans (from $6.85m. to $2.08m.); the amount of total net loans in 2001 was -$12.94m. An exception is Malta, a traditional beneficiary of Italian grants, together with Slovenia, which the OECD has excluded from the list of ODA beneficiaries as of January 2003 because of the high level of economic development now achieved. This relatively wealthy little country benefited from Italian grants until 2000 ($20.94m.).

along the same lines are also under way with Pakistan, the Philippines, Yemen and Djibouti.

An important tool for promoting the private sector is also found in concessional credit for the financing of joint ventures, but, as already noted, it has been used very little.[13] Overall, in the period between 1990 and June 2000, only 56 initiatives had been decreed under Section 7 of Law No. 49 of 1987, for a total of 170.9 billion lire (about $85m.), 28 of them for Asia, 11 for Latin America, 11 for the Mediterranean Basin and the Near East, five for Eastern Europe, and one for Africa. In 2001 only two new initiatives were approved, both for Tunisia, for a total of only 648 million lire (about $3.25m.).

With regard to the focus on least developed countries, they represented the bulk of Italian aid beneficiaries in the 1980s (around 60 per cent of total aid). Then, in the 1990s, Italian interests and the core mission changed – in line with the prevalence of neighbourhood, security and economic interests – with a larger amount of financial resources devoted to aid, but with the weight of the poorest countries declining. In the latter half of the 1990s, Italy renewed its interest in the poorest countries, as a way of operationalising the objective of poverty reduction and, again thanks to the effect induced by debt relief (which is basically targeted on poor countries), recent data can confirm this trend.

The level of concentration of Italian aid can be calculated by means of the percentage weight of the ten main beneficiaries (see Table 10.8, where the percentages are calculated over the total of bilateral ODA, excluding unallocated flows[14]). In the six years considered, the level of dispersion of the net total ODA flow has remained high: in 1979, 89 countries benefited, 13 of them with a negative net balance; in 1985, there were 97 beneficiaries, only one of which (Myanmar) had a negative net balance; in 1989 95 beneficiaries, four of them with a negative net balance; in 1995 112 beneficiaries, nine with a negative net balance; in 1999 107 beneficiary countries, 18 with a negative net balance; finally in 2001 there were 110 beneficiaries, 21 of them with a negative net balance.

---

13. Pursuant to Law No. 49 of 1987, these measures are included in the *Fondo Rotativo* (revolving fund) components, but statistically they are not included in the ODA figures.
14. This item records the amount of bilateral aid that cannot be referred to a single (specified) country.

TABLE 10.7
ODA TO LEAST DEVELOPED COUNTRIES, 1986–2001

Year	ODA to LDCs (US$m.)	Total bilateral ODA (US$m.)	ODA to LDCs (%)
1986	902.47	1486.77	60.7
1987	1074.28	1877.88	57.2
1988	1283.52	2407.83	53.3
1989	858.87	2189.14	39.2
1990	968.80	2112.09	45.9
1991	547.73	2244.87	24.4
1992	671.57	2430.45	27.6
1993	566.11	1930.12	29.3
1994	349.24	1834.38	19.0
1995	275.23	805.70	34.2
1996	244.37	811.08	30.1
1997	247.74	453.72	54.6
1998	480.96	697.45	69.0
1999	171.96	450.72	38.2
2000	240.09	376.80	63.7
2001	187.21	442.19	42.3

Source: Elaboration on OECD [*2003*] data (IDS online).

Furthermore, the relative weight should also be integrated with the absolute value of the financial commitment, also taking into account the different levels of commitment on the part of Italian ODA: net bilateral ODA rose from the initial US$38m. in 1979 to $660m. (1985), and $1,855m. (1989), only to decrease to $717m. (1995) and $450.72m. (1999) in the phase of 'aid fatigue'. The increase recorded in 2002 ($1,084.93m.) is mainly explained by the debt reduction.

As shown in Table 10.8, the ten main beneficiaries in the 1980s and mid-1990s represented, on average, 55 per cent of total ODA, rising to around 90 per cent in recent years (with regard to the 2001 data, it should be noted that the unallocated component of bilateral aid is extremely high). From this long-run analysis there also emerges a movement away from the 'African' orientation[15] to an economic-commercial tendency under Foreign Minister De Michelis,[16] and then

---

15. The Horn of Africa and Mozambique received 54.2 per cent of bilateral aid in 1979.
16. In 1989, the Horn of Africa still received 16.6 per cent of bilateral aid, but in the same period China and Argentina received 12.8 per cent.

to a phase of neighbourhood policy, emergency aid[17] and commitment to debt cancellation.[18] This trend was confirmed and accentuated in 2001 and 2002.

V. STRUCTURE AND MAIN COMPONENTS OF AID IN THE 1990S

The clearly different nature of the two main instruments of development co-operation – grants and soft loans – requires closer examination in terms of the trends seen during the 1990s.

At a resources level, as mentioned earlier, the grants channel was 'subsidised' by the loans channel during the last two years of the decade, as shown in Figure 10.1. What this means is that a minimum amount of resources was available for making grants, at the cost of a corresponding reduction in credit resources – as may be seen in the mirroring of the two curves – which determines an overall debit in credit expenditure for 1999 to 2002.

FIGURE 10.1
EVOLUTION OF BILATERAL ODA, 1997–2002

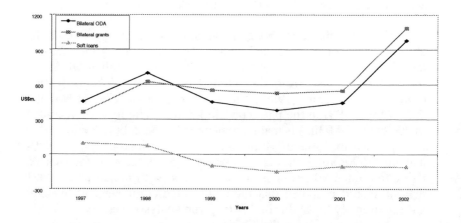

---

17. In 2001 Afghanistan for the first time enters the list of top beneficiaries due to the emergency following the war.
18. In 1999 Albania and Serbia/Montenegro received 27.3 per cent, while the debt restructuring of Congo and Tanzania was equal to 28.7 per cent of bilateral aid.

TABLE 10.8
WEIGHT OF TEN MAIN BENEFICIARIES 1979–2001 (IN US$M. AND AS % OF BILATERAL ODA EXCLUDING UNALLOCATED COMPONENT)

1979			1985			1989			1995			1999			2001		
Country	US$m.	%	Country	US$m.	%	Country	US$m.	%	Country	US$m.	%	Country	US$m.	%	Country	US$m.	%
Somalia	7.32	32.6	Ethiopia	81.60	12.5	Somalia	164.48	9.0	Ethiopia	116.12	16.9	Congo. Rep.	81.91	20.3	Eritrea	57.27	22.8
Ethiopia	5.02	21.6	Sudan	65.32	10.0	Ethiopia	138.40	7.6	Morocco	84.23	12.2	Albania	73.56	18.2	Serbia and Montenegro	32.08	12.8
Mozambique	2.15	9.2	Somalia	55.93	8.6	China	118.83	6.5	BiH	44.22	6.4	Serbia and Montenegro	36.82	9.1	Somalia	28.98	11.5
Brazil	1.74	7.5	Tanzania	34.82	5.3	Argentina	114.59	6.3	Ivory Coast	29.59	4.3	Tanzania	33.87	8.4	Afghanistan	22.11	8.8
Libya	1.45	6.2	Mozambique	27.96	4.3	Mozambique	80.0	4.4	Mozambique	26.90	3.9	Eritrea	30.27	7.5	Albania	21.66	8.6
Nicaragua	1.14	4.9	Chad	25.09	3.8	Peru	77.92	4.3	Uganda	24.14	3.5	Angola	24.98	6.2	Pal. adm. areas	15.84	6.3
Sudan	1.09	4.7	Tunisia	22.84	3.5	Senegal	72.29	4.0	Egypt	22.48	3.3	Malta	23.60	5.8	Ethiopia	13.56	5.4
Yemen	1.06	4.6	Turkey	22.22	3.4	India	71.03	3.9	Honduras	22.29	3.2	Somalia	19.60	4.9	Mozambique	13.08	5.2
Peru	0.90	3.9	Jordan	16.56	2.5	Cameroon	66.70	3.7	Gabon	20.29	2.9	Ethiopia	18.70	4.6	Honduras	12.76	5.1
Morocco	0.87	3.7	Angola	16.31	2.5	Tanzania	61.12	3.4	Jordan	17.84	2.6	Pal. adm.areas	18.53	4.6	Angola	11.89	4.7
Sub-total	22.74	98.8		368.65	56.5		965.45	53.0		408.10	59.2		361.84	89.5		229.23	91.4
Bil. aid	22.47			781.42			2189.14			805.70			450.72			442.19	
Bil. aid minus unallocated	23.28			652.45			1823.18			688.83			404.11			250.91	

Source: Elaboration on OECD [*2003*] data (IDS online).

A general consideration to be taken into account with regard to the Italian situation is a reminder about development financing, which was at the core of the UN conference in Monterrey, Mexico in March 2002. ODA is not in itself a motor for economic growth, but can be a useful catalyst for mobilising domestic and international resources. Foreign debt crises, although a real and serious danger, risk making people distrust the tool of credits (even those made on concessional terms) for developing countries, without finding any other solutions or mechanisms in the short term to replace them. In Italy, over recent years, the credit system has been widely used as a resource to draw on to keep grants going, thanks to the success of the Revolving Fund, which has seen the repayment of capital loans, and also to red tape, which has led to funds budgeted for not being spent. Another way of using the Revolving Fund set out in Law No. 49 of 1987 has been the financing of venture capital for small and medium-sized Italian enterprises in developing countries. Finally, the law on Italian participation in the stabilisation, reconstruction and development of the Balkans (Law No. 84 of 21 March 2001) states that part of the fund set up to help enterprises take part should be contributed by the Ministry for Foreign Trade, in terms of micro-credits (up to 200 billion lire, about US$100m.) and an increase in Revolving Fund credits. These funds are to be used for long-term investment in factories in countries outside the European Union, by providing free financial guarantees to enterprises damaged by the lack of payments from Yugoslavia after the war in that area.

The question which must now be asked is: How long can this process be sustained? In other words, will the funds paid into the Revolving Fund be used up quickly, requiring a new input from future Finance Bills, and in what scenario? Will this tool be used less in financing, meaning that only the grant channel will be used? Or will it be used to guarantee access to credit for countries with medium to low earnings, as is already the case?

It is easy to imagine an increasing use of the credit component in development aid, in terms of the interest already shown by Italian industry in international economic activities. In this sense, signs can be seen both in central government (the idea of the involvement of the 'Italian system' in the Balkans Stability Pact, and the strengthening of links via measures to promote foreign trade) and in the growing interest shown by regional and local government.

In any event, the data referring to the geographical destination of grants and soft loans identify priorities which do not coincide. With regard to grants, a preliminary remark about their volume is needed.

In view of the small size of the resources available, the increase in funds for grants in bilateral aid, as shown in Table 10.2, may seem interesting. However, when one subtracts the money destined for debt reduction (which in 2000 amounted to 38.4 per cent of the total of bilateral grants), one quickly realises that bilateral grants funds have been decreasing over recent years (and again in 2000, stand at approximately US$300–320m. of resources effectively paid out for development aid – an amount, in real terms, much lower than the bilateral grant total, minus the amount for debt reduction, paid out by any other country among the G-8, and most members of DAC).[19]

As regards the beneficiaries, in 2001 the geographical distribution of Italian grants confirms geopolitical motivations (Albania, Serbia/ Montenegro), interest in the Horn of Africa (Eritrea, Ethiopia and Somalia), Mozambique and Angola in sub-Saharan Africa, and the Palestinian Administered Areas in the Middle East, while Afghanistan becomes for the first time a major beneficiary due to the emergency crisis in late 2001 (see Table 10.8). Bringing up the rear are other countries where the importance of the grant channel is due more to other planning considerations currently under way (Argentina, El Salvador, Sudan, Uganda, India). Sub-Saharan Africa received 54 per cent of Italian bilateral grants in 2001, the top five beneficiaries in that year, Eritrea, Somalia, Mozambique, Ethiopia and Angola, confirming their position over the previous three years.

A particularly interesting item, in view of these considerations, is the weight given to emergency and food aid, when debt reduction is brought into the equation. This is not the place to debate the relationship between humanitarian or emergency aid, food aid and development aid. There are academics who make a clear distinction between them, and others who say that there must be complete continuity between the various phases and problems leading from emergency to development. In any event, as far as Italian aid is concerned, it cannot be denied that humanitarian and emergency aid – which makes up more than one-third of total grants to the top 15 beneficiary countries – is a relevant component.

Moving on to analyse soft loans, when the funds used for debt relief have been subtracted (i.e. when considering new credits), the picture that emerges shows that soft loans continue to be the main tool

---

19. While in 2001 the incidence of debt reduction on the bilateral grants total was very low, in 2002 it represented a very high percentage. Although the 2002 figure is not available, the impact of debt reduction on Italian bilateral aid in 2002 can be deduced from the fact that between October 2001 and October 2002 Italy cancelled external debt for a total amount of US$985.62m. (see Table 10.9).

of Italian ODA to Tunisia, Honduras, Albania and China (the top four beneficiaries). The same is true of other Latin American countries (Argentina and Brazil). In the Middle East the two main beneficiaries are Syria and Jordan. Sub-Saharan Africa no longer receives Italian credit (apart from a €777,555 loan to Eritrea and a €124,853 loan to Swaziland).

In conclusion, one can say that the tool of soft loans is declining in importance; the net amount for 2001 was -US$103.74m., which equals US$82.95m. in loans extended minus US$186.69m. in repaid loans (loans received). New credits are driven by aims that are clearly different from those underlying grants, and basically respond to economic policy interests.

VI. EXTERNAL DEBT REDUCTION AS THE NEW MAIN COMPONENT

If, on the one hand, we can see from the data in Table 10.1 a large reduction in Italian ODA resources (especially in terms of the bilateral channel), we can also see that this has been revised by the particular nature of operations put in place to deal with funds for development aid. In particular, from 1988 onwards, Italian bilateral aid has tended to reduce the importance of programme and project aid and moved over to the cancellation and rescheduling of developing countries' external debt.[20] Debt cancellation is categorised as a grant, whereas rescheduling is a soft loan. In Italy, there is no real additionality – recommended, among other things, by the World Bank and the International Monetary Fund's Heavily Indebted Poor Countries (HIPC) Initiative and subscribed to as an objective by most donors – in measures to reduce external debt. In fact, they become the principal element of its bilateral aid policies and represent the main distinction between rhetoric and performance. Since debt reduction is a non-structural component of bilateral aid (Law No. 209 of 25 July 2000 on debt cancellation, which represents a *una tantum* initiative, requires the Treasury to implement the debt reduction within three years – 2001 to 2004), Italian bilateral aid is expected to fall back after 2004, if no change in Italian policy occurs.

Data on Italian debt relief operations in 1998, 1999 and 2000 refer to the debt restructuring agreed at the Paris Club. In 2000 Law No.

---

20. Figures for 2001 go against this trend, for two converging reasons: (i) bilateral debt cancellation amounts to a mere US$0.33m., compared with $201.47m. in 2000; (ii) nevertheless, bilateral ODA shows an increase, over 2000, of $65.39m. (reaching $442.19m.), mainly due to increased disbursements in project and programme aid (approx. +$70m.) and technical co-operation (approx. +$65m.).

209 was passed, followed by its Implementing Regulation (No. 185 of 4 April 2001); implementation started in 2001 (22 October) when the first agreement (with Guinea) was signed.

Since official data for 2002 are not yet available, it is not possible to analyse the impact of Law No. 209 on bilateral ODA. A measure of the weight of Italian engagement in debt relief is provided by the data in Table 10.9, showing the amounts agreed from October 2001 to October 2002, as reported in the Treasury's Annual Report to Parliament.

The year 2000 was chosen for this analysis because it is the most recent year to show a strong impact of debt cancellation, in contrast to 2001, when debt forgiveness and rescheduling totalled only $14.24m. of bilateral aid (as it represented the interim period to bring Law No. 209 into operation).[21] In 2000 actions related to debt totalled $216.51m. (or, as already noted, 38.4 per cent of the bilateral grants total).

TABLE 10.9
HIPC COUNTRIES' FOREIGN DEBT CANCELLED (OCT. 2001–OCT. 2002)

Beneficiary	US$ m.	Euro m.	Date of signature
Total debt relief			
Mauritania	0.22	0.25	24/10/02
Tanzania	132.80	136.02	18/10/02
Malawi	0.23	0.26	17/6/02
Mozambique	525.30	557.30	11/6/02
Bolivia	69.85	74.25	3/6/02
Uganda	126.75	142.79	17/4/02
Sub-total	855.15	910.87	
Interim debt relief			
Mauritania	0.08	0.09	24/10/02
Mali	0.03	0.03	23/10/02
Cameroon	49.70	51.05	23/10/02
Benin	2.61	2.66	8/10/02
Chad	1.85	1.89	23/9/02
Guinea	15.93	17.87	22/10/01
Tanzania	45.10	50.48	10/1/02
Sub-total	115.30	124.07	
Pre-HIPC debt relief			
Ethiopia	10.31	10.99	5/6/02
Sierra Leone	4.86	5.53	22/3/02
Sub-total	15.17	16.52	
Total	985.62	1051.46	

Source: Zupi and Mazzali [2002].

---

21. With regard to 2001, it should be mentioned that Italy contributed to the multilateral HIPC Initiative with a disbursement of US$70m. to the newly constituted HIPC Initiative Fund.

During 2000, the only countries benefiting from Italian bilateral aid were those which had their overseas debt cancelled (Uganda, Cameroon, Zambia and Benin), or where debt cancellation was the main item (Bosnia-Herzegovina, Honduras, Senegal and Burkina Faso). Exceptions can be found – if one disregards the special agreement with Malta – namely those countries demonstrating the persistence of Italian interests owing to their proximity or the complex postwar emergency pay-out (Albania and the Federal Republic of Yugoslavia), or the emergency situation due to natural disasters in regions where Italy used to have interests such as the Horn of Africa (Eritrea and Ethiopia), and Mozambique. The Palestinian Authority is now a consolidated item where priority Italian action is required in development and technical assistance programmes.

An aspect directly affecting the overseas aid budget concerns the emphasis which Italy, more than other donors, has given to the cancellation of overseas debt in countries which traditionally have high debt levels. Italy's role as a member of the G-8, the European Union and the major multilateral institutions, as well as the particular interest of Italian public opinion in debt relief, make it an important actor in the international commitment to the reduction of poor countries' external debt. Moreover, Italy experienced two important civil society campaigns on debt cancellation: the *Campagna Sdebitarsi*, linked to the International Jubilee 2000 network, and the *Campagna della Conferenza Episcopale Italiana*, linked to the Vatican.

TABLE 10.10
FIFTEEN MAIN BENEFICIARIES OF BILATERAL AID DURING 2000 (US$m.)

Country	Total net aid	(a) Cancelled debt	(b) Rescheduled debt	(a)+(b) as % of total aid
Uganda	82.09	78.82	-	96
Bosnia-Herzegovina	32.97	25.03	-	76
Cameroon	26.96	24.41	1.92	96
Ethiopia	25.97	-	-	0
Zambia	24.02	22.83	-	95
Malta	20.83	-	-	0
Yugoslavia Fed. Rep.	19.30	-	-	0
Honduras	19.07	10.49	5.36	83
Benin	19.07	18.52	-	97
Eritrea	18.61	-	-	0
Albania	18.34	2.04	0.54	14
Senegal	15.21	10.70	-	70
Mozambique	13.09	-	-	0
Palestinian Nat.Auth.	11.78	-	-	0
Burkina Faso	8.20	6.95	-	85
Sub-total	355.51	199.79	7.82	58

Source: Zupi and Mazzali [2002].

TABLE 10.11
SHARE OF DEBT RELIEF COMPARED WITH DAC TOTAL, 2000 (US$m.)

	Net ODA	Net ODA debt relief	(as % of net ODA)	of which: bilateral	(as % of debt relief)	for HIPC	(as % of debt relief)
Italy	1,376	239	17.3	217	90.8	204	85.4
Total DAC	53,737	2,236	4.2	1,988	88.9	1,180	52.8

Source: Zupi and Mazzali [2002].

In this case, comparison is made between debt repayment (accounted for as part of grants) and rescheduling (part of soft loans) in the bilateral channel, and finance from the Trust Fund for HIPC activities in the multilateral channel. In the light of these data, the concrete problem to be found today at an Italian and – more generally – international level is the ratio between debt reduction/cancellation and development co-operation. The debt cancellation initiative is expressly based on the maintenance and reinforcement of current commitments to ODA, to which debt cancellation should be merely an adjunct. Unfortunately, this is not the case, and in view of the long time-lags in debt cancellation procedures, there is the danger that donors will budget for ODA initiatives over many years. This does not necessarily mean that they actually disburse the money: they merely budget for it. For the poorest countries, in contrast to the middle-income countries which are still paying their debt service to bilateral creditors, there is no additionality involved.[22]

In Italy, the figures for 2000 merely underline what had already been understood in 1999, when the two main ODA beneficiaries were Congo and Tanzania. They were not the countries for which most aid was destined, nor did they benefit from development co-operation budgets, but they had their external debts restructured by Italy ($78.16m. for Congo, $19.15m. for Tanzania).

Used correctly, external debt and poverty reduction policies via development aid should balance one another: when the debt crisis has been sorted out by cancellation, a new season of development financing should begin. In this way, developing countries would not start again on the downward spiral of debt, but would promote co-operation strategies for effectively reducing poverty. On the other

---

22. The relationship between debt cancellation and development aid is not simple, even from a purely accounting point of view. Total resources at multilateral level for debt reduction initiatives in poor countries correspond to the current net value of the foreign debt which will not be repaid. Accounting in donor countries, such as Italy, considers cancellation and rescheduling as part of ODA, although they do not generate new resources. This is because they were registered as ODA flows at the time of the original disbursement.

hand, incorrect use of debt cancellation will mean giving up playing an active role in development co-operation, which will then become a mechanism for reducing debt without offering any active measures for international co-operation.

It is also true that, for Italy, mention can be made of prevailing trends at the operating level of 'reducing' commitments. In other words, whereas Italian co-operation tends to be multilateral, this depends on the freeing up of commitments in the bilateral channel to the multilateral one, which thus become preponderant. The same may be seen in the bilateral channel: Italy's commitment to reducing developing countries' external debt becomes increasingly important, not because this item increases in absolute terms, but simply because this money, once budgeted for, becomes an expenditure commitment when other bilateral channel expenses are drastically reduced.

VII. DISTRIBUTION OF ODA BY OTHER SECTORS

As indicated above, Italian disbursements of development aid over recent years have noticeably decreased. This is especially true of bilateral aid, which owes its continued financial 'existence', in good part, to the 400 billion lire (US$200m.) subsidy paid from the Revolving Fund (pursuant to Law No. 266 of 1999) for allocating money to the Co-operation Fund for grants (2001). What is more, these reduced resources can be mostly attributed to forms of external debt rescheduling and cancellation. This means that analysis of resource distribution must start from this point. It also means that a great deal of discretion is needed when using the resources available – a conventional characteristic of Italian development aid, due in part to the fact that long-term budgeting is not binding and can indeed change from year to year.[23] The scaling down of resources means that more discretion has to be used, and that underlying trends are harder to identify. Indeed, change in any sort of plan from one year to the next is often enough for geographical and sectoral priorities to change. The time-lag between planning and actual disbursement means a delay of years in terms of the application of defined strategies.

It can reasonably be said, as we have done here, that the basic choice made by Italy in 1999 to 2002 was to espouse the cause of debt

---

23. Italy's new right-wing government has not moved away from this tradition: the brief period in office (May 2001 to January 2002) of Foreign Minister Renato Ruggiero, and the interim take-over of the office by Premier Silvio Berlusconi, are signs of potential discontinuity in strategy. Since November 2002, the Foreign Minister has been Franco Frattini.

cancellation. There was widespread agreement on this among Italian public opinion, but it tended to mean a reduction in development aid. At this point, any analysis of real trends in Italian ODA during the last few years becomes extremely relative, as is the case with any evaluation of its new planning direction, even though there are plenty of these. At a political level, the 1990s saw the falling off in resources coincide with the reappropriation of aid policies by the Foreign Ministry, in a fight back against what Foreign Minister Colombo defined as 'outside incrustations' [*see Rhi-Sausi (ed.), 1994; Rhi-Sausi and Zupi (eds), 2002*]. Development co-operation has become a powerful tool in Italian foreign policy, largely occupying the place once held by economic policy. Indeed, what remains of Italian ODA in South America and Asia confirms this view, as does the proximity to Italy of the main aid areas (the Balkans and the Mediterranean) in terms of security, stability and the control of immigration flows. The decision to take on the goal of external debt cancellation within the reduced aid resources has meant that the ties with foreign policy have been strengthened. Via its commitment to cancelling external debt, Italy has played a major role on the international co-operation stage, both during Holy Year 2000 and when it was President of the G-8 (2001). The growing use, over the last few years, of trust funds held by international organisations and the European Commission – especially when they are not tied up to funds from other donor countries – helps the Foreign Ministry in its policy of reappropriating development co-operation policies.

Table 10.12 compares data on Italian ODA per destination sector with reference to 1999, 2000 and 2001. In 2000 almost one-third went on operations linked to debt reduction – a figure which confirms and accentuates the trend first noticed in 1999, when debt reduction accounted for 17 per cent of ODA. Otherwise, social infrastructure and service expenditures (25.4 per cent), emergency aid (20.5 per cent) and commodity aid and general programme assistance (9.4 per cent) make up more than half of the total in 1999, a total amount of US$340m. In 2000 the predominant sectors were social infrastructure and service expenditures (20.6 per cent), production (especially industry, mining and construction, 15.4 per cent) and emergency aid (10.7 per cent). In 2001, the situation changes significantly: debt relief is irrelevant (2.2 per cent), as Law No. 209 has still to be implemented, and the first three sectoral components are social infrastructure and service expenditures (26.6 per cent), commodity aid and general programme assistance (16 per cent) and support to NGOs (15.7 per cent), while emergency assistance still holds a big share of the total (10.9

per cent). This figure is also mainly composed of grants: US$524.81m., as shown in Table 10.2, which means that credits amount to a mere $74.66m., which is even less, as noted above, than what is repaid on previous loans.

Infrastructure and social service payments could, in theory, show up in items which are considered *proxies* for commitments to poverty reduction, especially those on education and basic health care. However, if one goes on to break down these figures, it is easy to see how marginal they are to the day-to-day reality of bilateral co-operation.

In the education field, there is a preliminary problem with data availability, since the unspecified level share is very high (especially in 2001). As for the specified levels, the trend has been from tertiary (1999) to secondary education (2000), to both combined (2001), leaving very little over for spending on primary schooling, precisely the opposite of what a reading of intentions would lead one to suppose. The same is true in the health field.

The importance of the objective of reducing poverty, as the main focus of ODA, is nowadays recognised unanimously, and subscribed to by Italy. In fact, it continues to have a very ambiguous application in the field. An intrinsic risk concerns the general nature and dimensions of its definition, which can be used as a life-jacket to try to save everything and nothing (projects in the fields of finance, economics, the social sphere and politics). In this sense, what is actually done is particularly important for poverty reduction, where Italian ODA, over the past few years, has been concerned with providing a wide spectrum of priority strategies.

The 1999 DGCS guidelines outline the approach and content of an Italian initiative to be implemented with an initial allocation of US$12m. The initiative will consist of regional programmes in Central America, South America (Brazil and the Andean countries), the Maghreb, the Middle East, the Horn of Africa, the Sahel, Southern Africa and India, aimed to support poverty reduction strategies in two/three partner countries in each of those regions, within the framework of a consistent regional and international approach. Such programmes will build on the experience gained in previous initiatives, financed by Italian co-operation through the trust funds of international organisations, mainly the UNDP, and implemented by the UN Office for Project Services (UNOPS). In operational terms, it is a multi-bilateral initiative subject to direct monitoring by the DGCS.

TABLE 10.12
TOTAL ODA PER DESTINATION SECTOR 1999–2001 (US$m.)

Sectors	1999	2000	2001
I. Social infrastructure & services	*156*	*150*	*171*
Education, total	36	22	61
a) Education, level unspecified	26	8	38
b) Basic education	0	0	0
c) Secondary education	1	11	10
d) Post-secondary education	9	3	13
Health, total	32	39	16
a) Health, general	19	26	11
b) Basic health	13	13	5
Population programmes	1	7	2
Water supply & sanitation	19	52	15
Government & civil society	45	4	12
Other (social infrastructure & services, social services)	23	26	65
II. Economic infrastructure	*25*	*20*	*31*
Transport & storage	15	5	3
Communications	1	1	1
Energy	7	13	15
Banking & financial services	3	1	10
Business & other services	0	0	3
III. Production sectors	*39*	*112*	*54*
Agriculture, forestry, fishing	18	31	19
Industry, mining, construction	18	77	34
IV. Multisector	*45*	*50*	*60*
VI. Commodity aid/general prog. assistance	*58*	*37*	*103*
Food aid excluding relief food aid	44	32	76
VII. Action relating to debt	*105*	*217*	*14*
VIII. Emergency assistance	*126*	*78*	*70*
Relief food aid	0	0	0
Non-food emergency and distress relief	126	77	70
IX. DGCS expenses	*23*	*20*	*31*
X. Support to NGOs	*34*	*46*	*101*
XI. Unallocated/unspecified	*3*	*1*	*5*
Total	615	729	642

Source: Elaboration on OECD [*2003*] data (IDS online).

Thus, on a day-to-day basis, this effort should have been translated into a vast programme of multi- and bilateral-type projects, covering seven regions, with particular importance given to the *empowerment* of women and child care. What is new about the programme, launched in 1999, is how it has faced up to administrative difficulties which stopped it in its tracks until 2001. In terms of co-operation strategies, however, the most important element will be the effective definition of what is actually meant by 'poverty reduction' as the core mission of Italian development aid. The choice has been made, and it can be seen in the homogeneity of Italian aid programmes as compared with international commitments and co-operation (the DAC role is particularly important as a structure for stimulating and co-ordinating the main donor countries, including Italy). When one examines Italian policy in detail, it will be observed that there are some traditional strong points in co-operation practice, which can be directly linked to poverty reduction policies. In particular, social, health and children's programmes are a solid asset of experience in the field.[24]

VIII. TIED AID

Another element of development co-operation, which has been hotly debated in recent years, is tied aid, which is another term for loans or grants which imply the provision of goods or services by enterprises in the donor country.[25] In other words, with tied aid the tie is made at source, i.e. the beneficiary must use the loans and grants in the donor country. At international level, the practice of 'tying' a great deal of bilateral aid has spread, as have criticisms of the practice. Tied aid implicitly reduces the value of aid, because the tie reduces the beneficiary's choice of where the money may be spent. It also means an increase of 20 to 25 per cent in costs, and the principle of freedom of competition is violated (if the donor country has tied the aid given, the goods to be acquired may not be very competitive).

If, hypothetically, a number of countries speak out in favour of breaking the tie in this type of aid, in practice nobody actually wants

---

24. The Ministry of Foreign Affairs has recently brought out a book called *Italy for Children's Rights 2002, L'impegno dell'Italia per i diritti dei minori* (first edited in English in September 2001 to coincide with the UN General Assembly Special Session (UNGASS), 8–10 May 2002 in New York), which provides a great deal of information about Italy's aid activities on behalf of children.

25. 'Partially tied aid' is the term used when the supply is not limited solely to the donor country, but to several others as well. This may also include beneficiary countries. With untied aid, there is no obligation to suppliers in the donor country.

to take the first step: this would mean that the donor country could keep its own ties in place, while taking advantage of other countries' untying of aid. This is why there have been such long debates inside DAC about eliminating tied aid. On 14 May 2001, DAC published an agreement on untying aid to the 49 least developed countries (LLDCs). This deals with types of bilateral aid currently amounting to US$5 billion (of the $8 billion paid out bilaterally to the least developed countries, or 17 per cent of total bilateral aid).

Italy played an important role in the preparatory stages of the negotiations in favour of untied aid, but this goes against the current tendency in Italian jurisprudence. Law No. 49 of 1987, which governs Italian development co-operation policy, forbids the untying of loans. During the 1996–2001 Parliament a Bill to reform Italian aid policy was rejected, and implementation of the line adopted is not easy. Even the mechanism of international organisations' trust funds reinforces tying in the multi- and bilateral channels.

Final data on Italian bilateral co-operation during 2000 to 2001 show that, in terms of actual funds spent, and excluding technical assistance and administrative costs (which are necessarily tied), in 2000 38 per cent of aid was untied, but the remaining 62 per cent was tied. To be precise, 42 per cent of grants and 29 per cent of credits were not tied, while 58 per cent of grants and 71 per cent of credits was tied. The percentage of tied aid was high, but lower than it used to be. In 2001, when debt reduction had a very low incidence, the calculations show different results: only 8 per cent of total aid was untied, while 92 per cent was tied. In more detail, only 3 per cent of grants and 37 per cent of credits were not tied, while 97 per cent of grants and 63 per cent of credits were tied. In 2002, due to the volume of debt relief as the main component of Italian bilateral ODA (forecast as equal to about US$600m.), the percentage of tied aid will decrease substantially.

TABLE 10.13
BILATERAL ODA COMMITMENTS, EXCLUDING TECHNICAL ASSISTANCE AND ADMINISTRATIVE COSTS, 2000–2001 (US$m.)

	Grants		Soft loans		Total	
	2000	2001	2000	2001	2000	2001
Direct financing of imports	178.78	274.13	175.98	62.09	354.76	336.22
Untied aid	2.92	3.69	40.29	20.80	43.21	24.49
Partially tied aid	0	0	0	0	0	0
Tied aid (only to Italy)	175.86	270.44	135.70	41.29	311.56	311.73
Aid in kind (tied aid)	109.50	145.95	0	0	109.50	145.95
Debt relief (untied aid)	201.47	10.33	15.04	3.91	216.51	14.24

Source: Elaboration on OECD [2003] data (IDS online).

FIGURE 10.2
BILATERAL ODA COMMITMENTS, HISTORICAL TREND (1980–2001)

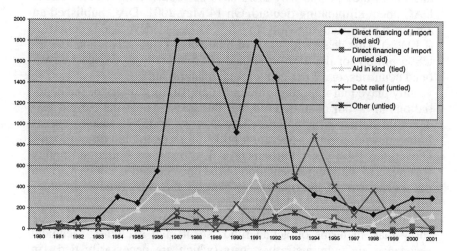

Source: Elaboration on OECD [*2003*] data (IDS online).

Once again, measures for the reduction of poor countries' external debt are a significant part of the overall result. In 2000 the high percentage of bilateral operations in debt reduction – which cannot be classified as tied – accounts for almost all the non-tied aid. Compared with the past, and especially before the late 1980s, when more than 90 per cent of all bilateral aid was tied and Italy was near the top of the list of countries offering tied aid, what has changed in the intervening period is that the amount paid out in aid has decreased, but measures for reducing external debt continue to increase.

IX. INSTITUTIONAL SET-UP, CHANGES AND PROBLEMS

Law No. 49 of 1987 created a Steering Committee (*Comitato Direzionale*), chaired by the Minister for Foreign Affairs (or his Under-Secretary), with the purpose of vetting all the DGCS's financing proposals. The DGCS comprises 12 offices, plus a new Central Technical Unit (*Unità Tecnica Centrale*, UTC), headed by a diplomat, to control the quality of the project cycle. In addition to the UTC at headquarters, Law No. 49 introduced Local Technical Units (*Unità Tecniche Locali*, UTLs) in the field, attached to the local embassies. These have been set up only very recently.

Management capability has always been an area of concern, particularly since the latter half of the 1980s, when the increasing amount of financial resources devoted to aid policy implied the need for better control of the effectiveness of interventions. After a series of political scandals in the early 1990s, the effects of general political turbulence plus a crisis of legitimacy affected the Ministry of Foreign Affairs and development co-operation policy itself. Bringing forward a major reinforcement of management capacity, in support of the work of DGCS, is considered to be the most urgent element in making Italian aid policy work better, as may be seen from the DAC Peer Review carried out by Canadian and Swedish reviewers between March and June 2000.

The Ministry of Foreign Affairs is still seeking to close the gap between the broad policy guidelines (the 1995 New Co-operation Policy Guidelines, updated annually under the Provisional Planning Act) and their practical implementation in the field. Italy still lacks any operative guidelines at intermediate level. A main administrative constraint is the lack of any methodical use of a 'Country-Programme' strategy: Italy began to implement a country programme in Albania and with the Ethiopian government at the end of the 1990s.

The gap between strategy and practice has also been caused by a lack of skilled professionals and the absence of a poverty reduction-oriented organisational and management structure in the MFA. This corresponds to a strongly sectoral approach and lack of poverty reduction mainstreaming, and implies a problem of coherence between the administration's human capital assets (its organisation) and the theory of how aid should work. Projects aimed at reducing poverty are not supported by specialised technical advisers and training activities.

Since 1995, a reform process has been set in motion. Following government approval of the Bill for reform at the end of 1997, it began to be discussed in Parliament. The main feature, in organisational terms, is the shift from a ministry-based management of ODA to the creation of an external agency, which should guarantee a more professionally oriented management structure. The main institutional problems may be summarised as follows:

(i) inadequate long-term professional staff and an unbalanced skill mix, both at headquarters and in the field, since priority was given to diplomatic and political control rather than to the technical management of ODA;
(ii) slow and confused administrative procedures, with the emphasis put on administrative accountability (particularly following the

corruption and political scandals) rather than on the effectiveness and adequacy of interventions;
(iii) limited decentralisation of decision-making authority to the field (UTLs), due to the need to guarantee political control at headquarters;
(iv) inadequate evaluation systems, which were seen as a tool for administrative purposes rather than to guarantee better project cycle management and stronger partnership;
(v) late introduction of a project cycle management system (1995) to appraise and evaluate ODA activities, in response to DAC and European Commission requests. Only in February 1996 was a Technical Evaluation Unit (*Nucleo di Valutazione Tecnica*, NVT) introduced to guarantee uniformity in appraisal methodology;
(vi) lack of effective poverty reduction mainstreaming, in terms of specific technical skills, country operations and incentive mechanisms at administrative level.

These problems lay behind the fact that the immediate change resulting from the reform of Law No. 49 of 1987 should have been the establishment of a new agency, with more adequate staffing, improved operational and administrative procedures, enhanced decentralisation, and the involvement of NGOs and local communities. The new legislation clearly implied increased resources and expansion of the bilateral channel. It stated explicitly that ODA should have increased continuously towards the internationally agreed 0.7 per cent target; and it introduced a Special Fund to manage soft loans, grants and food aid, aggregating all past funds.

A fundamental change should clearly have been induced by the introduction of this new development co-operation law, and the centre–left governments of the second half of the 1990s (under Romano Prodi, Massimo D'Alema and Giuliano Amato) searched for means of reforming the law. The Senate passed the law but the Chamber of Deputies did not approve it before the end of the 13th legislature in April 2001. Thus, Law No. 49 of 1987 still represents the legal point of reference, and the new political scenario, at national (the centre–right government under Silvio Berlusconi) and international (post-11 September 2001) levels, drives us to conclude that the only way to proceed is to reform aid policy within the context of Law No. 49, in continuity with it, through marginal reform. In particular, greater emphasis on decentralised co-operation at municipal and regional levels seems to represent the main direction to be followed.

X. THE MAIN ACTORS

NGOs, as private, voluntary, non-profit organisations, are important implementing bodies for Italian aid policy. But their importance is much more in terms of participatory and innovative approaches than in quantitative terms. A 1993 Special Commission Report [*Bottai and De Rita, 1993*] proposed separating the poverty reduction and the economic development objectives, giving private enterprises the implementing role for the latter, while NGOs are limited to ensuring solidarity and assistance. Since then, given the limited amount of resources devoted to bilateral aid, the lobbying pressure from private firms has been low, whereas NGOs have continued to ask for more support and improved effectiveness in aid's targeting of poverty reduction.

After the 1995 suspension of government funds to NGOs, the slow increase in NGO resources in 1996 was confirmed in 1997, and the revision and simplification of procedure mechanisms seemed to facilitate a trend for major support for NGO projects. However, the net reduction of bilateral resources and the persistent management problems hit the NGOs, especially in terms of the speed and complexity of procedures.[26] In 2000 the Italian Association of NGOs was set up, as a way of guaranteeing better co-ordination among the more than 150 existing NGOs.[27]

In terms of geographical distribution, the NGOs are 'universal', although historically their preferences have been towards Latin America for cultural, linguistic and religious reasons, together with the dynamic social-economic fabric of Latin American society. This trend has varied considerably, however, due to the new destinations for Italian aid, in particular those of the European Commission. In the coming decade the criterion of geographical proximity (the Balkans, the Mediterranean Basin and the Middle East) will probably be more significant than in the past. As regards sectoral distribution, education and training have continued to be the main fields of NGO interventions (representing some 60 per cent of total funds managed by NGO programmes), with health as the third main sector.

As already mentioned, a new component of Italian ODA is decentralised co-operation. It does not yet represent a significant part of Italian co-operation in absolute terms (in 2002, Regions and Auto-

---

26. It should be noted that data on recent years show an increase in support for NGOs: in terms of disbursement, in 2001 they received US$101m., as against $46m. in 2000 and $34m. in 1999; with a further increase reported for 2002.

27. Italian NGOs have been traditionally grouped into Catholic and lay federations, but some of the biggest ones do not belong to any federation.

nomous Provinces devoted about €34m. to development projects; aid activities carried out by the Municipalities are at present difficult to quantify). However, it seems to provide an interesting opportunity to realise the objective of involving different stakeholders from local communities via international partnership. The fact that most of the Regions – 18 out of 20 – enacted legislation to regulate development co-operation policy gives an important indication of this trend.

The complex links between aid and migration are becoming an important issue. Italy is attracting increasing flows of immigrants from the Balkans and the Maghreb, and this phenomenon will require a more comprehensive development co-operation approach in the future. The trend will further increase the role of decentralised co-operation. The migratory phenomenon can best be dealt with, in fact, at the subnational level, in terms of both institutional competence and operational policies.

The increased importance of decentralised co-operation will deeply affect the role of other actors in development co-operation. On the one hand, the trend towards the 'territorialisation' of the non-governmental sphere will probably be confirmed: NGOs will probably have subnational governments as their main reference points, rather than national governments or networks. Their international relations will also hinge on their own regional territory. On the other hand, enterprises, especially small and medium-sized ones (SMEs), will find a new space as agents of international co-operation, inasmuch as they develop territorial partnerships and strategic alliances with SMEs in developing countries.

XI. OPTIONS FOR THE FUTURE

Obviously, Italian co-operation policy will be substantially conditioned by the growing importance of international aid. The trends which seem to be emerging are, on the one hand, an approach whereby organisations are defined and established in terms of specific objectives to be pursued (target-oriented organisations), with the setting up of special funds for specific purposes (elimination of infectious diseases, E-government, food security, specific countries or geographical areas, etc.), supplied by voluntary contributions from the 'donors' club' – an approach which implies the predominance of the multilateral channel, both in co-ordinating activities and in managing funds – and, on the other hand, an approach towards neighbouring countries (in particular Balkan and Mediterranean countries), led by geopolitical priorities, which is in line with the EU framework.[28] The interest shown by the Italian government in the neighbourhood policy is translated in operative terms into a reciprocity-based strategy[29] which opens up new space for decentralised co-operation as well as for Italian entrepreneurship. As regards the latter, those which will benefit most from this trend will be the small and medium-sized enterprises, which are not able to gain access to the multilateral channel.

Regarding the first trend, two main factors come into play in choosing the organisation to be used according to objectives: a widespread intolerance of the poor quality and efficiency of international aid, which generates the need for objectives to be measured within a given time frame; and the search for a consensus in public opinion and among the contributors in donor countries which also pushes in this direction, in order that certain tangible results may be achieved. It is still too early to know whether these new trends in international co-operation will be able to be consolidated. Two main worries should be pointed out, however. On the one hand, the fragmentation and excessive rigidity of all development policies may increase. Streamlining co-operation could end up in pigeon-holing and fragmenting activities, casting a shadow over the theories of an integrated approach to development which have been put forward in the past decade. On the other hand, at a practical level, most international co-operation is still being carried out along traditional lines, by equally traditional organisations. It is difficult to see how the extremely formal methods currently being used in managing development projects will be flexible enough to handle the new organisation of development aid by objectives.

It is therefore quite probable that these new tendencies will be seen only in the management of new international co-operation funds. For donor countries and multilateral organisations, the new context will have greater or lesser repercussions depending on the characteristics of their policies, methods and management mechanisms. Structures such as the World Bank and some of the other UN specialist agencies may find it relatively easy to move over to a reorganisation of aid by

---

28. The reference here is to the so-called 'Wider Europe' approach, first stated in the Communication from the Commission, *Wider Europe – Neighbourhood* [*Commission of the European Communities, 2003a*] and to the EC's proposal to create a neighbourhood tool aimed at co-ordinating cross-border co-operation programmes – Interreg – and the various development co-operation programmes with the two above-mentioned regions – MEDA and Cards [*Commission of the European Communities, 2003b*].

29. An approach which differs from the traditional development co-operation approach, typically oriented towards the beneficiaries' interests.

objectives. Their specialisation and/or flexibility means that they can carry out co-ordination and management tasks. The European Commission, however, which experienced considerable difficulties in the past in leading and co-ordinating national co-operation policies, may face greater difficulties in meeting the new parameters, even though the reform currently under way in management structures should help to overcome these limitations. If the EC succeeds in developing and co-ordinating its neighbourhood policy towards the new neighbouring countries, its impact on Italian co-operation policy will be of great importance.

As far as donor countries' policies are concerned, the models which are least helpful in facing up to the new realities are those which are mainly bilateral, which have rigid management methods, and which are quantitatively very heavy. For beneficiaries, meanwhile, factors such as their commitment to development, their ability to absorb foreign aid, and the transparency and quality of their management structures will be increasingly decisive.

It is interesting to note how the recent evolution of Italian ODA has been more in tune – or less out of tune – with the more recent trends in international aid. As we have seen, Italy's activities have been organised according to different objectives, and the management mechanisms have been different, in the same way as the multilateralisation of aid is now a consolidated approach. Naturally enough, some differences do remain, both at an objectives and a content level, and in terms of the management methods used. We must remember, however, that harmonising Italian development aid with the new proposals for international co-operation is not the only method to be used. Some people are pushing for more bilateral aid, and this is also due to the relative insignificance that this channel enjoys in Italian development aid.

However, this new international context has reopened the debate on development co-operation. The two main topics for discussion are: the reform – whether legislative or otherwise – of aid policies, and the re-launching of co-operation activities, given that Italy is committed to raising its development aid from the current 0.2 per cent of GNP (US$2,313m. in 2002) to 0.39 per cent ($4,500m.) by 2006. As far as the reform of international co-operation is concerned, this was much debated during the 13th legislature, even though no reform law was approved, and despite the fact that the Senate had approved the drawing up of one. This shows how complex the idea of reform really is, so much so that we need to examine some significant points in the debate.

The first point is to be found in the fact that, although there is widespread agreement about reforming Italian development aid, there is no agreement that a new law is needed. This may seem to be paradoxical, in the light of the fact that during the previous legislature the various parties presented eight Bills for debate. But when it came to debating them, the chasm of difference of opinion became apparent, and crossed the party divide. This split in opinion can be explained by various factors, but the main reason was the division between those who argued that co-operation reform had to start with a new law, and those who maintained that Law No. 49 of 1987 only needed touching up to get over the impasse in Italian aid policy.

This explanation, however, has not specified what reform proposals should actually be made, i.e. whether future aid policy should be seen as an extension of current policy, quantitatively greater and qualitatively more efficient but in essence similar in terms of principles, orientation, priorities, objectives and management systems. If this is so, then the current laws need only be amended. Nowadays, this seems the most likely option.

In fact, the need for new legislation can be justified only if there is the political will to carry out bilateral co-operation, i.e. to express a direction which Italian development aid must follow; it must have sufficient financing and management, and must deal directly with developing countries (even when co-ordinated and consistent with the multilateral framework). If the view is not that of building up bilateral aid policies, but rather of consolidating the multilateral trend and the neighbourhood policy, the need for new legislation is much less pressing.

There are at least two arguments in favour of multilateral activity as the main plank in Italian aid policy. One line of argument (*multilateralist co-operation*) sees aid as part of Italy's commitments to the poorer members of the international community. The critical mass of resources necessary to meet the needs of these countries, the experience built up by European and international organisations in managing aid, the level of co-ordination and consistency that development co-operation requires (in international trade, external debt, environmental politics, etc.) make bilateral action by Italy almost an empty ambition. Italian interests and commitments, in both geographical and sectoral terms, can be catered for, according to this approach, by multilateral aid, without having to have a bilateral approach.

The priority given to the international agenda in Italy, particularly to the construction of a new global governance and Italy's active participation in it, is strengthened and increased by international political and economic factors. This vision, which tends towards the multilateralisation of Italian development co-operation, also reflects a deep distrust of the ability to manage bilateral policies. Naturally enough, explicit reference must be made to development co-operation during the 1980s, when bilateral aid was greater, both relatively and in real terms.

An important variant of the *multilateralist co-operation* model is that which proposes transferring all international aid policies to the European Union, the European Commission, or a mixture of Commission and national agencies. This is all part of the debate on EU foreign and security policy. European aid emerged as a concrete proposal during the lead-up to the Treaty of Maastricht, but the result has only been better co-ordination of, and consistency in, the policies of the individual member states. The question came up in the early 1990s in the Italian parliamentary debates, but none of the Bills during the 13th legislature mentioned the communitarianisation of Italian ODA. The new international situation could undoubtedly favour this proposal, but there are at least two obstacles still to overcome. One is the prevalence of bilateralism in the member states – apart from Italy – and the other is the lack of efficient management structures in the European Commission.

Another thread which can be inserted into the multilateralist argument is that which attempts to maintain bilateral alongside multilateral aid (*realistic co-operation*); in quantitative terms it is much smalller, but qualitatively it is much more highly oriented and better defined. This is the system which comes closest to the current Italian position. Bilateral aid, under this system, is useful and necessary for limited areas. It is indispensable, for example, when direct national security interests are in play, or where contractual commitments have been made, either formally or informally, with the international community. It is mostly the case with commitments to countries geographically close to Italy; geopolitical interests justify bilateral aid. Bilateral aid is also necessary for humanitarian and emergency aid, not merely because this is the usual practice for donor countries, but also because Italy has specialised in humanitarian aid which, among other things, produces political returns, both at home and abroad. Finally, another bilateral activity to safeguard is non-governmental activity (and perhaps decentralised co-operation activity as well). NGO participation in bilateral aid is an Italian tradition, and it does not require a particularly onerous financial commitment. This type of commitment, partly managed in both bilateral and multilateral modes, makes no substantial modification to the multilateral approach to Italian ODA.

The models for both *multilateralist* and *realistic* co-operation would require no legislative reform. Indeed, an internal amendment would suffice, plus a substantial reinforcement of the technical structure as laid down in Law No. 49 of 1987. In fact, in certain cases laws such as that governing Italian intervention in the Balkans can theoretically make up for lacunae in Law No. 49.

If during the current 14th legislature there were to be debate on a new development co-operation law, it would only be positive if one or both views of Italian co-operation were to be changed. In other words, new legislation would make sense only if the multilateral basis of Italian aid were to be severely reduced.

Meanwhile, what basis and what consensus can a consistent bilateral development policy count on, as *bilateralist co-operation*? If we consider the 30 years or so of Italian aid history, we shall quickly reach a simple conclusion. Bilateral development co-operation is now an affirmed policy, and has always been able to count on resources where the Italian economy was involved. Trends in Italian ODA follow a near-perfect parabola, according to whether or not Italian enterprises take part. Given that Italian ODA no longer represents a source of income for domestic industry, development co-operation has gone into steep decline.

This brings us to a second question: Is there any political future for Italian ODA which involves a special relationship with the economy? Or are there any other bases on which bilateral development co-operation can be built? To judge by the contents of Bills presented during the 13th legislature by the various parties, the first option would appear to be the one to discard. In the best of cases, there might be a secondary role for Italian enterprises, but it is more than likely that this role would be explicitly excluded. Even under the current bilateral system, the participation of the domestic economy is only residual. Current bilateral development co-operation is mainly guided by diplomatic interests, by the agendas of foreign and security policy, and the interests of economic policy enter the argument only marginally.

The signs of a return to economic participation in bilateral policy are contrasting. On the one hand, the international community is agreed that the main objective of development aid is the reduction of poverty in the world: this consensus was reached in part thanks to the active contribution of Italy, both during the sessions of the Development Assistance Committee of the OECD, and in the various UN agencies and at the European Commission. In the light of this, the role of enterprises in donor countries is limited. One need only think of the commitment to eliminating tied aid, which is the mechanism still in quite frequent use by the various aid agencies. On the other hand, as was hinted at above, the new proposals for international co-operation assign a leading role to the economic sphere, even in donor countries. The European neighbourhood approach towards the new southern and eastern neighbours also seems to be moving in this direction. Here the question is still open, and it will be decided by the position the current government – and the parties in alliance with it – take. The return of economic interests as a significant (if not deciding) factor in Italian aid will doubtless be a strong incentive for *bilateralist co-operation*.

Finally, one must remember that Italian enterprises, especially the larger ones, have been in favour of multilateral aid for some time now, even where this preference was born out of necessity. The volume of multilateral financial resources makes this channel more attractive to large enterprises, as is the case with the mechanisms used in managing it.

Concerning other motives for sustaining a bilateralist approach, immigration has been quoted as a new rationale for development aid. The debate, although useful in understanding the new dimensions of international co-operation, has provided no credible arguments for breathing fresh life into Italian policies for co-operation with developing countries. Furthermore, the duality of multilateral and bilateral approaches is very clear here, as in other areas.

One conclusion that may be drawn when dealing with models for *bilateralist co-operation* is that there are many structural difficulties in reopening a bilateral phase in Italian development co-operation. The factors which might determine the outcome of this type of policy are many and varied (humanitarian, economic, geopolitical, security-related) and difficult to broaden out on to the current international stage. In the past, Italian ODA achieved an unsteady balance between them only when resources were plentiful. For balance in the future, the economic commitments already undertaken will have to be maintained – and redistributed. What is certain is that balance cannot be achieved when resources are scarce.

Those who propose an internal reform of Law No. 49 of 1987, and in particular those who favour the model of *realistic co-operation*, therefore have many elements in their favour. This model is even compatible with a scenario of relatively abundant resources, as well as one where they are scarce. If a wide-ranging reform of development co-operation is not necessary (or not possible), the need to restore some measure of vitality in international aid policy does remain. Development aid as it currently stands in Italy is extremely weak, and continues to pay the price of management problems, increasingly exacerbated by cuts in technical staffing levels over the past few years. Some sort of reform is therefore necessary, and the current debate must shed light on where it can take place.

REFERENCES

Bottai, B. and G. De Rita (eds), 1993, *Libro bianco della cooperazione italiana allo sviluppo negli anni Ottanta* (White Book on Italian Development Co-operation), 2 vols., Roma: CENSIS.

Commission of the European Communities, 2003a, *Wider Europe – Neighbourhood. A New Framework for Relations with our Eastern and Southern Neighbours*, Communication from the Commission, Brussels, 11.3.2003, COM (2003) 104 final.

Commission of the European Communities, 2003b, *Paving the Way for a New Neighbourhood Instrument*, Communication from the Commission, Brussels, 1.7.2003, COM (2003) 393 final.

Italian Ministry of Foreign Affairs, 2001, *Italy for Children's Rights 2002* (ed. Tosca Barucco), First edn. Rome, September.

Italian Ministry of Foreign Affairs, 2003, *Relazione annuale sull'attuaazione della politica di cooperazione nel 2001* (Annual Report to Parliament on Italian ODA), Rome.

OECD, 2000, *DAC Peer Review of Italy*, Paris: OECD.

OECD, 2003, *The DAC Journal Development Co-operation 2002 Report*, Vol.4, No.1, Paris: OECD DAC.

OECD, International Development Statistics (IDS) online, http://www.oecd.org/dataoecd/50/17/5037721.htm, 2003.

Rhi-Sausi, J.L. (ed.), 1994, *La crisi della cooperazione italiana: Rapporto del CeSPI sull'aiuto pubblico allo sviluppo* (The Crisis of Italian Co-operation. CeSPI Report on Italian ODA), Rome: Edizioni Associates.

Rhi-Sausi, J.L. and M. Zupi (eds), 2002, *Ripensare la cooperazione. Rapporto CeSPI sull'aiuto pubblico allo sviluppo* (Rethinking Co-operation. CeSPI Report on Italian ODA), Rome: Memoranda.

Stocchiero, A. (ed.), 2001, *I Comuni italiani e la cooperazione internazionale* (Italian Municipalities and International Co-operation), Rome: CeSPI Working Paper, No. 6.

Zupi, M. and A. Mazzali, 2002, *Il debito estero dei paesi poveri. Un manuale per l'orientamento e l'analisi* (Poor Countries' Foreign Debt. A Handbook for Orientation and Analysis), Rome: CeSPI.

Zupi, M. and A. Mazzali (eds), 2003, *Mobilitare le risorse finanziarie a livello locale per lo sviluppo del sud del mondo. Alcune esperienze europee* (Mobilising financial resources at local level for development. Experiences from Europe), Rome: CeSPI-UCODEP.

11

A New Member of the G-0.7: Luxembourg as the Smallest and Largest Donor

PAUL HOEBINK

In the first speech ever to be delivered by the leader of a foreign government to the National Assembly of the Cape Verde Islands, the Prime Minister of Luxembourg, Jean-Claude Juncker, stated on 31 January 2002 that his country might not be a member of the G-7, but that it was proud to be part of the Group of 0.7. It demonstrates that small countries can be large in some things. Luxembourg is a relative newcomer on the donor scene, but it has quickly catapulted itself among the donors that fulfil the UN General Assembly's 1970 promise that rich countries should be giving 0.7 per cent of their GNP to foreign aid. What is more, Luxembourg has recently promised to reach 1.0 per cent of GNP in 2005. This relatively high volume of aid also means that Luxembourg can be an important donor in some developing countries.

How did Luxembourg, a country that at the beginning of the 1990s was still criticised by the Development Assistance Committee of the OECD for its lack of interest in development co-operation, arrive at this situation? What kind of example does it present for those newcomers from Central and Eastern Europe that are entering the donor scene? Luxembourg is a rich but small country with a surface area of 2,586 km^2 and 440,000 inhabitants. It can thus become at one and the

---

Note of acknowledgement. Apart from an analysis of the document presented in the literature list, this chapter is based on a series of interviews in April and June 2002 in Luxembourg and a visit to the Cape Verde Islands in the last weeks of May 2002. The author would like to thank officials of Lux-Development and the Directorate for Development Co-operation of the Ministry of Foreign Affairs, in particular Philippe Aschman, co-ordinator of the Ministry's Bureau in Praia, Cabo Verde, and also representatives of Luxembourg's NGOs and political parties who were willing to share their views and ideas with him. Special thanks go to Mike Mathias of ASTM for collecting a series of (old) articles and documents on Luxembourg's development co-operation.

LUXEMBOURG: A NEW MEMBER OF THE G-0.7

same time the largest as well as the smallest OECD donor. Its small size challenges its development co-operation programme with some specific problems, illustrated in its programme in Cape Verde, from which these new donors can learn. This chapter will focus on the aid rationale of this small donor, and at the same time try to analyse some of the problems of 'smallness'.

I. FROM A SMALL TO A LARGE DONOR

In the 1970s and 1980s the governments of Luxembourg did not show much interest in development co-operation. Aid hovered at between 0.1 and 0.2 per cent of GNP, consisting mainly of (obligatory) contributions to multilateral organisations.[1] The Luxembourg governments of those days emphasised its neutrality with regard to development problems and the absence of moral responsibility, since Luxembourg had not been a colonial power.[2] Furthermore, the members of the government stressed, as did the then Minister of Foreign Affairs, Gaston Thorn, in Parliament in 1975, that in reality Luxembourg's aid could be considered to be higher, since no strings were attached to it and it was not tied.[3] The sporadic bilateral projects, mainly in health, had their origin in official visits, for instance, by the Grand Ducal family to Rwanda, Tunisia and Togo.

From 1977 the pressure on political parties gradually grew to raise the aid volume and to come to a 'coherent development policy'.[4]

---

1. Two-thirds of the budget went to international organisations (EDF, IDA) at that time. There were some early bilateral projects such as a nursing school and a hospital in Rwanda, a (psychiatric) hospital (Togo) and seven family planning/maternal care centres (Tunisia). Furthermore, a few private investment projects, such as a livestock development project in Zaïre/Congo, were supported. Several parties, such as the Socialist Party, supported this emphasis on multilateral aid and there was little political pressure to increase the aid volume. See the interview with Minister of Foreign Affairs Gaston Thorn and the discussion with political parties in the first Week of Solidarity with the Third World [*Brennpunkt Drëtt Welt, 20, 1976*].
2. Although quite a few Luxembourgers had been active in the Belgian Congo, and they had officially an equal status as Belgians there.
3. From the end of the 1970s, however, with the Tunisia project as a first example, the Ministry of Foreign Affairs definitely tried to engage Luxembourg firms in projects.
4. In the first Week of Solidarity with the Third World a petition was signed by 1,500 people, demanding among other things an increase of the volume of aid to 0.7 per cent of GNP and concentration of aid on the 25 poorest countries. A year later the same organisations demanded a 'coherent policy' *vis-à-vis* the Third World from the Luxembourg government [*Brennpunkt Drëtt Welt, 23, 1977*]. It should be noted that at the beginning of the 1970s, several MPs from the Democratic and Socialist parties pleaded for a higher volume of aid, but this was not followed up by the government nor by the MPs themselves. At the beginning of the 1980s, there were already about

NGOs, like the *Action Solidarité Tiers Monde* (ASTM), did not hesitate to link these demands with the analysis of Luxembourg as a financial market-place and an investor in developing countries, pointing to examples of corruption and mis-investment [*Delcourt, 1980*].[5] Luxembourg, they claimed, was the second richest of the DAC member countries in terms of GNP per capita, but last but one in relative ODA volume (in relation to GNP): 'The state of Luxembourg is one of the most egoistical in the world' [*ibid.*]. They reminded the government regularly that it had subscribed in 1969 to the International Development Strategy of the United Nations, promising to reach the aid target of 0.7 per cent of GNP by 1980. The group of NGOs further complained that a regulation regarding volunteers and *coopérants*, civil servants and teachers who were willing to work for several years in developing countries, although promised as early as 1971, had still not come into force after ten years. They further demanded the breaking of the growing link between development assistance and Luxembourgian exports, visible in a law of October 1981 which permitted tied loans to developing countries.

The growing pressure of an increasing number of NGOs led to three laws being put forward by the Santer government of Christian Democrats and Socialists. In December 1985 these laws came into force, creating a Development Aid Fund (FAD) and a Development Co-operation Fund (FCD), and establishing a co-financing arrangement with NGOs.[6] The three laws were criticised by the NGO community among others for their lack of interconnection and general policy vision and the absence of a policy on development education [*Brennpunkt Drëtt Welt, 64/65, 1985*]. The laws were thus more the expression of an active NGO community than of a real interest on the

---

30 NGOs active in Luxembourg. They had collected already in 1978 LUF50 million (approximately $1.3m.), about a quarter of the amount of the official development assistance.
5. The activities of Jean Ziegler in Switzerland clearly served as an example here. Delcourt concluded that Third World issues did not raise interest in Parliament and also not among Luxembourg's population. See also Delcourt's speech as representative of Luxembourg's NGOs at a conference organised by the Luxembourgian United Nations Association (24 October 1981) [*Brennpunkt Drëtt Welt, 45, 1981*].
6. The FCD was intended to pay for bilateral projects, the co-financing of projects of NGOs and for *coopérants*. Because of its special constitution it falls outside the annual budgetary constraints. Since 1991 it can also co-finance projects with the multilaterals. The FAD was meant to collect money from private citizens to finance NGO projects. Gifts to the FAD were tax-deductible. It was discontinued in 1996 due to lack of success.

part of the government regarding development co-operation.[7] Nevertheless, the laws led to the first major debate in Parliament on development co-operation policy; furthermore, an increase of ODA with an intermediate objective of reaching 0.35 per cent by 1995 was foreseen. The second Santer government, which arrived in power in 1989, had, however, very limited ambitions with regard to development co-operation and adopted the laws and policy as they were.[8] In 1991 the government published its first annual report on development co-operation[9] and under this government Luxembourg finally became a member of the DAC in December 1992. It was also under this government that Lux-Development became implicated almost entirely in development projects.

The Luxembourg government produced its own particular development policy via the laws of 1985, 1989 and 1996, and also through its annual reports. A law of 1982 had first regulated the status of civil servants and others willing to work as development workers in developing countries. This was replaced by the law of 1989 which improved the position of the 'agents de la coopération' and the 'coopérants'. The three laws of December 1985 instituted the Development Aid Fund, the Development Co-operation Fund and the co-financing mechanism. The law of 1996 defined the central goals of Luxembourg's development co-operation, and also further delineated the roles of the various organisations, the co-financing arrangements and the priority sectors. Other laws of that year specified in greater detail the co-financing arrangement and the 'leave of absence for development co-operation'. No further policy documents were published over the years.

The Development Co-operation Law of November 1995, which came into force on 6 January 1996, spelled out the three major objectives of Luxembourg's development co-operation as:

- to contribute to a lasting economic and social development of the least developed countries, in particular;
- to contribute to the gradual and harmonious integration of developing countries into the world economy;
- to contribute to the reduction of poverty in developing countries.

---

7. Although there were some fears inside the NGO community that the installation of a common development fund would lead to a diversion of private aid to public projects [*Brennpunkt Drëtt Welt, 66, 1986*].
8. Interview with Georges Wohlfahrt, Secretary of State [*Brennpunkt Drëtt Welt, 94, 1989*].
9. Which it was obliged to do by the 1985 law, but it took five years to produce the first report.

These goals come very close to the goals of European development co-operation as stipulated in the Treaty of Maastricht. NGOs had a large influence on the formulation of this law, produced finally by Secretary of State Georges Wohlfart. It embedded development co-operation firmly in Luxembourg's foreign policy, but also made it a special part of it. The law, as well as development co-operation in general, was broadly supported by all political parties. In parallel with it the Directorate for Development Co-operation was also created inside the Ministry of Foreign Affairs in 1998.

Public statements by the Prime Minister, the Minister of Foreign Affairs, or the Minister of Development Co-operation and Humanitarian Action were mostly not of great length or special content, and primarily addressed questions of aid volume. Thus it was to the surprise of many that Prime Minister Santer announced at the UN Conference on Environment and Development in Rio de Janeiro in 1992 that Luxembourg would embrace the 0.7 per cent of GNP target and would achieve it by 2000. His successor, Jean-Claude Juncker, repeated this promise in his government declaration of February 1995 [*Lux-Development, 1995*]. In the Government Declaration of 12 August 1999 Juncker announced that Luxembourg would reach an ODA target of 1.0 per cent of GNP in 2005, a promise repeated by Charles Goerens at the Finance for Development Conference in Monterrey in March 2002.[10] The emphasis has thus been mostly on volume.

TABLE 11.1
LUXEMBOURG'S DEVELOPMENT CO-OPERATION: AID VOLUMES

	1985	1990	1995	2000	2001	2002	2003
ODA ($m.)	10.00	25.00	65.00	123.00	141.00	147.00	189.00
% of GNP	0.17	0.21	0.36	0.71	0.82	0.77	0.80
Multilateral aid ($m.)	8.00	11.00	22.00	24.00	35.00	31.00	n.a.
% of ODA	80.00	44.00	33.80	19.50	24.80	20.40	
Technical co-operation ($m.)	..	1.00	2.00	2.00	5.00	3.00	n.a.

Source: OECD/DAC, *Development Co-operation*, Efforts and Policies, various years.
Note: Not yet available.

---

10. *Discours Prononcé par M. Charles Goerens, Ministre de la Coopération et Action Humanitaire. Conférence International sur le Financement du Développement.* Monterrey, 18–22 March 2002. See also *Statement on Foreign Policy*, by the Minister of Foreign Affairs, Lydie Polfer, in the Chamber of Deputies on 7 March 2002.

In conclusion one could say that politicians for a long time were not particularly interested in development co-operation. There was obviously no pressure from the business community, which had little interest in trade with developing countries. The only force for change was a growing NGO community with close links to the powerful Christian Democrats and the Socialists, which for 15 years from 1984 to 1999 had formed the coalition government. Among the OECD member states Luxembourg has probably the highest number of NGOs per capita, connected with the Catholic Church as well as with secular organisations. Their growing influence manifested itself in a change in politicians' attitudes towards development co-operation. Prime Minister Juncker demonstrated this already in his first speeches. He showed a keen interest in development co-operation 'not to be just a manager of wealth',[11] and in this sense he initiated a change. It is as if this new generation of politicians seems more eager to show Luxembourg's face in this sector of international co-operation, where a small country can be great and have influence. One cannot ignore, however, the fact that this 'human face' gives Luxembourg a chance to divert attention away from its 'banker's face' with the scars of money laundering and the secret bank accounts of criminals and dictators.

In the years before the Juncker government neither the Ministers of Foreign Affairs nor their Secretaries of State spent much of their time on development co-operation; theirs were the 'important' international questions like European unification, international security and other global governance issues. As a country that exported the major part (up to 90 per cent) of its iron and steel, leather and chemicals production, Luxembourg had many interests to defend in more than 60 international organisations.[12] As a small nation and a small market, Luxembourg has to derive its welfare and growth from internationalism. It is only able to grow through international co-operation. Furthermore, this country with its old industries ran into a major crisis in the second half of the 1970s, which consumed all the political energy of its governments at the time. Its balance of payments turned to the negative and it had to readjust. In this climate the Secretary of State was supposed to keep an eye on development co-operation.

Only when the post of Secretary of State came into the hands of the Socialist Lydie Err, until then chair of the parliamentary Foreign

---

11. Interview with Jean-Claude Juncker [*Brennpunkt Drëtt Welt, 148/149, 1995*]. As also shown by his visit to Cape Verde.
12. According to Newcomer [*1984:273*], Luxembourg already in 1965 was a member of 66 international organisations. See also Kirt [*2000*].

Affairs Committee, was a change made. In the eyes of many in the Luxembourg 'aid business' she was the first member of the government who was really interested in development co-operation and who tried to change it into a more or less self-growing aid system.[13] She was probably also the first to spend more time on development co-operation than on her other government tasks.[14] According to her own calculations, three-quarters of her time was devoted to co-operation issues [*Interview, June 2002*]. In the new Juncker government that emerged following the defeat of the Socialists and Christian Democrats in 1999 the new status of development co-operation was expressed in the fact that there is now a Minister for Development Co-operation (and Defence), the conservative Liberal Charles Goerens.

II. THE ORGANISATION OF LUXEMBOURG'S DEVELOPMENT CO-OPERATION

A series of ministries and administrative bodies is involved in Luxembourg's development co-operation. Several ministries contribute to the development co-operation budget, mostly with very small amounts. With regard to the Ministries of Health and Education this concerns the payment of the salaries of teachers or health workers working as *agents de la coopération* in developing countries and the contributions to WHO and UNESCO. The Ministry of Finance was responsible for 6 per cent of the budget in 2000 (5.5 per cent in 2001), being the part of multilateral aid which is going to the World Bank and the IMF and the International Fund for Agricultural Development. Its share of the overall budget has fallen from over 30 per cent at the end of the 1980s to this 5.5 per cent, parallel to the increase of the bilateral budget. The Ministry of Foreign Affairs is now responsible for about 85 per cent of the budget. There is a separate budget line, of more than 8 per cent of total aid, for aid channelled through the European Union that is part of the general EU budget.[15]

The Ministry of Foreign Affairs is thus the administrative body which has the final responsibility for the major part of Luxembourg's

---

13. She did bring 'wind in the windmill', as the German expression used in *Brennpunkt Drëtt Welt*, 178/179, March 1998, put it.
14. The Minister of Foreign Affairs as well as the Secretary of State in his department also have other government tasks. Lydie Polfer, for example, was also Minister of Administration and Administrative Reform. For more general background information to Luxembourg politics see Fehlen [*2000*].
15. The aid to the ACP countries via the European Development Fund is extra-budgetary aid and comes within the general budget.

development co-operation. Its Directorate of Development Co-operation, created only in 1998, is the largest directorate within the Ministry, and is responsible for bilateral and humanitarian aid. Since 1996 there has been a clear effort to increase its staff. Of the Ministry's total professional staff of 67 persons 23 work in this Directorate, whereas 13 work in the Directorate of Political Affairs and 13 again in the Directorate of International Economic Relations.[16] In its other four smaller directorates 23 staff members are working. The 23 staff working in the Directorate of Development Co-operation are distributed over six bureaus: budget, bilateral co-operation (including an Africa and a Latin America Bureau), multilateral co-operation, co-operation with NGOs, humanitarian action and technical assistance.

The Ministry is responsible for multilateral aid within the UN family and EU aid, the bilateral programme and the co-financing mechanism with NGOs. Officially it is the Ministry which has to identify and select projects, but until recently the Ministry did not have any representatives in the field, so much of the identification was also left to Lux-Development. In January 2001 a first development co-operation office was opened in Dakar and later that year a second one in Praia, Cape Verde. New offices are to be opened in the years to come in South-East Asia and Central America. The Ministry is also responsible for evaluation, but little progress has been made in this respect, although the Government Declaration of 12 August 1999 clearly stated that an evaluation policy should be developed.

Lux-Development is the implementing agency. It prepares identified projects and sends them for approval to the Ministry. After approval it is responsible for their implementation and management. It also has an evaluation task of its own, but up until now only a few of the projects have been evaluated. Originally the *Société luxembourgeoise pour la Coopération internationale* (Lux-Development) was set up in 1978 in the midst of the steel crisis as a private limited liability company to promote exports from Luxembourg's companies to developing countries in order to create new jobs. Since most of the smaller companies were directed only to the regional market and the larger companies had their own export offices, it turned out that there was little for Lux-Development to do in this field. So it gradually became involved in the preparation and implementation of development projects, stimulated in particular by its second director, who more or less rebuilt the company. With the rapidly growing aid

---

16. These figures exclude secretaries and other administrative staff. For the Directorate of Development Co-operation this means that four administrative staff are excluded. Total staff in this Directorate is now 27, up from 17 in 1999.

portfolio Lux-Development's role also grew. It now handles about 30 per cent of Luxembourg's total ODA with a total project portfolio of about €41 million at the beginning of 2002. Its staff has grown from 11 members in 1996 to 48 in 2002, about 20 of whom are responsible for projects. It has offices in Praia (Cape Verde), Ouagadougou (Burkina Faso), Windhoek (Namibia), Managua (Nicaragua) and Hanoi (Vietnam). Its staff at headquarters in Luxembourg as well as in its field offices is international, resembling the general situation in the country where 60 per cent of those employed do not have Luxembourg nationality.[17]

Due to its presence in the field and to its growing funds, Lux-Development has been able to play an increasingly important role in Luxembourg's development co-operation. The agency is allowed to devote 8 per cent of the expenditure of projects which it handles to its own administrative costs. It has thus been able, with its growing portfolio, to expand its staff, in contrast to the Ministry, which has a more or less fixed number of functionaries. Thus Lux-Development has taken over certain responsibilities, in particular with regard to the identification of projects, which should have been in the hands of the Ministry. By doing so the division of tasks between the two organisations has become more or less blurred. In view of its increasing functions in bilateral co-operation the Secretary of State decided in 1998 to transform Lux-Development into a public liability company with the state as the major shareholder.[18] Since then its board of 18 members consists of a mix of representatives from ministries, NGOs, trade unions and industry, presided over by Prince William. There has also been an attempt by the opening of field missions by the Ministry, to delineate the roles between Lux-Development and the Ministry better, in particular leaving the identification of projects more in the hands of the Ministry. This seems to be put into practice slowly.

A third major player in Luxembourg's development co-operation is the NGOs. The Ministry has a co-financing arrangement with NGOs whereby it pays up to 85 per cent of their project costs and 75 NGOs with an official status are able to receive money from this budget line. 59 of these NGOs are members of the umbrella organisation, the *Cercle de Coopération des ONGDs de Luxembourg*.[19] The total budget line in 2000 for NGOs was €18m. or 13.1 per cent of total

---

17. Interviews with Lux-Development's staff in April and June 2002.
18. Following a ruling of the European Commission, Luxembourg's government took 99 per cent of the shares of Lux-Development in 2003. In a reaction, the director, Eugène Rausch, decided to leave (*d'Letzeburger Land*, 5 December 2003).
19. Another six NGOs which do not have an agreement with the Ministry, are also members of the *Cercle* (as at mid-2002).

ODA, and in 2001 €21.5m., having increased by more than 87 per cent since 1998. The four largest NGOs, such as Médecins Sans Frontières (MSF) or Caritas Luxembourg, receive amounts between €1m. and €2.5m. a year, the smallest several thousand euros. Seven NGOs receive about €11m. on a programmatic basis. In addition, 190 project proposals were given a positive answer in 2000, the following year 147 for an amount of €8.5m.

Seventy to 75 per cent of the NGOs work only with volunteers, but even the larger ones may still depend on them. Only 21 have project staff, ranging from one project officer in SOS Sahel to eight project officers in MSF Luxembourg. Since 1999 the government has allowed the NGOs the possibility of directly claiming 4 per cent of the volume of their activities as administrative costs or of asking for a reimbursement of actual costs up to a maximum of 10 per cent of total project costs. Furthermore, the government is trying to professionalise the NGOs by offering technical support and by entering into long-term agreements with some of the larger NGOs, on the basis of either sectoral or geographical specialisation or of the NGO's particular approach. Eight of these agreements have been concluded up until now (2002). Since 2001 a Technical Assistance Bureau, subsidised by the Ministry, has been functioning in the office of the *Cercle*. Among the first things it organised was a training course in project planning, in the use of the logframe method. About two NGO projects a year are evaluated externally, and since 2001 the Auditor's Office has been auditing the NGO fund on a regular basis.

In conclusion, one could say that of the three main players in Luxembourg's development co-operation the Ministry seems the weakest partner. Obviously it is the one that sets the rules and distributes the money. But its operations are clearly hampered by the lack of staff and administrative resources (we shall return to this below). Apart from the legislation it has no policy documents to guide it. But the NGO community also does not come out very strong. It consists of a large number of very small NGOs with no or few professional staff. Only recently have activities been undertaken to professionalise the organisations. Lux-Development emerges as the strongest partner in this triple alliance in terms of its growing number of staff, its presence in the field and its other resources. But its somewhat technocratic emphasis on project preparation and implementation and its lack of policy guidance do not make it a very strong partner either.

III. COUNTRIES AND SECTORS

With the rapid increase in Luxembourg's aid volume, the number of aid-receiving countries also grew fast. A first step towards concentration in 1990 ('with the goal of not dispersing the financial means too much') led to the establishment of 14 programme countries. Luxembourg's development co-operation originally started in francophone countries and in Cape Verde, because of the presence of an important Cap Verdean immigrant community. Rwanda, Togo and Tunisia were the first countries to receive assistance, followed by Senegal, the Comoros and Equatorial Guinea later on. Already in its first annual report on development co-operation in 1990 the government announced the intention to concentrate its aid on ten countries, seven of them in Africa. Respect for human rights as well as a low level of military expenditure would be among the criteria in the selection of countries. The list expanded quite rapidly, however, to include also Latin American countries like Chile, El Salvador, Ecuador and Nicaragua as well.

Wohlfahrt's successor, Secretary of State Lydie Err, tried to streamline the choice of countries more by removing four from the list – Burundi, Mauritius, Tunisia and Ecuador. The criteria for selection were: regional concentration, in particular on (West) Africa with Mali as the only new country; the level of poverty, with those countries at the bottom of most lists being chosen; and good governance, with an emphasis on the protection of human rights. This resulted in a selection of programme countries in West Africa (Burkina Faso, Mali, Niger, Senegal and Cape Verde), in Central America (El Salvador, Nicaragua) and in South-East Asia (Laos, Vietnam) plus Namibia and the Palestinian Territories. Excluded are countries in East Africa and South Asia, mainly due to the linguistic aspect of the selection. Within countries often the most poverty-stricken regions are targeted, such as Okavango in Namibia, the island Santo Antão in Cape Verde and the 'shadow zones' (*zones d'ombres*) in Tunisia.

The recipients receiving the largest volumes of aid (Burkina Faso, Cape Verde, Mali, El Salvador, Laos and Vietnam) received in 2001 from €5.0m. to €5.7m.). The smallest recipients among these 11 target countries (the Palestinian Territories, Senegal) got around €2.0m. In total they received €49.4m. in 2001, about 31.5 per cent of total ODA.[20] About 20 other countries were receiving aid from Luxem-

---

20. Part of this was aid through the multilaterals or NGOs, leaving €30.9m. for bilateral assistance to the 11 programme countries (or a little less than 60 per cent of bilateral aid).

bourg in that year, in total €21.1m. This means that more than 40 per cent of bilateral aid was still going to non-programme countries. The major part of Luxembourg's aid goes to the least developed countries, although their share has been decreasing in recent years (see Table 11.2).

TABLE 11.2
AID DISTRIBUTION BY COUNTRIES AND SECTORS (IN % OF TOTAL ODA)

	1990–91		1995–96		2000–1	
Major recipients	Cape Verde	3.7	Cape Verde	7.3	Yugoslavia	5.4
	Senegal	3.2	Nicaragua	3.8	Cape Verde	5.2
	Rwanda	3.2	Viet Nam	3.1	Nicaragua	4.9
	Burundi	3.0	Senegal	3.1	Burkina Faso	4.2
	Chile	2.4	Mauritius	2.9	El Salvador	3.9
Major recipient	LDCs	63.6	LDCs	37.8	LDCs	45.1
groups	Other LICs	6.9	Other LICs	27.9	Other LICs	19.2
	LMICs	16.8	LMICs	23.9	LMICs	31.4
	UMICs	11.0	UMICs	10.4	UMICs	4.3
Sectoral distribution	(not reported)		Social sector	40.7	Social sector	69.8
			Agriculture	5.4	Agriculture	4.0
			Industry	0.5	Industry	3.2
			Program.ass.	1.1	Program.ass.	1.6
			Other	22.3	Emergency	11.4

Source: DAC, Development Co-operation. Efforts and Policies, various years.

This means that the concentration process has not progressed very far.[21] Even in 2001 aid spending in non-programme countries like Morocco and Tunisia was often higher than in programme countries. Also countries with a relatively high income, such as Chile and Ecuador, were still receiving aid. New projects were still being approved in non-programme countries or even with new aid recipients. In addition, politicians in Luxembourg like to travel with presents in their hands and for developing countries these are often development projects. China and Mongolia are examples of countries where new projects were started after official visits and in violation of the official selection criteria.

From the onset Luxembourg's development aid programme has been concentrated on the social sectors. The major proportion of its bilateral ODA goes on health, water and sanitation. Education and rural development have always been runners-up. In 2001 more than 80 per cent went to the social sectors, the productive sector receiving a mere 15 per cent. Health stood at 28 per cent of money spent on bilateral aid projects, education at 25 per cent, water and sanitation at

---

21. The first 11 recipients received 53.6 per cent of bilateral aid in 1991/92 and 57.4 per cent in 1996/97 [OECD/DAC, 1999].

11 per cent. With a doubling of the budget by the second Santer government, Secretary of State Georges Wohlfahrt[22] tried to concentrate on a series of projects which 'would benefit the local population to the maximum': housing, small hospitals and dispensaries, schools, water and electricity supply. This has been the trend in the years that followed. Small hospitals, dispensaries, secondary schools and water projects received, and are still receiving, the most aid. Nearly all the money comes in the form of projects.

Only a very small proportion of Luxembourg's aid is tied. Luxembourg's economy, mainly oriented towards the service (banking, insurance) sector, has limited opportunities to deliver goods used in development co-operation. The introduction of tied loans in 1982 led to several projects and deliveries under aid being criticised by the NGO community.[23] At the beginning of the 1990s two-fifths of ODA was still tied [*OECD/DAC, 1993:3*]. Since then this has been reduced, but there are also some more recent examples of aid-tying.[24] First of all in a series of projects, not managed by Lux-Development, refrigerators for medical use were delivered by Electrolux, which has a factory in the electoral district of the former Secretary of State, Georges Wolhlfart. Factory halls (hangars) and steel constructions were delivered by Astron to, among others, Kosovo and Cape Verde. Food aid (wheat from the north of Luxembourg) has been delivered to the Balkans and the Sahel. Furthermore, several consultancy firms have been involved in project preparation and implementation.[25] In the banking sector the Agency for the Transfer of Financial Technologies has a training project in Vietnam. But in general the major part of Luxembourg's aid is spent on financing local costs, much of it going to local construction companies. There are no statistics on the

---

22. Interview with Georges Wohlfahrt [*Brennpunkt Drëtt Welt, 132/133, 1993*]. Wohlfart was himself a medical doctor, which partly explains his interest in the social sectors.
23. One example was a palm oil project in Ecuador in which a Luxembourgian together with a Belgian firm were to deliver the oil mill to the plantation 'Palmeras de Ecuador'. The NGO *Iwerliewen* denounced the project, because it would lead to the disappearance of at least another 25,000 ha of tropical forests, in addition to the 20,000 ha already occupied by the plantation. Several newspapers had stories on this project [*Brennpunkt Drëtt Welt, 66, 1986 and 76, 1987*]. Another example was the delivery of truck-tyres to Turkey. The then Secretary of State Goebbels declared that they were 'old corpses in the cellar' of the former government. Another so-called 'state credit' to China was also under fire [*Interview with Georges Wohlfahrt, Brennpunkt Drëtt Welt, 94, 1989*].
24. *Inauguration de la 79e Fil. Discours de M. Charles Goerens*, 6 October 2001.
25. There is also a fund to involve Luxembourg consultants at the World Bank.

return flow, but they were estimated for the second half of the 1990s at a mere 5 per cent [*OECD/DAC, 1999:38*].[26]

Multilateral aid fell relatively and very fast during the 1990s, due to the steady increase in ODA. At the beginning of the 1990s it still amounted to more than 40 per cent of total ODA. It went down to a little more than 35 per cent in 2000 [*OECD/DAC, various years*]. Luxembourg sticks here to its official and obligatory contributions, which means that by far the major part (in most recent years about half) of it goes to the EU. The IDA contribution comes second, and Luxembourg raised its contribution in its most recent replenishment. Voluntary contributions to the UN agencies are relatively low, but have been increasing in recent years. The WHO has always been the largest recipient, but recently the UNFPA has caught up with it. Contributions to UNDP, UNICEF and UNHCR are relatively small. Luxembourg is not a member of the regional development banks, apart from the European Bank for Reconstruction and Development.

IV. LUXEMBOURG'S DEVELOPMENT CO-OPERATION WITH THE CAPE VERDE ISLANDS

A case study of Cape Verde as the major recipient of development assistance from Luxembourg can illustrate the programmatic content and background of this relatively new donor, and can demonstrate its strengths and weaknesses. The Cape Verde Islands are an archipelago of islands of mostly volcanic origin off the coast of Senegal (off Cabo Verde). The archipelago consists of ten larger islands, nine of which are inhabited, and a series of uninhabited smaller islands. Its total area is not more than 4,033 km^2, about the size of a medium-sized district or province in Europe or, indeed, one and a half times the size of Luxembourg. The islands were uninhabited when the Portuguese arrived in 1462. They acted as an entrepôt in the centuries of the slave trade and a fuelling station for coal in the times of the steamships and for kerosene until recently, in particular for South African planes. More than this the history of the islands has been characterised by erosion, droughts and famines, during which thousands of people starved, often through the negligence of the Portuguese colonial power. After these disasters many thousands emigrated to the United States, Portugal, France and the Netherlands, or, forced by the colonial regime, to São Tomé and, to a lesser extent, Príncipe. The

---

26. Figures for 1996 and 1997. This seems a bit too low though, but it will not be much higher.

430,000 or so inhabitants (slightly fewer than in Luxembourg) are said to be at least equalled by Cape Verdian emigrants abroad, but this emigrant figure is probably too high.[27] Only 11 per cent of the islands' soil is cultivable, so that 80 to 90 per cent of the food that is consumed has to be imported. Its main export products are music and literature.[28] The main source of income, apart from aid and the remittances of emigrants, is tourism.

The Cape Verde Islands had, so to speak, the 'bad luck' to be colonised by an underdeveloped European nation. At independence (5 July 1975) the new government inherited a country with almost no physical or social infrastructure. The literacy rate was 21.5 per cent and there were only two secondary schools.[29] Mindelo's coal harbour was constructed (and abandoned) by the British and the international airport at Sal by an Italian air company.[30] Woods had vanished under Portuguese (absence of) rule. Erosion and desertification were almost not countered at all. This left the young state with the tremendous task of developing the country from scratch, and this in a period of severe droughts (1973–85). Since independence hundreds of kilometres of paved roads have been constructed. The surface of woodlands has gone up from about 3,000 to more than 50,000 ha. Enrolment in primary education rose to 95 per cent, with a gradual extension of compulsory attendance to six years. Health statistics show that life expectancy has gone up to 71 years and that child mortality is now one-quarter of what it was in 1975. All these figures show the Cape Verde Islands performing much better than other countries in the Sahel.

This impressive progress was to a large extent made possible by foreign aid. The Cape Verdian government used food aid in particular very cleverly in food for work programmes to enable road construc-

---

27. These figures should be treated with caution, like most of the figures regarding the islands. Here one has to take into consideration who one still wants to call a 'Cape Verdian' (immigrants of the third or fourth generation?). Head counts of immigrants are erratic, in particular in countries with high immigration rates and a policy of acculturation. Lesourd [*1995:273*] counted for 1990 between 420,000 and 450,000 emigrants, and, with a stricter definition, between 250,000 and 300,000.
28. Its exports in 2000 were estimated at $40m., creating a trade deficit of $250m. Shoes and clothing exports have been increasing in recent years.
29. In fact, under the Estado Novo more than half of the existing primary schools were closed. Due to the mix of its population and the use of Cape Verdians as colonial civil servants, the illiteracy rate was not so disastrous as in the other Portuguese colonies, where illiteracy reached 99 per cent.
30. It was sold in 1947 to the Portuguese state. It goes without saying that the road infrastructure was in a similar state of underdevelopment.

tion, reforestation and the building of dams and terraces.[31] In addition, there were donors' knowledge and capabilities with regard to water management and the fight against erosion.[32] Donors also assisted the country in its strategy of investing in human resources. Education and health always received a percentage of development aid, which was clearly higher than in most other African countries. Small proportions of foreign aid went to other productive sectors, apart from agriculture, partly because these sectors were in the hands of state companies until the end of the 1980s.

The down-to-earth government policies[33] clearly inspired donors to invest in the things most needed in the country. One could add that, in view of the small market, no major donors' interests are involved in Cape Verde, apart from some possible strategic interests in controlling a major shipping route.[34] An almost impeccable human rights record and a move to full democracy at the beginning of the 1990s stimulated this all the more. With a prudent macroeconomic and monetary policy this means that all the conditions of 'good governance' are almost fulfilled. It is therefore amazing to see that in recent years several donors have been leaving the country, due mainly to a domestic policy of rationalising their overall aid programmes rather than out of a concern about the effectiveness of the co-operation programmes in Cape Verde.[35] Aid dependency has remained large, however. In the DAC's 25th anniversary report Cape Verde was quoted as the most aid-dependent country in the world, net ODA in 1982 to 1983 being 58 per cent of GNP [*OECD/DAC, 1985:128*]. This percentage clearly diminished fast over the years, but still hovered between 25 and 30 per cent in the second half of the 1990s, but then it

---

31. These FAIMO (Frentes de Alta Intensidade de Mão de Obra – High-Intensive Labour Systems) were originally to be discontinued by the MdP government, but later on were again mobilised in public works.
32. As Lesourd [*1995: ch.6*] clearly elaborates in his chapter about these programmes.
33. Lesourd [*1995: passim*] in his major study on *Cabo Verde* insists on the fact that Cape Verdian government policies were, also under the PAIGC/PAICV, first and foremost practical, notwithstanding the Socialist vocation of the party, inspired by its historical leader Amilcar Cabral, a Cape Verdian killed in January 1973 in Conakry by the Portuguese secret service. The policies clearly differed from those of their counterparts in the other former Portuguese colonies.
34. But here the Azores are clearly more important. In recent years there have been some concerns about drug trafficking and the impossibility of the government being able to control this.
35. This was, for example, the case with the Netherlands which tried to cut down the number of aid recipients and mainly seems to have excluded the Cape Verde Islands, because they would not be able to absorb a large amount of aid. The Netherlands will return with an environmental programme in the coming years.

dropped to 17.1 per cent in 2000 and 13.1 per cent in 2001.[36] This was due mainly to the fact that a large donor, Germany, ended its aid programme, but World Bank loans were also at a minimum in 2000. This means that normally small donors (Portugal, Luxembourg, Austria) are now playing a large (or larger) role.

Luxembourg's aid programme in Cape Verde falls well in line with this analysis. The programme started at the end of the 1980s with the reconstruction of a slum area in Mindelo, São Vincente, paying for the imports of building materials and, a little later, the construction of a health centre. The programme extended to the other side of the channel to the island of Santo Antão, which became a concentration region for Luxembourg's aid to Cape Verde. Initially the projects there were in health (a dispensary in Porto Novo and a small hospital in Ribeira Grande), but education followed with the construction of a boarding-house for grammar school students in Ribeira Grande. From 1993 onwards a quite large amount, about €8.5 million, was spent on rural electrification.[37] In 1996 three new projects were added: water and sanitation in urban centres, house building,[38] and the enhancement of goat-derived products,[39] all on Santo Antão. A year later Luxembourg co-financed with the European Union an important road construction project. From then on the programme spread to other islands: boarding-houses on São Vicente and Praia, health projects in Praia, waste collection in São Nicolau, and nationwide projects on health in schools and against drugs abuse.

Education, health and basic infrastructure, together with food aid, are thus the main components of Luxembourg's development co-operation programme with Cape Verde. Food aid started in 1991 and amounted to 2,000 tonnes a year of Luxembourg wheat and a similar amount in money to finance other food imports, most recently rice. The total amount involved was about €1 million a year over the period 2000 to 2002. Food aid is still very necessary in a country that even in years of good rainfall has to import 90 per cent of its grains [*Metz, 2002; UN System, 2001:31–8*]. The major problem with food aid is the use of the counterpart funds. There were continuous discussions

---

36. This was due to a huge fall in net ODA, from $136.6m. in 1999 to $94.1m. in 2000.
37. Luxembourg's projects are mostly rather small in size, although a lot larger than the $100,000 to $200,000 reported by the DAC in its first Aid Review of Luxembourg in 1993. The average project size between 1990 and 1993 was $413,000. Although figures were not available for all the projects in Cape Verde, a sample of 13 projects gave an average of about €2 million.
38. Houses were built for higher executives to encourage their return; the income of the sales was used for the construction of apartments in a social housing project.
39. A breeding centre as well as cheese and sausage production.

(with several donors) with the government on the use of these funds and the poor transparency the government was providing on this use. In particular during the last two years before the election of 2000 the Movement for Democracy government used these funds on a spending spree, creating also an enormous budget deficit. This brought some donors to the decision to end their food aid [*Metz, 2002*].[40] Furthermore, there were severe delays in payments by the implementing agencies. Luxembourg faced the same problems. Its counterpart funds should be, and are, mainly used for construction works in the municipalities, most recently for kindergarten. Austria, which is also a major food aid donor, discontinued giving its food aid in kind in 1996 and since then has provided the money value instead. Efficiency considerations were at the root of this: higher cost-effectiveness, better timing and more flexibility [*ibid.*]. Other donors such as the European Commission and Switzerland also discontinued giving food aid in kind and since then have given it in monetary form. Luxembourg's wheat is in practice one of the few components of its co-operation with Cape Verde that involves tied aid. The wheat is clearly overpriced;[41] Luxembourg food aid thus loses a lot of its efficiency.

Luxembourg has financed several health projects in the last ten years or so. Among the latest are the already mentioned equipment of a health centre in Porto Novo[42] and Paúl, and a regional hospital in Ribeira Grande, all on Santo Antão, and the construction and equipment of a surgical unit in Praia. The surgical unit had already been on the drawing-board for some years, but was delayed because the hospital finally wanted a larger construction so that a diagnostic centre could be placed in the basement, to bring idle-standing equipment under the same roof.[43] The Agostinho Neto Hospital is at the pinnacle of the health pyramid of the country (together with, but to a lesser

---

40. Italy, France, Germany, Switzerland and Sweden were said to have ended their food aid in recent years. This had, for example, severe consequences for the school canteen project in the cities (the programme in the rural area does get food from the WFP) [*Metz, 2002*].

41. Austrian wheat could be bought in 2001 for €140 a tonne [*Metz, 2002*], whereas Luxembourg's wheat did cost an estimated €250 a tonne. C.i.f. prices for Cape Verde in 2000 and 2001 were respectively €132 and €140 a tonne. Lux-Development indicated that the f.o.b. price in 2001 for Luxembourg's wheat was €137. As from 2002 Luxembourg stated that rice would no longer be delivered but 6,000 tons of wheat yearly. This was due to the poor performance of the public marketing agency EMPA, which is heavily indebted, in contributing to the counterpart funds.

42. The buildings for this centre were paid for by the Dutch, but were standing idle due to lack of equipment. On the health projects see also Estrela and Reitmaier [*1998*].

43. For example, a scanner donated by the Fundação Calouste Gulbenkian and some other Portuguese equipment.

extent, the Hospital in Mindelo), but it also serves as a general health centre for Praia and a general hospital mainly for the island of Santiago. Up until now it had only one surgical room, the roof of which collapsed, so that it had to resort to emergency measures. The new unit thus fulfils a clear need, but with its four rooms it seems too large in relation to the number of operations carried out in recent years. The construction looks sober and adequate. The equipment delivered comes mainly from France, and after-sales service is part of the contract. Technicians have been sent to Senegal for training. But in this respect there still seems to be a need for more training.

Luxembourg's education projects consisted mainly of boarding-houses and a technical school in Assomada/Santa Caterina on Santiago Island. Boarding-houses should enable children from rural villages to go to a secondary school in the (small) towns. The children eat, sleep and study in the house and visit their parents every other week during the school year of 36 weeks. Parents have to pay 6,000 CV Escudos a month for their stay.[44] As one comes down into the valley of Assomada in central Santiago the two buildings of the technical school and the boarding-house opposite really stand out, with their bright yellow colours in an overwhelmingly grey town. The new boarding house caters for 87 students aged between 11 and 19, roughly half boys and half girls. It has a large restaurant and well-equipped kitchen, which also sells meals to other students and teachers. The Grão Duque Henri Technical School was also opened in 2002 and had 650 students in its first year, expected to rise to more than 1,000 in the second year. Although nearly all the teachers have the required qualifications, which is said by the director to be 'super bom',[45] those particularly in the technical classes suffer from lack of training in dealing with machinery. For this reason some machines remain unused, even though the school does not seem over-equipped. The electricity class, however, is also used in the evenings for adult students who pay for lessons in electricity construction work.

To sum up: Luxembourg has grown over the years into a major donor in Cape Verde, the third largest of the bilaterals, with a portfolio of $8.3 million in 2000. Its aid consists mainly of projects, for the most part in the social sectors. The emphasis has been on construction and 'visibility'. Only in food aid and in some of the early

---

44. For comparison, the daily wage of a worker in the FAIMO scheme is Esc.400; a better educated construction worker would earn around Esc.1,000.
45. In most schools only 50 per cent are qualified. The school has some foreign staff from Portugal and France, paid for under Technical Co-operation, but also some African staff under local contracts.

projects could domestic trade interests be found; by far the major part of the equipment in health and education projects comes from other European countries, in particular from France. The programme seems more or less self-developing; there are no country strategy papers, no sector papers. One project seems to follow another; identification is dependent on the presence of a Lux-Development representative and recently on a field office of the Ministry of Foreign Affairs. The lessons learned up until now are not collective, but are learned individually. There are no signs of a regular exchange of experiences and knowledge with other donors or other forms of close collaboration.

The Luxembourg programme has difficulty in reaching the poorest parts of the population. All educational projects appear to be catering mainly for the middle (and upper) classes and their children. Fees for the boarding-houses and (technical) schools are not within the reach of the poorest parts of the population. Apart from the food-for-work programmes, some water projects and the health posts, reaching the poorest or targeting their aid programmes on them seems to be a major problem for most donors.

There are also clear weaknesses in the donors' way of handling aid. Donors have more or less carved up the country, so that every donor had or has its 'own' island (the Germans Fogo and Brava, the Swiss Boa Vista, the French São Nicolau, the Dutch Santo Antão, the Belgians Maio). Luxembourg joined in by concentrating on Santo Antão and recently also on Praia. This seems to have prevented cross-project learning. Co-ordination among donors is still very weak. The last donor Round Table organised by the UNDP was in 1999. Once every three weeks there are EU meetings, but these do not seem to go further than some straightforward exchange of information. Only now is the Department for International Co-operation trying to set up a database, with the support of Luxembourg, of all donor projects on the islands.[46] A general overview of the aid projects and programmes and lessons learned is lacking. Sector programmes are only now being formulated, the World Bank and the Netherlands being the only donors to support them. Increasingly the country seems to be left with one, more than major, donor – Portugal, which does not have a long history of development co-operation, has other private interests[47] and

---

46. The UNDP, which is mostly responsible, although also mostly deficient (due partly to insufficient reporting by donors), has left this task to one side in the last five years.

47. In the recent privatisations of state companies it was mostly (only) Portuguese companies which won the deals; for instance, in the most recent privatisation under way of the water and electricity company Electra, of which the Portuguese EDF will be the major shareholder.

seems to lack clear strategies. The number of donors is diminishing: Switzerland and Belgium have left, the islands did not become a programme country in Dutch development co-operation, from Germany only the GTZ has stayed but it is now also phasing itself out. Luxembourg will be left with only a few major partners.

IV. MAIN CONSTRAINTS AND WEAKNESSES OF LUXEMBOURG'S DEVELOPMENT CO-OPERATION

For a small country the outside world is large. This means that it has to cover a whole set of issues, countries and organisations with a small number of staff. The 18 staff members of the Development Directorate in the Ministry of Foreign Affairs not only have to follow important conferences and meetings, they also have to approve projects and programmes and to follow international developments in the conceptualisation of development problems. In larger aid administrations it is often a heavy burden for staff to keep up with recent thinking and theorising; it is therefore easy to imagine that for a very small office this is almost impossible. The problem is exacerbated by the large number of countries in Luxembourg's bilateral programme and the large number of NGOs that qualify for co-financing. The staff seem overstretched and overloaded.

Two further issues are adding to this. First, the major part of the staff consists of career diplomats with no academic background in development co-operation. Development-specific expertise is lacking. The *agents de la coopération* working in the Directorate, who have often been working in development projects for a larger number of years, do not have a clear status inside the Ministry and are working only on a temporary basis.[48] In Lux-Development several staff members have experience of some kind in developing countries, but a clear balance of expertise relevant for poverty reduction is lacking there also. They draw heavily on outside consultants for expertise.

Second, inside the Ministry staff changes are frequent. In the last three years up to 2002 the whole staff was renewed, apart from the director. Also the Development Co-operation Directorate in this Ministry does not seem to be attractive to career diplomats, who often focus more on the political department.

Lux-Development has been totally reorganised since 1992. Up until then it had the traces of an export promotion agency. There was no

---

48. Since my visits to Luxembourg in the first half of 2002 three of them did get a permanent employment contract.

systematic annual budgeting, there were almost no project documents [*Interviews 2002. See also Golooba-Muteba, 1999:7*]. After the reorganisation more and more specialised staff were recruited, made possible by the rapidly growing projects portfolio, 8 per cent of which can be spent on administrative costs. This gave the agency more or less a monopoly position in Luxembourg's development co-operation, with an overloaded Ministry, on the one hand, and small and professionally weak NGOs on the other.[49]

As noted, Lux-Development still shows traces of the export promotion agency it once was.[50] Its emphasis is clearly on projects and hardware. Preparation of projects is extensive but seems to be mainly technical. Already in the distant past this emphasis led to projects which were not well embedded in the social context, an example then being the hospital in Tsévié, Togo [*Brennpunkt Drëtt Welt, 40, 1980 and 73, 1987*]. Luxembourg's projects in Cape Verde also show some of these characteristics. This is not to say that they are outside the range of social and economic development in Cape Verde, but that the emphasis has been on construction and delivery of machinery and instruments, not on training, teaching programmes and institution building. Or, to put it in different words, the emphasis is still very much on the 'hard' and less on the 'soft'.[51]

The government's policy statements during the past decade have been largely confined to planning on the achievement of new ODA targets. The development of policy has been slow and the programme seems to develop in a more or less pragmatic way. Of course, programme countries have been chosen and a clear sector concentration

---

49. This was also because the division of tasks between the Ministry and Lux-Development was not clearly delineated in an official document, as the first audit of Lux-Development concluded [*See the overview of the audit in Brennpunkt Drëtt Welt, 166/167, April/June 1997*]. Added to this was the lack of capacity at the Ministry which caused the appraisal of projects to be inefficient (according to the first audit of Lux-Development) [*ibid.*].

50. Also in a judicial sense, since major companies and banks still hold an important part of its shares and are also represented on the Board of Directors (6), together with NGOs (1) and trade unions (2). The government is the majority shareholder and has the largest number of representatives (7) on the Board. In its statutory legislation export promotion is still mentioned as a major goal.

51. As was mentioned in the first audit of Lux-Development (see above), where it was concluded that 84 per cent of the expenses of the preceding year (1994) went on investments in infrastructure. In the project proposals seen for this study this emphasis on 'hardware' was clearly visible. It could also be concluded that the technical handling of the projects and of the deliveries attached to them was impeccable. A recent critique of the establishment of a milk factory in Montenegro bears the same sort of remarks: 'Wo die Milch aus den Brunnen fliesst – Luxemburger Entwicklungshilfe in Montenegro' [*d'Lëtzebuerger Land, 3 August 2001*]. In this article the 'technizistischen Herangehensweise' was seen as a major obstacle to the success of this factory.

has been stipulated, but from time to time they are lost from sight. Furthermore, the link between these country and sector choices is made in principle, but the links with the overarching goals of Luxembourg's development co-operation are not very clear. The emphasis on the social sectors seems to indicate the importance of poverty reduction goals within Luxembourg's development co-operation, but there has been no mainstreaming of poverty reduction within these sectors. Guidelines on gender issues and the environment are also missing. Staff shortages and a lack of experience in development co-operation are probably felt most in this field of policy development.[52]

The DAC review of 1993 concluded that the Ministry had 'not established any formal mechanisms for programming and evaluation'. A lot of progress has been made since then, in particular in Lux-Development. Lux-Development had been looking for ISO certification since 1997, but discovered that the old standard was not well adapted to service organisations of a rather small size. In 2000, however, Lux-Development initiated a project to be certified by the new standard [*Annual Report, 2000*].[53] Since 1998 NGO projects have been audited and in 2000 a technical assistance bureau was set up. All these efforts do not, however, seem to be part of an overall strategy to enhance the quality of aid with clear programming and planning, targeting, monitoring and evaluation.[54]

Only a very limited number of projects have been evaluated up until now. Evaluation is another weak spot in Luxembourg's development co-operation. An *agent de coopération* was recruited to set up a monitoring and evaluation system, but he left the Ministry after a short time. Part of the first development co-operation days in July 2000, organised for Luxembourg's development workers, was devoted to evaluation, and in 2001 a manual was produced [*Ministère des Affaires Étrangères, 2001*], but a clear policy and planning are still lacking. Money has been set aside and a consultant hired to assist the Ministry in elaborating a clearer line in this field, but a lot still has to be done here. This means that the learning and feedback capacity is still low and individualised. Institutional memory is largely limited to the brains of a few 'old' hands.

---

52. In the first audit of Lux-Development it was also concluded that policy guidelines were missing [*Brennpunkt Drëtt Welt, 166/167, April/June 1997*].
53. Neither the 2001 Annual Report nor the website give an outcome as to whether the organisation has yet been certified.
54. In the DAC *Review* [*1993: 28*] it was stated that 'these measures have been taken piecemeal rather than within a framework of an overall strategy' and 'quality control remains a major challenge'.

V. CONCLUSIONS

Luxembourg is an interesting case as a relative newcomer in development co-operation. Although the country is small and therefore has limited human resources in terms of the numbers of staff as well as expertise, it set out to copy the development co-operation programmes of the large donors. Its almost total emphasis on projects makes it exceptional in comparison with those donors, but in most other characteristics its aid programme tends to follow them: a relatively large number of aid-recipient countries, a relatively large number of (small) projects, and a comparatively huge number of NGOs. It is easy to imagine that its staff is overstretched and overloaded. This leads to a clear undervaluation of policy development and a neglect of embedding all activities in an aid strategy. Evaluation and learning lessons have the same low value. Development co-operation issues other than aid are more or less lost sight of.[55]

This leads to the question of how within ten years Luxembourg as a 'non-donor' within a Europe full of donors could have become one of the most generous aid donors, and within a few years probably the most generous one. How was Luxembourg able to quadruple its aid between 1994 and 2003? As we have seen above, the rapid change from being an iron and steel producer to being a service provider had only marginal influence on Luxembourg's development co-operation. Lux-Development emerged from this crisis as the aid agency, but aid itself was only distantly and indirectly the result of economic-commercial pressure from related interest groups, as clearly shown by the case study of Cape Verde. In addition, Parliament with its rare debates on development policy cannot be seen as an important force for change. In Luxembourg's consensus policies it seems to make little difference (with maybe the Social Democrat Lydie Err as the one exception) whether the ministerial position is occupied by a Conservative Liberal or a Social Democrat. The general public supports the recent generosity and is active in a large number of mostly small NGOs, but knowledge and interest do not run very deep.

It was first and foremost a few active and vocal NGOs who put a lot of pressure on politicians and political parties to increase the aid budget. But this activity already existed in the 1970s and it took a

---

55. This is most clearly seen with regard to 'European coherence issues', in particular the Common Agricultural Policy, where Luxembourg defends the position of its peasantry, 80 per cent of whose income depends on Brussels (see an interesting series of articles on Luxembourg's agriculture that appeared in *d'Lëtzebuerger Land* from December 1999 to January 2003), and the CAP, plus pleading for the creation of regional agricultural markets in the South.

long time before it ultimately became successful. The growing number of NGOs and their close links with the government parties are a first explanation of the changes in Luxembourg's development co-operation. One cannot escape the impression, however, that outside pressure together with a sense of embarrassment and moral responsibility was the combination that made politicians change direction. Criticism from the OECD that such a rich country could be such a small donor was a first sign of this pressure. Embarrassment at the large inflow of 'black money' into Luxembourg's banks, frequently coming from the political leaders of developing countries, was a second driving force for change.[56] A feeling of international responsibility and the chance to make a small country big in the important international field of development co-operation were also significant in recent years.[57] One can wonder, though, why it took Luxembourg so long to arrive where it is now, for a country that has been so versed in international co-operation, just because it is a small country. The ambiguity could be that this particular field of international co-operation remained out of Luxembourg's sight, because it was so busy in other fields of international co-operation and because it was so small and already had so many international tasks.

Luxembourg as a donor never seems to have questioned what niche it could occupy in the world of donors. Instead it set out to develop itself as the twenty-first or twenty-second DAC donor, with the whole range of activities attached to that position. It is clear that Luxembourg's aid programme is quite impeccable in comparison with other DAC members with its more than serious emphasis on the social sectors and its lack of important political or economic interests.[58] Luxembourg (and its rulers and politicians) likes to show itself in its aid efforts as a 'humane internationalist' nation that can excel in this sector of foreign policy (and by way of this 'excellence' is also building up influence in other sectors of foreign policy).[59] But there

---

56. Luxembourg's sensitivity on this issue could be seen in particular in its immediate and vocal reaction to the French National Assembly's Peillon-Montebourg report on money laundering which was presented in January 2002 in Paris. Luxembourg rejected instantly the accusations in the report. Minister of Justice Luc Frieden declared at a press conference: 'Le rapport constitue un ensemble de clichés et de préjugés méconnaissant le droit et la réalité luxembourgeoise.'
57. As shown above in Prime Minister Claude Juncker's keen interest in development policy and to a lesser extent also in the Grand Ducal family's activities and visits.
58. Only 15 per cent of the financial volume of contracts awarded by Lux-Development in recent years are with engineering, consultancy and construction firms [*OECD/DAC, 2003*].
59. At the press conference presenting the Annual Report on development co-operation of 2001 Minister of State Goerens insisted that Luxembourg was clearly gaining

would still have been possibilities to look for a special place for Luxembourg in the donor community. Instead of extending the number of NGOs and aid recipients, it could have been more selective, choosing, for example, only a limited number of small programme countries, only land-locked or island states where the number of donors is limited and where Luxembourg could have taken a more prominent place (as it has in Cape Verde).

What is probably most amazing is that Luxembourg has developed its bilateral programme in 'splendid isolation'. There are no signs – apart from the fact that it is part of the 'normal' donors' coteries, the DAC and the Development Committee of the EU – that Luxembourg has tried to learn lessons from other, older donors' experiences. It most probably did not have the manpower for it, but no planning seems to have been undertaken on institutional and policy development, when the decision was taken to allow the aid budget to grow to 0.7 per cent of GNP. Policy documents and guidelines are absent, which gives the impression of Luxembourg's development co-operation programme as self-aggrandising, extending on all sides. It does not even always look like copying the institutional paraphernalia of the larger donors, because some essentials (such as planning and policy development) are missing.

If 'semi-copying' DAC donors becomes the trend for the new Central and Eastern European donors that will come into the 'aid market' after the EU enlargement, it will exacerbate the already huge co-ordination problems that development co-operation and aid-recipient countries are now facing. Aid recipients are not capable of co-ordinating the ever growing number of donors, and many donors do not seem ready yet to be co-ordinated. From this perspective, Luxembourg should be an example of what the new donors should not do in terms of policy development, country selection and co-financing arrangements. In terms of volume and untying of aid and of sector choice Luxembourg could be a donor to follow. Recent adjustments are said to be streamlining the programme better. It needs more changes to make the central slogan of the public relations campaign started in November 2002 more realistic: *'La coopération ça bouge!'*

---

in influence by means of its development efforts, citing as an example its recent actions in the EU [*d'Lëtzebuerger Land*, 27 July 2002].

REFERENCES

Almeida, G., 1999, *Estórias Contadas* (Histories Told), Mindelo: Ilhéu Editora.
Bayani, R., 2000, *Étude sur l'évolution de la politique de coopération au développement luxembourgeoise de 1986 à 1999* (Study on the Evolution of the Luxembourgian development co-operation policy between 1986 and 1999), Luxembourg: Ministère des Affaires Étrangères, Direction de la Coopération au Développement, September.
Bourgain, A., P. Piereti and G. Schuller, 2002, *Compétivité de l'économie luxembourgeoise. Rapport 2001* (Competitiveness of the Luxembourgian economy. Report 2001), Luxembourg: Cahier de CREA-Cunlux-Statec no. 93-02.
Brennpunkt Drëtt Welt, a series of articles between 1974 and 2002.
Delcourt, C., 1980, 'Dossier: La contribution à l'aide au développement du Tiers Monde, prêtée par le Grand-Duché de Luxembourg', *Brennpunkt Drëtt Welt*, No.40.
Estrela, Y. and P. Reitmaier, 1998, Rapport d'Évaluation des Projets de Santé au Cap Vert financés par la Coopération Luxembourgeoise, Luxembourg, December.
Fehlen, F. (dir.), 2000, *Les élections au Grand-Duché de Luxembourg – Rapport sur les élections du juin 1999. Étude réalisée pour la Chambre des Députés du Grand-Duché* (The elections in the Grand Duchy of Luxembourg – Report on the elections of June 1999. Study realised by the Chamber of Deputies of the Grand Duchy), Luxembourg: Centre de Recherche Public George Lippmann, October.
Goerens, Ch., 2003, Déclaration sur la politique de coopération au développement et d'action humanitaire du Luxembourg, Luxembourg: Ministére des Affaires Étrangères, 12 February.
Golooba-Muteba, F., 1999, *DAC Informal Network on Poverty Reduction Scoping Study; Country Study: Luxembourg*, London: Overseas Development Institute.
Kirt, R., 2000, *Kleine Nation, Grosse Union: Über Möglichkeiten und Grenzen kleinstaatlicher Aussenpolitik im geopolitische neu parzellierten Europa. Ein Essay,.* Echternach: Éditions Phi.
Klonski, S., 2001, *Lutte contre la Pauvreté – La politique de coopération luxembourgeoise* (Fight against poverty – The Luxembourgian development co-operation policy), Luxembourg: Ministère des Affaires Étrangères, Direction de la Coopération et de l'Aide Humanitaire.
Lesourd, M., 1995, *État et société aux îles du Cap-Vert*, Paris: Karthala.
Lux-Development, 1998–2002, *Rapport Annuel/Annual Report*, Luxembourg: Agence luxembourgeoise pour la Coopération au Développement.
Metz, M., 2002, *Evaluation of Austrian Food Aid*, Berlin: CODEPLAN, May.
Ministère des Affaires Étrangères, 1993 to 2002, Lëtzebuerger Entwéklungszesummenaarbecht/La coopération Luxembourgeoise au Développement – Rapport Annuel, Luxembourg: Ministère des Affaires Étrangères, Direction de la Coopération.
Ministère des Affaires Étrangères, 2003, La politique luxembourgeoise de coopération au développement. Mémorandum présenté au Comité d'Aide au Développement à l'occasion de l'examen des pairs le 18 mars 2002, Luxembourg, January.
Murat, H. and M. Cabral Sanchez, 2000, L'aide française au Cap Vert 1991–1999. Retour sur l'évaluation-pays (1975–1991). (The French aid to Cape Verde 1991–1999. Return to a country evaluation (1975–1991)), Paris: Ministère des Affaires Étrangères, Direction Internationale de la Coopération Internationale et du Développement, July.
Newcomer, J., 1984, *The Grand Duchy of Luxembourg: The Evolution of Nationhood 963 A.D. to 1983*, Lanham, MD: University Press of America.
OECD/DAC, 1993, *Aid Review 1993/94: Luxembourg* (DCD/DAC/ AR (92) 2/13. Paris: OECD, September.
OECD/DAC, 1999, *Development Co-operation Reviews: Luxembourg*, Paris: OECD, No. 32.

OECD/DAC, 2003, *DAC Peer Review of Luxembourg*, Paris: OECD, *The DAC Journal*, Vol.4, No.2.
OECD/DAC, various years, *Development Co-operation: Efforts and Policies of the Members of the Development Assistance Committee*, Paris: OECD.
OECD, 2001, *OECD Economic Surveys: Luxembourg*, Paris: OECD.
Trausch, G. *et al.*, 1996, *Le Luxembourg face à la construction européenne*, Luxembourg: Centre d'études et de recherches européennes Robert Schuman.
United Nations System/Système des Nations Unies au Cap Vert, 2001, Bilan Commun du Pays (CCA), Praia: October.

12

All in the Name of Quality: Dutch Development Co-operation in the 1990s

LAU SCHULPEN

1. INTRODUCTION

The post-1989 period was a stimulating one for Dutch development co-operation, with a strong focus on improving the quality of aid. Starting with a substantial critique at the end of the 1980s [*for example, Hoebink, 1988; IOV, 1988*], the 1990s saw for the first time a much more clearly worked out poverty reduction policy [*DGIS, 1990*] and, under different ministers for development co-operation, specific strategies to achieve a higher impact of aid. In contrast to the often expressed idea of being a 'front-runner in the field of development co-operation', the Netherlands might better be viewed as adhering closely 'to international thinking on development as expressed primarily by the World Bank and the UN organisations' [*Hoebink and Schulpen, 1998: 39*]. Naturally, there are some differences in emphasis and there are certainly some original ways in which the Netherlands has shaped its development policy. Five guiding issues are taken to provide an overview of the changes in thinking and practice of Dutch development co-operation since the end of the 1980s. This chapter will concentrate on the period from the mid-1990s onwards, but will occasionally go back a little further in time.

The chapter starts with an overview of the general thinking on development co-operation, particularly from a poverty perspective, based on the policy of two Ministers (Pronk and Herfkens) who have, each in his and her own way, shaped development co-operation in the 1990s. To this will be added some of the recent policy guidelines of Van Ardenne, who succeeded Herfkens in 2002.[1] In the second part,

---

Note of acknowledgement. The author wishes to thank Louk de la Rive-Box and Rob Visser for their comments on an earlier draft of this chapter.
1. Officially, Van Ardenne was a state secretary during the first Balkenende Cabinet which lasted for only a few months. It was the first time since 1965 that development

and following a short overall description of Dutch aid volumes, the history of Dutch development co-operation in the 1990s will be traced following the five issues placed centre-stage here. This covers, first, the selection of countries or, more precisely, the effort undertaken particularly from the end of the 1990s to concentrate Dutch bilateral aid on a limited number of developing countries. Second, it is concerned with the change in strategy and the ultimate emphasis on the sector-wide approach based on such notions as ownership and Poverty Reduction Strategy Papers (PRSPs). Third, it deals with the role of the private sector in development (co-operation), which was reluctantly taken up by Herfkens but which forms a major part of the strategy of Van Ardenne. Fourth, it turns to the private aid channel for which an entirely new subsidy system was set up in the late 1990s. Finally, attention is paid to coherence, an important topic already under Pronk but made central under Herfkens.

II. POVERTY REDUCTION: CHANGING PERCEPTIONS

Only two Ministers, both from the Labour Party, shaped Dutch development co-operation in the 1990s: Pronk from 1989 to 1998 and Herfkens from 1998 to the beginning of 2002. Although their policies were largely comparable, there is nevertheless a clear distinction between them. Whereas Pronk will most likely be remembered as a 'philosopher' (exemplified by two major policy papers which were quite positively received within the Dutch aid community), Herfkens is definitely more of a 'do-er'. No major policy papers were presented during her time in office, and she not only managed to antagonise a large part of the development co-operation community in the Netherlands but also prepared the ground for substantial changes in Dutch development co-operation. The Christian Democrat Van Ardenne succeeded Herfkens in 2002. Although she did emphasise different issues from those of her predecessors, it is too early to see what will be their outcome. Despite the differences between the Ministers, all three clearly started from the overall objective of Dutch development co-operation already established in the 1970s: poverty reduction (whether or not preceded by 'sustainable').

---

co-operation was no longer headed by a minister. For many, this meant a downgrading of development co-operation within the political spectrum. New elections held in January 2003 eventually led to the second Balkenende Cabinet comprising Christian Democrats (CDA), Conservative Liberals (VVD) and Progressive Liberals (D66). Van Ardenne, who earlier openly declared that she wanted to return, indeed did so. Development Co-operation is thus again headed by a minister.

In November 1989, the third Lubbers Cabinet came into office, a coalition of the Christian Democrats (CDA) and the Labour Party (PvdA), with Jan Pronk as Minister for Development Co-operation, having already held this post in the 1970s (1973–78). In 1994, Pronk continued as Minister in the first Kok Cabinet (also known as the Purple Coalition, of the PvdA, the Conservative Liberals (VVD) and the Progressive Liberals (D66)). Several events made this a dynamic period in Dutch development co-operation. For one thing, there were the continuing attempts to reduce the aid budget. In absolute terms, Dutch aid continued to increase year by year, but aid levels fell sharply as a percentage of GNP. In fact, there was a constant decline over the period 1989 to 1994. In the latter year, Dutch aid reached its lowest level (i.e. 0.76 per cent of GDP) since 1976. Thereafter, it slowly recovered to reach 0.83 per cent in 1996. Indonesia's cancellation of Dutch aid may also be regarded as an important event, although the aid relationship was resumed full swing from 1999 onwards (see below). More important for our discussion, however, have been the publication of the White Paper *A World of Difference* and the reverification exercise in the mid-1990s.

September 1990 saw the publication of the lengthy report entitled *A World of Difference* [DGIS, 1990]. Under the Minister's leadership, a team of civil servants had worked for months on the 385-page report, which quite rightly bears the subtitle 'a new framework for development co-operation in the 1990s'. It attempts to analyse the place of development co-operation in a decade distinguished by the end of the Cold War, and stated that 'sustainable poverty alleviation' should be the central goal of the Netherlands' development policy.

1. Poverty Reduction Defined

Although the long history of poverty reduction within Dutch development policy suggests otherwise, it was only with *A World of Difference* that a more structured vision was elaborated with regard to poverty and poverty reduction. The White Paper generally followed international thinking, particularly with regard to the multi-dimensional definition of poverty presented. Three dimensions were distinguished: economic, social and political. An effective strategy for poverty reduction should tackle all three. Essential, then, were: (i) investments in poor people and improving their productive assets; (ii) meeting basic needs; and (iii) enlarging the participation of the poor in political decision-making by strengthening local civil society organisations. Also important was the commitment by the developing country's government to the objective of poverty reduction and, in general, to good governance and good policy.

Sustainable development also consisted of three elements: growth of production, equitable distribution and the maintenance of the 'ecoscope'.[2] Poverty reduction in the 1990s would, according to the White Paper, be different from previous periods. First, because the economic aspect has to be linked explicitly to the social and political aspects, thereby making social conditionality an important aspect of donors' policy. Second, because a macro approach has to be linked to a micro approach, meaning that a healthy macroeconomic policy together with realistic exchange rates, a sound fiscal policy and price stability are absolute prerequisites for the success of sustainable poverty reduction.

A major problem remains with regard to the operationalisation of the policy. Poverty reduction has recently been the subject of an extensive OECD study covering all Northern donors. The study comes to the conclusion, among other things, that donors are 'more determined than ever before to ensure that development co-operation benefits the poor', but also that 'the multi-dimensional definitions [of poverty] ... boil down to little more than extended taxonomy of the characteristics of poverty with few operational implications'. In other words, bringing their poverty focus into practice seems to be extremely difficult. This is definitely true for the Netherlands [*Schulpen and Hoebink, 1999*].

The optimism expressed in *A World of Difference* with regard to the possibilities for tackling poverty in the era following the fall of the Berlin Wall was comparable to the optimism expressed by such institutions as the World Bank and the UNDP. Only a few years later, it seemed to have disappeared in a new White Paper, *A World in Dispute* [*DGIS, 1993*]. Despite the fact that increased tensions between and within states formed the central theme of this paper, poverty reduction received considerable attention. However, the paper mainly reiterated what was already said in *A World of Difference*, and simply concluded that poverty reduction 'is an absolute prerequisite for reaching sustainable development and combating environmental degradation and conflicts' [*ibid.: 139*].

The implications for policy of the renewed attention to poverty reduction were seen as a focus on the poorest countries, support for

---

2. The term 'ecoscope' is used in a *World of Difference* to sum up all the economic functions of the environment. 'At a given stage of its evolution an ecosystem is able to supply a limited quantity of resources and process a similarly limited quantity of waste substances without coming to any harm' [*DGIS, 1990: 79*].

grassroots processes and NGOs, an increase in the effectiveness of poverty projects, greater attention to the environment and urban poverty, a more gender-specific aid focus, and decentralisation [*DGIS, 1990*]. However, the translation into concrete proposals for action is perhaps the weakest part of the first White Paper. In the last two chapters on the implementation of the new policies, the Minister was hindered by the absence of evaluations of several aid instruments, the lack of integration of existing evaluations from other donors and his own naïveté on the practical workings of aid. However, this does not detract from the fact that, for the first time in years, a White Paper had been produced that was of sufficiently high quality to provoke a serious discussion on development co-operation policy. Moreover, the lack of evaluations of several aid instruments has been at least partly overcome in recent IOV/IOB studies and in a few studies written specifically for the reverification of foreign and development policy in 1995 [*Hoebink, 1995; IOV, 1992, 1994; Jepma, 1995*].

2. Reverification of Foreign Policy

The reverification of Dutch foreign policy was the direct result of the failure, at the beginning of the first Kok government, of the three coalition partners (i.e. PvdA, D66 and VVD) to reach agreement on the budget for development co-operation. This reverification was a political as well as an organisational exercise. Given the changing international situation following the fall of the Berlin Wall, it was felt that development co-operation was no longer the stepchild of international relations but instead should be viewed as an integral part. *A World in Dispute* [*DGIS, 1993*] concluded that there was increased attention to such issues as emergency aid in conflict situations, stability and peace, ecological sustainability, democracy, human rights and good governance. As such, development co-operation increasingly touched upon other policy fields [*DGIS, 1996: 69*]. This called for 'decompartmentalisation' between Ministries such as Development Co-operation, Foreign Affairs and Economic Affairs, aimed at 'improving policy coherence, better coordination between the Ministries involved in foreign policy matters, and improving working links with Dutch society at large' [*OECD, 1998a: 13*].[3]

---

3. As part of the reverification exercise, there is from 1997 onwards one expenditure group for foreign policy known as the Homogeneous Group for International Co-operation (HGIS) with a total budget of 1.1 per cent of GNP. This then includes, among other things, all ODA (around 0.8 per cent of GNP), peace operations, programmes for Central and Eastern Europe, export instruments of the Ministry of Economic Affairs, embassies and consulates, and contributions to international organisations such as NATO and the OECD. With regard to the ODA part, the call for

The reverification exercise had two main outcomes: a fixed ceiling for official development assistance (0.8 per cent of GNP) and an organisational restructuring. From September 1996, a new matrix structure was introduced, with regional directorates co-ordinating the regional tasks of the three Ministries. As one of the four Directorates General, International Co-operation (DGIS) was 'responsible for developing policies for various themes of Dutch co-operation and assisting DGRB (Directorate General Bilateral Relations) to draw up the medium-term regional programmes on the basis of the thematic policies' [*OECD, 1998a: 19*]. A major element of the reorganisation was the shift of the main responsibility for the bilateral aid programme from The Hague to the various embassies, which, from January 1997 onwards, were the first entities responsible for the selection, assessment, monitoring and evaluation of development projects and programmes, receiving only general policy guidelines from The Hague. The preparation and implementation of development co-operation were made more efficient by shortening the decision-making lines. Also a greater feeling of ownership on the part of the developing countries was promoted by interaction between the various implementing departments and line-ministries and the embassies in those countries.

Along with the reverification exercise, the Minister for Development Co-operation also carried out his own review known as the small reverification. Together with a call for more recipient ownership, a simplification of procedures, better monitoring and a systematic restoration of evaluation findings, this led to a call for fewer projects. In order to improve the quality of Dutch aid, programme aid was increasingly to replace projects, seen in the 1996 policy paper as the activities of the donor and not of the target group. Programme aid includes balance-of-payments support, sectoral budget support and debt relief, but also programmes to strengthen the institutional capacity of the recipient country to set up policy on its own and to

---

more effective aid led to a restructuring of the budget. Country programming is replaced by thematic programming, thereby aiming at a more flexible approach. In the 1997 budget, 12 so-called thematic categories or themes were distinguished. In the 1998 budget these were reduced to 11, including one non-ODA category: (1) economy, employment and regional development; (2) the environment; (3) social development; (4) education, research and culture; (5) human rights, conflict prevention and good governance; (6) humanitarian aid; (7) macroeconomic support and debt relief; (8) Surinam, Netherlands Antilles and Aruba; (9) multilateral programmes; (10) other programmes (e.g., expert assignments, cost of refugees and administrative expenses); and (11) non-ODA expenditures. Themes 1–7 are the main themes and, although not new for Dutch co-operation, are regarded as constituting 'a clearer picture of priorities than previously' [*OECD, 1998a: 15*].

implement projects. Finally, the call for fewer projects also reflected a rethinking on technical assistance (see also Box 12.1). The 1996 budget even went so far as to state that projects would remain important but mainly (or only) after it was clear that more general support, institutional support and technical assistance had failed to lead to the desired results. Sectoral budget support was seen as a 'new' instrument with which it was hoped to overcome some of the inherent problems which too great an emphasis on the project approach might bring.

BOX 12.1
TECHNICAL ASSISTANCE

Rethinking technical assistance (TA) was already part of the reverification exercise of the mid-1990s. However, the elaboration on this was left to Herfkens. In 1999, she announced that she wanted to limit the number of Dutch experts. As in the case of the MFOs (see below), the Minister could fall back on a so-called IBO research which concluded that there certainly was room for improvement in the effectiveness and efficiency of technical assistance [*Hoebink and van der Velden, 2001: 81*].[4] Technical assistance was increasingly researched during the 1990s. Although the different studies provide a mixed picture, a relatively critical assessment prevails. Problems are mainly seen in such fields as knowledge transfer and training, the expert-counterpart system, the high costs of TA, crowding out local expertise, and institutional development. Herfkens initiated the discussion on TA by presenting a number of cliché-images (for example, well-paid, high-living experts who spend a lot of time by their swimming pools and take over the jobs of local experts). This critique was not well received in Parliament nor by the Dutch organisations active in this field, many of which had already started changing their strategies. The impact of the changes proposed by Herfkens (i.e., a stop to sending out experts by the Ministry itself) was considered marginal [*Hoebink, 2002a*].

---

4. Interdepartmental Policy Researches (IBOs) date from the time of economising, when highly placed civil servants would brainstorm over possible savings. Within development co-operation, there have been IBOs for technical assistance, international education and the co-financing organisations.

3. Herfkens

All in all, the changes in thinking that guided the period up to 1998 provided a perfect 'breeding ground' for the ministerial ambitions of Herfkens. Soon after she got her job (returning from a long sojourn abroad in Washington and Geneva) a new word was added to the Dutch vocabulary: de-Pronking (meaning strong deviation from the lines set out by Pronk). For part of her policy this is certainly true (as for the country selection – see below), but in more general terms Herfkens mainly continued along the lines set out earlier, although definitely at a higher speed. Poverty reduction remained the central objective and Herfkens made quite clear from the start that all Dutch aid is imbued with poverty reduction. More decisively than Pronk, she laid the emphasis on ownership in contrast to donorship, and in the wake of this the Netherlands became a strong believer in the Poverty Reduction Strategy exercise initiated by the IMF and the World Bank (see below). At the same time, the doubt already expressed in Pronk's time about the impact of project aid and the necessity to move towards sector or budget support now became the norm. The same may be seen with regard to coherence. Pronk had already moved the issue of coherence to the forefront, but it was Herfkens who made it a much more central theme in her policy (see below).

In other fields, the policy of Herfkens indeed broke significantly with that of her predecessors. As already mentioned, this holds particularly true for the concentration of Dutch bilateral aid on a smaller number of countries. But in other fields as well, such a deviation can be witnessed. This is the case for technical assistance, but also, for the first time since the early 1980s, a radical change was agreed upon with regard to the so-called co-financing scheme or, more generally, the way NGOs are subsidised by the Dutch Ministry for Development Co-operation (see below). Finally, and although Herfkens remained somewhat reluctant, for the first time a policy paper on private-sector development (PSD) was issued in 2000. It was left to Van Ardenne, however, to press more seriously for the role of private companies and PSD in her policy.

In one of the few policy papers produced after 1998, Herfkens made clear that a broad integrated approach to poverty is needed [*DGIS, 2001a*]. Central here is the so-called micro-macro approach which stipulates that 'poverty reduction must start from the situation of the poor themselves (micro level)' but that action is then to be taken at the national and international level (macro level), as that is where the causes of poverty have to be sought. 'Removing the causes of poverty at these levels will ultimately improve conditions for the poor on a long-term basis, without the need for intervention directly targeting them or their immediate surroundings, as in the case of direct poverty reduction.' In this way not only the move away from projects could be justified, but the approach also meant that three key (macro-level) principles almost automatically emerged: ownership, coherence and global governance. Although the poverty policy as such may be regarded as linking to new insights in the field of poverty (mainly also on the basis of World Bank studies) and to an emerging international consensus (as exemplified by the Millennium Development Goals), there is an important catch. In 1998, Hoebink and Schulpen not only concluded that Dutch policy in the field of poverty 'did not get much further than what had already been set in motion internationally', but also that 'the most important weakness of the central Policy Papers has been ... insufficient attention to implementation' [*Hoebink and Schulpen, 1998: 40*]. The poverty policy of Herfkens was no exception to this. It remains unclear what this conceptualisation of poverty in practice means, and guidelines from the Ministry do not, to put it somewhat bluntly, go much further than a call for sector approaches and 'the use of [PRSPs] as a primary focal point for field strategy, implementation, donor co-ordination, and monitoring and evaluation' [*OECD, 2001: 1–21*].

Although at the end of the 1990s a poverty mainstreaming exercise was undertaken at the Ministry guided by the poverty desk, its outcome remains unclear. Moreover, since the decentralisation of the major part of bilateral development co-operation to the embassies, one may wonder whether such a mainstreaming exercise is of much use without their active involvement. Even if one departs from the logic that attention to the implementation aspects will be greater at the individual country level, there still remains the problem of a lack of institutional memory particularly at that level, as well as an absence of clear guidelines from The Hague and the lack of country policy papers.[5] The latter should be regarded as a particularly important problem, certainly in the light of the absence of clearly worked out PRSPs which can form the basis for Dutch aid. The DAC thus concludes that 'until such a PRSP strategy can become a reality ... it would seem necessary and useful for the Netherlands to develop some form of simple statement of strategy and rationale for Dutch development assistance at the country level' [*OECD, 2001: 1–40*]. Nine

---

5. The DAC, for instance, notes 'an apparent problem with communication between headquarters and the field' and adds that 'country-level development strategy documents are no longer generated, and have only been replaced by annual programme and budget submission documents' [*OECD, 2001: 1–40*].

pages later, the DAC examiners are even more outspoken: 'like most donors, the Dutch are grappling with the principle of poverty reduction, without fully understanding the dimensions of the problem, without a clear sense of documented priorities, and without a clear sense of how to organise around the problem.' Such problems are not easy to overcome and there are indeed few signs that the Ministry managed to do just that.

This also holds for the short period of Van Ardenne, who for a large part of the time followed along the same lines, emphasising the integration of poverty reduction with other foreign policy instruments, coherence ('an absolute priority') and a focus on Africa. After her installation as minister in the second Balkenende Cabinet in 2003, she announced a total of ten policy emphases. Besides the three already mentioned these are: (i) a regional approach, (ii) co-ordination and harmonisation, (iii) a greater role for the private sector (see below), (iv) partnership, (v) sustainable development, (vi) migration and development, and (vii) millennium goals with an emphasis on education, environment, AIDS and reproductive health [*DGIS, 2003a*].[6] None of these aims or emphases have been truly worked out yet, however, not even in the official White Paper of October 2003 under the title *Mutual Interests, Mutual Responsibilities* [*DGIS, 2003b*].

All in all, the above shows that the Minister is an important determining factor in Dutch development co-operation. In particular Pronk and Herfkens have had a major influence on policy. Naturally, they were guided not only by international thinking on development co-operation but also by internal Dutch political developments and pressure from the domestic constituency. Hoebink [*1999a: 198–9*], for instance, attributes the changes in the aid programme 'to a complex overlap of socio-economic, bureaucratic and economic factors'. The economic factors at the time include such elements as the 'interaction between the structure of the Dutch economy, changes in the world economy and changes in the economies of aid-receiving countries'. The latter, for instance, 'made it impossible for most aid-receiving countries to implement new projects', so 'programme aid became an

---

6. Interestingly, a regional approach and co-ordination and harmonisation were also part of Bukman's policy (Christian Democrat Minister for Development Co-operation 1986–89), while attention to the role of the private sector figured prominently in the policies of Christian Democrat Minister De Koning (1977–81) and conservative liberal Minister Schoo (1982–86). The emphasis on the Millennium Goals was already there under Herfkens. The fact that Van Ardenne wants to concentrate on only a few of these (mainly in the social sectors) is new, however. Truly new is the aim of migration and development which most likely stems from the increasing discussion on migration in the Netherlands following the terrorist attacks in the United States.

important instrument'. In addition, pressures from private aid agencies and other stakeholders in the Netherlands should not be forgotten. The DAC [*OECD, 2001: 1–28*], for instance, concluded that 'the NGO community plays a strong role in Dutch development co-operation. They directly impact on public awareness of Dutch development programmes and on the politics and strategic thinking related to development co-operation.' At the same time, this NGO community has close links with political parties in Parliament, which in the end formally establishes 'national development co-operation goals and directions ... based largely on policy initiatives presented to it by the Minister for Development Co-operation' [*ibid.: 1–20*].

III. VOLUME TARGETS

Since the early 1970s, the Netherlands has managed to surpass the 0.7 per cent target as set by the United Nations. Over the past decade around 0.8 per cent of GNP has been set aside for official development assistance (ODA) in accordance with the 1995 political compromise of the reverified foreign policy. Although this is a decrease compared to the 1980s (when the ODA/GNP ratio was nearer to 1 per cent), in absolute terms there has been an almost constant growth of ODA (see Figure 12.1 for the period 1988 to 2000). The discussion about the target volume is probably the most frequent one on development co-operation within the Netherlands. Particularly during periods of increasing economic problems, the call for downsizing the development aid budget (for instance, to the internationally agreed level of 0.7 per cent of GNP) grows stronger. Not surprisingly, this call is especially strong from the right-wing political parties, whereas the left-wing parties are generally in favour of an increase (often up to 1 per cent of GNP). Table 12.1 shows the positions of the parties with regard to the volume for ODA in 2002. Obviously, there seems to be a clear link between an increase in the budget and the (then) felt need for the re-establishment of a Minister for Development Co-operation. In this regard, it is noteworthy that a recent poll showed 26 per cent of the Dutch public to be in favour of such an increase (with 53 per cent in favour of maintaining the present volume) [*NCDO, 2002*]. Interestingly, the political preference of those in favour of a higher budget is largely in line with the viewpoints of those political parties themselves.

FIGURE 12.1
TOTAL NET ODA (IN CURRENT US$ AND % OF GNP)

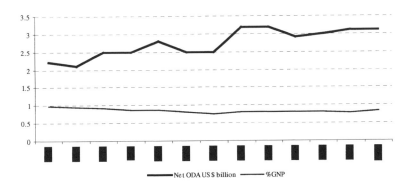

Source: [OECD, *1989, 1992, 1995, 1998b, 2002*].

TABLE 12.1
POLITICAL PARTIES AND THE PUBLIC ON THE AID BUDGET AND THE MINISTER

Political party	Proposed budget (% of GNP)	Minister for Development Co-operation	Public opinion on budget[b]	
			> 0.8% GNP	0.8% GNP
CDA (Christian Democrats)	0.8	Not mentioned	24%	65%
VVD (Conservative Liberals)	0.7	No	16%	62%
LPF (Conservative populists)	0.7[a]	No	12%	39%
PvdA (Labour Party)	1.0	Yes	38%	53%
GroenLinks (Green Left)	Increase	Yes	54%	41%
SP (Socialist Party)	1.0	Yes	53%	36%
D66 (Liberal Democrats)	0.8	Yes	36%	54%
Christian Union (Christian)	1.0[c]	Yes	38%[d]	46%[d]
SGP (Conservative Christian)	0.9	Not mentioned	38%[d]	46%[d]
Leefbaar Nederland (Conservative populists)	0.7	Not mentioned	22%	24%

Source: Election programmes of different political parties [*NCDO, 2002*].
Notes:
[a] Two weeks before the 22 January elections of 2003, the LPF proposed an extra cut in the budget to 0.6 per cent.
[b] In percentages of those who voted during the last general elections for that particular political party.
[c] According to the Christian Union, the budget should increase to 1.0 per cent in 1996 and then even further to 1.3 per cent.
[d] The Christian Union and SGP are taken together under 'small Christian parties' in the NCDO poll.

The distribution of Dutch ODA between multilateral and bilateral channels (the latter including assistance through the private aid channel) has not changed since the end of the 1980s. Although fluctuating somewhat, around 30 per cent of all ODA is channelled through multilateral organisations. As such, the preference of Herfkens for this channel obviously did not materialise.[7] In addition, there are no signs for the near future that there will be a major change in this situation, although the budgets for structural bilateral development co-operation are under strain. Recently, Minister Van Ardenne announced that due to unforeseen disbursements (e.g. following humanitarian aid to Iraq), the beginning of an economic crisis in the Netherlands (leading to a decreased budget for development co-operation) and sanctioning of long-term commitments for such issues as the Global Health Fund and the co-financing organisations (see below), the budget for structural bilateral aid will be hit hardest. Whereas in 2002 24 per cent of the total budget still went to the Dutch concentration and thematic countries (see below), it is expected that this will decrease to 22 per cent in 2003.[8] Despite the reasons given, the need for these reductions seems to have come largely as a surprise to the Ministry, indicating a lack of proper financial planning and of anticipating the effects of granting long-term funding. The constantly growing budget (at least in absolute terms) during the 1990s, combined with the fact that the DGIS from the 1970s onwards had found it difficult to disburse all available funds, could also be said to have made the Ministry inattentive to administrative management.

Within the multilateral channel itself, there have been some earlier changes, however. These were mainly due to the appreciation of individual multilateral organisations. In a number of papers (on the EU, the international financial institutions and the UN as channels for development co-operation), Herfkens clearly showed a much more critical attitude towards the multilateral channel than any of her predecessors. The EU, in particular, was severely criticised for its inefficiency and ineffectiveness (lack of a clear poverty focus, lack of coherence, problems in management and organisation, staff problems, lack of sustainability and of an independent evaluation process [*Box and Tromp, 2001: 194–6*]). Hardly any funding consequences could

---

7. And despite the fact that Herfkens was in favour of increasing the share of the multilateral channel in development co-operation if only 'to place Dutch assistance more in line with the DAC average' (being 33%), but also as 'an obvious way to help disburse a steadily expanding aid budget in a context of organisational capacity constraints' [*OECD, 2001: 1–27*].

8. In effect, total disbursements for target and thematic countries will be reduced from €901m. in 2002 to €828m. in 2003.

be attached to this, however, because of the contractual obligations of the Netherlands as an EU member. The Dutch contribution to the EU decreased from US$248m. in 1991 to US$233m. in 2000.

With regard to the UN, Herfkens [*DGIS, 1999a*] saw four primary causes for the 'failings in the way UN agencies function': (i) 'inconsistencies in the policies pursued by Member States'; (ii) 'the UN's worldwide mandate for operational activities is out of date'; (iii) the 'failing management' of UN agencies; and (iv) 'too little co-operation … among UN agencies and between the UN and the World Bank'. The agencies emerging best from the screening operation were UNICEF, UNFPA, UNAIDS, UNCDF and WHO. 'Least impressive are FAO, UNESCO and UNDP.' In effect, the Dutch contribution to the UNDP (still the major UN beneficiary of Dutch aid) was kept at almost the same level, while the contribution to UNFPA more than doubled between 1998 and 2000. In addition, the Netherlands' contribution to the IDA increased marginally throughout the 1990s.

FIGURE 12.2
TOTAL BILATERAL NET ODA TO RECIPIENT COUNTRIES
(US$m., IN CURRENT AND CONSTANT DOLLARS)[a]

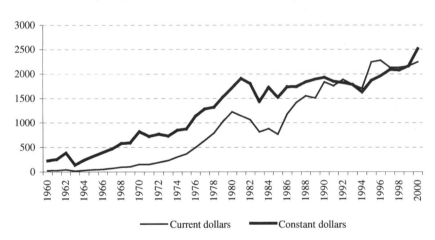

Source: Calculations on the basis of OECD/DAC (cd-rom).
Note: [a] Using the deflator for the Netherlands (1999=100).

The major channel for Dutch ODA remains the bilateral one (including aid provided through the private aid channel).[9] Figure 12.2 provides an overview of the growth of bilateral aid over the period 1960 to 2000. A steep increase in the 1970s is then followed by a fluctuation in the 1980s at around the same level. From the mid-1990s onwards bilateral aid (in constant as well as current US$) again increases, due to economic prosperity in the Netherlands combined with a fixed ODA/GNP ratio of 0.8 per cent.

Since 1997, loans have been abolished entirely within Dutch bilateral ODA due to an (economising) decision taken already in 1991 by Pronk. In effect, all ODA now consists of grants. It is interesting that earlier loans are still being repaid by developing countries and net ODA is therefore negatively influenced. If, for instance, all earlier development loans were to be cancelled, total net bilateral ODA over the period 1997 to 2000 would have been not US$8,671m. but US$9,319m. (or 7.5 per cent more). For particular countries this difference would have been substantially higher, for example, for India. Over the period 1997–2000, India received a total of US$250m. in Dutch aid. At the same time, India paid on earlier loans an amount of US$204m. to the Netherlands, with the effect that in 1999 and 2000 net ODA to India was negative.

Finally, a closer look at the sectoral division of aid commitments shows that commitments to activities in 'economic infrastructure and services' and 'production', which showed a steady decline over the years, declined even further throughout the 1990s. Bad experiences with industrial projects such as in cement and food processing as well as fertiliser exports all contributed to the decline of the production sectors in Dutch aid commitments. In contrast, 'social infrastructure and services' (for example, education, drinking water, health) remained an important sector for Dutch aid, reflecting the commitment by Pronk to the so-called 20/20 initiative.[10] General programme

---

9. An exact calculation of the NGO element in this bilateral aid is not readily available for a longer time period. According to the DAC [*OECD, 2001*], core support to NGOs during 1998 to 1999 amounted to US$355 million or 21 per cent of total bilateral ODA. A statistical overview by the DGIS [*2002a*] over the period 1998 to 2001 showed that total expenditure increased from €418.1m. in 1998 to €560.2m. in 2001, excluding emergency aid and international education and research funds. The majority of these funds went to Dutch NGOs (particularly the so-called co-financing organisations – see below) and a relatively small part went to international NGOs such as the International Planned Parenthood Federation and the World Conservation Union (IUCN).

10. In 1995, during the World Social Summit in Copenhagen, the so-called 20/20 initiative was adopted stipulating that an expenditure of 20 per cent of the national budget for basic social services in developing countries would be matched by 20 per cent of the budget for development co-operation of donor countries being used for the

aid (i.e. funds made available for general development purposes without any sectoral allocation), which was still insignificant in the 1970s and early 1980s but increased in the 1980s, fell in the early 1990s. The end of the 1990s again showed an increase, due mainly to disbursement problems. These same problems also partly explain the changes in debt relief. All in all, however, debt relief decreased under Herfkens. Most striking is the steady increase in unallocated or unspecified aid commitments. Whereas in the 1970s and 1980s only some 5 to 8 per cent of all aid commitments fell into this category, in 2000 it had increased to more than 38 per cent.

IV. COUNTRY SELECTION

One of the few issues that over the years has led to sometimes heated public and parliamentary discussions is the selection of countries eligible for Dutch aid. Since the late 1960s, following the substantial increase in bilateral aid, the Netherlands has been familiar with the concept of concentration countries (in the 1980s called programme countries and at present target countries).[11] The history of Dutch policy towards concentration countries (see also Box 12.2) shows that: (i) a poverty criterion has always formed part of the selection criteria for concentration countries, although it has been expressed mainly in terms of a GNP per capita figure; (ii) besides concentration countries there have (therefore) always been many other countries that have received Dutch bilateral aid and everything points to this continuing to be the case in the future; (iii) the criteria in the past have not been executed to the letter, as is already clear from the fact that certain countries (for example, Indonesia) have remained on the list despite the fact that they did not meet all the criteria set or that Macedonia arrived on the list recently (although too rich); and (iv) other criteria

---

same services. These 'basic social services' include expenditures for basic health, basic education, water and sanitation, nutrition, and reproductive health and population.

11. Obviously, concentration countries are not a specific Dutch phenomenon. In essence, all donors have always concentrated their (bilateral) aid to specific countries, if only owing to the limited funds available, since it was impossible to provide aid to all developing countries. For the selection of these concentration or programme countries, donors have used different criteria, not in the final instance, political criteria or conditionalities. This is not to say that these criteria were always adhered to and that applying them did not cause substantial problems as other, often not openly mentioned, criteria also played their role. The human rights criterion used by the Netherlands already in the 1970s, for instance, did lead to suspending aid to Uganda 'because of the reign of terror of Idi Amin [and] to Chile ... after the *coup d'état* of General Pinochet' [*Hoebink, 1999a: 191*]. At the same time, this criterion did not stop the Netherlands providing large amounts of aid to such human rights violators as Indonesia.

have played a role, such as the economic and political relationships with the Netherlands.

BOX 12.2
HISTORY OF THE DUTCH CONCENTRATION COUNTRY POLICY (MID-1960S TO MID-1990S)

The main reason for concentrating Dutch bilateral aid was the opinion that the Netherlands, in view of its limited funds, could hardly be expected to be active in all developing countries. Moreover, by concentrating its efforts on specific countries it was expected to increase the effectiveness and impact of Dutch aid. Whereas in the 1960s the main criterion for selecting concentration countries was the question of whether there was a donor consortium in place, in the 1970s more specific criteria were developed: (i) the poverty situation (based primarily on GNP per capita); (ii) the need for aid; and (iii) the commitment of the recipient government to promoting policies directed at improving the position of the poorest groups. In practice, other criteria also played a role such as the state of economic relations between the aid recipient and the Netherlands and their political relations (for example, in the case of the Middle East). These specific criteria, however, did not lead to the removal of countries from the list, but only to adding some new countries such as Cuba and Jamaica.

At the end of the 1970s, Minister De Koning changed the criteria used by his predecessor Pronk. The new criteria were: (i) GNP per capita of less than US$550; (ii) the extent to which a country implements a policy aimed at the distribution of economic welfare; and (iii) the human rights position. In effect, the first criterion of Pronk was reduced to the GNP question, the second criterion was removed, the third was (only slightly) played down, and the human rights criterion (which was part of Pronk's third criterion) was added. This all led to the removal of five countries from the list, while no new countries were added.

In the 1980s, under Minister Schoo, three criteria were brought forward for the selection of what were then called programme countries: (i) GNP per capita of maximum US$795; (ii) a socio-economic policy aimed at 'development'; and (iii) the question of whether a lasting developing relationship could be built up with the country. In effect, three more countries were removed from the list (leaving ten programme countries in 1986), but they received further aid under regional programmes. In addition to these programme countries so-called regions were also identified (i.e. Southern Africa, Sahel, Central America and Andes) in which the Netherlands would be active mainly through its rural and industrial programmes.

In *A World of Difference*, Pronk again announced new criteria for the selection of countries: (i) the need for foreign aid (which is deductible in accordance with the degree of poverty, the amount of aid received from other donors, and the foreign debt situation); and (ii) socio-economic policy, including policies on women and on human rights, the latter being substantiated by declaring that what is essential is the commitment of the national government to tackling the problem of poverty. In addition to these two criteria, the experience with the existing aid programme would be taken into account as well. Three different categories of recipients were distinguished: (i) programme countries (i.e. India, Indonesia, Pakistan and Bangladesh) which would be eligible for all kinds of Dutch aid; (ii) regions (i.e. clusters of countries in a specific region – East Africa, Southern Africa, Sahel, Nile and Red Sea, Central America, Andes and Mekong – a total of 45 countries); and (iii) four sector countries, which are eligible only for specific sectoral aid.

From *A World in Dispute* onwards, there was a new division of countries: (i) regular co-operation countries (eight countries and six regions covering another 28 countries); (ii) three countries under sustainable development contracts; (iii) seven countries in a state of conflict or rehabilitation; and (iv) 13 countries in transition. In 1996, Pronk abolished the list of programme, region and sector countries entirely and replaced country programming by thematic programming, thereby aiming at a more flexible approach.

In 1996, the number of Dutch bilateral aid recipients reached a peak of 136 countries, due mainly to Pronk's idea that it is important to 'sit at the table' even with quite small sums at one's disposal. Confronted with this wide dispersion of Dutch aid funds, one of Herfkens' first steps when she took office at the end of 1998 was to announce a substantial reduction in the number of countries.[12] This reduction was thought necessary not only because fewer funds were available, but also (and mainly) in order to improve the management of the aid and to increase its effectiveness. The criteria used by Herfkens to determine which countries would be eligible for long-term Dutch bilateral aid covered, in addition to the poverty situation (combined with the need for foreign aid), criteria under the umbrella of good policy and good governance.[13] Herfkens therefore started from the general notion that the policy and governance in developing countries determine to a large extent the effectiveness of aid. Moreover, the precise criteria used under the general notions of good policy and good governance remained unclear, as also the manner in which these criteria were judged.[14] The selection of countries creates the impression that in reality quite different criteria have been decisive, such as the personal preferences of the Minister and the

---

12. With this, Herfkens also followed in the footsteps of the National Advisory Council which a few years earlier had pressed Pronk to drastically reduce the number of recipient countries.

13. The World Bank report *Assessing Aid [World Bank, 1998]* was important here. It stated that aid should be targeted to low-income countries with sound economic management (i.e. with good macroeconomic policies and the right institutions to implement these policies). With only minor reference to possible other explanations, the report thus takes the domestic situation in developing countries as decisive for the effectiveness of aid. The reasoning is relatively simple: economic growth (expressed in per capita incomes) is the most important element for development, simply because 'incomes and social indicators tend to improve together' and 'the greatest improvement has been in fast-growing economies'. At the same time, economic growth 'does not solve all ills'. For curing such ills as sex discrimination, social exclusion and environmental degradation, one has to rely on good policies and good institutions. In short, good policies and good institutions are the key to development and where these are in place, aid can have a significant contribution.

14. Only the criteria to determine the poverty situation in a country have been clear from the start (i.e. eligibility for IDA credits). Under good governance and good policy so many different elements have been mentioned over the months that the concentration exercise kept the Ministry busy. To give some examples: good policy embraced the macroeconomic, economic structure and social policy. In the social policy, the composition and poverty orientation of public spending are considered, as also the effort in the fields of the environment and gender. Good governance covered such issues as the integrity of government service, the prevention of corruption, transparency in the management of public money, the separation of powers, legal security, democratisation and the human rights situation.

assessment of the existing bilateral relationships with specific countries.[15]

After an initial selection on the basis of these criteria, a further selection was made on the basis of an assessment of the ongoing aid programme, the country's active role in support of the legal order in its region and/or its relationship with the Netherlands in the socio-cultural or economic fields. In the end, 19 (later reduced to 17, and then restored to 19) true 'concentration countries' remained. With these, the Netherlands planned a long-term relationship. However, this is in no way the entire story. Besides these 19 target countries, another three countries (currently South Africa, Egypt and Indonesia – the so-called +3 countries) have been added, with which a structural relationship for a limited time-period (i.e. five years) is planned.[16] Next to these, three so-called thematic groups of countries are distinguished: (i) the environment (12 countries); (ii) human rights, peace building and good governance (17 countries); and (iii) the private sector (12 countries). Finally, there are also still three so-called DOV (Sustainable Development Treaties) countries. Because of some country overlap between these (thematic) categories, the total number of 'concentration countries' still reaches 49, or 50 if Surinam is included. Table 12.2 provides an overview of these 'concentration countries' of Dutch bilateral aid.

Figure 12.3 shows not only the substantial increase in the number of recipient countries up to 1998, but also the decrease to 105 countries in 2000. This decrease, notwithstanding the slow pace at which it is moving, may be regarded as the first results of the 'concentration effort' of Herfkens. Perhaps even more important, the figure shows that on average of more than one-third of all aid recipients received less than US$1 million (in current dollars) a year during the period 1989 to 2000 and another one-third less than US$10 million. Only a handful of countries ever received more than US$100 million per year during the entire period 1960 to 2000 (i.e. India, Indonesia, the Netherlands Antilles, Surinam and ex-Yugoslavia).[17]

---

15. However, the new policy of Herfkens with regard to a concentration of bilateral aid on good policy countries is clearly a deviation from the policy of her predecessor Pronk. Under Pronk bilateral aid spread even further over the Third World mainly for political reasons. Herfkens obviously took quite seriously the conclusion in the World Bank study *Assessing Aid* that 'efforts to "buy" policy improvements ... have typically failed'.

16. Initially, there were four countries added. The Palestinian Administered Areas, however, were in later years removed and put on the thematic list (under 'human rights, peace building and good governance').

17. India received more than US$100m. in 1980, 1981, 1986, 1987, 1988, 1990 and 1991, while Indonesia received such a sum in 1982, 1987, 1988, 1989, 1990, 1991

DUTCH DEVELOPMENT CO-OPERATION IN THE 1990S

TABLE 12.2
TARGET COUNTRIES OF DUTCH BILATERAL AID (2003)

	Target countries	*+3 countries*	*Thematic countries*			*DOV*
			Environment	Human rights, etc.	Private sector	
1	Bangladesh	Egypt	Brazil	Albania	*Bosnia*	Benin
2	*Benin*	Indonesia	Cape Verde	Armenia	*China*	Bhutan
3	Bolivia	South Africa	*China*	*Bosnia*	*Colombia*	Costa Rica
4	Burkina Faso		*Colombia*	Cambodia	*Ecuador*	
5	Eritrea		*Ecuador*	*China*	El Salvador	
6	Ethiopia		*Guatemala*	*Colombia*	*Guatemala*	
7	Ghana		Mongolia	*El Salvador*	Ivory Coast	
8	India		*Nepal*	Georgia	Jordan	
9	Macedonia		Pakistan	*Guatemala*	*Moldavia*	
10	Mali		*Peru*	Guinea-Bissau	Nigeria	
11	Mozambique		*Philippines*	Honduras	*Peru*	
12	Nicaragua		Senegal	Kenya	*Philippines*	
13	Rwanda			*Moldavia*		
14	Sri Lanka			Namibia		
15	Tanzania			*Nepal*		
16	Uganda			Palestinian territ.		
17	Vietnam			Zimbabwe		
18	Zambia					
19	Yemen					

Source: website DGIS [*www.minbuza.nl*].
Note: Countries in more than one category are italicised.

FIGURE 12.3
NUMBER OF RECIPIENT COUNTRIES (POSITIVE NET ODA) – 1960–2000

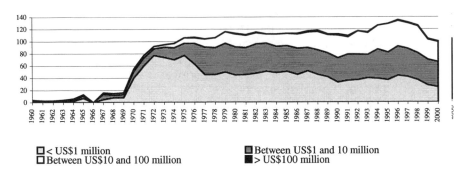

☐ < US$1 million
☐ Between US$10 and 100 million
▩ Between US$1 and 10 million
■ > US$100 million

Source: Calculations based on OECD/DAC (cd-rom).

---

and 2000. The Netherlands Antilles received more than US$100m. annually only during 1996 to 1999, while Surinam received the largest sum directly following its independence from the Netherlands in 1976 and ex-Yugoslavia only in 1993.

425

Moreover, the existence of concentration countries does not automatically mean that the majority of geographically allocable bilateral aid indeed finds its way into these countries. In its most recent review of the Netherlands, the DAC [*OECD, 2001*], for instance, noted that 'the share of bilateral ODA reaching the most important beneficiaries has actually decreased over the years'. In effect, during 2000 even the 19+3 target countries received only 33.2 per cent of all bilateral aid,[18] while 11.9 per cent went to the 28 different thematic countries.[19] Whether in the near future the number of aid recipients will be even further reduced remains to be seen. In 2002, Van Ardenne announced that she supported the policy of selecting a limited number of countries for long-term partnerships, if only because the Netherlands ought to be a 'reliable partner'. At the same time, she considered that the country policy focused too strongly on individual countries, while many development issues (for example, water management, conflict management, economic co-operation) were not restricted to individual countries but required a more regional approach.

Although it still remains unclear exactly how this regional approach is to be linked to the country concentration policy, and thus what the call for a more regional approach will mean in practice, Van Ardenne in June 2003 had already announced several changes in the Dutch country policy. From the Mutual Interests, Mutual Responsibilities policy paper onwards [*DGIS, 2003b*] the 'existing 19+3, environment, and good governance, human rights and peacebuilding lists [have been merged] into a single list of "partner countries"'.[20] The remaining 26 partner countries have been selected using the same criteria as earlier (i.e. poverty, need for aid, good governance and sound policy) but also foreign policy considerations have played a role[21] as well as criteria of 'complementarity' and 'concentration'. According to the Minister, the latter means 'focusing knowhow and resources on a limited number of countries in order to boost effectiveness by obtaining sufficient critical mass. This critical mass will also be achieved through complementarity, i.e. the added value that can be generated through cooperation with other donors, whether they are governments, multilateral organisations or NGOs' [*DGIS, 2003a*].

---

18. This includes a negative net ODA for India, which was probably the most important recipient of Dutch bilateral aid for decades.
19. The DAC [*OECD, 2001: 1–22*] attributed this to 'the fact that only a part of overall aid is allocated as "delegated" funds to embassies in both the [then] 17+4 partnership countries and the 29 thematic countries'.
20. Interestingly, this would only leave the 'private sector' list of countries as a separate one.
21. 'The decision to continue providing support to the Palestinian National Authority, Yemen and Egypt is partly based on the unstable situation in the Middle East' [*DGIS, 2003b*].

FIGURE 12.4
DISTRIBUTION OF BILATERAL ODA PER INCOME CATEGORY (CURRENT US$) – 1960–2000

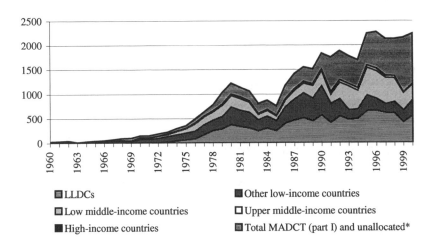

Source: Calculations based on OECD/DAC (cd-rom).
Note: *MADCT (More Advanced Developing Countries and Territories) covers those countries that during the 1990s were removed from part I of the DAC list (i.e. those countries eligible for ODA) and transferred to part II.

Finally, and despite all the dispersion of Dutch aid and the struggle to achieve further concentration, it should be clear that the majority of Dutch bilateral aid has always been used to finance development activities in the world's poorest countries. Schulpen and Hoebink [*1999: 21*], for instance, showed that 'net disbursements of Dutch ODA to in total 121 aid receiving countries over the period 1969–1996 show that 71.6 per cent of bilateral ODA went to those countries grouped under Low Income Countries (LICs). If the Human Development Index (HDI) is taken as "dividing line" this percentage is 77.1 per cent for countries grouped under the heading of Low HDI. Countries under High HDI received over the same period 5.2 per cent of Dutch aid, whereas the remaining 17.7 per cent went to the Medium HDI countries. In the case of the GNP index, high income countries received less than 1 per cent, upper middle income countries some 9 per cent, and lower middle income countries 19 per cent.' In addition, the DAC [*OECD, 2001: I–22*] concluded that 'more than 60 per cent of Dutch bilaterally allocated ODA benefited the least developed and other low-income countries in 1999. This exceeded the

DAC average of 52 per cent.' Figure 12.4 provides an overview of the division of Dutch bilateral net ODA per income category for the period 1960 to 2000.

V. SECTORAL APPROACH

One of the central issues in the effort to improve the quality of Dutch aid in the 1990s was the sectoral approach. As mentioned already, the basic ideas for such an approach were set out in 1996 under Pronk [*DGIS, 1996*]. In this policy paper, sectoral budget support (defined as the provision of foreign exchange, the value of which in local currency is used to finance specific sectoral or sub-sectoral programmes)[22] was seen as a way of overcoming the problems with project aid (considered to lead frequently to unsustainable 'islands of development'). In order to make such a sectoral approach workable, it is essential that there is not only sufficient implementation capacity on the recipient side but also a broad convergence between the donor's priorities and the recipient's goals. As under Herfkens, it was in essence those countries that were considered to have good policies and good governance that were selected, and it was thought possible to extend the sectoral approach to all these countries. The major difference between Pronk and Herfkens with regard to the sectoral approach was therefore that the latter introduced this approach to all countries with which the Netherlands has a long-term bilateral relationship.[23]

The general idea is that Dutch aid will be concentrated on two to three sectors within a country and that aid should preferably be provided in the form of direct budget support.[24] With this, the ownership of the country in question would be taken as the point of departure, as the country was to decide on how the money would be spent. The country itself and not the donor would thus be in the

---

22. In contrast to general programme aid (aimed at general economic adjustment), sectoral budget support has a specific poverty alleviation focus.

23. In doing so, Herfkens was (as in the case of country selection criteria) strongly influenced by the World Bank study *Assessing Aid* [*World Bank, 1998*] which stipulated that a sector approach is seen as a possibility for those countries that do 'things reasonably well, but [continue] to spend on many things that are inefficient'.

24. The term 'preferably' is used here deliberately. Although it remains the objective in the end, the Netherlands is not so blind that it does not acknowledge that even within a sectoral approach projects remain possible and that it will take a while before the project form of aid will disappear (if ever). It was not therefore surprising that the DAC in the latest review of the Netherlands concluded that project financing is still dominant if only because it is not so easy to change practice as it is to change policy [*OECD, 2001*].

driver's seat. For true ownership it is, of course, necessary not only that the government alone decides but also that this is done in consultation with other stakeholders in society (for example, civil society organisations, the private sector, labour unions, academia, etc.). This shows not only the importance attached to democratisation as almost a precondition for a sectoral approach, but also why the Netherlands attaches so much importance to Poverty Reduction Strategy Papers (initiated by the World Bank and the IMF). In the meantime, it is well known that experience with these PRSPs demonstrates that precisely in the consultative sphere a lot still needs to be done and that the poverty strategies set up in more and more countries are not necessarily a reflection of, nor an answer to, the true poverty problems. Perhaps for lack of a better instrument, the Netherlands maintains its faith in the PRS process, however. All this automatically leads to a call for further co-ordination as 'critical to successful use of sector approaches' and for coherence in order to ensure that 'ODA-funded programmes are not compromised by inappropriate policies in other key areas (e.g. trade, agriculture, intellectual property, arms)' [*OECD, 2001: 1–21*]. Policy-wise, this brings together the four central aspects of Dutch development co-operation (namely, poverty reduction, ownership, sector approaches and co-ordination). Whether it also provides sufficient operationalisation is open to doubt, however (see also section II above).

The Netherlands is by no means the only donor to attach such high values to ownership and to instruments such as PRSPs to achieve it. Nor is it the only donor to introduce a sectoral approach in recent years as an answer to the need for more co-ordinated efforts at ownership. It is, however, practically the only donor that has made the sectoral approach the norm. This is in sharp contrast to such donors as Sweden and Norway, which introduced a sectoral approach even earlier than the Netherlands, but which followed a step-by-step introduction. According to Hoebink [*2002a*], there are very good reasons to prefer the Swedish and Norwegian way of working. In view of the fact that it is very likely that a recipient is not able to plan, implement and/or monitor a realistic programme for a specific sector, it is logical that donors are reluctant to provide budget support. Hoebink therefore concludes that the Netherlands is moving far too fast in introducing a sectoral approach simultaneously in all its target countries. Some of the realism obviously required was clearly expressed by Van Ardenne, who stated that she would keep a close watch on the performance of developing countries in using Dutch aid. Nevertheless, also for Van Ardenne, 'the sector-wide approach will remain the guiding principle governing bilateral development cooperation'. At the same time, she announced her intention to restrict the number of sectors in each country 'to two, or at most three', while adding that 'various activities aimed at improving the quality of governance will also be launched in each country' [*DGIS, 2003a*].

VI. PRIVATE-SECTOR DEVELOPMENT (PSD)

According to Van Ardenne, greater co-operation between the public and the private sector is needed and can be achieved. One of her 'key points' thus refers to this public–private partnership and to a greater involvement of the private business sector (but not restricted to the Dutch private sector). In this, she deviated from the standpoints of both Pronk and Herfkens. Both Labour Ministers supported in theory the need for developing the role of the private sector in developing countries (also in the light of poverty reduction) but were in practice quite reluctant to give it much weight within development co-operation, and certainly if it involved Dutch private enterprises (if only because they feared a greater say of the Ministry of Economic Affairs over (part of) their budget). Nevertheless, it was once again Herfkens who provided the room for discussion about PSD.

So, during the budget discussions in December 1998, Minister for Development Co-operation Herfkens stated that 'the private sector is the most important source of employment and sustainable poverty reduction'. She added that particular attention should be paid to creating the preconditions in developing countries themselves for an adequate functioning of the private sector. Furthermore – and this was meant specifically for Dutch business – she argued that the involvement of the Dutch corporate sector in development co-operation was perfect as long as this sector could show itself to be a useful instrument for developing the private sector in the Third World. With these statements, Herfkens set the tone for the discussion on the private sector, development and the programmes of the Ministry for Development Co-operation.

Programmes aimed at the private sector, and in particular the Dutch private sector, have existed for quite some time in the Netherlands. Already in the 1960s, the Netherlands had budget lines to promote the private investments of Dutch companies in developing countries by financing pilot projects or feasibility studies. Furthermore, insurance against political risks was made possible already at an early stage. In the 1970s and 1980s, Dutch aid money was used substantially for financing goods of Dutch origin (for example, fertiliser, trucks) as part of balance-of-payments or import support. Although several Ministers, particularly in the 1980s, tried to create more room for the Dutch private sector in development co-operation, they did not succeed. Apart from the fact that it made some of them express their dissatisfaction with the lack of response from the sector, the involvement of Dutch business actually decreased in the 1980s. Nevertheless, there are still many instruments aimed at the Dutch private sector in particular, a few of which were started during the 1990s when Pronk was Minister for Development Co-operation.

Whatever the categorisation of instruments one uses,[25] the Dutch private-sector armamentarium shows a large diversity in programmes, the implementation of which falls under the responsibility of several different organisations and agencies. There was certainly not 'one window' for private-sector programmes, as has been, and still is, advocated by private enterprises in the Netherlands. More important, however, is the fact that a large part of these programmes are principally aimed at Dutch private firms and also that the majority of funds are reserved for these programmes. This holds true, for instance, for the ORET programme, but also for guarantees and insurances for Dutch investments. Although some of the programmes are described as such, there are in fact no Dutch aid programmes aimed specifically at private-sector development in developing countries [*van den Bosch, 1998: 1*].

For a long time there has been no integrated policy for the private sector, and certainly not in relation to the main objective of Dutch development assistance: (sustainable) poverty reduction. The Social and Economic Council (SER) [*1997*] pointed to this lack, and its plea for a key role for employment creation (through the development of the private sector) as a central notion in Dutch development policy should (also) be seen in the light of this. Pronk was very frank about this absence of an integrated policy. In 1997, he stated that the wide range of instruments for the Dutch private sector aimed at investment or export promotion were 'insufficiently tuned to each other and are managed and deployed in a fragmented manner'.

---

25. The categorisation of Dutch instruments aimed at the private sector that has been used covers (i) export promoting instruments (ORET/MILIEV); (ii) facilitating instruments (e.g. funding of studies into investments, trade possibilities and the setting up of pilot projects under the PSOM programme); and (iii) investment-promoting instruments (e.g. the Dutch Investment and Matching Fund (NIMF) and the loans and risk-capital provisions of the Financing Company for Developing Countries (FMO)). A similar division was used by Pronk in a letter to Parliament on promoting Dutch investments in developing countries in December 1997. Other categorisations are also used, however. In the 1997 Budget, for instance, the private-sector programmes are divided into two categories: (i) instruments in support of an enabling environment, and (ii) instruments in support of the Dutch private sector.

Part of the business community in the Netherlands was convinced that the lack of a clear integrated PSD policy within Dutch development assistance was due to a negative attitude within the Ministry with regard to PSD and, in particular, the involvement of Dutch private enterprises in aid [*Dalmeijer and Schulpen, 1999: 13–14*]. Pronk was seen as a prime example of this attitude, perhaps because as a Social Democrat he regards development as first and foremost a task for the government or because he felt that poverty reduction would be better aimed at the informal sector which was of less interest to foreign firms. The latter is said to be the main reason why the Ministry directs its attention with regard to employment mainly at microenterprises in the urban areas. Finally, it is also possible that Pronk (and many at the Ministry with him) felt that the Dutch private sector ought to be able to function without assistance from the government. The negative attitude was, however, not only 'visible' within the DGIS but also obviously worked the other way round. The relationship between the private sector and the DGIS has been described as one of 'institutionalised distrust'. Whether Herfkens succeeded in dispelling this distrust may be open to doubt.

One of the clearest examples of the focus on Dutch firms, and at the same time one of the hottest issues in Dutch development policy, is the Development-related Export Transactions Programme (ORET). According to most critics, ORET is a purely Dutch-oriented export promotion programme with no connection with the development co-operation budget. Backed by a motion in Parliament, Minister Herfkens announced her intention to move away from exports to investments. Investments are said to offer better guarantees for long-term development in aid-recipient countries, while at the same time they are thought to be less sensitive to corruption. Moreover, such a move would meet the wish, expressed for years within Dutch aid circles, to achieve further untying of aid. Even the SER (in which private-sector organisations are also represented) had spoken in 1997 against tied aid. However, a distinction should be made between the exporting and the non-exporting private sector in the Netherlands. Not surprisingly, the exporting sector, and particularly that part which makes use of the ORET programme, was less charmed by this untying story of the SER.[26]

---

26. Dalmeijer and Schulpen [*1999*] show that the controversy between exporting and non-exporting companies goes beyond the issue of tying or untying of aid. Several representatives of non-exporting firms expressed their dissatisfaction with the ORET programme and particularly with the role of the exporting firms in the programme. The latter were accused of lacking a long-term vision and of being interested in short-term financial gain only. As such, these critics were also convinced that the ORET

Whether or not the ORET programme should be seen as a pre-investment programme, the fact remains that promoting Dutch foreign investment in developing countries through aid is not an easy task. And it becomes all the more difficult if one has to do it without a clear and integrated policy (or any policy at all) that also links these investments to the central objective of sustainable poverty reduction. The 2000 policy paper on 'economy and development' was therefore warmly welcomed by the private sector [*DGIS, 2000*]. For the first time, an effort was being made to link the existing 'PSD' instruments to overall development policy. Although the enthusiasm from the private sector was understandable, the policy paper was characterised by others as nothing new. Van Dijk [*2001: 152*], for instance, concluded that it 'provides a systematic overview of existing instruments but without presenting anything new: no new insights, instruments or approaches'.

In essence, the policy paper follows the line of thinking on PSD as this was already expressed in the mid-1990s by the DAC (for an overview of PSD thinking, see Schulpen and Gibbon [*2002*]). Trade, investment, debt relief, primary products and coherence are presented as important elements at the international level for PSD. At a more national level (i.e. in the individual developing countries), the main elements are in the field of macroeconomic policy, political stability and good governance, physical and social infrastructure and the protection of humans and the environment. So far, there are certainly few new insights. The same holds true for the micro-level (or the level of direct support to the private sector). Three types of obstacles are distinguished: knowledge, profitability and risks. The already existing PSD instruments are then grouped under these three headings: training in management and investment instruments under the heading of 'knowledge', financial support for investments and exports under 'profitability', and guarantee schemes naturally fall under the heading of 'risks'. Perhaps the most interesting point refers to stronger centralisation by making (as far as possible) one agency responsible for the implementation of each set.[27] In this way it is expected that

---

programme was not 'equal to investments' particularly because it is concerned only with commercially non-viable activities and in most cases there is no 'aftercare'. Other representatives of Dutch firms in this study were convinced that under the ORET programme not only in most cases were capital or investment goods delivered but also that these exports had to be seen as pre-investments. In other words: exports improve the relationship between the Dutch private sector and the recipient country and can thus, in the longer run, lead to investments.

27. The FMO (Financing Company for Developing Countries) will become the principal agency for the implementation of programmes falling under the heading of profitability, and the NCM (Dutch Credit Insurance Company) for those under the

greater clarity and transparency will be created and the criticisms expressed by Dutch enterprises over the years that there was no single window can be tackled.

As mentioned earlier, Van Ardenne made PSD one of her central policy issues. She emphasised public–private co-operation and development of the private sector in developing countries (which was also the viewpoint of her predecessors). In the short time between and after elections, there was of course little possibility to elaborate this policy. We thus have to wait to see how Van Ardenne will operationalise her aim of giving a greater role to the private sector and thereby to the creation of productive employment and the promotion of trade and investment. The history of Dutch aid obviously does not provide for a glowing perspective in this regard.

VII. PRIVATE AID CHANNEL

Within the private channel for development co-operation in the Netherlands a distinction is often made between 'recognised development organisations' (referring mainly to the so-called co-financing organisations – termed MFOs following the Dutch abbreviation) and 'categorial development organisations' (see, among others, Kruijt et al. [1983] and Schulpen and Hoebink [2001]). The latter heterogeneous category covers hundreds of (often small) organisations and includes such bodies as solidarity and country committees, church-related organisations, humanitarian institutions, women's organisations and trade unions. More recent are cultural organisations and institutions related to the business sector. An attempt by Schulpen and Hoebink [2001] to briefly describe the diversity within the categorial Dutch NGO sector showed that three types of activities prevail. First, this covers financial support to development projects often implemented by local organisations; in some cases, the funding is provided initially by the Dutch government but in most cases it comes from donations from the general public. Second, these NGOs are active in the field of

---

risks heading. An exception is made for programmes under the heading of knowledge. Although major knowledge programmes (e.g. PESP (Programme Economic Co-operation Projects); PSO (Programme Co-operation Eastern Europe), PSOM (Programme Co-operation Emerging Markets)) are brought together under the Senter agency, the activities of CBI (Centre for the Promotion of Imports from Developing Countries) and PUM (Netherlands Management Co-operation Programme) are regarded of having such an individual character that integration will not be advantageous. Within the Ministry a special co-ordinating desk was created with three sub-desks covering the three fields for the stimulation of PSD: international markets, national enabling environment in developing countries and the micro-level.

awareness-raising in the Netherlands and, third, in exchange programmes (twinning, volunteer services, etc.).

Another classification of the private aid channel was brought forward by the DGIS in 2001. The first group then consists of those NGOs with a broad all-encompassing approach to structural poverty reduction (i.e. the MFOs). The second category includes those NGOs working towards poverty reduction from a specific thematic point of departure (for example, women, the environment, health, the media, peace, human rights). Third, there are organisations with a different primary objective but which are also active in the field of poverty reduction, such as labour unions, the peace and environmental movement, and the Association of Netherlands' Municipalities. The final category consists of private initiatives of (groups of) individual citizens [*DGIS, 2001a*]. It should be noted that this classification was mainly used in the light of the changes in the subsidy system for the private aid channel that were discussed at the end of the 1990s (see below).

Not only in these subsidy discussions but also in earlier discussions on the private aid channel, the MFOs predominated as the main beneficiaries. This makes these co-financing organisations the most interesting to discuss. They are the biggest organisations, and all have a strong (financial) relationship with official Dutch aid. Significant changes occurred in the funding and policy agreement between these MFOs and the Ministry for Development Co-operation during the second half of the 1990s and the early 2000s.

The MFOs have been part of Dutch development co-operation since December 1964 when funds were set aside in the budget for 'co-financing by the government of development projects of non-commercial private organisations'. Since then, the MFO programme experienced almost constant growth, both in funds and in the room for manoeuvre they created for themselves. In this regard, a crucial decision to provide annual block grants to the MFOs was taken in the early 1980s. This block grant system was introduced without an evaluation of the programme. Apparently Dutch politicians were so convinced of the MFOs' qualities that a formal assessment was considered unnecessary. The MFOs fuelled this conviction by promoting themselves as the primary channel for poverty reduction.

Nevertheless, by the end of the 1980s even the MFOs could no longer escape the growing criticism of official aid. In 1989, the MFOs initiated an independent impact study that was presented in 1991 [*SIM, 1991*]. Although this was quite critical of their work, the MFOs persisted in presenting themselves as a perfect aid channel, though in a more modest way. After an initially defensive reaction to the unexpected criticism, they also embraced some of the study's recommendations.[28] Following the policy lines set out in *A World of Difference*, Pronk announced in 1991 a structural increase in the funds for the MFOs.[29] The co-financing organisations were once again on track with more money at their disposal. The critical period of the late 1980s was over, or so it seemed. It all changed when Herfkens spoke at Novib's New Year's reception in January 1999.

She criticised the MFOs for forming a 'closed front', for their financial dependence on the government and for the fact that in her eyes they provided little in the way of alternatives. Herfkens was convinced that there is a distinct difference between the roles of the government and the NGOs; the latter should concentrate on providing countervailing power to the poor and not take over the responsibilities of governments in such fields as health and education. Backed by a new subsidy law, the Minister set out to fundamentally change the relationship with the MFOs.[30] Heated discussions followed on the respective roles of government and NGOs in development, and on a new funding system for the MFOs. Even before such a new system was worked out, the Minister decided in December 1999 to grant Foster Parents Plan Netherlands the status of an MFO and thus access to funding through the Ministry. For Herfkens, this was one of the ways to open up the closed MFO system.[31]

---

28. The MFOs came forward with an official reply in February 1992. Changes in a number of areas were promised: more attention would be paid to the sustainability of the activities supported, the process of making their policy more explicit would be continued, Africa would receive greater attention, co-operation among the MFOs would be strengthened, and more attention was promised for institutional support to NGOs. Four years later they presented their final report in which they extensively documented the measures taken [*GOM, 1995*].

29. One of the main reasons for this development is that NGOs are regarded as being closer to the target groups of the poor and of following a more participatory approach, thereby forming an important channel in the field of poverty reduction. Thus, by structurally increasing the MFO budget in the 1990s, Dutch politicians actually reaffirmed their conviction that the MFOs were a good aid channel. Others had a different opinion. Quarles van Ufford [*1991: 41*], for instance, saw it as a lack of political interest in the MFO programme.

30. The 'attack' on the NGOs (and more particularly the MFOs) was in line with Herfkens' efforts to streamline bilateral aid (concerning countries, interventions, sectors) and discussions with the four co-financing organisations (MFOs) in order to arrive at a division of tasks between the DGIS and the MFOs thereby maximising the efforts in the fight against poverty. In order to enhance the quality of aid rendered, it was considered important that these two channels should be geared to one another, something that had often been lacking in the past.

31. The decision to grant Foster Parents Plan (in 2002 renamed Plan Netherlands) MFO status was heavily criticised mainly by the 'old' MFOs (i.e. Novib, Icco, Hivos and Cordaid). They argued that the organisational structure, objectives and pro-

FIGURE 12.5
GROWTH OF THE CO-FINANCING PROGRAMME (€m.)

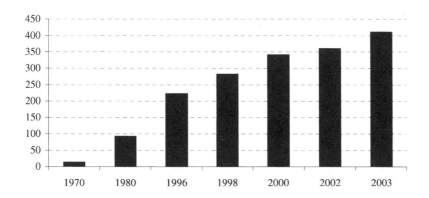

Source: Based on Commissie-Box [*2002*].

In April 2001, and following a broad consultation involving a large part of the Dutch NGO community, the MFOs and the Ministry reached an agreement. It was decided to increase the funding of the MFOs from 10 per cent to (maximum) 14 per cent of the budget for development co-operation, depending on the quality of the proposals (see Figure 12.5 for an overview of changes in the budget for MFOs). The subsidy would be granted for a period of four years, after which the MFOs had to apply again for another four years. For the first time, activities in the Netherlands itself (in such fields as education and lobbying) could be financed by funding from the Ministry directly under the co-financing programme (MFP). In addition, the MFO programme would be open to other organisations, although Herfkens kept repeating that she could think of no other organisation in the Netherlands that would qualify. With this agreement, both the Ministry and the MFOs may be regarded as winners. The MFOs were satisfied because the importance of the private aid channel was reconfirmed, the continuity of their work was (more or less) guaranteed and activities in the Netherlands itself were brought under the MFO subsidy.

---

gramme of Foster Parents Plan did not correspond to the official criteria. Moreover, they were afraid that it would open up the MFO programme to other organisations. Added to this, no independent evaluation of Foster Parents Plan had been undertaken prior to the decision to grant it the status of MFO. The latter issue was particularly interesting as simultaneously a public discussion was held on the quality of the programme of Plan International (e.g. in Colombia and Haiti), through articles in the press, documentaries on TV and by two organisations of critical foster parents.

The Minister was satisfied because in principle the MFOs' 'closed front' was broken and a start was made with a more qualitative appraisal of the MFOs.

Whereas the above-mentioned MFO system (known under the name 'MFP-wide') was meant only for those organisations with a broad, all-encompassing strategy and programme for poverty reduction,[32] another subsidy window ('MFP-small', later renamed as 'MFP-thematic') was opened for those organisations that work towards poverty reduction from a specific approach (for example, the environment, women, health, the media, human rights). Finally, a third window was proposed (initially called the front office) meant for small-scale activities in the field of development co-operation.[33] In 2003, this last window was still being further developed mainly by the MFOs and the National Committee for Sustainable Development (NCDO) which intend to be responsible for it. The other two windows ('MFP-wide' and 'MFP-thematic') became operational from January 2003 onwards.

In essence, this meant that the responsibility for judging the quality of proposals of organisations applying for funding under one of these two windows was in the hands of Herfkens' successor, Van Ardenne. The decisions taken by the Ministry at the end of 2002 with regard to the MFP-thematic proposals (on the basis of an internal DGIS assessment) did not lead to an open public discussion, although there were rumours that the decisions taken had been sharply criticised by the organisations and that discussions are still continuing. Also the decisions taken in the 'MFP-wide' programme did not lead to a public debate, although the media paid some attention to them and one more fundamental critique was brought out (but did not receive substantial news coverage). In a way, this was strange as the assessment by the advisory commission was quite critical.

Not only did the five existing MFOs presented their plans to the independent advisory commission, but also another organisation (Terres des Hommes) did the same and was admitted. From 2003

---

32. In practice, this meant that the organisations which want to apply for subsidy under the 'MFP-wide' programme had to be active in three different fields: direct poverty reduction activities, lobbying and activities with the purpose of strengthening civil society.

33. This means that the fourth category of NGOs distinguished by the DGIS [*2001b*] (i.e. the private initiatives of groups of citizens) are referred to this front office. For the third NGO category (i.e. those organisations for which poverty reduction is not the primary objective but which are also active in this field) no new and specific window was opened, although these organisations remain entitled to programme funding for parts of their work (e.g. through the already existing labour union co-financing programme).

onwards the MFO programme thus includes six organisations. The advisory commission proposed a ceiling of 11 per cent of the budget for the six MFOs, mainly because their plans and proposals were not considered of sufficiently high quality to justify a larger budget. Novib in particular was criticised by the commission: its plans were considered inadequate. It therefore received an insufficient mark and should, according to the advisory commission, receive substantially less subsidy than it applied for. Plan Netherlands also scored below average and was to receive less than it asked for. The other four organisations were to receive precisely the amount they requested.

Van Ardenne followed the recommendations of the advisory commission. At the same time, she did not open up the criticism addressed to the commission itself. Hoebink [*2002b*] was the most outspoken in his criticism. According to him, the work of the commission should be judged as arbitrary and subjective. He particularly criticised the failure of the commission to follow the criteria set in the MFO policy but also the fact that hardly (or no) attention had been paid to the experiences and previous effectiveness of organisations.[34] If the latter had been taken into consideration, Novib could never have ended up with such a low mark, according to Hoebink. Finally, he also criticised the fact that Terres des Hommes had been added without any knowledge about the quality of this organisation in the past.

Although the commission's report thus led to a minor stir in the Dutch development community (and particularly at Novib), Van Ardenne gave both Novib and Plan Netherlands the opportunity to present new plans within two years. If these were judged more positively, both organisations would be entitled to an increase in subsidy.

All in all, the period from 1988 onwards has indeed been a period of ups and downs for the MFOs in general and for particular organisations. For the second time since its inception in 1964, the MFO programme became the subject of public debate, which, like the first time (in the late 1970s/early 1980s), led to a new subsidy structure. The starting point of this new system seems somewhat harsher than in the 1980s, but it remains a system that in essence carries the backing of both the MFOs and the Ministry. There seems to be a general feeling within the NGO sector that it should be given the opportunity to prove itself, although Van Ardenne obviously does not share this feeling. In her policy paper, she repeated her wish to integrate 'theme-based cofinancing into the Broad-based Cofinancing Programme with effect

---

34. This particularly concerned the fact that a two-year-long evaluation of the MFO programme on several themes was not taken into consideration.

from 2007', thereby effectively killing the newly established subsidy structure almost in its infancy [*DGIS, 2003b*].

VIII. COHERENCE

All the major political parties in the Netherlands are in favour of opening up markets to the goods of developing countries and some explicitly mention the often disastrous effects of the European Common Agricultural Policy on these countries. We are thus in the midst of the so-called coherence debate which has been going on (although not necessarily under the same name) since the 1970s but which received important reinforcement in the 1990s. It has already been mentioned above that Pronk emphasised coherence, but that it was under Herfkens that the coherence issue was placed centre-stage. Van Ardenne also stated that policy coherence remained a crucial issue.

In defining policy coherence ('the non-occurrence of effects of policy that are contrary to the intended results or aims of policy'), Hoebink [*1999b: 181*] distinguishes between internal and external coherence. Internal then refers to development policy in the narrow sense and means that 'the effects of a certain part of development policy should not be contrary to the intended results or aims of the same or other parts of development policy or foreign policy'. Put differently, reference is made here to issues that were part of the reverification exercise of the mid-1990s (see above), and it therefore 'implies coherence in the rationale behind development co-operation, in goal-setting and prioritisation, between dialogue and implementation, between different types of aid, between various donor programmes, and between donor's policies, multilateral policies, aid recipient and NGO policies' [*ibid.*]. Many of the issues brought forward during the 1990s aimed at improving the quality of development aid could well be placed within this internal coherence definition. It holds for the emphasis on ownership, for the struggle over the tuning of the activities of NGOs (i.e. MFOs) and the Ministry, and for trying to bring private-sector instruments more in line with the overall objective of development co-operation.

External coherence, in contrast, refers to a broad perspective 'in which goals and activities in a given policy sector should not be at odds with policies in another sector, in this case development policy'. Such issues as 'market policies, policies that privilege certain economic sectors ... migration policy, and environmental policies' are important [*ibid.*]. This external coherence was already part and parcel of Dutch development policy in the early days, although it was then referred to by such terms as 'structural policy', 'integral' and 'integrated'. For instance, already in 1974 the Dutch government stated that 'trade and aid issues should be dealt with simultaneously', exemplified as this was in the formal support for the New International Economic Order (NIEO). In later policy documents, the need for an integral (or integrated) development policy was emphasised by opting for a 'coordinated approach towards policy on the transfer of financial flows, debt, raw materials, trade, industrialisation and agriculture' [*ibid.*].

The 1980s saw a narrowing down of development policy in favour of aid. In budget papers important problems facing the developing countries (such as the debt burden) were ignored, and the Minister for Development Co-operation in fact became a minister for development aid. This changed once more at the end of the 1980s when Pronk took over for the second time. Development co-operation again covered major non-aid issues and the necessity of an integrated policy encompassing aid, trade, international finance, etc. was again stressed. The policy paper *A World in Dispute* proved to be the stepping-stone for a concentrated effort to abolish incoherence. Under the term 'decompartmentalisation', Pronk called for bringing down the walls between different approaches, policies, ministries and departments with regard to foreign policy at large. Development co-operation should no longer, according to Pronk, be viewed as the soft sector of foreign relations.[35]

In March 2002, as one of her last efforts as Minister, Herfkens brought out a policy paper on coherence. Ever since she started in 1998, coherence had been on the tip of her tongue. In the policy paper, four types of coherence are distinguished: (i) within development policy (i.e. between different policy themes and target groups and the overall objective of poverty reduction); (ii) within the broader foreign policy (i.e. between development co-operation and other objectives of Dutch foreign policy); (iii) the 'big' coherence (i.e. between poverty reduction and those areas covered by other ministries such as Economic Affairs (trade), Finance (IMF) and Agriculture (the EU Common Agricultural Policy); and (iv) international coherence

---

35. Naturally, such a call for 'decompartmentalisation' ought to work both ways. In effect, 'Pronk (the aid budget) had to pay for the growing influx of asylum seekers, for Dutch language courses and for a transport aircraft for military interventions. Overall, these austerity measures and budget cuts brought the volume of Dutch aid back [from] about 1 per cent of GNP in 1989 to around 0.8 per cent in the mid-1990s. As ex-minister De Koning stated: "It is nice to talk about decompartmentalisation, but when you go into the cabinet you have to remind yourself that you enter into the company of pickpockets'" [*Hoebink, 1999a*].

(i.e. between the policies of organisations such as the World Bank, the IMF, the WTO and the UN). It was already noted above that Herfkens has been active in all these areas during her time in office. Take, for instance, the assessments of the EU, the UN and the World Bank in which she called for more co-operation and consultation, or the discussions with the MFOs. In this sense, it is understandable that Hoebink [2002a], when discussing Herfkens' four years in office, concluded that she could certainly not be reproached for paying insufficient attention to coherence. According to the DAC [OECD, 2001], which feels that 'policy coherence has always been a key concern of Dutch development policy', the Netherlands has very actively sought coherence in the field of trade. Reference is then made to its strong support for the 'Everything but Arms' agreement in the EU,[36] as well as for reform of European agricultural policy.

In the meantime, Van Ardenne also set coherence centre-stage in her policy. She even announced that she would like to become known as 'Miss Coherence'. Already in her first short term as state-secretary, she managed to take the first step towards it by publishing (together with the Minister for Agriculture) a paper on policy coherence with special reference to development co-operation and agriculture [DGIS, 2002b]. This paper calls, *inter alia*, for trade liberalisation and further reform of the agricultural sector in Europe. Moreover, within the Ministry a special Coherence Bureau was established. In her June 2003 letter, coherence is among the ten central policy emphases. Presented as crucial for poverty reduction, coherence in the policy of the Dutch government, the European Union and the international organisations is considered 'essential' [DGIS, 2003a].

IX. CONCLUDING REMARKS

The 1990s was indeed a stimulating decade. Driven by critical comments and papers on such issues as the role of the private sector, the relationship between NGOs and the Ministry, technical assistance and the weaknesses and strengths of multilateral organisations, the public debate on development co-operation was revitalised. In this, the Min-

---

36. 'The EU's 2001 "Everything But Arms" (EBA) initiative was originally intended to provide immediate free market access for all non-military exports from LDCs. However, following a concerted campaign by European producers and traditional Caribbean exporters, who feared that they would lose market share to LDC exporters, the proposal was modified so that free LDC market access for three important products (rice, sugar, and bananas) will be delayed for up to eight years' [OXFAM, 2002: 101].

isters played an important role as did the NGO community and, in part, the private sector. Not in all cases are the outcomes of these debates yet clear. Take, for instance, the multilateral organisations. Despite the critical studies brought out by the DGIS, the changes in funding have up until now been marginal and any future changes seem to be more dependent on budgetary considerations and not on policy decisions. In addition, the discussion on the relationship between NGOs and the Ministry is likely to continue for some time, although probably with less concentration on the MFOs which have dominated this discussion in the past.

In the meantime, public commitment to development co-operation remains high in the Netherlands and obviously did not suffer from the growing criticism from the end of the 1980s onwards. This public commitment is at the same time an indication of what Hoebink [*1999a*] called the 'humanitarianisation' of Dutch foreign policy in the 1990s, an important feature of which was the further de-commercialisation of Dutch aid. In this light, it is interesting to keep track of more recent moves towards a greater emphasis on the role of the commercial sector in development co-operation. Notwithstanding the fact that the discussion centres on the development of the private sector in developing countries, it should not be forgotten that the majority of PSD instruments are directed first and foremost at the Dutch private sector.

During the 1990s important steps were taken towards greater understanding of poverty and poverty reduction (brought forward even more vigorously as the main objective of Dutch aid). International thinking on development was an important guiding force for many of these steps, as may be seen in the fields of the conceptualisation of poverty, the sector-wide approach, the country concentration efforts and private-sector development. In other fields, specific developments (as in the area of the private aid channel) should be judged as being guided more by internal Dutch considerations. All in all, however, the Netherlands does not deviate from mainstream thinking on development co-operation and in fact never has done so. In some cases, as with the sector-wide approach and the efforts towards coherence, it even seems that the Netherlands is aiming to be at the top of the class.

This also means that as a donor it is susceptible to the same kind of problems as other donors experience. This holds true, in particular, for the operationalisation of its policy. Whereas on paper Dutch policy seems to be quite logical and consistent, with a strong link between ownership, sector approaches, co-ordination and coherence, the question remains whether this logic also extends to daily practice. The DAC [*OECD, 2001*] has already been quoted as concluding that the Netherlands lacks a clear understanding of the dimensions of poverty, of priority-setting, and of organisation. Added to this is that 'the internal operations ... are not yet organised around the concept of poverty' [*ibid.: 1–49*]. In effect, one could say that the 1990s did provide a somewhat better elaborated poverty policy, but that this policy as yet provides insufficient guidelines for implementation. The emphasis on (internal) coherence only contributes to the problem, as it brings in the need for a poverty focus outside the DGIS (while it is already quite problematic to arrive at a poverty focus within the DGIS itself). In addition, the decentralisation to the embassies is not necessarily an advantage in this, particularly in view of the fact that communication between The Hague and the embassies is far from perfect. Moreover, due to frequent transfers of personnel, the institutional memory at field level is even more restricted than at headquarters.

In the meantime, the reliance on PRSPs as a proxy for ownership and as a guiding force for Dutch development co-operation has led to the final dismissal of the idea of setting up country strategy plans. In theory, this is quite consistent as it is seen as odd to opt for recipient ownership, on the one hand, and at the same time draft your own policy as a donor. In practice, however, the PRS process is not necessarily running as smoothly as was hoped. The reliance on the PRSPs even creates the impression of a 'wait-and-see' attitude at embassy level. It would perhaps be wiser to search for a clear combination of ownership and donorship, in which the latter is not excluded from thinking through its own policy at country level.

REFERENCES

Bosch, F. van den, 1998, 'Trends in het ontwikkelingssamenwerkingsbeleid – De ontwikkeling van de particuliere sector – beleid, organisatie en programma's in de DAC donorlanden' (Trends in development co-operation policy – the development of the private sector – policy, organisation and programmes in DAC donor countries), The Hague: Ministry of Foreign Affairs (internal report).

Box, L. and S. Tromp, 2001, 'Variaties op een thema: Nederlands beleid betreffende Europese ontwikkelingssamenwerking' (Variations on a theme: Dutch policy with regard to European development co-operation), in Schulpen (ed.).

Commissie-Box, 2002, *Breed uitgemeten – advies van de Commissie Medefinancieringsprogramma-breed inzake toetreding en toewijzing van middelen 2003-2006* (Wide measured – Advice of the Commission Co-financing programme-wide with regard to accession and allocating funds 2003–2006), The Hague: Commissie Medefinancieringsprogramma-breed.

Dalmeijer, R. and L. Schulpen, 1999, 'Armoede (m/v), particuliere sector en ontwikkelingssamenwerking – een inventariserend, beleidsvoorbereidend onderzoek' (Poverty (m/f), private sector and development co-operation – policy preparing research), Zeist/Nijmegen: internal DGIS report.

DGIS, 1990, *A World of Difference – a new framework for development co-operation in the 1990s*, The Hague: DGIS, September.

DGIS, 1993, *Een wereld in geschil – de grenzen van ontwikkelingssamenwerking verkend* (A World in Dispute), The Hague: DGIS.

DGIS, 1996, *Hulp in uitvoering – ontwikkelingssamenwerking en de herijking van het buitenlands beleid* (Aid in Progress – Development co-operation and the reverification of foreign policy), The Hague: DGIS.

DGIS, 1999a, *De kwaliteit van de VN als kanaal voor ontwikkelingssamenwerking* (The quality of the UN as a channel for development co-operation), The Hague: DGIS.

DGIS, 1999b, *De kwaliteit van de Europese hulp* (The Quality of European Aid), The Hague: DGIS.

DGIS, 2000, *Ondernemen tegen armoede – notitie over economie en ontwikkeling* (In business against poverty – note on economy and development), The Hague: DGIS, October.

DGIS, 2001a, *Poverty Reduction: Dutch Policy in Brief* (Development co-operation: a global task), The Hague: DGIS, December.

DGIS, 2001b, *Beleidskader MFP-breed* (Policy framework MFP-wide), The Hague: DGIS, May.

DGIS, 2002a, 'Ontwikkelingssamenwerking en Civil Society, bestedingen Particulier Kanaal (exclusief noodhulp en internationaal onderwijs en onderzoek) – Statistisch overzicht' (Development Co-operation and Civil Society, spending private channel (exclusive of emergency aid and international education and research) – statistical overview), The Hague: DSI/MY, February.

DGIS, 2002b, *Notitie beleidscoherentie ontwikkelingssamenwerking-landbouw* (Policy coherence development co-operation-agriculture), The Hague: DGIS/EZ.

DGIS, 2003a, 'Development cooperation long-term outlook', letter to the House of Representatives of the States General, 17 June.

DGIS, 2003b, *Mutual Interests, Mutual Responsibilities – Dutch development co-operation en route to 2015*, The Hague: DGIS, October.

Dijk, M.P. van, 2001, 'Hulp en/of handel versus trade rather than/no aid' (Aid and/or trade versus trade rather than/no aid), in Schulpen (ed.).

Forster, J. and O. Stokke (eds), 1999, *Policy Coherence in Development Co-operation*, London: Frank Cass (EADI Book Series 22).

Goei, K., 2002, 'Agnes van Ardenne: Miss Coherentie en makelaar in armoedebestrijding' (Agnes van Ardenne: miss Coherence and broker in poverty reduction), *Hivos Magazine*, Vol. 8, No. 4.

GOM, 1995, *Met het oog op kwaliteit – eindrapportage over de maatregelen van Cebemo, Hivos, Icco en Novib naar aanleiding van de Impactstudie* (With an eye on quality – final report on the measures taken by Cebemo, Hivos, Icco and Novib following the Impact Study), Oegstgeest: GOM, March.

Hoebink, P., 1988, *Geven is nemen: de Nederlandse ontwikkelingshulp aan Tanzania en Sri Lanka* (To give is to take: Dutch development aid to Tanzania and Sri Lanka), Nijmegen: Stichting Derde Wereld Publikaties.

Hoebink, P., 1995, *De effectiviteit van de hulp: een literatuuroverzicht van macro- en microniveaus* (The effectiveness of aid: a literature review of macro and micro levels), The Hague: DVL/OS.

Hoebink, P., 1999a, 'The Humanitarianisation of the Foreign Aid Programme in the Netherlands', *European Journal of Development Research*, Vol. 11, No. 1, June.

Hoebink, P., 1999b, 'Aid and Trade Coherence and Development Policies in the Netherlands', in Forster and Stokke (eds).

Hoebink, P., 2002a, 'Van wervelwind tot nachtkaars? Vier jaar Eveline Herfkens op ontwikkelingssamenwerking' (From whirlwind to nighlight? Four years Eveline Herfkens on development co-operation), *Internationale Spectator*, Vol. LVI, No. 4, April.

Hoebink, P., 2002b, 'Beoordeling van het rapport "Breed Uitgemeten Advies van de Commissie Medefinancierings programma-breed inzake toetreding en toewijzing van middelen 2003–2006"' (Judging the report 'Wide measured – Advice of the Commission Co-financing programme-wide with regard to accession and allocating funds 2003–2006'), Nijmegen: CIDIN (unpublished report).

Hoebink, P. and L. Schulpen, 1998, *Netherlands' Aid Policies for Poverty Reduction*, ODI Working Paper 115, London: Overseas Development Institute, December.

Hoebink, P. and F. van der Velden, 2001, 'Van tropenarts en vrijwilliger tot institutieopbouw: Nederlandse technische assistentie in verleden en toekomst' (From doctor in the tropics and volunteer to institution building: Dutch technical assistance in past and future), in Schulpen (ed).

IOV, 1988, *Uitbesteed, goed besteed? – een onderzoek naar het uitbestedingsbeleid voor bilaterale ontwikkelingsactiviteiten met nadruk op rurale en regionale ontwikkeling* (Contracting out – research on the contracting out policy for bilateral development activities with an emphasis on rural and regional development), IOV Report No. 245, The Hague: IOV/DGIS.

IOV, 1992, *The Sector Programme for Rural Development: A Programme Evaluation with Special Reference to Indonesia, Sudan, Rwanda, Tanzania and Nicaragua*, The Hague: IOV/DGIS.

IOV, 1994, *Evaluatie van de Nederlandse hulp aan India, Mali en Tanzania (samenvattend rapport)* (Evaluation of Dutch aid to India, Mali and Tanzania), The Hague: IOV/DGIS.

Jepma, C.J., 1995, *On the Effectiveness of Development Aid*, The Hague: DVL/OS.

Kruijt, D. and M. Vellinga (eds), 1983, *Ontwikkelingshulp getest: resultaten onder de loep* (Development aid tested: results scrutinised), Muiderberg: Coutinho.

Kruijt, D., K. Koonings and M. Vellinga, 1983, 'De Nederlandse ontwikkelingsgemeenschap' (The Dutch development community), in Kruijt and Vellinga (eds).

NCDO, 2002, 'NCDO-draagvlakonderzoek 2002' (NCDO public support research 2002), Amsterdam: NCDO.

OECD, 1989, *Development Co-operation in the 1990s – Efforts and Policies of the Members of the Development Assistance Committee (1989 Report)*, Paris: OECD.

OECD, 1992, *Development Co-operation – Efforts and Policies of the Members of the Development Assistance Committee (1992 Report)*, Paris: OECD.

OECD, 1995, *Development Co-operation – Efforts and Policies of the Members of the Development Assistance Committee (1994 Report)*, Paris: OECD.

OECD, 1998a, *Development Co-operation Review Series – The Netherlands 1997*, No. 24, Paris: OECD/DAC.

OECD, 1998b, *Development Co-operation – Efforts and Policies of the Members of the Development Assistance Committee (1997 Report)*, Paris: OECD.

OECD, 2001, 'Development Co-operation Review of The Netherlands', *DAC Journal 2001 – International Development*, Vol. 2, No. 3: I–7 – I–69.

OECD, 2002, 'Development Co-operation – Efforts and Policies of the Members of the Development Assistance Committee (2001 Report), *The DAC Journal Development Co-operation*, Vol. 3, No. 1, Paris: OECD.

OXFAM, 2002, *Rigged Rules and Double Standards: trade, globalisation, and the fight against poverty*, London: OXFAM.

Pietilä, T. (ed.), 2000, *Promoting Private Sector Development – Issues and Guidelines for Aid Agencies*, Helsinki: Institute of Development Studies, Policy Papers 1/2000.

Quarles van Ufford, P., 1991, 'Kwaliteitszorg in de ontwikkelingssamenwerking – Novib in de jaren negentig' (Quality care in development co-operation – Novib in the 1990s), in Quarles van Ufford and Wils (eds).

Quarles van Ufford, P. and F. Wils (eds), 1991, *Kwaliteitszorg in de jaren '90 – twee lessen aan Novib voor haar beleid* (Quality care in the 1990s – two lessons for Novib's policy), The Hague: Novib.

Schulpen, L., 1999, Private *Sector* Development, Poverty Reduction and Bilateral Donors: The Case of the Netherlands, Nijmegen: CIDIN (unpublished report).

Schulpen, L., 2000, 'Private Sector Development and Poverty Reduction: An Inbuilt Incoherence?', in Pietilä (ed.).

Schulpen, L. (ed.) 2001, *Hulp in ontwikkeling – bouwstenen voor de toekomst van internationale samenwerking* (Aid in development – building blocks for the future of international co-operation), Assen: Van Gorcum.

Schulpen, L. and P. Gibbon, 2002, 'Private Sector Development – policies, practices and problems', *World Development*, Vol. 30, No. 1.

Schulpen, L. and P. Hoebink, 1999, 'DAC Scoping Study of Donor Poverty Reduction Policies and Practices: The Netherlands', Nijmegen: Third World Centre (unpublished report for the DAC scoping study).

Schulpen, L. and P. Hoebink, 2001, 'Ontwikkelingssamenwerking via particuliere ontwikkelingsorganisaties – de MFO's in perspectief' (Development co-operation via private development organisations – the MFOs in Perspective), in Schulpen (ed.).

SER, 1997, *Advies De particuliere sector in internationale samenwerking* (Advisory report on The Private Sector in International Co-operation), The Hague: Sociaal-Economische Raad (November).

SIM (Stuurgroep Impactstudie Medefinancieringsprogramma), 1991, *Betekenis van het medefinancieringsprogramma: een verkenning* (Meaning of the cofinancing programme – an exploration), Utrecht: Libertas.

World Bank, 1998, *Assessing Aid – What Works, What Doesn't, and Why – a World Bank Policy Research Report*, New York: Oxford University Press.

13

Norwegian Aid Policy: Continuity and Change in the 1990s and Beyond

OLAV STOKKE

I. INTRODUCTION

What concerns are forming and driving Norway's aid policy – basically altruistic ones or (enlightened) self-interests of various kinds? What does Norway want to achieve by way of aid and why? And how does it proceed? To what extent is the policy formed through endogenous policy processes or through adaptations to internationally driven processes? To what extent and how have the transformation of the international system which took place in the early 1990s and the evolving shifts in international development and aid paradigms during recent years (see Chapter 1 of this volume) affected the main features of Norway's policy towards developing countries: towards international convergence?

These are core questions to be addressed in this chapter. The answers are sought in the main features of Norway's aid policy – the main justifications, the stated objectives and guidelines and how they have been implemented. Emphasis will be given to the stated policy, because intentions and ambitions are most clearly expressed there. However, the follow-up in terms of volume performance, the distribution of ODA among various aid channels, the main recipients and purposes will also be part of the analysis, as will the institutional framework.

Norwegian aid policy has been formalised through a series of government White Papers and proposals and Parliament's response to these initiatives.[1] These major policy documents, reflecting endogen-

---

Note of acknowledgement. A draft of this chapter benefited from the scrutiny of the EADI workshop in Nijmegen in July 2002. I am grateful to participants, Helen O'Neill and Lau Schulpen in particular, for valuable comments and to Paul Hoebink for comments on a later draft. The usual disclaimers apply.
1. The first comprehensive White Paper on aid was presented to Parliament in 1961 by the Gerhardsen government (Labour). The next major review came in 1967 by the

ous processes although also responding to impulses from outside, will constitute the main sources.²

In the following sections the main features of Norway's development co-operation policy will be outlined and discussed, based on the norms established by these major policy papers and the way they have been implemented. A brief presentation of the patterns up to the end of the 1980s will constitute a baseline for a description and analysis of the main policy changes of the 1990s and beyond. The focus will be on justifications and overall objectives (section II), components of development and aid strategies (section III), guidelines (section IV), volume targets and performance (section V) and the aid administration (section VI), before the concluding section (section VII).

Borten government (non-socialist coalition). In 1972, the Bratteli government (Labour) introduced the White Paper that has probably wielded the strongest influence on aid policy followed, in 1975, by a White Paper on North–South relations (the NIEO). In 1984, the Willoch coalition government (Conservatives, Christian Democrats, the Centre Party) undertook a new major review of aid policy, with the first-ever Minister of Development Co-operation, Reidun Brusletten (Christian Democrat), in charge. Three years later, the Brundtland (II) minority government (Labour) added to this White Paper, before Parliament determined the policy. In 1992, the Brundtland (III) minority government presented a new major White Paper, emphasising broader North–South relations, including aid. The government then appointed a commission to consider these issues, which reported in 1995. On this basis, the Brundtland government came up with a new White Paper later that year and Parliament set the policy in 1996. Since then, the major policy proposals have been introduced in Parliament in the annual statement on development policy by the Minister of International Development and followed up in the government's budget proposal later on. On 30 April 2004, the Bondevik II government presented a new major White Paper dedicated to 'A Joint Fight Against Poverty'.

Since 1990, there have been a series of (minority) governments: Brundtland III (Labour) (1990–96), Jagland (Labour) (1996–97), Bondevik I (Christian Democrats, Centre Party, Liberals) (1997–2000), Stoltenberg (Labour) (2000–1) and Bondevik II (Conservatives, Christian Democrats, Liberals) (2001–). In the Bondevik (Christian Democrat) governments, a Christian Democrat was (is) Minister of International Development.

The Minister of International Development has, with only one exception, been a woman. Kirsti Kolle Grøndal (Labour) (1986–89) succeeded Ms Brusletten, and was followed by Mr Tom Vraalsen (Centre Party) (1989–90), Ms Vesla Vetlesen (Labour) (1990–92), Ms Kari Nordheim-Larsen (1992–97), Ms Hilde Frafjord Johnson (Christian Democrat) (1997–2000; 2001–) and Ms Anne Kristin Sydnes (Labour) (2000–1).

2. This chapter draws on earlier studies of Norway's aid policy [*among others, Stokke, 1975, 1979, 1984, 1989, 1992ab*] as well as a monograph in the process [*Stokke, forthcoming*].

In mid-2003, three volumes on the the history of Norway's development assistance appeared, resulting from a research project funded by the Ministry of Foreign Affairs and organised by the Research Council of Norway. It involved several historians over a three-year period – Jarle Simensen [*2003*] covered the period 1952 to 1975, Arild Engelsen Ruud and Kirsten Alsaker Kjerland [*2003*] the period 1975 to 1989, and Frode Liland and Kirsten Alsaker Kjerland [*2003*] the period 1989 to 2002. For an excellent review of these volumes from outside Norway, see Wohlgemuth [*2003*].

II. JUSTIFICATIONS AND OVERALL OBJECTIVES

1. Justifications

The justifications given for a policy by the political authorities (government and Parliament) represent a primary source where the motives are to be found. Hidden agendas will always be there and ideal (altruistic) arguments may camouflage selfish interests. The stated motives are, therefore, to be modified and contradicted by the policy implemented. If the stated justifications are to be trusted, what are the motives driving the aid policy?

In an earlier study, I concluded that successive governments and Parliament have laid emphasis on altruistic motives for development co-operation. The most powerful incentive over the years has been the moral obligation to help human beings in distress. Development assistance has been considered an instrument to promote international common goods such as international stability and peace and internationally shared interests such as economic stability and growth, improved health, and environmental concerns. Although self-centred interests have been discernible, they have been of secondary importance [*Stokke, 1989: 169ff.*]. This assessment holds true also for the 1990s and beyond, although instrumental arguments came strongly to the fore in the mid-1990s.

From the start, development assistance was considered an instrument in promoting *peace*: economic and social progress was considered a precondition for peace and stability [*Engen Committee, 1961: 5, 7*]. The main White Papers up to the late 1980s repeated this argument.[3] However, the ethical imperative, to assist those in need, has been the predominant justification. Governments based on different political parties have repeatedly emphasised altruistic motives.[4]

---

3. The argument has followed two main tracks: aid channelled through the UN would help strengthen the capability of the UN system to resolve conflicts, and development assistance would in itself contribute to stability in a world characterised by growing interdependence.
4. The Engen Committee [*1961: 7*] stated that the motives were based on 'the principle of equality of human beings and a feeling of solidarity with all countries and races.' In the 1975 White Paper, the government (Labour) stated that the moral obligation to help human beings in distress was the most powerful incentive behind Norway's aid contributions, and this obligation did not stop at national borders: 'our people firmly adhere to such fundamental precepts of human values as the Christian philosophy of universal brotherhood and a general feeling of solidarity between all human beings' [*NORAD, 1975: 13–14*]. In 1984, the coalition government (Centre Party, Christian Democrats and Conservatives) argued in the same vein, stating that 'Norwegian aid originates from humanitarian and Christian traditions and the philosophy of solidarity. Development assistance is an extension to the international

In the early 1990s, the international environment for aid changed dramatically (see Chapter 1). To what extent and how did the transformation of the international system and the emerging Washington–New York consensus affect the justifications given for Norwegian aid?

In the major White Papers of the 1990s and beyond, the old justifications were repeated. However, the emphasis gradually changed, and adjusted to the post-Cold War international discourse: security concerns, as related to an extended security concept, and environmental concerns came to the fore, although international solidarity remained a fundamental pillar. The 1992 White Paper made a stronger link between the effects of aid in recipient countries and the long-term and short-term interests of the industrialised countries (Norway included) than previous White Papers.

In 1995, the North–South Commission took human rights as the point of departure.[5] Nine years later, the Bondevik II government followed up this challenge in the 2004 White Paper. The main thrust of the (Labour) government's 1995 White Paper, in contrast, was on the instrumentality of international solidarity: Norway had won the trust of other countries, due to solidarity expressed through many years of Norwegian development assistance, humanitarian aid and political support for poor countries, the government argued. 'This has given Norway influence in international co-operation. The Government wishes to maintain and strengthen the instruments Norway has at its disposal to demonstrate solidarity with the weakest members of the world community' [*Report No. 19, 1995: 5*].

---

level of the efforts to create social justice, characteristic of the Norwegian welfare state' [*Report No. 36, 1984: 19*].

5. The notions of development and basic human rights were considered from a human rights perspective. Development was understood as 'a positive process in society, where the basic needs and rights are secured for all, at the same time as new opportunities for further improvements in living conditions and quality of life are created' [*NOU, 1995: 5: 94*]. In 1996, Parliament came out with an almost identical definition [*Recommendation No. 229, 1996: 8*]. The commission summarised the motives driving Norway's engagement for development issues abroad: 'Solidarity, justice, equality and benevolence [nestekjærlighet] have for years been given as basic motives for Norway's development assistance, the long-term development co-operation as well as disaster aid. Lately, a concern for the environment of the Earth and future generations has also entered as an important justification for why we engage ourselves. Our values are increasingly related to both human beings and the environment, given a perspective where the Earth and all living beings are integral parts of a whole. The increasing ecological and social crisis of the global society affects all and has to be solved in common. There is here a question *about both common interest and self-interest*' [*NOU, 1995: 5: 94–102; quotation p. 99, emphasis in original*].

One important reason given for Norway's involvement in North–South issues was that sustainable development in the South was of common interest to people in both the North and the South. Co-operation to achieve development must be based on the recognition of a common interest in promoting certain basic goals. This applied to trade, aid and other international co-operation, the government argued, adding that the main task for Norway's foreign policy was 'to promote Norway's interests in relation to foreign countries, which includes contributing to the world community finding common solutions to international problems' [*ibid.: 10*]. However, Norway's interests, as perceived, were the decisive ones.[6] Parliament, in turn, laid the emphasis on long-term common interests[7] – bringing the justification back to square one.

In the 2004 White Paper, the development policy was anchored in a human rights perspective. Fighting extreme poverty was considered the major challenge of our time.[8]

*

During most of the 1980s and early 1990s, the government took pride in Norway being the OECD country that provided the most development assistance in relative terms. In 2001, the Minister for International Development, Anne Kristin Sydnes (Labour), gave this pride an extra twist: the development policy represented an important part of the image of Norway that was formed outside the country, where Norway's oil fortune and welfare were well known. 'The development policy also counteracts a picture of Norway as a country where self-interest, materialism and egoism rule the ground' [*Proceedings of Parliament, 2000: 8 May*].

---

6. These interests 'are linked to a number of factors which affect our lives, national identity and international position in various ways. In the Government's view, an active South policy is an important element of our overall approach to save these interests' [*Report No. 19, 1995: 10*].

7. The majority of the Parliamentary Committee on Foreign Affairs (all members, with the exception of one, representing Frp) 'underscore long-term common interests as the fundamental motivation for Norway's aid policy towards the developing countries in the South. Aid must primarily be about securing rights for the weakest, founded on the basic philosophy that a better distribution of the benefits will create a better world to live in for all'. It found the fact that more than one billion people lived in absolute poverty the most important global challenge. 'From a Norwegian perspective it is of crucial importance to work for greater international concern for the fight against poverty' [*Recommendation No. 229, 1996: 7–8*].

8. 'Development policy is not charity. It means to contribute to realising human rights for all. ... Norway's development policy, the way it is expressed in this White Paper, rests on the fundamental precondition that all human beings have the same human value. Also development policy, therefore, becomes a rights agenda. It also means that human rights should be promoted – the economic, social and cultural rights as well as the civil and political ones. Fighting poverty is a case about justice' [*Report No. 35, 2004: 5*].

In the stated policy, the moral obligation to assist poor developing countries in improving the conditions of their poorest people emerges as the driving force in Norway's development assistance. Although the emphasis on international solidarity is basically altruistic, this justification has been framed as the enlightened self-interest of a small country. Whereas international common goods – such as peace and international economic and political stability – were given the main emphasis in the earlier decades, the instrumental aspects of development assistance in a foreign policy context came increasingly to the fore in the 1990s.

2. Overall Objectives

To what extent have the overall objectives set for aid been attuned to the ideal norms reflected in the justifications? The general direction was pointed out in 1961 to 1962: aid should promote economic, social and cultural development in developing countries [*Report No. 23, 1961; Recommendation No. 75, 1962*].[9] Ten years later, fairer income distribution was added. Poverty alleviation and social justice emerged as primary tasks. A long-term goal was also given, namely, '[c]o-operation not involving direct aid, but rather based on mutual exchange of goods and services between the countries and increased cultural and political intercourse and co-operation with the people of the partner countries' [*Report No. 29, 1972; Recommendation No. 135, 1973*].

A decade later, the Willoch (coalition) government confirmed these concerns, adding two dimensions: democratic development within (recipient) countries and liberation from economic dependence in external relations [*Report No. 36, 1984*].[10] These overall objectives were agreed to, in 1987, by the Brundtland II (Labour) government, which added that they might be contradictory. It emphasised five subsidiary objectives: responsibility in the administration of natural resources and the environment; economic growth; improvements in the conditions of the poorest, particularly women; promotion of

---

9. The Engen Committee [*1961: 13*] had laid emphasis on 'economic and social development'. In hindsight, it is interesting that the government added that aid 'should be provided without the intention of promoting particular political, economic or religious interests' [*Report No. 23, 1961: 1*].

10. Poverty alleviation was emphasised more strongly than ever. The overall objective was 'to assist in creating lasting improvements in the economic, social and political conditions of people in developing countries. Development assistance shall be used in a way that maximises its development effects for the poor sections of the population. Aid shall in the first place be directed to the poorest developing countries and be provided in a form that creates as little dependence on continued aid as possible' [*Report No. 36, 1984: 20*].

social, economic and political human rights; and the promotion of peace between nations and regions. Promotion of peace, previously among the motives, was established as an objective in its own right [*Report No. 34, 1987*].[11]

To what extent and in what way did the post-Cold War international situation and the international convergence on development aims in the 1990s affect these overall objectives? One feature is characteristic of the major White Papers of the 1990s: they had a broader focus than development co-operation, aiming at formulating overall objectives for Norway's total policy towards developing countries. In such a setting, it became important to strike the connection between the overall objectives for development co-operation and those set for the broader co-operation.

The 1992 White Paper confirmed the overall objectives set for Norway's development co-operation in 1987, which Parliament endorsed [*Report No. 51, 1992: 27; Recommendation No. 195, 1993: 37–8*]. The 1995 White Paper, in contrast, anchored the objectives in the context of Norwegian foreign policy interests. It provoked a heated debate that involved the very identity of aid (international solidarity vs national interest). The Christian Democrats even issued a counter-report [*Christian Democrats, 1995*]. Parliament restored the old order [*Recommendation No. 229, 1996: 1*].

According to the 1992 White Paper, the objectives for Norway's North–South policy and those set for development co-operation were basically the same: 'to contribute to a more rational and just distribution and use of resources internationally and to contribute to development and increased welfare in developing countries'.[12] The North–South Commission agreed and so did the government [*NOU, 1995: 5: 122; Report No. 19, 1995: 10*].

---

11. Parliament agreed to the overall objectives as proposed by the Willoch (coalition) government. The five subsidiary objectives proposed by the Labour (minority) government should be assessed according to the extent to which they would contribute in fulfilling the overall objectives; alone they could not justify aid projects [*Recommendation No. 186, 1987*]. At the end of the 1980s, therefore, promoting economic, social, cultural and political development, with particular emphasis on social justice and poverty alleviation, constituted the core of the 'traditional' objectives set for development co-operation.

12. The following distinction was made between the two policy areas: '*The North–South policy* directs itself mainly towards the international framework conditions for economic and social development, while *the development assistance* aims directly to facilitate conditions in order to create economic growth and social welfare in the separate developing countries. The government therefore consider our North–South policy and our development co-operation policy to be two parts of a greater whole' [*Report No. 51, 1992: 37*].

Parliament agreed that Norway's aid policy had to be considered in a broader North–South context connected, in particular, to international conditions that affected the economy, debt and trade. The need for coherence in the policy towards developing countries should have consequences not only for the aid policy, 'it must have consequences for what we are doing in foreign policy and trade policy.' The long-term and altruistic objective of the development co-operation policy should remain an important point of departure for Norway's policy towards the countries of the South [*Recommendation No. 229, 1996: 7–8*].

The White Papers of the 1990s developed the five subsidiary objectives identified in the 1987 White Paper. The order of their presentation, however, indicates changes in priorities. In the 1992 White Paper, 'environmentally sustainable development, economic growth and social justice' came first. Development that was environmentally sustainable was set as a crucial objective. Economic growth was considered a necessary precondition for developing countries to be able to reduce poverty, improve the welfare of the population, and become independent of continued aid. Then followed the promotion of human rights and peace, equal opportunities for women and men in all areas of society, and humanitarian relief [*Report No. 51, 1992: 27, 175*].

The 1995 White Paper changed this order, most probably reflecting a change also in the relative importance given to the objectives. Efforts to promote peace, human rights and democracy came first, followed by economic and social development for poor countries and population groups, sound management of the global environment and biological diversity, promoting equal rights and opportunities for men and women, and preventing and alleviating distress arising from conflict and natural disasters.

The following year, Parliament set the priorities, placing the traditional objectives for development assistance at the top of the list [*Recommendation No. 229, 1996: 16*].[13] During subsequent years, the

---

13. The majority (not including the Progress Party (Frp)) set the main objectives in the following order:
    1. To fight poverty and contribute to lasting improvements in living conditions and the quality of life, and in this way promote more social and economic development and justice nationally, regionally and globally. Employment, health and education are areas that must be centrally placed in order to attain such a development.
    2. To contribute to the promotion of peace, democracy and human rights.
    3. To promote a responsible administration and utilisation of the environment and biological diversity of the Earth.
    4. To contribute to preventing and alleviating distress in connection with conflicts and natural disasters.

1996 accord on aid policy between the political parties, with particular reference to the revival of the previous 'centre–left' alliance, was almost ritually confirmed in the annual statements on development policy and the discussion of these statements in Parliament. Towards the end of the 1990s, the internationally agreed objectives [*OECD, 1992, 1996; UN, 2000*] were integrated into the stated policy, with particular reference to poverty eradication [*Proceedings of Parliament, 1998: 5 May; MFA, 2000: 32; Proposition No. 1, 2001: 60–1*]. In the 2004 White Paper, the government followed up. It anchored its policy in the UN millennium development goals and emphasised the fight against poverty [*Report No. 35, 2004: 19–35*].

*

The 1990s brought new concerns to the fore and added emphasis to the old ones. The overall objectives were increasingly adapted to internationally agreed objectives, with a primary concern for poverty eradication, much in line with the international discourse. They were adapted to the international security situation of the post-Cold War era – emphasising conflict prevention and humanitarian assistance. Liberal political values, in terms of human rights, democracy and good governance, also came to the fore – much in line with international trends. Both the traditional objectives and the new concerns are well attuned to the motives given for development assistance, with a particular emphasis on improving conditions for poor developing countries and poor and disadvantaged people.

III. MAIN STRATEGIES

The development philosophy behind Norway's development assistance has changed over time. Strands belonging to various development paradigms have been intertwined. Until the publication of the1984 White Paper, the main policy documents had not related their policy directly to development theory or development paradigms, although reflecting shifting paradigms. However, humanitarian assistance aside, two main strategies and a further policy thrust may be inferred from the White Papers and from aid implementation up to the late 1980s, namely:

– *a welfare strategy* that sought to assist in improving welfare within developing countries with an emphasis on the improve-

---

5. To contribute to the promotion of equal rights and opportunities for women and men in all areas of society.

ment of social services and such services as will directly improve the ability of local communities to meet their own basic needs (including support for rural development, improvements within the primary sectors (agriculture, forestry, local fisheries), improvement of drinking water in rural and urban areas, education, health and nutrition, etc.);
- a *strategy for industrial development and the promotion of trade* that sought to assist in building up the productive capacity and exports of developing countries, with an emphasis on the industrial sector and export-oriented production, export promotion, improvement of the infrastructure (telecommunications, main roads and transportation systems), and energy, particularly water energy.
- There was also a *cluster of policies designed to advance particular Norwegian (private-sector) interests*. Although operating within the context and under the guise of development co-operation, their primary objective was different from those set for development assistance.

The two main strategies co-existed: although they differ in emphasis, they might be combined within the same programme. In 1987, Parliament attempted to combine the two by emphasising that Norway's development assistance should 'give strong emphasis to increasing the productivity of the poor part of the population', adding that Norway should be a front runner in giving priority to an ecologically responsible basic-needs strategy for development assistance [*Recommendation No. 186, 1987*].[14]

The changed international environments of the 1990s affected the development strategy quite profoundly. As already noted, the White Papers of the 1990s discussed development beyond development co-operation, involving North–South relations such as trade policy, investments and the environment, and also the peace–conflict dimension. The debt burden and ways to overcome this crisis were high on the agenda of the 1980s. Structural adjustment and *economic* liberalism emerged as main recipes – with which the Norwegian government only reluctantly and at a late stage had associated itself. In the 1990s, *political* liberalism came increasingly to the fore, involving

---

14. The two main strategies identified, along with the cluster of policies (involving a set of mechanisms) with the primary objective to advance Norwegian private-sector interests, are described in greater detail and analysed elsewhere (see, in particular, Stokke [*1989: 174–88; forthcoming: Ch. 2*]).

democracy, human rights and good governance – about which the government had few if any inhibitions.

In its description and assessment of the changing international reality, the 1992 White Paper made extensive use of reports emerging in the early 1990s, in particular reports by the World Bank, the UNDP and the OECD (plus the WCED 1987 report) – indicating the strong influence they exerted on policy formation. The government argued that a broad international consensus had emerged both on development objectives and on the ways in which they might be attained. This perception was echoed both in the report of the North–South Commission and in the 1995 White Paper. It represented a new trend in Norway's policy within this area.

The 1992 White Paper removed the NIEO strategy from the national political agenda, although respect was paid to the aspirations that had been driving the policy.[15] The strategy had failed because a precondition on which it rested was not met, namely that fundamental systemic change could be achieved through international negotiations. Several industrialised countries disagreed with the objectives set, among them the most powerful, including the United States, increasingly so after new governments, signalling a fresh ideological course for the 1980s, had taken over in Washington and London, the White Paper argued.[16]

The major policy papers of the 1990s adopted a holistic perspective on development and insisted on policy coherence. The various components of the strategy outlined for the 1990s and beyond, as sketched out below, were systematically interwoven with each other.

(i) Promotion of economic liberalism and economic growth. According to the 1992 White Paper, the structural adjustment programmes had two basic objectives, namely to stabilise the economic situation in developing countries by restricting the demand side and through balance-of-payments support, and to change their economic policy 'in order to ensure a more *sustainable* economic development in the longer term'. As noted by the government, this approach was in stark contrast to the policy pursued two decades earlier, emphasising economic self-sufficiency, non-intervention and a belief in the state as the

---

15. Norway's NIEO policy was based on the 1975 White Paper, as broadly endorsed by Parliament [*Report No. 94, 1975; Recommendation No. 192, 1976*]. Together with a group of like-minded states, the Norwegian government was spearheading this policy among the OECD countries. The policy is presented and discussed in Hveem [*1977, 1989*], Løvbræk [*1990*], Stokke [*1979, 1980*], and Stokke, ed. [*1980*].
16. For a discussion of Norway's NIEO policy in the early 1980s and the way the 1992 White Paper put an end to it, see Stokke [*forthcoming: Ch. 4*].

dynamo in the development process. With reference to the critique of the reform programmes, the government committed itself to 'adjustment with a human face' [*Report No. 51, 1992: 49*]. Parliament, in contrast, was more critical of the conformity of the adjustment policy [*Recommendation No. 229, 1996: 14–15*].[17]

The major policy papers of the 1990s saw economic growth in poor developing countries as a precondition for their ability to tackle the major problems confronting them – poverty and environmental problems, in particular. Economic growth was regarded as dependent on predictable framework conditions, just distribution and investments in human capital – health and education. Economic growth and social development were seen as mutually dependent and reinforcing. Trade and environmental concerns were also integrated into the growth discourse, as was the debt problem. Private-sector development was increasingly highlighted.[18] However, as indicated in Table 13.1, the money followed other priorities: in the 1990s, the shares allocated to economic infrastructure, agriculture, industry and other production declined drastically.

(ii) Promotion of political liberalism: democracy, human rights and good governance. In the 1990s, liberal political values came to the fore as objectives to be promoted by means of development assistance. They were also part of a strategy to attain development as so defined. This represented a new policy course, and was recognised as such by the government,[19] although the values had also been part

---

17. The ambivalence is reflected in policy statements combining an added emphasis on market mechanisms with a social liberal/social democrat chord that related to the development strategies of the 1970s, although now given an economic growth justification: 'Poverty reduction, satisfaction of basic needs, and improvements of conditions for women were ... seen as contributions to economic growth and not only as issues related to justice' [*Report No. 51, 1992: 49–50*].

18. A plan for debt reduction, directed towards the poorest, most debt-ridden countries, was presented in the budget proposition for 1999 [*Proposition No. 1, 1998: 166–71, and (unprinted) enclosure*]. In order to engage the Norwegian private sector more strongly in development activities in developing countries, NORFUND was established in 1997, followed by a strategy for Norwegian support for private-sector development in developing countries [*MFA, 1999a*]. In the first half of 1999, the government also set up a Forum for Productive Sector Development in the South with representatives from Norwegian industry and business, institutions and voluntary organisations.

19. As stated in the 1992 White Paper, '[d]emocracy and human rights, social equality and popular participation in the development process, good government and effective administration are all elements that have been given stronger emphasis in the 1980s. Earlier, such issues were rejected as interference in the internal affairs of developing countries, but clear recognition has gradually emerged: positive development within these areas constitutes an important incentive for economic effectiveness

of the previous policy. Democracy, human rights and economic and social development were closely interlinked.[20]

This policy was continued. In 1999, the Minister presented a four-pillar strategy to meet the two main challenges in the countries concerned, namely *popular mobilisation* and the building up and *consolidation of new, fragile democracies*.[21] A plan of action for *human rights* was also presented to Parliament in 1999, addressing issues related to human rights in a broad perspective, including the integration of four central human rights conventions into Norwegian law. All development assistance was cast in a human rights perspective: development and human rights were considered two sides of the same coin. Parliament considered the White Paper one year later [*Report No. 21, 1999; Recommendation No. 23, 2000*].[22] In the stated policy, human rights-based aid became increasingly important.

Fighting corruption was a central aspect of the government's good government pillar. In the 1999 statement on development, the Minister launched 'a new offensive' in the *fight against corruption* [*MFA, 1999b: 9–11*]. Parliament agreed to make NORAD a front organisation in this fight, emphasising, however, that actors in the North who

---

and social justice, and the national authorities have the responsibility for the development' [*Report No. 51, 1992: 50*].

20. Popular participation in the development process was seen both as an objective in its own right and as a means to attain economic and social development. 'Participation includes economic, social, political and institutional dimensions of society and affects economic policy, access to services and development of human resources, participation in decision-making processes, strengthening of the legal system and democratic institutions, improvement of the position of women, growth of voluntary organisations, etc. The question of human rights is interwoven in all this. Gradually, popular participation in development has come to include good government with transparency and responsibility in the public administration and rational use of resources' [*Report No. 51, 1992: 31–2*].

21. The four-pillar strategy included [*MFA, 1999b: 11–13*]:
    - support for economic and social development to give people the experience that democracy gives them a better life. Education was emphasised;
    - assistance in building up functioning public institutions. Good governance was emphasised, with particular reference to the fight against corruption;
    - assistance to strengthen civil society, to give the people themselves an opportunity to influence development. The emphasis was on voluntary organisations;
    - assistance to strengthen political structures, political parties and free media to ensure that active and competent control mechanisms *vis-à-vis* the established elite were in place.

22. Human rights were considered to be universal. Bilateral and multilateral aid should be channelled to recipients according to their ability and willingness to improve such rights. Positive measures were considered more effective than aid conditionality [*Report No. 21, 1999: 109ff*]. Parliament agreed. Central to a human rights focus would be 'efforts to improve democracy, but also a poverty orientation, with efforts for women, children and the handicapped' [*Recommendation No. 23, 2000: 8*].

promoted corruption in developing countries should also be confronted [*Recommendation No. 28, 1999*].

(iii) Promotion of human development constituted another component of the strategy. The emphasis was on education and health.[23] In 2003, a strategy for assistance in the field of education was presented as 'Job Number One'. Education as a human right was the point of departure. It should help combat poverty and give the poor the possibility of economic, social and political participation. Education for girls was to be given priority and primary education should be free and promote democracy and human rights (*MFA, 2003ab*).

(iv) Promotion of sustainable development was a fourth component, involving both global and local aspects – an integral part of the development strategy since the mid-1980s.[24]

(v) Promotion of peace and prevention of conflict was considered both as an objective in its own right and a means to enhance development. The major policy papers of the 1990s saw stability and peace as a precondition for development. Aid for these purposes was considered within a humanitarian relief context. The 1995 White Paper saw poverty as a threat to peace, national reconciliation and global

---

23. In the stated policy, human development was considered part of a holistic approach: it was related to economic growth that, in turn, was linked to social justice. Both were linked to good government in the widest interpretation of the concept. And the social dimension was integrated into the human development concept. Women and children were identified as particular target groups. In 1996, Parliament emphasised that the building of competence and capacity and institutional development should be more strongly prioritised [*Recommendation No. 229, 1996: 12, 26*]. In the 1998 statement on development policy, the Minister, Hilde Frafjord Johnson (Christian Democrat), emphasised that the highest priority would be given to education and health, with education as job number one. In the 2000 and 2001 statements, her successor, Anne Kristin Sydnes (Labour), followed up with particular emphasis on basic human services, the fight against HIV/AIDS and other diseases, including support for the global vaccination programme (GAVI) by the WHO, and the follow-up of the ILO Convention against the worst forms of child labour [*MFA, 2000: 6; Proceedings of Parliament, 2001: 24 April*].

24. In the 1992 White Paper, the government pledged itself to mobilise 'increased resource transfers to developing countries to assist them in their efforts to solve local, national and regional environmental problems. Furthermore, additional resources have to be mobilised, beyond the regular development assistance transfers, as compensation to cover the extra costs inflicted on developing countries by their choice of alternatives that contribute to protect the global environment' [*Report No. 51, 1992: 55*]. A strategy for environmental aid was worked out in 1997, based on the major Norwegian policy papers and international commitments (Agenda 21 in particular), emphasising an integrated approach that included broader policies towards the developing countries, and stressing policy coherence [*MFA, 1997a*].

security.[25] Parliament agreed, pointing out that the transition between long-term development assistance and humanitarian aid should be improved [*Recommendation No. 229, 1996: 22*].

(vi) A strong gender perspective. Since the mid-1970s, a gender perspective has been firmly entrenched in Norway's aid policy and more generally in its policy towards developing countries. Strategies and plans of action for the promotion of women in development and gender equality were prepared in the early 1980s and reflected in the two major White Papers of that decade. This policy was continued into the new millennium.[26]

(vii) Responsibility for own development. International development assistance has from the very beginning been cast as 'help for self-help', placing the main responsibility at the doorsteps of the authorities of the countries concerned. In the context of Norwegian aid policy, this has been reflected in the guideline prescribing recipient-oriented aid. In the 1990s, it was followed up by a complementary guideline, prescribing recipient responsibility (see below, section IV).[27]

---

25. The government would maintain a high degree of involvement in the UN with a view to helping make the organisation an effective body for conflict resolution and peacekeeping activities, human rights, the environment, and economic and social development, the White Paper stated, indicating the interrelation between these various efforts. Peace, reconciliation and democratic development were considered important also for economic and social development. Support for conflict resolution was regarded as an integral part of Norway's humanitarian and relief aid and an important part of long-term development co-operation as well [*Report No. 19, 1995*].
26. In the stated policy, this priority was systematically followed up in one policy area after the other. When alleviation of poverty through economic growth and social development was discussed, the gender perspective was integrated into it. When the government discussed private-sector development, it argued that a strategy for the further development of the private sector 'must emphasise strengthening women's control of and access to the means of production', involving the right to own land and the same access to credit, training and technology as men. Such support for women was justified in its own right, but also considered instrumental for development in general; thus, improvements in the health of women and children 'may have particularly favourable ripple effects for society as a whole' [*Report No. 19, 1995: 21–3*]. In the 1998 statement on development policy, the Minister observed, *inter alia*, that about 70 per cent of the poor in the world were women (the 'feminisation' of poverty): 'Poverty-oriented aid is therefore women-oriented aid' [*Proceedings of Parliament, 1998: 5 May*].
27. The 1995 White Paper argued that aid could only support a country's own efforts – it 'is not intended to replace the mobilisation of internal resources, for instance in the form of collection of taxes and duties'. Previous warnings about the negative consequences of aid dependency were repeated. Parliament, in turn, agreed [*Report No. 19, 1995: 6, 10, 16; Recommendation No. 229, 1996: 10, 16, 18–20*].

(viii) Differentiating between developing countries. The development trends and the actual situation vary extensively from one developing country to another. Internationally, these differences led to a country-by-country approach in the effort to meet the development challenges. The 1992 White Paper therefore found it less natural to go for the type of universal solutions, characteristic of the NIEO philosophy: different means towards different groups of countries were found to be more appropriate.[28] Parliament, critical of the blessings of the globalisation process, particularly as it involved the poorest countries, found it of the utmost importance that Norway's policy towards developing countries should take the increased regional differences as a point of departure, emphasising that Norway's own assistance be concentrated on the poorest countries, and sub-Saharan Africa in particular [*Recommendation No. 229, 1996*]. In the stated policy, the poverty orientation of Norwegian aid continued to dominate the scene in the years that followed. One aspect of this policy was that a large share of the development assistance should go to Africa.[29] However, the policy as implemented stands out in strong contrast to these policy prescriptions: in the 1990s, the share of ODA going to sub-Saharan Africa – and to the least developed and other low-income countries – declined.[30]

(ix) An active role within multilateral organisations. In the policy papers of the 1990s, an active role within multilateral organisations was advocated as a means to attain the development objectives prioritised by Norway.[31] However, during these years the share of multilateral

---

28. 'For the most advanced developing countries, improved trade conditions and increased private-sector co-operation will be the most interesting. For the poorest developing countries, development assistance will remain important for many years to come' [*Report No. 51, 1992: 55*].
29. In the debate on the 2000 statement on development policy, Einar Steensnæs (Christian Democrat) argued that the poverty orientation was the most fundamental principle for Norwegian aid. In the debate on the 2001 statement, Anne Kristin Sydnes (Labour) stated that the fight against poverty 'is and will be the very foundation of Norway's development policy' [*Proceedings of Parliament, 2000: 16 May; 2001: 8 May*]. In 2002, this policy was followed up in a systematic way by an action plan for fighting poverty [*MFA, 2002a*].
30. See *infra*, section IV, particularly notes 45 and 47, which offer explanations as well.
31. In Report No. 51[*1992: 35*] the government argued that, through active efforts within multilateral organisations, 'Norway will be able to exert a far greater influence on development, particularly in co-operation with the other Nordic countries, than we can hope to achieve through bilateral efforts.' Similar views were also expressed by the 1995 White Paper [*Report No. 19, 1995: 6, 12, 19*], arguing, *inter alia*, that Norway in this way had contributed to a development where 'the development banks, in addition to being finance institutions, have increasingly become development agencies whose main goal is the alleviation of poverty' [*ibid.: 12*]. This parallels Den-

aid decreased (Table 13.2). The follow-up is discussed later (see below, section IV).

(x) Voluntary organisations as development agents. The major policy papers of the 1990s considered voluntary organisations complementary channels to government-to-government bilateral aid, at times even as alternative channels, particularly where intergovernmental aid was not considered feasible – especially involving humanitarian aid. In the late 1980s and early 1990s, this type of aid increased substantially, leading the 1992 White Paper to observe that Norwegian NGOs played a crucial role as partners in the administration of development assistance.[32] In the 1995 White Paper, the government followed this up, committing itself to facilitate a continued high involvement of NGOs in development co-operation, 'with increasing emphasis on ensuring that they contribute towards strengthening civil society in Norway's partner countries, i.a. through organisational development' [*Report No. 19, 1995: 9, 11–12, 19*]. Parliament argued in the same vein, and subsequent governments continued to cherish the so-called 'Norwegian model' – the close co-operation between Norwegian authorities and NGOs in the field of development co-operation and relief.[33]

(xi) Aid conditionality. Traditionally, Norway's relations with its partners in development were governed by guidelines prescribing non-interference (recipient orientation of aid) in stark contrast to the first-generation conditionality initiated in the early 1980s related to

---

mark's so-called active multilateralism, initiated in the mid-1990s (see Chapter 5, this volume). In both countries, the domestic audience was the primary one.

32. 'The organisations can supplement the total Norwegian development assistance since they can reach out to prioritised target groups and work within sectors that are not equally well taken care of by the government-to-government development assistance.' The government also committed itself to support local NGOs in recipient countries 'in order to strengthen popular participation in development' [*Report No. 51, 1992: 33*].

33. In 1996, Parliament noted that almost one-third of bilateral aid was channelled through voluntary organisations. It found NGOs to be flexible and effective bodies for getting through to particular target groups for aid, and also well placed to 'contribute to developing the civil sector and to striking a balance between the public and the voluntary sector in developing countries' [*Recommendation No. 229, 1996: 27*]. In the 1998 statement on development policy, Hilde Frafjord Johnson (Christian Democrat) identified the actors of civil society – voluntary organisations, churches, universities and other institutions – as the 'social capital' that should be made better use of in the fight against poverty. In the 2001 statement, Anne Kristin Sydnes (Labour) argued that in recipient countries with good governance, voluntary organisations might represent a useful supplement to other forms of aid; in countries with bad governance they might represent a good alternative [*Proceedings of Parliament, 1998: 5 May; 2001: 24 April*].

the structural adjustment policy referred to earlier. However, in the mid-1980s, Norway, along with the other Nordic countries, fell into line. When it came to first-generation conditionality, the emphasis remained on the so-called positive measures: providing assistance for activities that were assumed to lead in the desired direction, but without coupling them to ultimate demands for policy reform. The objectives associated with the second-generation conditionality of the early 1990s (promotion of human rights, democracy and good governance) have, in contrast, been given high priority in Norway's stated policy, along with social justice and poverty alleviation. Nevertheless, the attitude remained ambivalent with regard to the use of aid conditionality.[34] Although the ambivalence continued, particularly involving brinkmanship with aid as a tool, shifting governments became gradually tougher, especially when high-priority policy issues were involved. In the budget proposition for 2001, Zimbabwe was no longer included among the main partner countries.[35]

(xii) Policy coherence. In the policy papers of the 1990s and beyond, the strategy components outlined briefly above ((i)–(xi)) are presented as interconnected. The policy papers prescribe increased coherence in Norway's aid policy, in its total policy *vis-à-vis* developing countries and in the policies of donors towards these countries. The strong emphasis on policy coherence belongs to the 1990s, and is well attuned to the prescription, initiated by the high-level meeting of DAC in December 1991, that policies towards developing countries be more coherent [*Forster and Stokke, 1999; Forster and Stokke, eds, 1999; Stokke, 1999*]. The 1992 White Paper stated the importance of

---

34. I have previously concluded that '[t]orn between competing values, Norway has been a somewhat hesitant free-rider in the conditionality drive, particularly with regard to the first generation but also to the second generation aid conditionality' [*Stokke, 1995b: 197*]. The North–South Commission, however, was clear on the normative question involved: international society had both a right and a duty to intervene in countries where the authorities grossly suppressed parts of their population, committed genocide or similar grave injustices, created extensive flows of refugees, and threatened peace and security in other countries. Donors of development assistance 'have a right and duty to set conditions *vis-à-vis* recipients of aid involving both the administration of such transfers and the economic and political framework for the development policy in recipient countries' [*NOU, 1995: 5: 177*]. The 1995 White Paper was clear on the conditionality involving recipient responsibility. However, when it came to the extended policy, involving trade, the environment and debt relief, 'positive measures' were recommended [*Report No. 19, 1995: 16, 27–8, 42*]. In Parliament, the Labour government obtained the strongest support from the Conservatives [*Recommendation No. 229, 1996: 13–15, 32*].
35. In the 1999 statement on development policy, announcing an offensive against corruption, the Minister stated that sanctions would also be applied in particularly grave cases [*MFA, 1999c: 10*]. For the Zimbabwean case, see MFA [*2000: 9–11*].

policy coherence over again, as did the North–South Commission. The 1995 White Paper considered it an essential instrument for positive, lasting change.[36] In the 1999 statement on development policy, international alliance building was set out as an instrument to co-ordinate international development assistance. The World Bank initiative – the Comprehensive Development Framework (CDF) – was welcomed. Parliament, in turn, agreed [*MFA, 1999b: 6; Recommendation No. 28, 1999*].

*

The follow-up of the intentions of this strategy and its separate components is not easily traced by following the aid disbursed. In the 1990s, major changes took place in the distribution of bilateral aid on the main purposes (Table 13. 1). The strong and growing emphasis given to social and administrative infrastructure, increasing from 18 to 58 per cent, represents the most striking change. Given the high priority of education and health in the stated policy since the mid-1990s, their shares were surprisingly low (6 and 3 per cent, respectively, in 2000).[37] Support for 'government and civil society' made up the largest share (18 per cent in 2000), also reflecting the prominent role

---

36. See, *inter alia*, Report No. 51 [*1992: 11, 22*], NOU, 1995: 5 [*10–11, 14–15*], Report No. 19 [*1995: 6, 10,12–13, 15–17*].

37. According to DAC statistics [*OECD, 2002: Table 19; 2003: Table 19; 2004: Table 18*], the aid for education, 6.1 per cent of total bilateral commitments in 2000, included students and trainees. Only 1.8 per cent went to basic education (1.4 per cent in 2001, but 5.5 per cent in 2002). 2.7 per cent of total bilateral commitments in 2000 went to health; only 1.6 per cent went to basic health (1.5 per cent in 2001, but 4.1 per cent in 2002). The shares of health-related sectors such as population (including reproductive health) and water supply and sanitation were 1.9 per cent and 2.0 per cent, respectively, in 2000. It goes without saying that support for education through multilateral aid agencies is not included.

NORAD portrays a somewhat different picture [*MFA, 2002b: 73*]. In the years 1999 to 2001, 9.5, 8.4 and 9.9 per cent, respectively, of bilateral ODA went to education, and 4.2, 4.8 and 6.2 per cent, respectively, to health. The statistics published by the MFA may be politically 'doctored' by also including activities outside those defined by DAC as 'education' and 'health', respectively, in order to adapt reality to prescriptions – although, in principle, the same statistical standards apply. A possible explanation may be that aid for education within other sectors, defined as sector assistance in the DAC statistics (such as agriculture) has been included in the NORAD overview.

However, there may be other explanations as well: the DAC statistics for 1999 record 10.4 and 5.8 per cent of total bilateral ODA for education and health, respectively; the statistics for 2001 record 7.0 and 10.7 per cent, respectively, for education and health; and for 2002, these percentages were 11.3 and 6.6, respectively.

The stated policy is briefly outlined above. In the 1995 White Paper, the government committed itself to the 20/20 proposal of the UN Social Summit in Copenhagen [*Report No. 19, 1995: 21*], targeting 10 per cent of *total aid* to health and 10 per cent of *total aid* to education by the year 2000; in 1998, a new target of 15 per cent of total aid was set for education. The emphasis was on *basic* education and *basic* health [*Proceedings of Parliament, 1998: 5 May*]. The 2003 strategy [*MFA, 2003ab*] identifies education as 'Job Number One'.

TABLE 13.1
NORWAY'S BILATERAL COMMITMENTS AS DISTRIBUTED ON MAIN PURPOSES (%)

	1979–80[a]	1990	1994–95[a]	1999–2000[a]	2001	2002
Social and administrative infrastructure	12.9	17.5	25.2	41.5	44.4	58.4
Economic infrastructure	30.5	22.2	17.0	8.1	13.8	6.7
Agriculture	19.7	9.3	6.2	5.5	4.1	5.3
Industry and other production	12.9	7.3	3.0	1.4	2.9	0.7
Commodity aid and programme assistance	8.0	2.1[b]	1.7	1.2	2.2[d]	3.4[d]
Emergency aid	8.1	14.3[c]	23.2	23.2	11.9	11.9
Other	7.8	27.3	23.7	19.1	20.6	13.6
	100	100	100	100	100	100

Sources: OECD DAC – Annual reports (various) and calculations based on these sources.
Notes: [a] Averages; [b] Programme assistance and food aid; [c] Other than food aid; [d] Programme assistance.

attributed to aid through NGOs. Support for economic infrastructure and for industry and other production was drastically reduced, and the aid for agricultural development declined from an already low level. Emergency aid increased from a high level in 1990 (14.3 per cent) to 25.6 per cent in 2000 [*OECD, 2002: Table 19*].

How can this be explained? The increases are more easily explained than the decreases. Since the mid-1980s, emphasis in the stated policy has been on institutional development, with the purpose of strengthening the capability of recipient authorities to plan, prioritise and implement their own development. It was also justified by the new emphasis on recipient responsibility. The large share of emergency relief may be explained by the pressing needs for such aid, also in Central Europe, but equally well by the government's determined efforts to give Norway a strong humanitarian foreign policy profile – for ideal as well as instrumental reasons.

IV. GUIDELINES

Some of the guidelines established for Norway's development assistance prescribe how aid is to be distributed between the various aid channels, recipients and purposes. Others set norms for the quality of aid. Most of these norms were established in the late 1960s and early 1970s and formalised in the 1972 White Paper.[38]

---

38. These guidelines are only sketched here. Fuller accounts and analyses are provided elsewhere (see Stokke [*1975, 1979, 1984, 1989, 1991, 1992b, forthcoming*]).

1. A Large Multilateral Component

One guideline prescribes that *ODA should be divided equally between multilateral and bilateral aid channels*. It was established in 1969 and formalised in the 1972 White Paper. In the 1960s, multilateral aid was considered altruistic, effective, organised in the best interests of recipients, free from the strings associated with bilateral aid, including procurement tying, and with the national identity of the donors removed. It was well tuned to a major Norwegian foreign policy concern: to strengthen global institutions, the UN in particular. The main function of the guideline has been to protect a large multilateral aid component. The guideline was confirmed in the major policy papers of the 1970s and 1980s, although given a flexible interpretation: a 45 (minus)/55 (plus) distribution in favour of bilateral aid was considered to be within the agreed frame.

How did the changed international environment after the Cold War affect this guideline and the perceptions on which it rested? The major policy papers of the 1990s continued to see efforts through international co-operation, the UN system in particular, as a means of promoting the objectives set for the Norwegian North–South policy and Norway's foreign policy more generally (see above, section III). It was logical, therefore, that the 1992 White Paper confirmed the guideline, particularly since the multilateral share had shown a downward trend in the 1980s and dropped below 40 per cent in the early 1990s. Norway (jointly with the other Nordic countries) should make multilateral agencies, particularly UN organisations, more effective, it was argued. Multilateral aid should be distributed according to 'the effectiveness of the organisation concerned and the extent to which its activities fulfil the objectives set for Norwegian development co-operation' [*Report No. 51, 1992: 180ff.*].

The 1995 White Paper added an instrumental justification: by 'nudging the large, dominant multilateral organisations in the direction desired', Norway influenced the utilisation of resources that were vitally important to the political and economic development of recipient countries. The strategic use of earmarked funds had also proved effective in a multilateral context, the government argued, adding that it would make 'more conscious use of the possibilities offered by the combination of participation in governing bodies, maintaining general contributions and targeting earmarked funds to steer policy in the desired direction'. The report was vague, however, when it came to the volume target [*Report No. 19, 1995: 36–7*]. The majority in Parliament argued strongly for increasing the share of multilateral aid [*Recommendation No. 229, 1996: 16–17*].

TABLE 13.2
NORWAY'S ODA AND ITS DISTRIBUTION ON BILATERAL AND MULTILATERAL AID CHANNELS,
1980–2002

	1980	1990	1992	1994	1995	1997	1999	2000	2001	2002
ODA (US$m.)	473	1,205	1,273	1,137	1,244	1,306	1,370	1,264	1,346	1,696
as % of GNI (GNP)	0.82	1.17	1.12	1.05	0.87	0.85	0.90	0.80	0.80	0.89
Bilateral ODA (US$m.)	271	756	811	828	907	916	1,007	934	940	1,145
as % of total ODA	57.3	62.7	63.7	72.8	72.9	70.1	73.5	73.9	69.8	67.5
Multilateral ODA (US$m.)	202	449	462	309	337	390	363	330	406	551
of which:										
IDA (US$m.)	42	79	79	77	86	71	53	31	80	73
Regional development banks (US$m.)	19	44	55	8	13	49	42	49	47	62
The UN system (US$m.)	130	239	308	209	230	238	231	208	245	303
Multilateral ODA as % of total ODA	42.7	37.3	36.3	27.2	27.1	29.9	26.5	26.1	30.2	32.5
ODA in deflated prices (1999=100)	–	102.63	105.44	94.33	108.29	104.28	100	102.06		

Sources: OECD, DAC – annual reports (various) and calculations based on these sources.

The follow-up of this guideline is shown in Table 13.2, where the distribution in the main multilateral systems is also given. In spite of the strong stated support for the guideline, by governments and Parliament alike, the bilateral (including the multi-bi) component increased continuously in the 1990s, reducing the multilateral share to less than 30 per cent. This, most probably, reflects a will to decide what purposes were to be supported. A large and increasing multi-bi component strengthens this explanation, although the concern for a higher return flow may also explain this particular pattern.[39] Despite an expressed concern about the inefficiency of the UN system (and the perception of an efficient World Bank), the largest share of the multilateral aid component has consistently been channelled through the UN system – reflecting the importance Norway continues to place on the UN as an arena for its foreign policy. It may also reflect the influence Norway, jointly with its fellow Nordic countries, has traditionally exerted on the development policies of these agencies, the UNDP in particular.

2. Geographical Concentration

Another guideline, established in the late 1960s, prescribed that *bilateral aid should be concentrated geographically on a few regions and*

---

39. The 'multi-bi' component, aid earmarked for specific projects run by multilateral agencies and development finance institutions, is defined as part of bilateral aid because the end use (purpose) is identified before the commitment is made. Its share of total ODA increased from 6 per cent in 1990 to 13 per cent in 1995, and 16 and 13 per cent, respectively, in 1999 and 2000.

countries.[40] It was operationalised through the selection of a fairly small number of main partner countries with which extensive, broad development co-operation was established on a long-term basis, involving a four-year revolving country programme and an office of a NORAD resident representative. The main rationale was a concern for aid effectiveness. In practice, however, the share of bilateral aid being channelled to the group of main partner countries declined from 78 per cent in 1970 to 61 per cent five years later. It remained at around 60 per cent for the next ten years before declining slightly towards the end of the 1980s.

In the 1990s, less emphasis was given to the principle of geographical concentration and, accordingly, less importance given to the main partner countries. This change did not come as a response to the new environments of international politics: it reflected changed perceptions at the level of domestic politics and a stronger emphasis on national (private-sector) interests. The authorities opted for more flexibility, giving emphasis to regions and special programmes rather than individual countries, apart from the rhetoric to the contrary. This process had been initiated in the late 1980s, and was formalised in the 1992 White Paper.[41] It was carried further by the 1995 White Paper, emphasising flexibility and competition for aid money rather than continuity in old development co-operation relations. The rationale offered for the change was that it would ensure that activities receiving support were of the highest quality possible. The new approach represented a fundamental breach with the previous system, established to ensure predictability in development co-operation relations and to make long-term planning possible for the recipient authorities through long-term commitments. Most emphasis was now given to the regional allocations, where a large number of countries, in addition to

---

40. Several activities have been excepted from the principle of geographical concentration, such as support for family planning, projects run by Norwegian enterprises or NGOs, some elements of technical assistance and, of course, humanitarian aid. Some of the exceptions were motivated by development concerns (such as family planning), others by particular Norwegian interests (private-sector, NGOs).

41. The share of bilateral ODA allocated to the main partner countries had declined to 46 per cent in 1991 [*Stokke, 1992b*]. The 1992 White Paper removed the distinction between main partner countries and partner countries, and established 12 so-called programme countries – adding Namibia, Nicaragua and Zimbabwe to the previous main partner countries (Botswana, Mozambique, Tanzania and Zambia in Africa, Bangladesh, India, Pakistan and Sri Lanka in Asia, with Uganda on the waiting list). Development co-operation with Botswana, India and Pakistan was to be phased out. The aid was to be concentrated on three regions: southern and eastern Africa, South Asia and Central America. Major emphasis was to be given to southern and eastern Africa [*Report No. 51, 1992: 228–34; Recommendation No. 195, 1993: 43–5*].

the programme countries, were eligible for development assistance. In 1996, Parliament, by and large, followed the government.[42]

However, in 2000, the (Labour) government signalled a turn towards the old system: to concentrate bilateral development co-operation more strongly on a small number of main partner countries. The co-operation with these countries was to be based on a broad dialogue on development policy and long-term efforts to reduce poverty [*Proposition No. 1, 2001: 74–5*].[43] Although the stated intention was to concentrate bilateral assistance geographically, the new reality did not differ much from the previous one.[44]

The main justification for the guideline of geographical concentration – combined, as it was, with a set of criteria for the selection of main partner countries and another guideline prescribing long-term commitments in aid relations (to be outlined in the following two subsections) – was a concern for aid effectiveness that was to be attained through (mutual) insights and a good planning and follow-up system, operationalised via the country programming process. In the 1990s, the emphasis fell increasingly on flexibility, which in reality confronted the old system – this, too, justified by a concern for aid effectiveness. The guideline, as a result, became weakened. *How* weakened is shown in Table 13.3, although this table does not convey the full story since it is based on *total* bilateral aid to the countries concerned, not just the part included in the country programmes (subject to systematic planning and negotiation). The share of total aid to the main partner countries that was provided through the country programmes declined progressively, a process which had started already in the late

---

42. The number of programme countries remained the same (12) after the removal of Botswana, India, Namibia and Pakistan: war-ridden Ethiopia and Eritrea were newcomers together with Malawi and Nepal [*Recommendation No. 229, 1996: 17, 26*]. However, as noted, inclusion in the group meant less in the mid-1990s than before.

43. The group of 11 programme countries (Zimbabwe had already disappeared) was reduced to seven: Malawi, Mozambique, Tanzania, Uganda, Zambia, Bangladesh and Nepal remained, while Eritrea, Ethiopia, Nicaragua and Sri Lanka were no longer included. However, in addition to the new group of main partner countries, co-operation was to be developed with a limited number of so-called special partner countries. For this purpose, four regional allocations were to be established, each with earmarked budget lines with regard to themes and areas. The government proposed 18 countries to be included among the special partner countries [*Proposition No. 1, 2001: 74ff.*].

44. Although somewhat reduced compared with the previous list of 11 priority countries and 31 partner countries, the combined number of 'main partner countries' and 'special partner countries' added up to 27 countries and one territory. To illustrate the point further: the budget for 2002 earmarked NOK970m. for the 'main partner countries' and NOK1,1447m. for regional allocations. The allocation to 'main partner countries' represented less than 7.2 per cent of the ODA budget.

TABLE 13.3
THE PROGRAMME COUNTRIES' SHARE OF TOTAL BILATERAL ODA, 1970–2001 (%)

1970	1980	1975	1985	1987	1991	1995	1999	2000	2001
78	61	64	59	56	41	30	27	24	23

Sources: Report No. 51 [*1992: 230*], MFA [*1997: 51; 2002: 70*] and Stokke [*1989: 226*].

1980s: it decreased from 81 per cent in 1985 to about 60 per cent in 1991 [*Report No. 51, 1992: 230*]. This raises the question of whether the principle of geographical concentration and the notion of main partner countries still apply in a meaningful way, rhetoric aside.

The weakening of the position of the programme countries – and of the rationale behind the system, the country programming – may be illustrated by another major trend of the 1990s: the increasing share of aid going to Europe, primarily ex-Yugoslavia, and to Palestine.[45]

Characteristic of this 'political' assistance is that it has been exempted from the principle of geographical concentration and other norms guiding the implementation of aid (the aid has been tied, recipient responsibility has not been practised, its time perspective has been short, and the recipients were not among the least developed countries) – and, accordingly, is outside the normal planning procedures established to secure the quality and effectiveness of bilateral aid. It was justified as a contribution to relieve human suffering, promote peace and build democracy. In the late 1990s, similar concerns became decisive even in the selection of main partner countries (Eritrea, Ethiopia).

Several explanations may be offered for this dramatic change, which is in contrast to repeated policy statements prescribing greater concentration. The aid effectiveness justification cannot be totally discarded. However, the main explanations may be sought elsewhere: the increased emphasis given to humanitarian assistance, conflict resolution, and (Norwegian) private-sector interests. The latter led to a duality in development co-operation policy: a concern for poverty alleviation was to guide the co-operation with the main partner countries while a concern for developing broader co-operation with economically and technologically more advanced developing countries (involving Norwegian private-sector interests and institutional interests) was to guide much of the remaining aid.

---

45. According to DAC statistics, the share of total bilateral ODA being channelled to Europe increased from 0.2 per cent in 1989/90 to 21.0 per cent in 1999/2000 (down to 13.6 for 2000/01) (even larger than is shown in Table 13.4, based on Norwegian statistics), while the share channelled to sub-Saharan Africa (given priority in the stated policy) was reduced from 64.7 to 42.8 per cent (same share for 2000/01) [*OECD, 2002: 272; 2003: Table 28*].

TABLE 13.4
BILATERAL ODA TO EX-YUGOSLAVIA AND PALESTINE, 1990–2001 (%)

	1991	1992	1993	1994	1995	1997	1999	2000	2001
Europe	0.1	0.6	5.7	12.1[a]	8.4[a]	7.3[a]	17.4	16.4	11.6
of which ex-Yugoslavia		0.5	5.6	12.0	8.3	7.1	14.7	12.4	8.8
Middle East[b]				-	-	-	7.1	6.7	7.1
of which Palestine				16.4	4.7	4.8	3.0	3.2	4.3

Notes: [a] Europe and Oceania; [b] Middle East is reported under Asia, 1994–98.
Sources: MFA [*1997b: 61; 1999b: 54, 59; 2002b: 70*], Report No. 49 [*1991: 104–5*], Report No. 9 [*1993: 133*] and Report No. 8 [*1994: 141*].

3. Criteria for the Selection of Main Recipient Countries

The principle of geographical concentration, when first established, was put into operation through the selection of a few countries as main partners in development. Criteria were established for their selection: they were to be among the poorest in the world (among the least developed or the low-income countries) and their governments should pursue a development-oriented and socially just policy. In 1976, Parliament added that they should be expected to respect human rights as laid down in the UN declaration and convention. Most of the main partner countries up to the early 1990s had, however, been selected before these criteria were formalised.

What happened to these criteria in the 1990s? The 1992 White Paper maintained the poverty criterion and extended and even sharpened the regime criteria, in line with the emphasis given to democratic development, human rights and good governance in both the overall objectives set for aid and in the strategy. Democracy was added to the list and pursued both in first- and second-generation aid conditionality [*Stokke, 1995ab*]. This fits well into patterns established by other donor governments and may, at least partly, be explained by the changed international political environment after the end of the Cold War.

However, these criteria were not alone in deciding to whom aid should be provided – 'Norwegian foreign policy and, in some cases, particular geopolitical concerns can influence decisions concerning the selection of programme countries and the volume and orientation of development assistance' [*Report No. 51, 1992: 231*]. The 1995 White Paper followed up on the regime criteria involving priority countries. However, they were not necessarily to apply to the allocations for 'extended co-operation'; there, the promotion of broader areas of contact with Norway was the primary purpose [*Report No.

19, 1995]. Parliament, in turn, laid great emphasis on the poverty criterion [*Recommendation No. 229, 1996*].

The actual selection of priority countries in 1996 was well in line with the poverty criterion. It was less attuned, if indeed at all, to the established regime criteria – the recipient governments' respect for human rights, pursuance of a socially just distribution policy, promotion of democracy and qualities associated with good governance, and the stated expectations with regard to effectiveness and efficiency. In 2001, when a restructuring of the groups of recipient countries was contemplated, the main focus was on the recipient government's ability to take control of its country's destiny. Good governance was established as one of the two major regime criteria; the poverty criterion remained.[46]

Still, the poverty orientation of Norway's development assistance (operationalised in terms of GNP per capita of recipient countries) was dramatically weakened during the 1990s, in stark contrast to the strong emphasis given to fighting poverty in the stated policy (job number one). In 1989 to 1990, 84.7 per cent of bilateral ODA went to the least developed (60.5 per cent) and other low-income countries. A decade later, this share was reduced to 59.4 per cent (of which 46 per cent went to the LLDCs).[47] The main explanations are given in the previous subsection.

4. Development Co-operation: Still a Long-term Commitment?

Another guideline, established in the early 1970s, prescribed that development co-operation should be a long-term commitment. As already indicated, this guideline might conflict with the political criteria set for the selection of partner countries in cases where a shift of regime on the recipient side takes place. This possibility was anticipated in the 1972 White Paper: in cases where a new government pur-

---

[46]. 'The countries must show a clear political will to solve important national challenges themselves. We want to co-operate with those who themselves take control and promote good governance' (Anne Kristin Sydnes (Labour) in the 2001 statement on development policy [*Proceedings of Parliament, 2001: 24 April*]).

[47]. For 1999/2000, although slightly above the DAC average of 58.3 per cent, Norway ranked 13th among the DAC countries on this account. During this period, the group of lower middle-income countries (LMICs) had increased their share from 9.8 to 38.0 per cent [*OECD, 2002: Table 32*]. For 2000/2001, the share of bilateral aid to the least developed countries and other low-income countries increased to 67.1 per cent (of which 47.4 per cent to the LDCs). Nevertheless, Norway ranked 13th among the DAC countries on this account (but fourth on the latter score) [*OECD, 2003: Table 26*]. However, the calculations for 2001/2002, that included imputed multilateral ODA, recorded 69.1 per cent to the LDC-LIC group of countries (49.7 per cent to the LDCs, ranking third), 25.9 per cent to the LMICs and 5.0 per cent to the UMICs [*OECD, 2004: Table 26*].

sued a policy contrary to the (political) criteria established, the aid should be concentrated on sectors in which, with some certainty, it would attain the objectives aimed at, with particular reference to social justice.

What happened to this guideline in the 1990s has already been indicated: as from the late 1980s, there was an increasing demand for flexibility in aid relations. Nevertheless, the rhetoric of the 1992 White Paper was clear: development co-operation was to remain a long-term commitment. At the same time, the long-term perspective was toned down: several of the policy changes involving the development strategy, other guidelines and administrative mechanisms pointed in the opposite direction, opting for greater flexibility in aid relationships. The 1995 White Paper took these trends further. However, the reorganisation initiated in 2001 indicated a change towards a stronger commitment to more long-term aid relations. Nevertheless, this applied only to a limited (and declining) part of the bilateral aid, namely the aid to the main partner countries, and not so much to the second group of special partner countries and other types of bilateral aid.

For years, this guideline had been given preponderance *vis-à-vis* the political criteria for selecting main partner countries when conflicting. In the 1990s, rhetoric aside, it was weakened both in its core area (involving partner countries) and generally: new programmes and the so-called broader co-operation took a larger share of the aid budget – governed by a concern for flexibility in aid relations. The strong emphasis given to humanitarian aid also worked in this direction.

5. Recipient Orientation of Aid and Recipient Responsibility

Norwegian development assistance should be recipient-oriented: integrated into the development plans and political priorities of the recipient governments. This guideline, established in 1972, presupposed partner countries with governments that met the criteria set for their selection. The main rationale was aid effectiveness. In the 1984 White Paper, the government modified the guideline slightly in order to overcome a potential conflict between the overall objectives set for Norway's development assistance and this particular guideline: equal weight should be given to both.

In the 1990s, the perspective was gradually changed, although the core of the old guideline survived. In 1990, NORAD combined this guideline with a new, complementary guideline of *recipient responsibility*: the responsibility for planning and implementing all development activities should rest with the authorities and institutions of the recipient country. The implications were made crystal clear: 'In cases where NORAD is in disagreement with or lacks confidence in important national priorities and goals, we shall have to assess other forms of co-operation with the relevant country, and in special cases propose to the Norwegian authorities a reduction in the government-to-government aid.' It was essential that recipients kept their part of the agreement [*NORAD, 1990: 13–15*].

The 1992 White Paper confirmed the old guideline of recipient orientation and, in a somewhat modified way, agreed to the new one. In the 1995 White Paper, recipient responsibility became the key concept, and now more uncompromisingly than ever. It became an integral part of a development co-operation approach geared to increased aid effectiveness, involving greater flexibility for the donor in aid relationships [*Report No. 19, 1995*]. Parliament agreed, emphasising the need to strengthen the institutional capacity on the recipient side by way of development assistance [*Recommendation No. 229, 1996: 18*]. However, during subsequent years, the willingness to use the axe has not been extensively demonstrated.

6. Aid as Gifts and Untied

A long-established guideline prescribes that aid should be provided on a *grant* basis. By and large, it was followed up during the 1970s and 1980s, and was also confirmed in the major policy documents of the 1990s.[48] In terms of the grant share of total aid, as defined by DAC, this was also followed up.[49]

Another guideline states that development assistance should *not be tied* to procurement of Norwegian commodities and services. This guideline has been controversial from the very beginning, reflected in the way it was formulated and practised since it was first introduced in 1967: there were both ambiguities and modifications to the general principle. When the Norwegian economy, particularly certain export industries, came under pressure in the second half of the 1970s, this norm came under pressure as well.[50] Little changed in the 1980s. The

---

48. In the 1992 White Paper, the justification was quite brief and simple: 'The development assistance shall mainly go to the poorest developing countries and should not contribute to increase the debt burden of these countries' [*Report No. 51, 1992: 174–5*].

49. The grant element of total ODA was 99.6 per cent in 1990, 99.3 per cent in 1995 and 99.6, 99.7 and 99.8 per cent, respectively, for the years 1998 to 2000 [*OECD, DAC annual reports (various)*].

50. In 1981, with the intention of correcting a 'flexible' implementation of the guideline, Parliament recommended that NORAD, as a general rule, should buy at market prices, though with a unique modification that reflected the political cross-pressures at work [*Recommendation No. 255, 1981: 6*]: 'However, in cases where particular

1984 White Paper confirmed the guideline as defined three years earlier.[51] Towards the end of the 1980s, when the Norwegian economy was again under strain, systematic efforts were made to ensure a larger return flow of aid in order to serve the needs of the private sector – Norwegian industries and services in particular [*Stokke, 1991*]. Special mechanisms were already in place. Their financial frames, however, remained at a modest level – increasing from 1.3 per cent of total aid in 1984 to 3.3 per cent in 1991 – and the actual use was even less.

What happened to the guideline in the 1990s and beyond? In spite of the strong pressure for a higher 'return flow', the 1992 White Paper restated the principle of untied aid, without the reservations previously attached to it: the exception of the special arrangements for the private sector already referred to. This may be explained by the way it was defined: it was limited to the *grants* and did not, as before, involve total aid [*Report No. 51, 1992: 175*]. Although remaining a theme in the development discourse during subsequent years, the 1992 position remained in force – and changing governments argued for the untying of aid internationally. In the second half of the 1990s, stimulated by the unilateral decision of the UK (Labour) government, the Bondevik (minority) government argued that Norway should follow the UK example in order to increase the value of aid for recipients and make it more effective. However, it did not succeed in this.

In domestic politics, this issue has been controversial along an altruism–self-interest axis. It did not fully follow a right–left axis, although the parties to the right (and the private sector) favoured aid forms that could be directly or indirectly tied to Norwegian commodities or services. The Socialist Left Party and the parties in the middle, especially the Christian Democrats, have consistently argued

---

reasons justify it, NORAD may favour Norwegian suppliers with regard to commodity, programme and project aid. In such cases ... the concept "not ... on appreciably less favourable terms" may allow the choice of Norwegian supplies even if they are up to 10 per cent more expensive than what could be obtained elsewhere.' The implication was that, in actual practice, these forms of aid included a 10 per cent margin to the benefit of Norwegian producers.

51. The justification given is worth highlighting [*Report No. 36, 1984: 26ff.*]: 'We have to see to it that as little as possible of aid disappears as a result of bad planning, corruption and inefficient administration in the recipient country. Similarly, the aid administration must see to it that the value of the aid is not reduced on the Norwegian side because of too high a price or a quality that is not adapted to the purpose.' Leakage on the donor and the recipient side is placed at the same level, with a reference to corruption. However, it appears from the context that the major concern was to obtain fair prices *in the Norwegian market*, not to ensure international competition to obtain favourable prices. The subsequent Labour government agreed to this position [*Report No. 34, 1987*].

for untied aid (and so have several NGOs, including the Church of Norway). However, Labour has been ambivalent, and unwilling to take unilateral steps. The position of Labour may be explained by the convergence of (private-sector) export interests and trade unions' concern for employment opportunities, particularly when the national economy was under strain.

DAC statistics on aid tying do not capture all forms of tied aid, let alone all efforts to ensure the main purpose with aid tying, namely to ensure a high return flow. However, they tell part of the story and may indicate trends. While 39 per cent of Norway's bilateral aid was tied in 1990, the share was reduced to 23 per cent in 1995 and to 1.1 per cent in 2001.[52]

7. Poverty-oriented Aid

Poverty orientation of development assistance has been an overall norm since the early 1970s and even before. As noted, the norm has been integrated into the stated objectives for aid, with particular reference to the target groups that have been identified. As we have seen, it has also been integrated into the development strategy and the criteria set for the selection of main partner countries.

The 1992 White Paper turned poverty orientation of aid into a separate principle, claiming that it was time-honoured, and rightly so. The principle was established at a time when the poverty orientation of aid had been weakened in practice for a variety of reasons. The way it was defined underscores its well-established character [*Report No. 51, 1992: 174*].[53] In 1993, Parliament agreed [*Recommendation No. 195, 1993: 38*].

In 2002, the Bondevik government presented its Plan of Action for Fighting Poverty in the South up to 2015. The plan represents an effort to operationalise, at the level of national politics, the UN millennium goals. It systematises Norway's various policies that directly

---

52. In 1990, US$358m. was untied and 226m. tied (including technical assistance but not administrative costs). In 1995, US$541m. was tied and 162m. untied (excluding administrative costs *and* technical assistance). By the end of the 1990s, tying of bilateral aid (excluding administrative costs and technical assistance) was almost non-existent – US$8m. (0.9 per cent) in 1999, 15m. in 2000, 8m. in 2001 and 8m. (0.9 per cent) in 2002 [*OECD, DAC annual reports (various)*].

53. It meant 'partly, that Norwegian aid shall be channelled to the poorest developing countries, partly that Norwegian development assistance shall be used to promote social and economic development for the broader layers of the population and particularly for the groups that were most problematically situated. In this connection, it is important to direct attention towards the mechanisms that affect the development of poverty. In this context, the socio-economic structure is of importance and the extent to which the authorities promote a policy that is socially just and development-oriented' [*Report No. 51, 1992: 174*].

or indirectly affect developing countries in a way geared towards poverty reduction, with particular reference not only to development cooperation, but involving trade policy, debt relief, energy development and other policy areas as well. Development policy was defined as a core element of Norway's foreign policy. Policy coherence – in the widest sense of the notion, involving Norwegian national policies as well as Norway's policy in international organisations (in particular the UN and the Bretton Woods institutions) and *vis-à-vis* other donors and recipients of aid – was prescribed as the core strategy, necessary to attain the main objective. The strategy took the international objectives, as defined by the high-level meeting of DAC in 2001 and by the so-called Millennium Assembly of the UN in 2000, as the point of departure [*MFA, 2002a*]. The 2004 White Paper followed up on this strategy, as already noted: fighting poverty was what development was all about.

Both the plan and the new White Paper aimed at a wider set of policy areas than the aid policy, such as the trade policy and finance, including debts. And the mechanisms involved cover a broader spectrum than the national systems to which resources are channelled; the policy of recipient governments as well as the purposes within the systems to which aid is channelled are also considered to be important. Nevertheless, it remains a paradox that during a period when poverty alleviation has been emphasised more strongly than ever in the stated policy, the countries that need assistance the most have been receiving an increasingly smaller share of Norway's bilateral aid, as documented above (subsections 2 and 3, and fn. 47).

V. THE AID VOLUME: TARGETS AND PERFORMANCE

In 1973, Parliament set 1 per cent of GNP as the target for Norway's official development assistance, to be attained four years later, accompanied by a plan for stepped-up budgetary allocations. All political parties in Parliament except the Conservatives endorsed this. The target was reached in 1982 and ODA remained above this level for the rest of the 1980s – peaking at 1.17 per cent of GNP in 1986 and 1990. It stayed above the 1 per cent target during the following four years. In 1995, however, it dropped below the target (to 0.87 per cent of GNI),[54] and remained below it for the rest of the decade and

---

54. The sudden decline in 1995 had a 'technical' rather than a political explanation: Norway was one of the first OECD countries to introduce a new system of national accounts. This revision raised GNI substantially: the nominal growth between 1994

into the new millennium. Nevertheless, ODA continued to increase in absolute terms, and remained at a high level also in relative terms, as a share of GNI (see tables IN.1, IN.3 and 13.2).

Most political parties, including those forming the present government (Christian Democrats, Conservatives, Liberals), have committed themselves to restore ODA to 1 per cent of GNI by 2005. In the budget for 2002, the outgoing Labour government had proposed 1 per cent of GNI as a target, but gave no date for its attainment. The incoming government (Bondevik II) set the date, but did not follow up with a plan for stepped-up budget allocations, as had been done in the mid-1970s. In the 2004 White Paper, the government repeated its commitment to this target, although it was vague with regard to when it would be attained.[55] Although the volume has been increasing both in absolute and relative terms, the upward move has been slow, making the last step up the ladder a difficult one.[56]

This broad agreement does not imply that the volume target has been kept above party politics: the Progress Party (Frp) has for years opposed most official aid except for humanitarian purposes, and the Conservatives have been holding back, although (since the mid-1970s) they have accepted the international 0.7 per cent of GNP target. The Socialist Left Party and the Christian Democrats (along with the Centre Party and the Liberals) have been pressing for higher ODA volumes. Labour has taken a middle position – and over time determined the actual volume level.[57]

Decisions on the volume target have, since the mid-1970s, been the outcome of endogenous processes. Since the early 1980s, Norway has

---

and 1995 was 33 per cent. The aid budget had been fixed in current NOK as a percentage of a calculated (much lower) GNI for 1995.

55. 'The government aims at increasing ODA to 1 per cent of GNI. Once the target is attained, ODA will be maintained at at least 1 per cent for the next parliamentary period (2005–2009)' [*Report No. 35, 2004: 86*].

56. The performance 1980–2002 is provided in Table 13.2. In 2001, ODA amounted to NOK12,312m., increasing to NOK13,939m. in 2002 and NOK14,392m. in 2003 (revised budget). In its budget proposal for 2004, the government increased ODA to NOK15,294m. In the budget proposals, ODA amounted to 0.92 per cent of the calculated GNI for 2002, 0.93 per cent for 2003 and 0.94 for 2004 [*Proposition No. 1, 2002: 52; 2003: 46*].

57. It may seem strange that governments under Conservative prime ministers reached the peak of 1.17 per cent of GNP in ODA (in 1986 and 1990). However, these were coalition governments that also included, *inter alia*, the Christian Democrats, for whom development co-operation has been and remains a high priority concern. The Conservatives had to accept a high ODA volume in order to establish the coalition governments. This also applies to the new commitment of attaining the 1 per cent of GNI target by 2005 – shortly before, the party (in opposition) had proposed large cuts in the ODA budget, a position it also defended in the 2001 election campaign.

remained among the top providers of ODA in relative terms (as a share of GNP/GNI). The decision to reach the 1 per cent target was made before the country became an oil-rich nation, and fits well into the stated justifications for aid. However, as noted above, keeping ODA at a fairly high level has also been part of the image building of changing governments – for both domestic and external consumption – and sustained also for this reason.

VI. THE AID ADMINISTRATION

For most of the time, the political responsibility for development co-operation has been vested in the Ministry of Foreign Affairs (MFA).[58] In 1984 a Ministry of Development Co-operation (MDC) was established with overall responsibility for most aid issues and a shared responsibility (with the MFA) for some others; in 1990, the MDC merged with the MFA. However, since mid-1983, the political responsibility for most issues within this policy area has been vested in a Minister for International Development.

In the early years, the executing agency for most bilateral aid (NORAD) was given an independent position, governed by a board of directors appointed by the government and an advisory council appointed by Parliament, thus making aid a special case. One justification for creating a distance from the MFA was that development objectives might conflict with the self-centred foreign policy objectives traditionally associated with a ministry of foreign affairs. However, in the first half of the 1970s, the MFA brought NORAD under tighter control, insisting on a final say on 'political' issues. Nevertheless, a large degree of independence prevailed – at times also with tensions. Thus, the country programming, involving the main partner countries, remained the prime responsibility of NORAD and the offices of resident representatives in these countries continued to report to NORAD. In 1984 NORAD was, for a few years, integrated as a department (an 'internal directorate') of the MDC.

In the 1990s, the end of the Cold War constituted a basic precondition, internationally, for a strong commitment to coherence in policies towards developing countries. To what extent did this commitment result in institutional reform and restructuring at the national level?

---

58. The responsibility for relations with the development banks (including the World Bank) was transferred from the Ministry of Trade to the MFA in 1979; for most practical purposes, however, the MFA has had the final say in issues involving aid since 1971. Nevertheless, the World Bank govenor was drawn from the Bank of Norway up until the mid-1990s.

A series of institutional reforms were initiated with the purpose of integrating development co-operation and the more traditional tasks of the MFA. The preparation of the 1992 White Paper was part of this process, involving policy co-ordination. The insistence on policy coherence, both within the aid policy and with regard to Norway's policies towards developing countries in general, may also be explained from this perspective. Institutional reforms also took place in order to ensure co-ordination of foreign policy, including development co-operation. In the main partner countries, so-called integrated embassies were established in which the office of the NORAD resident representative became an integral part.[59]

Although the former 'MDC' departments, following the merger, continued for some time as separate departments, reporting to 'their' Minister, major efforts were made later on to achieve a fuller integration by means of institutional restructuring. The lion's share of humanitarian aid (the 'political', not the 'natural disaster', part) remained the responsibility of the Minister of Foreign Affairs, making it possible to pursue an active humanitarian foreign policy, involving traditional humanitarian aid as well as efforts to resolve conflicts. This may be part of the explanation of why humanitarian aid made up such an extraordinarily large share of Norwegian ODA throughout the 1990s and beyond (see Table 13.1).

A new reform of the aid administration was initiated in 2002. The MFA commissioned an evaluation of the present system with a focus on relations between the ministry and NORAD. The government's programme for the modernisation and simplifying of the public sector, making it more effective, was one point of departure and the government's strategy for fighting poverty in developing countries up to 2015 the other [*ECON, 2003: 71 ff.*]. In the budget proposition for 2004, the government concluded: a combination of centralisation of policy decisions, the formulation of policy and strategy, state-to-state bilateral aid, and information to the MFA, with an increased decentralisation of the implementation to the (integrated) embassies in recipient countries. The integration of the former offices of NORAD's resident representatives into the MFA was finally settled:

---

59. In these embassies, the 'NORAD' people were in a clear majority. The first integrated embassy was established in Windhoek (Namibia) in 1991, with the former Secretary-General of the MDC as ambassador. In the following years, a few senior NORAD administrators were appointed ambassadors in some of these embassies, and in most of the remaining ones, the 'NORAD representative' served as second-in-command to the ambassador.

administratively, they were to be the responsibility of the Minister of Foreign Affairs.[60]

VII. CONCLUDING OBSERVATIONS

Public attitudes have been in favour of providing development assistance at a high level. This is reflected in opinion polls taken at regular intervals; towards the end of the 1990s, almost 90 per cent of respondents were in favour.[61] Traditionally NORAD, in addition to its own information service, has financed information activities about developing countries, development and development co-operation through NGOs. This information (and education) was filtered through the value system of the organisations concerned. The probability is that the positive attitude to development assistance, as reflected in the opinion polls, results to a great extent from this strategy and the active involvement of a variety of NGOs in public education.

NGOs have increasingly been used as channels for the ODA, both in pursuit of long-term development and as channels of humanitarian aid, especially in the 1990s, when about one-third of bilateral ODA was channelled through Norwegian, international and local NGOs. In the 1990s, the major role involving long-term development assigned to NGOs was that of a development agent *vis-à-vis* civil society in recipient countries, with particular emphasis on organisational development based on common values or interests.

Whereas private-sector transfers to developing countries have traditionally been at a low level compared with most DAC countries (and skewed to countries that are well-off compared with the traditional recipients of Norwegian ODA), Norwegian NGOs have themselves

---

60. NORAD was stripped of many of its functions. It was to remain an independent directorate, but its mandate was to be changed to that of providing technical advice for the MFA and the embassies, evaluation, and the administration of application-based aid through voluntary organisations, the private sector, media, etc. Several man-years were transferred to the MFA and the integrated embassies [*Proposition No. 1, 2003: 77ff.*]. The implementation of the reorganisation started early in 2004. It is therefore far too early to assess how it will work out in practice.

61. Since 1972, Statistics Norway has carried out opinion surveys for NORAD at regular intervals. In 1972, 72 per cent of the respondents were in favour of the aid, 19 per cent against and 9 per cent did not know. Since then, the trend was favourable until it dropped again in 1990 (77 per cent in favour, 17 per cent against and 6 per cent did not know). In the 1990s, the support increased again – to 85 per cent in 1993, 80 per cent in 1995, 84 per cent in 1996, and 88 per cent in 1999 and 2001. In 1999, 53 per cent found the volume level of aid satisfactory, while 11 per cent wanted to increase the volume and 30 per cent to reduce it (3 per cent wanted to end aid and 2 per cent did not know) [*Rønning, 2000: 12; http://www.ssb.no/uhjelphold/main.html*].

contributed quite extensively.[62] This active engagement, it must be assumed, improves knowledge of the situation prevailing in a multitude of different developing countries and generates empathy with fellow human beings, affecting attitudes in the wider society as well.

Basically, Norway's North–South policy, including the aid policy, has emerged from endogenous political processes. Among the political parties, the Christian Democrats have exerted a particularly strong influence, directly and indirectly, both on the stated and the implemented policy. Development co-operation has been singled out among the top priority issues of that party. However, Labour, traditionally the strongest single party in Parliament and more often than not in government position, has struck the balance and usually decided major policy outcomes. Also NGOs carry strong influence on policy outcomes, particularly the 'big five' humanitarian aid organisations (the Norwegian Church Aid, the People's Aid (Trade Union), the Norwegian Red Cross, Save the Children and the Norwegian Refugee Council) – along with the Church of Norway, the trade union movement, and the interest organisations for industry, farmers and exporters. Civil servants in central positions in the aid administration, the MFA and NORAD, have been well placed to combine experience from the implementation of aid with impulses from international and bilateral aid agencies and development finance organisations and signals from the major actors in the national arena – being themselves major actors within this policy area. Contrary to what has been the case in many other donor countries, the aid policy has been a high-profile issue on the national political agenda.

It emerges from the previous sections, however, that, in the 1990s and beyond, Norway's stated development co-operation policy has also been increasingly based on what may be termed 'the Washington–New York consensus' (see Chapter 1 of this volume). The international agenda – set by global conferences on specific themes, such as social development, women, children and the environment, in particular, and by major reports by the UN (especially the human development reports) and the Bretton Woods institutions – has strongly influenced the national agenda. The transformation of the international system in the late 1980s and early 1990s and the new emphasis on liberal political ideas (human rights,

---

62. According to DAC statistics (various annual reports), aid through Norwegian NGOs amounted to 0.09 per cent of GNI in 1984–85, 1989–90 and 1994–95, increasing to 0.11 per cent of GNI in 1999–2000. In 1999–2000, Norwegian NGOs ranked first among the DAC countries (the DAC average was 0.03 per cent of GNI); in 1994–95, they ranked second (with the Netherlands; Ireland peaked with 0.10 per cent); and in 1989–90 they peaked (with the Netherlands).

democracy, good government) that followed in its wake have both contributed strongly to the new approach – adapting Norway's development policy to internationally agreed objectives, strategies, norms and mechanisms.

This represents a new trend. In the 1970s and the early 1980s, policy-makers in Norway – along with those of a group of like-minded countries, the Netherlands and the fellow Scandinavian countries, in particular – took pride in their own norms and achievements within this policy area. The demand for greater donor co-ordination – argued for also during those years, not least in the DAC – was fended off: donor co-ordination was seen to imply adapting Norway's policy to those of the major (Western) powers. In contrast to these policies, policy-makers perceived Norway to be pursuing different values and objectives: its aid was perceived as basically altruistic, oriented to meet the needs of the poor in poor developing countries, not to serve the strategic or economic interests of the donor; it was generous with regard to both volume (as a percentage of GNP) and the conditions on which it was provided (grants, not loans, and mainly untied); it did not interfere in the internal affairs of the recipients, but was recipient-oriented; bilateral development co-operation was considered a long-term commitment, organised in a way that allowed the authorities of a small group of main partner countries to plan their development for years based on available resources, aid included (four-year revolving country programmes); and a large share of the aid was channelled through multilateral agencies, the UN in particular, removing the strings and the flag of the donor. These features of the aid policy were attuned to predominant values in Norwegian society [*Stokke, 1989*]. They were inspired by and in line with norms set for development assistance in the UN strategies for the first two development decades.

In the 1990s, many of these qualities became less pronounced. Basically, however, altruism prevailed, involving both justifications and objectives – although instrumental justifications were also brought forward: generous aid was good for Norway's self-identity as well as for its image abroad. Aid was increasingly seen as a tool in the foreign policy of Norway, although the main focus here continued to be on international common goods[63] and humanitarian intervention. It also continued to be generous: most aid was provided as grants and

---

63. Here and earlier in the text, the reference has been to the broad concept of 'international common goods' (such as international stability and peace, economic stability, the environment, etc.). It may also include elements of the more specific current concept of international public goods (e.g. health, education, sound drinking water). See, in particular, two publications that have emerged from the UNDP Human Development Report Office [*Kaul* et al.*(eds), 1999, 2003*].

became increasingly untied. Although 'new' objectives came to the fore – human rights, democracy, good governance – the core objectives of the 1970s and the 1980s, social and economic development in the South, emphasising poverty alleviation, continued to be task number one, along with a concern for the environment.

The main changes *vis-à-vis* the old model relate to the non-intervention principle, multilateral aid, and the (lack of) stability in aid relations involving main partner countries. The new attitude to the non-intervention principle, most strongly expressed in the guideline of recipient responsibility, was ambivalent. Non-interference was, on the one side, emphasised more strongly than ever: in the stated policy, the responsibility for prioritising, planning and implementing aid-financed activities was placed firmly on the doorstep of the recipient government. But the recipient authorities were expected to deliver according to the agreement signed by both parties: if not, future development co-operation might be negatively affected. The dilemma involved, however, relates to the asymmetric power relations typical of aid relations, particularly those involving poor, aid-dependent recipient governments (the primary targets of Norwegian development assistance). This also applies to the broader issue of aid conditionality – both first- and second-generation conditionality.

The new strategic approach, emphasising international co-operation and co-ordination of development policy as well as development assistance, has been justified by instrumental arguments: such co-ordination would make aid more effective. This argument is not new. The crucial questions have been and remain what difference this co-ordination makes to aid – what kind of development is aid promoting? And how is the donor–recipient power balance affected? Is the influence of the recipient government over its own country's development, all rhetoric about participation, ownership and it being in the driver's seat aside, reduced even further?

The objectives set for aid answer the first of these questions: Norway has increasingly associated itself with the international objectives set out by the UN Millennium Assembly, emphasising poverty reduction. The answer to the second set of questions, the rhetoric about putting the recipient authorities in the driving seat aside, is more complex. The argument that this strategy makes it possible for Norway to exert stronger influence on the policy of international agencies towards developing countries (implying that this is to their benefit) through large contributions and an active participation in the steering bodies of these organisations, than would be possible through its bilateral aid, may be meant for home consumption – although it may contain a grain of truth, as well.

There is, of course, the other side of the coin, admitted in the policy documents: Norway has to adapt its policy to what is agreed in these international bodies – where governments resisting reform traditionally hold a strong position. In an area where values cherished at home are brought to the fore – human rights, democracy, good governance, the environment, along with the reduction of poverty – this may be less of a problem.

In this chapter, the main focus has been on the stated policy in order to capture the philosophy and intentions driving the development co-operation and changes that have taken place over time, particularly in the 1990s and beyond. The stated intentions have not always been followed up: contradictions have been identified, particularly related to the actual allocation of development assistance.

This focus has resulted in important aspects only being touched upon. This applies, first of all, to the implementation of the stated aid policy: the problems involved when the intentions, norms and values imbedded in objectives and strategies are to be operationalised, particularly those related to poverty reduction, equity and economic and social justice, but also those related to the promotion of liberal political values (good governance, democracy and human rights). The operationalisation of such values posed a huge challenge in the 1970s and 1980s, when most of the aid was channelled to projects or more circumscribed programmes. This challenge has not diminished with the new strategic approach of the 1990s; on the contrary, for a small donor, new dimensions have been added.

The aid strategy since 1990 has had four core components: the principle of recipient responsibility, combined with programme aid (including sector programmes) that was to be anchored in the recipient country and co-ordinated by the aid providers. The first three components were rooted in previous aid strategies; the fourth represented the main novelty. This strategy represented a challenge for an aid administration that was, basically, accustomed to handling 'projects': how to get to grips with the policy-makers at various levels in the recipient country in order to influence central aspects of their policy in accordance with objectives and norms adopted for the aid policy? For a small donor, this challenge is no less *vis-à-vis* the major donors, bilateral as well as multilateral ones.

Rhetoric aside, these challenges have only to a very limited extent been met so far. If influence on outcomes in accordance with stated intentions is what matters, then the answer for a small donor is probably to concentrate on niches considered to be of particular importance, combined with developing its own competence within these niches, presence where decisions are arrived at – and alliance building with like-minded donors. Even so, for a small actor with high ambitions with regard to exerting influence, the discrepancy between rhetoric and reality is doomed to persist.

REFERENCES

Budget Recommendation No. 3, 2000 (Budsjett-innst. S. nr. 3 (2000–2001)), Innstilling fra utenrikskomiteen om bevilgninger på statsbudsjettet for 2001 vedkommende Utenriksdepartementet m.v. (Recommendation by Parliament's Standing Committee on Foreign Affairs and Constitutional Matters [later: Committee on Foreign Affairs] on Allocations in the Budget for 2001 Concerning the Ministry of Foreign Affairs and Others), Oslo.
Christian Democrats, 1995, KrFs motmelding (1995-96), Solidarisk sør-politikk. Hovedtrekk i norsk politikk overfor utviklingslandene (The Counter-Report of the Christian Democrats, 1995. A Solidarity South Policy. Main Patterns of Norway's Policy towards Developing Countries), Oslo: Christian Democrats.
ECON, 2003, *Evaluation of the Norwegian Development Policy Administration*, Report 19/03, Oslo: ECON (April).
Engen Committee, 1961, Innstilling fra Utvalget for utredning av spørsmålet om Norges hjelp til utviklingslandene (Recommendations by the Committee Appointed to Review the Issue of Norway's Aid to Developing Countries), Oslo: Utenriksdepartementet (Ministry of Foreign Affairs).
Forster, Jacques and Olav Stokke, 1999, 'Coherence of Policies Towards Developing Countries: Approaching the Problematique', in Forster and Stokke (eds).
Forster, Jacques and Olav Stokke (eds), 1999, *Policy Coherence in Development Co-operation*, London: Frank Cass (EADI Book Series).
Hveem, Helge, 1977, *En ny økonomisk verdensorden og Norge* (A New International Economic Order and Norway), Oslo: Universitetsforlaget.
Hveem, Helge, 1989, 'Norway: The Hesitant Reformer', in Pratt (ed.).
Kaul, Inge, Isabelle Grunberg and Mark A. Stern (eds), 1999, *Global Public Goods: International Cooperation in the 21st Century*, New York: Oxford University Press.
Kaul, Inge, Pedro Conceico, Katell Le Gulven and Ronald U. Mendoza (eds), 2003, *Providing Global Public Goods: Managing Globalization*, New York: Oxford University Press.
Liland, Frode and Kirsten Alsaker Kjerland, 2003, *Norsk utviklingshjelps historie*, Bind 3, *1989–2002: På bred front* (The History of Norway's Development Assistance, Vol. 3, 1989–2002: Approaching Broadly), Bergen: Fagbokforlaget.
Løvbræk, Asbjørn, 1990, 'International Reform and the Like-Minded Countries in the North–South Dialogue 1975–1985', in Pratt (ed.).
MFA, 1997a, Strategi for miljørettet utviklingssamarbeid (Strategy for Environment in Development Co-operation), Oslo: Utenriksdepartementet (April).
MFA, 1997b, Norsk bistand i fokus, Bistandspolitisk redegjørelse, Rapport om norsk bistandsvirksomhet i 1996 (Norwegian Aid in Focus, Statement on the Development Policy, Report on Norwegian Development Co-operation in 1996), Oslo: Utenriksdepartementet.
MFA, 1999a, *Strategi for støtte til næringsutvikling i Sør* (Strategy for the Support of Productive Sector Development in the South), Oslo: Utenriksdepartementet (February).
MFA, 1999b, Norsk bistand i fokus, Utviklingspolitisk redegjørelse 1999, Rapport om norsk utviklingssamarbeid i 1998 (Norwegian Aid in Focus, Statement on the

Development Policy 1999, Report on Norwegian Development Co-operation in 1998), Oslo: Utenriksdepartementet (May).

MFA, 1999c, *Strategi for styrking av forskning og høyere utdanning i tilknytning til Norges forhold til utviklingslandene* (Strategy for the Strengthening of Scientific Research and Higher Education in Connection with Norway's Relations with Developing Countries), Oslo: Utenriksdepartementet (March).

MFA, 2000, Norsk bistand i fokus, Utviklingspolitisk redegjørelse 2000, Rapport om norsk utviklingssamarbeid i 1999 (Norwegian Aid in Focus, Statement on the Development Policy 2000, Report on Norwegian Development Co-operation in 1999), Oslo: Utenriksdepartementet (June).

MFA, 2002a, *Kamp mot fattigdom. Regjeringens handlingsplan for bekjempelse av fattigdom i sør mot 2015* (Fight against Poverty. The Government's Plan of Action for Fighting Poverty in the South towards 2015), Oslo: Utenriksdepartementet (March).

MFA, 2002b, *Norsk bistand i fokus. Utviklingspolitisk redegjørelse 2002, Rapport om norsk utviklingsarbeid 2001* (Focus on Norwegian Development Co-operation), Oslo: Utenriksdepartementet.

MFA, 2003a, *Education – Job Number 1, Norwegian Strategy for Delivering Education for All by 2015*, Oslo: Norwegian Ministry of Foreign Affairs.

MFA, 2003b, Education for all through Norway's Development Cooperation, Technical Basis for the Strategy for Education of the Norwegian Ministry of Foreign Affairs: 'Education – Job number 1', Oslo: Norwegian Ministry of Foreign Affairs (January) (mimeo).

NORAD, 1975, 1979 (3nd (*sic.*) edition), *Norway's Economic Relations with Developing Countries* (off-print of Report No. 94, 1995), Oslo: NORAD.

NORAD, 1990, *Strategies for Development Cooperation – NORAD in the Nineties*, Oslo: NORAD.

NOU, 1995:5, *Norsk sør-politikk for en verden i endring* (Norwegian South Policy for a Changing World), Report by the North–South/Aid Commission, Oslo: Statens forvaltningstjeneste/Statens trykning.

OECD, 1992, *Development Co-operation, 1992 Report*, Paris: OECD, DAC.

OECD, 1996, *Shaping the 21st Century: The Contribution of Development Co-operation*, Paris: OECD, DAC.

OECD, 2001, *The DAC Journal Development Co-operation, 2000 Report*, Paris: OECD, DAC.

OECD, 2002, *The DAC Journal Development Co-operation, 2001 Report*, Paris: OECD, DAC.

OECD, 2003, *The DAC Journal Development Co-operation, 2002 Report*, Paris: OECD, DAC.

OECD, 2004, *The DAC Journal Development Co-operation, 2003 Report*, Paris: OECD, DAC.

Palme Committee, 1962, 'Rapport rörande det svenska utvecklingsbiståndet' (Report on Sweden's Development Assistance), Appendix to Prop. 1962:100.

Pearson, Lester B., 1969, *Partners in Development. Report of the Commission on International Development*, New York: Praeger.

Pratt, Cranford (ed.), 1989, *Internationalism under Strain*, Toronto and London: Toronto University Press.

Pratt, Cranford (ed.), 1990, *Middle Power Internationalism*, Kingston and London: McGill-Queens University Press.

Proceedings of Parliament, 1998 (Stortinget (1997–1998), (1998–1999)), Oslo.

Proceedings of Parliament, 2000 (Stortinget (1999–2000), (2000–2001)), Oslo.

Proceedings of Parliament, 2001 (Stortinget (2000–2001), (2001–2002)), Oslo.

Prop. 100, 1962, Kungl. Maj:ts proposition nr 100 år 1962, Kungl. Maj:ts proposition till riksdagen angående svenskt utvecklingsbistånd; av den 23 februari 1962 (The Swedish Government's Proposition to Parliament on Swedish Development Assistance), Stockholm.

Proposition No. 109, 1967 (St.prp. nr. 109 (1966–67)), I. Om den videre utbygging av Norges bistand til utviklingslandene. II. Om opprettelsen av 'Direktoratet for utviklingshjelp' m.m. (I. About the Continued Development of Norway's Aid to the Developing Countries. II. About the Establishment of 'The Directorate for Development Co-operation', etc.), Oslo: Utenriksdepartementet.
Proposition No. 1, 1998 (St.prp nr. 1 (1998–99)), For budsjetterminen 1999. Utenriksdepartementet (For the Financial Year 1999. Ministry of Foreign Affairs), Oslo: Utenriksdepartementet.
Proposition No. 1, 2001 (St.prp. nr. 1 (2001–2002)) For budsjetterminen 2002. Utenriksdepartementet (For the Financial Year 2002. Ministry of Foreign Affairs), Oslo: Utenriksdepartementet.
Proposition No. 1, 2002 (St.prp. nr. 1 (2002–2003)) For budsjetterminen 2003 (For the Financial Year 2003), Oslo: Utenriksdepartementet.
Proposition No. 1, 2003 (St.prp. nr. 1 (2002–2004)) For budsjetterminen 2004 (For the Financial Year 2004), Oslo: Utenriksdepartementet.
Recommendation No. 75, 1962 (Innst. S. nr. 75 (1961–62)), Innstilling fra utenriks- og konstitusjonskomitéen om Norges hjelp til utviklingslandene (Recommendation by the Committee on Foreign Affairs on Norway's Aid to Developing Countries), Oslo.
Recommendation No. 167, 1967 (Innst. S. nr. 167 (1967–68), Innstilling fra utenriks- og konstitusjonskomitéen om den videre utbygging av Norges bistand til utviklingslandene (Recommendation by the Committee on Foreign Affairs on the Continued Development of Norway's Aid to Developing Countries), Oslo.
Recommendation No. 135, 1973 (Innst. S. nr. 135 (1972–73)), Innstilling fra utenriks- og konstitusjonskomitéen om enkelte hovedspørsmål vedrørende Norges samarbeide med utviklingslandene (Recommendation by the Committee on Foreign Affairs on Some Major Issues Related to Norway's Co-operation with Developing Countries), Oslo.
Recommendation No. 192, 1976 (Innst. S. nr. 192 (1975/76)), Innstilling fra den forsterkede utenriks- og konstitusjonskomité om Norges økonomiske samkvem med utviklingslandene (Recommendation by the enlarged Committee on Foreign Affairs on Norway's Economic Co-operation with Developing Countries), Oslo.
Recommendation No. 255, 1981 (Innst. S. nr. 255 (1980–81)), Innstilling fra utenriks- og konstitusjonskomiteen om Noregs samarbeid med utviklingslanda i 1979 (Recommendation by the Committee on Foreign Affairs on Norway's Co-operation with Developing Countries in 1979), Oslo.
Recommendation No. 186, 1987 (Innst. S. nr. 186 (1986–87)), Innstilling fra utenriks- og konstitusjonskomiteen om Norges hjelp til utviklingslandene (Recommendation by the Committee on Foreign Affairs on Norway's Aid to the Developing Countries), Oslo.
Recommendation No. 195, 1993 (Innst. S. nr. 195 (1992–93)), Innstilling fra utenriks- og konstitusjonskomiteen om utviklingstrekk i Nord–Sør-forholdet og Norges samarbeid med utviklingslandene (Recommendation by the Committee on Foreign Affairs on Trends in North–South Relations and Norway's Co-operation with the Developing Countries), Oslo.
Recommendation No. 229, 1996 (Innst. S. nr. 229 (1995–96)), Innstilling fra utenrikskomiteen om hovedtrekk i norsk politikk overfor utviklingslandene (Recommendation by the Committee on Foreign Affairs on Major Features in Norwegian Policy vis-à-vis Developing Countries), Oslo.
Recommendation No. 28, 1999 (Innst. S. nr. 28 (1999–2000)), Innstilling fra utenrikskomiteen om utviklingspolitisk redegjørelse fra utviklings- og menneskerettighetsminister Hilde Frafjord Johnsen (Recommendations by the Committee on Foreign Affairs on the Statement on Development Policy by the Minister of International Development and Human Rights, Ms Hilde Frafjord Johnsen), Oslo.
Recommendation No. 23, 2000 (Innst. S. nr. 23 (2000–2001)), Innstilling fra utenrikskomiteen om menneskeverd i sentrum. Handlingsplan for menneskerettigheter

(Recommendations by the Committee on Foreign Affairs on Placing the Value of Human Beings in the Centre. Plan of Action for Human Rights), Oslo.

Report No. 23, 1961 (St.meld. nr. 23 (1961–62)), Norges hjelp til utviklingslandene (Norway's Aid to Developing Countries), Oslo: Utenriksdepartementet.

Report No. 29, 1972 (St.meld. nr. 29 (1971–72)), Om enkelte hovedspørsmål vedrørende Norges samarbeid med utviklingslandene (On Some Major Issues Related to Norway's Co-operation with Developing Countries), Oslo: Utenriksdepartementet.

Report No. 94, 1975 (St.meld. nr. 94 (1974–75)), Norges økonomiske samkvem med utviklingslandene (Norway's Economic Co-operation with Developing Countries), Oslo: Utenriksdepartementet.

Report No. 36, 1984 (St.meld. nr.36 (1984–85)), Om enkelte hovedspørsmål i norsk utviklingshjelp (On Some Major Issues in Norwegian Development Assistance), Oslo: Departementet for utviklingshjelp (Ministry of Development Co-operation).

Report No. 34, 1987 (St.meld. nr. 34 (1986–87)), Om enkelte hovedspørsmål i norsk utviklingshjelp. Tilleggsmelding til St.meld. nr. 36 (1984–85) (On Some Main Issues in Norwegian Development Assistance. Supplementary Report to Report No. 36, 1984), Oslo: Departementet for utviklingshjelp.

Report No. 49, 1991 (St.meld. nr. 49 (1990-91)), Om Norges samarbeid med utviklingslandene 1990 (On Norway's Co-operation with Developing Countries in 1990), Oslo: Utenriksdepartementet.

Report No. 51, 1992 (St.meld. nr. 51 (1991–92)), Om utviklingstrekk i Nord–Sør forholdet og Norges samarbeid med utviklingslandene (On Trends in North–South Relations and Norway's Co-operation with the Developing Countries), Oslo: Utenriksdepartementet.

Report No. 9, 1993 (St.meld. nr. 9 (1993–94)), Om Norges samarbeid med utviklingslandene i 1992 (On Norway's Co-operation with Developing Countries in 1992), Oslo: Utenriksdepartementet.

Report No. 8, 1994 (St.meld. nr. 8 (1994–95)) Om samarbeidet mellom Noreg og utviklingslanda i 1993 (On the Co-operation between Norway and the Developing Countries in 1993), Oslo: Utenriksdepartementet.

Report No. 19, 1995 (St.meld. nr. 19 (1995–96)), En verden i endring (A World in Change), Oslo: Utenriksdepartementet.

Report No. 21, 1999 (St.meld. nr. 21 (1999–2000)), Menneskeverd i sentrum. Handlingsplan for menneskerettigheter (Placing the Value of Human Beings in the Centre. Plan of Action for Human Rights), Oslo: Utenriksdepartementet.

Report No. 35, 2004 (St.meld. nr. 35 (2003–2004), Felles kamp mot fattigdom (Joint Fight against Poverty), Oslo: Utenriksdepartementet.

Ruud, Arild Engelsen and Kirsten Alsaker Kjerland, 2003, *Norsk utviklingshjelps historie*, Bind 2, *1975–1989: Vekst, velvilje of utfordringer* (The History of Norway's Development Assistance, Vol. 2, 1975–1989: Growth, Good Will and Challenges), Bergen: Fagbokforlaget.

Rønning, Elisabeth, 2000, *Holdninger til og kunnskap om norsk utviklingshjelp 1999* (Attitudes to and knowledge about Norwegian development assistance 1999), Oslo: Statistics Norway.

Simensen, Jarle, 2003, *Norsk utviklingshjelps historie*, Bind 1, *1952–1975: Norge møter den tredje verden* (The History of Norway's Development Assistance, Vol. 1, 1952–1975, Norway Approaches the Third World), Bergen: Fagbokforlaget.

Stokke, Olav, 1975, *Norsk utviklingsbistand* (Norway's Development Assistance), Uppsala: Scandinavian Institute of African Studies.

Stokke, Olav, 1978, *Sveriges utvecklingsbistånd och biståndspolitik* (Sweden's Development Assistance and Development Co-operation Policy), Uppsala: Nordiska afrikainstitutet.

Stokke, Olav, 1979, *Norge og den tredje verden* (Norway and the Third World), Oslo: Universitetsforlaget.

Stokke, Olav, 1980, 'En ny økonomisk verdensordning – program og realiteter (A New International Economic Order: Programme and Results), in Stokke (ed.).

Stokke, Olav, 1984, 'Norwegian Aid: Policy and Performance', in Stokke (ed.).
Stokke, Olav, 1989, 'The Determinants of Norwegian Aid Policy', in Stokke (ed.).
Stokke, Olav, 1991, 'Norsk bistandspolitikk ved inngangen til 1990-tallet' (Norwegian Aid Policy on the Doorsteps to the 1990s), *Norsk Utenrikspolitisk Årbok 1990*, Oslo: NUPI.
Stokke, Olav, 1992a, 'Nord–Sør-meldingen: Nye signaler?' (The North–South White Paper: New Signals?), in Stokke (ed.).
Stokke, Olav, 1992b, 'Mål, strategi og prinsipper for norsk bistand: Old bottles?' (Aims, Strategy and Guidelines for Norwegian Development Assistance: Old Bottles?), in Stokke (ed.).
Stokke, Olav, 1995a, 'Aid and Political Conditionality: Core Issues and State of the Art', in Stokke (ed.).
Stokke, Olav, 1995b, 'Aid and Political Conditionality: The Case of Norway', in Stokke (ed.).
Stokke, Olav, 1999, 'Development Co-operation and Policy Coherence: The Case of Norway', in Forster and Stokke (eds).
Stokke, Olav, forthcoming, *The Dilemmas of Policy Coherence. Aspirations and Realities in Norway's Policies Towards Developing Countries*.
Stokke, Olav (ed.), 1980, *En ny økonomisk verdensordning?*, Oslo: NUPI.
Stokke, Olav (ed.), 1984, *European Development Assistance*, Vol. 1, *Policies and Performance*, Vol. 2, *Third World Perspectives on Policies and Performance*, Tilburg: EADI (EADI Book Series 4).
Stokke, Olav (ed.), 1989, *Western Middle Powers and Global Poverty. The Determinants of the Aid Policies of Canada, Denmark, the Netherlands, Norway and Sweden*, Uppsala: Scandinavian Institute of African Studies.
Stokke, Olav (ed.), 1992, *Norsk Nord–Sør-politikk: Lever den opp til sitt rykte?* (Norwegian North–South Policy: Up to its Reputation?), Oslo: Norwegian Institute of International Affairs/Research Report No. 163 (December).
Stokke, Olav (ed.), 1995, *Aid and Political Conditionality*, London: Frank Cass (EADI Book Series 16).
UN, 2000, *United Nations Millennium Declaration* (ResolutionA/RES/55/2 8 September 2000) Millennium Summit, New York, 6–8 September 2000, New York: UN Department of Public Information (DPI/2163).
Wohlgemuth, Lennart 2003, 'The History of Norwegian Aid: A Case for Contextualisation', *Forum for Development Studies*, Vol. 30, No. 2, Oslo: NUPI (December).
http://www.ssb.no/yhjelphold/main.html

14

Spanish Foreign Aid: Flaws of an Emerging Framework

JOSÉ ANTONIO ALONSO

I. INTRODUCTION

In merely two decades, Spain has radically changed its position in the international community. As recently as 1979, Spain was still an aid recipient, yet in 2003 it was ranked as the world's eleventh largest donor, contributing 2.9 per cent of the total aid of the OECD's Development Assistance Committee (DAC). In that time, it had to create the human resources, instruments and institutions necessary to develop a framework for foreign aid. As a result there were great improvements in the Spanish aid system, but remarkable shortcomings remain, such as a lack of integration between financial and technical co-operation instruments, unsuitable geographical and sectoral orientations of aid disbursement and limited technical capacity and management ability. Lack of experience and the weakness of political commitment to aid policy are the main reasons for some of these defects. This chapter briefly explains the evolution of Spanish aid by means of a critical analysis of its main characteristics. Thus, it points out some of the reform tasks involved that the new government, elected in March 2004, will need to face.

II. BUILDING THE INSTITUTIONAL FRAMEWORK OF SPANISH AID

In the mid-1980s the Spanish government took on the task of creating a specialised institutional framework for the management of aid. In 1985, the State Secretariat for International Co-operation and Latin America (SECIPI) was set up as part of the Ministry of Foreign Affairs, with a mandate to direct foreign aid policy. In 1986, the Interministerial Commission for International Co-operation (CICI) was established with the objective of co-ordinating co-operation activities inside central government departments. Three years later, within the

SECIPI, the Planning and Evaluation Office (OPE) was also created to assist the SECIPI technically in programming and evaluation of aid – a mission that would become effective only in the latter half of the 1990s.

Prior to these new institutions, a foreign aid loan instrument existed, known as the Development Aid Fund (FAD). Set up in 1976, its purpose was to make soft loans to developing countries, tied totally or partially to the acquisition of Spanish goods and services. The fact that this instrument was introduced almost a decade before the creation of an institution responsible for foreign aid policy indicates the commercial objective of export promotion that inspired it. An Interministerial Commission on Development Aid Fund (CIFAD) was created to manage applications for FAD loans.

A further important institutional advance was made through the creation, in 1988, of the Spanish Agency for International Co-operation (AECI), under the auspices of the SECIPI, to manage bilateral grants. The AECI in effect became the institution responsible for managing development programmes, projects, technical co-operation and humanitarian aid, and has benefited from the network of 29 Technical Co-operation Offices (OTC), located in the priority recipient countries. Thus, Spanish foreign aid adopted a dual system, with one body (SECIPI) given the political oversight, while the other (AECI) was entrusted with the management of bilateral grants.

This evolving system was consolidated by Spain's incorporation into two influential international bodies: the European Union and the Development Assistance Committee of the OECD. Attaining EU membership in 1986 allowed Spain not only to take part in the decision-making process of an important multilateral actor, but also to learn from the experiences and good practices of other EU donors. Furthermore, its incorporation into the DAC, in 1991, stimulated the convergence of Spanish aid policy with international criteria and standards – a convergence that might well have been slower and more incomplete without the stimulus afforded by its international connections.

The second half of the 1980s brought new actors into the aid system. First of all were the NGOs, which, despite their lack of experience in the development co-operation arena, have secured a growing presence both in managing aid projects and in political advocacy. The technical and managerial shortcomings of the central and decentralised administrations have favoured an increasingly important role for NGOs, which administered close to one-third of bilateral ODA in 2002. Second, regional and municipal governments became more involved, developing aid programmes from their own budgets. From modest beginnings, the involvement of this mode of public co-operation – called 'decentralised co-operation' – is now a substantial part of the Spanish system, contributing 14 per cent to total ODA in 2002 (nearly 25 per cent of all bilateral aid).

In 1995, as a result of public pressure, the Development Co-operation Council was created to assist aid policy by allowing for extensive participation by the aid community – government, NGOs, trade unions, employers' organisations and academic experts. The aim of this consultative body was to create a forum in which the various participants could make reports and recommendations aimed at improving the quality and effectiveness of aid policy.

Finally, the consolidation of this aid system was furthered by the Law on International Co-operation for Development, approved in July 1998. This law defines the principles, objectives and priorities of aid policy, as well as defining the institutional framework and its instrumental resources. The law, which received broad public and parliamentary support, aligned the Spanish system with the dominant doctrine among donors, declaring poverty to be the central objective for aid policy. As a result of the law, the *Strategic Guidelines for Spanish Co-operation Policy 2001–2004* was approved in 2000, becoming the first planning document of the Spanish aid system.

III. THE EVOLUTION OF SPANISH OFFICIAL DEVELOPMENT ASSISTANCE (ODA)

Four stages may be identified in the evolution of Spanish ODA over the past two decades (see Table 14.1). The first, up to 1988, was somewhat unstable; the volume of resources handled in the first two years of the 1980s was not repeated until the end of the period, after undergoing a severe decline in 1983. The second stage is characterised by a rapid expansion of resources, tripling the volume of the previous period. This bullish approach lasted until 1992, the year in which the volume of ODA peaked at US$1.5 billion. The third stage, until the end of the 1990s, began with a relative stagnation in resources, at an annual sum of approximately $1.3 billion. Finally, in 2001 aid underwent a remarkable spurt of growth, reaching $1.7 billion as a result of a one-off triangular cancellation of debt with Nicaragua and Guatemala. In 2002 and 2003, the most recent years for which data are available, aid resources amounted to $1.7 and 2 billion, respectively, including a notable expansion in loan components (FAD and micro-credits) in 2002 and an increase in emergency aid (linked to Iraq) in 2003. Overall, the data from recent years tend to confirm the high volatility of Spanish aid, indicating the lack of both a clear model for operations and a solid budgetary base.

TABLE 14.1
THE DISTRIBUTION OF ODA, 1981–2002 (US$m. AND %)

	1981–84	1985	1986	1987	1988	1989	1990	1991	1992	1993	1994	1995	1996	1997	1998	1999	2000	2001	2002
Volume (US$m.):																			
Bilateral aid	134	129	59	114	158	265	633	761	1100	936	854	816	888	765	838	829	720	1150	998
Grants	61	53	52	85	77	122	227	243	250	190	257	533	563	540	666	653	603	966	769
Loans	73	76	7	28	81	143	407	518	849	746	597	283	325	225	171	176	117	184	229
Multilateral	37	40	145	118	90	277	332	501	419	367	450	532	364	469	538	534	475	588	714
Total ODA	171	169	203	232	248	541	965	1262	1518	1304	1305	1348	1251	1234	1376	1363	1195	1737	1712
ODA/GDP	0.1	0.1	0.09	0.08	0.14	0.14	0.2	0.24	0.27	0.28	0.27	0.24	0.22	0.24	0.24	0.23	0.22	0.30	0.26
Percentages:																			
Bilateral aid	78.3	76.3	29	49.1	63.7	48.9	65.6	60.3	72.4	69.8	65.1	60.3	70.9	62.0	60.9	60.8	60.2	66.2	58.2
Grants	35.6	31.3	25.6	36.6	31	22.5	23.5	19.2	16.4	15.8	18.9	39.3	45	43.7	48.4	47.9	50.4	55.6	44.9
Loans	42.7	44.9	3.4	12	32.6	26.4	42.1	41	55.9	53.9	46.2	20.9	25.9	18.2	12.4	12.9	9.8	10.6	13.3
Multilateral	21.6	23.6	71.4	50.8	36.9	51.2	34.4	39.7	27.6	30.2	34.9	39.5	29.1	38.0	39.1	39.1	39.7	33.8	41.7
Total ODA	100	100	100	100	100	100	100	100	100	100	100	100	100	100	100	100	100	100	100

Source: DAC.

As noted, the most fundamental increase in the amount of Spanish ODA came from 1988 to 1992, a period in which Spanish aid was consolidated from both an institutional and a budgetary point of view. However, most of the progress associated with that period was a result of decisions taken outside the national aid system. The increase in aid was due, first, to the contributions resulting from Spain's entry into the European Union – to the EU budget and the European Development Fund (EDF) – and, second, to the allocations of FAD credits resulting from a more active commercial policy. Hence, it is the reduced strength of these two components – and not so much the internalisation of supposed 'aid fatigue' – which explains the stagnation in the amount of Spanish aid in the second half of the 1990s.

In the first period, the ODA/GDP ratio remained at a modest 0.10 per cent. In 1988 the ratio rose sharply, reaching 0.28 per cent in 1992; in 1994 a steady decline began, reaching the lowest point of 0.22 per cent in 1996. From 1997 to the end of the decade, it fluctuated between 0.23 and 0.24 per cent. In the past three years, the ratio has risen to its highest level of 0.30 per cent, in 2001, before falling to 0.26 per cent in 2002 and 0.25 per cent in 2003 – though this is still higher than the average recorded in the late 1990s.

In any case, as a result of the agreements adopted by the EU at the Monterrey Conference, Spain will be required to devote at least 0.33 per cent of its GDP to aid in 2006. Given the trends in the flow of aid in the last few years, achieving this goal will require a remarkable effort on the part of the Spanish aid system. There is so far no strategy outlining how this may be achieved.

Despite the recent growth in its aid resources, Spain was ranked eleventh among donors in 2003. The position changes, however, if the size of the aid effort is related to GDP or to population. According to the ODA/GDP ratio, Spain (at 0.25 per cent) ranks fifteenth out of the 22 DAC members. Its quota is similar to the weighted DAC average, although perceptibly below that of the EU average. When its efforts are expressed in terms of aid per capita, Spain (with $39 per person) ranks eighteenth, ahead of Greece, Portugal, Italy and New Zealand. Yet in this case the coefficient is below both the DAC and the EU averages.

IV. THE COMPOSITION OF AID

As indicated above, factors like integration into the European Community in the second half of the 1980s, increasing multilateral cooperation, and the policy of support for the internationalisation of business at the start of the 1990s, strengthening commercial loans, played a key role in the evolution of Spanish aid. As a consequence, the components of Spain's foreign aid changed remarkably during the period, in contrast to the majority of DAC members, whose main aid components were kept largely stable. In fact, these variations highlight Spain's lack of an established aid policy model and consolidated political and budgetary practices.

Despite the variability, however, it is possible to detect some traits that characterise Spanish ODA. Priority has not been given to grants in the bilateral aid. Data from the 1990s show that only 19 per cent of this aid was provided as grants while the DAC average was 58 per cent in the first half of the decade, and 45 per cent compared to 62 per cent in the second. In 2002 – the year with the most recent data – the corresponding quotas were 45 per cent and 61 per cent respectively. Although the differences seem to get smaller with time, they are still perceptible, revealing the lower weight of the most generous component of aid (grants).

This characteristic may be further clarified by analysing the relative influence of the main component of bilateral grants aid. Comparisons with the DAC average in the last two trienniums (1994–96 and 1997–99) reveal the relative lack of importance Spain has attributed to 'projects, programmes and technical co-operation', a component more generally associated with the donor's technical capacities in the management of aid. Furthermore, this component amounted to 27 per cent of Spanish bilateral grants aid in 1994–96, as opposed to an average of 38 per cent in the DAC, and to 21 per cent as opposed to 40 per cent, respectively, in 1997–99. This reveals the relative weakness of the Spanish system in confronting the formulation and management of direct intervention in beneficiary countries.

The technical weakness of Spanish co-operation is shown also by the scarce use of new forms and instruments of aid, such as direct budget support, sectorial programmes, basket funds or sector-wide approaches (Swaps). Up to now, Spanish co-operation is based, mainly, on development projects, with limited capacity to participate in a more demanding international co-ordination, such as Swaps or Poverty Reduction Strategies.

A second trait is the significant role played by commercial loans. In fact, bilateral aid in the past has focused on FAD credits, such as in 1992–93, when more than half of total ODA was provided as commercial credits. This is a characteristic that seems to have diminished with time, at least when aid is expressed in net terms. FAD credits amounted to 48 per cent of Spanish aid between 1990 and 1994, as opposed to 11 per cent for the DAC. These quotas were, respectively, reduced to 18 per cent and 5 per cent between 1995 and 1999, with an even more pronounced levelling off at 11 and 4 per cent, respectively, in 2002.

This reduction in the relative influence of FAD credits is a consequence of the increasing difficulties of making new allocations, as well as the increasing importance of repayments. On the one hand, the OECD's so-called 'Helsinki Agreement' has entailed a more rigorous selection of operations and beneficiaries for this type of aid. This has had a direct effect on Spanish aid, forcing the exclusion in the middle of the 1990s of some Latin American countries like Mexico, Argentina, Chile, Uruguay and Venezuela from the list of eligible recipients. Difficulties in making FAD allocations have also arisen from the more favourable terms of access to private capital markets for some developing countries in the second half of the 1990s. On the other hand, as the FAD credits mature, the amount repaid becomes more important in relation to total annual allocations, in turn reducing the contribution of this instrument in net ODA.

The traditional weighting of bilateral aid in favour of FAD credits has had important consequences. First, it has caused a certain deviation from aid's primary objectives, given the preference that this policy instrument gives to the donor's commercial objectives over and above developmental needs. Second, it has made aid more costly for the recipient because it generates a larger external debt. Third, it has lessened the potential room for manoeuvre of the beneficiary, by tying resources partly to the acquisition of Spanish goods and services. Lastly, to the detriment of the poorest, it has upset the selection of beneficiaries, giving greater leverage to middle-income countries with more promising markets. Furthermore, the presence of some higher-income recipients like China or Indonesia has no other explanation than the attractions of their potential markets for Spanish business.

For these reasons there has been considerable criticism of this type of instrument, with questions raised over its excessive influence on aid policy, its obvious commercial connections, and its dissociation from the rest of the objectives and instruments of development co-operation. In addition, the DAC has insisted on the need to promote closer integration of such credits into general aid policy, and to orient resources to the service of a clearer objective to fight poverty. As stated in the last peer review of the DAC, 'Spain should continue its review of FAD loans in a comprehensive manner to reinforce their poverty reduction orientation' [*OECD, 2002: 16*] – an observation that seems highly pertinent given the level of autonomy in the management of credits with regard to the rest of aid policy. However, the government has strongly resisted changing current practice.

In another respect, the weight of reimbursable aid has risen since 1998 as a result of the creation of a Micro-credit Concession Fund (CM), which from a budgetary perspective enjoys considerable financial backing. The establishment of this instrument in Spanish international co-operation was welcomed as a positive step by the aid community, although questions were raised over the fact that financing is exclusively through credits (with hardly any donations) and over its partial orientation towards the poorest sectors of the recipient populations. Indeed, the main objective of the programme appears to be the strengthening of local financial sectors rather than improving the poorest people's access to credit (see Alonso *et al.* [2002]).

The weight of the third basic component of aid – multilateral aid – has suffered slight variations. As already indicated, the progress ODA experienced in the second half of the 1980s appeared to be associated with the contributions brought about by Spain's integration into the European Community. Throughout those years, there was therefore a remarkable increase in the relative importance of multilateral aid, which between 1986 and 1991 accounted for 47 per cent of total ODA, a quota markedly higher than the DAC average of 28 per cent.

Since 1991, the distribution between bilateral and multilateral aid has moved towards the 'norm': the proportion of multilateral aid has diminished, although it is still above the DAC average. The weight of this multilateral component does not reflect a particular slant in Spanish aid, which traditionally abstains from reinforcing multilateral commitments, but shows the existing imbalance between aid resulting from obligatory contributions and more autonomous aid components. Nevertheless, in the last few years an effort has been made to enhance the Spanish presence in international financial institutions (the World Bank, the Inter-American Development Bank, the Andean Development Corporation and the Central American Bank for Economic Integration). Yet this seems to be a policy aimed at improving the access of Spanish business to markets promoted by the development banks, and not an action intended to reinforce the multilateral commitment of Spain's aid policy. In fact, Spanish voluntary contributions to non-financial institutions are very low. And so, Spain has run a passive policy towards most multilateral institutions, hardly proposing any initiatives in this regard. However, Spanish participation has been slightly better within the EU, particularly with regard to issues that have involved a more active commitment with Latin American countries.

V. THE GEOGRAPHICAL ALLOCATION OF AID

The geographical distribution of Spanish aid suggests that it is relatively dispersed, highly conditional on foreign and commercial policy, and oriented mainly to middle-income countries, with little attention paid to the poorest countries.

The first characteristic, the relative dispersion of aid, is especially pronounced in the large number of countries with which Spain has cooperated in the last few years – an average of 110 countries (or territories). Given the limited amount of resources available, it is assumed that in many cases the amount of aid allocated would be minimal. In fact, during the triennium 1997 to 2000, half of the recipients received less than $1 million, such a meagre quantity that its effectiveness must have been limited. Thus, in order to become more efficient, it will be necessary to develop a more precise definition of what Spanish aid should concentrate on. This is the theoretical objective of the *Strategic Guidelines* in listing the 28 priority countries for Spanish aid, in addition to the Palestinian territories and the Western Saharan population (see Table 14.2). However, the *Strategic Guidelines* does not define the criteria needed to translate these geographical preferences into operative terms, including the allocation of resources between countries.

TABLE 14.2
PRIORITY COUNTRIES ACCORDING TO THE *STRATEGIC GUIDELINES*

	Africa	*Asia*	*Europe*	*Latin America*
LLDCs	Angola Cape Verde Equatorial Guinea Guinea Bisseau Mauritania Mozambique Sao Tomé and Principe			
OLICs	Senegal	China Vietnam		Honduras Nicaragua
LMICs	Algeria Morocco Namibia South Africa Tunisia	Philippines	Albania Bosnia-Herzegovina Yugoslavia F.R.	Bolivia Dominican R. Ecuador El Salvador Guatemala Paraguay Peru

The level of dispersion is overshadowed by a high concentration of resources on the main recipients. Half of Spanish bilateral aid is concentrated on ten countries; and if this is extended to the top 20 recipients, the quota of bilateral aid rises to 75 per cent. More remarkable still is the relationship Spain has had with the main recipients during recent years. While countries defined as priority recipients in the *Strategic Guidelines*, such as Tunisia, cannot be found among the 30 top aid recipients, Indonesia, one of the top five, is not even given priority status. This seems to be a result of the autonomy in the allocation of FAD credits and the lack of a commitment to establish and enforce the priorities of the *Strategic Guidelines* for the allocation of aid.

Analysis of the geographical distribution of aid reveals that the criteria of regional preference concur with Spain's foreign policy priorities (see Table 14.3). At the beginning, there was a marked interest in Latin America, to which almost two-thirds of bilateral aid was directed at the end of the 1980s. Equatorial Guinea, the only African ex-colony that has maintained links with Spain, was also a member of this privileged group of important aid recipients, in spite of its frequent human rights violations. After entry into the EU, Spanish aid extended the scope of its geographic involvement, incorporating some countries in North Africa, like Morocco, and in sub-Saharan Africa, especially the Portuguese-speaking ex-colonies, along with some Asian countries like the Philippines, that have shared historical ties. Even so, other countries have also been added to the preferred list due to the weight of FAD credits and business interests.

TABLE 14.3
REGIONAL DISTRIBUTION OF NET ODA (WEIGHT AVERAGE)

	Spain				DAC 2000
	1988–91	1992–95	1996–99	2000	
Sub-Saharan Africa	17.9	10.5	21.6	16	29
North Africa	13.0	12.2	7.3	12	7
Latin America	40.2	46.5	40.8	41	12
Middle East	1.1	1.5	3.9	3	4
Far East	13.5	20.8	9.3	18	25
South and Central Asia	2.7	0.5	1.3	1	13
Europe	0.3	0.4	4	9	7
CEEC/NIS	1.2	-	1.1	1	1
Oceania	-	-	-	-	2
Total	100	100	100	100	100

Source: DAC.

In Latin America, attention has been concentrated on the poorest countries of Central America (El Salvador, Guatemala, Nicaragua and Honduras), the Dominican Republic and the South American Andean countries (Colombia, Venezuela, Bolivia and Peru). In North Africa, Morocco is the main recipient, and within sub-Saharan Africa, the countries where the greatest volume of resources is concentrated are Mauritania, Angola, Mozambique and Senegal. In Asia, China and Indonesia are the principal destinations for Spanish aid, followed at some distance by the Philippines and Vietnam. Finally, within Europe aid is concentrated on Bosnia and Yugoslavia (including Kosovo).

A final characteristic of the distribution of Spanish aid is its inadequate focus on the socio-economic status of the beneficiaries, with a perceptible shift in favour of middle-income countries (MICs) and against those in the lowest socio-economic bracket (see Table 14.4). In fact, half of the 1996 to 2001 Spanish aid went to lower middle-income countries (LMICs), a percentage which is much higher than that of the DAC average, which allocated 36 per cent to that group of countries. Furthermore, Spanish aid to low-income countries (LICs) is notably lower, at 49 per cent as opposed to 59 per cent for the DAC. The least developed countries (LLDCs) receive less than half of what the DAC dedicates to that group of countries – 13 per cent as opposed to 28 per cent. According to the available data, Spain has decidedly failed to fulfil the intention, agreed upon at the latest Conference on Least Developed Countries, to dedicate at least 0.15 per cent of the donor GDP in the form of aid to this group. In 2000, the quota for Spain barely reached 0.02 per cent. In fact, of the DAC members, Spain is next to Greece and the United States in making the least effort to realise this goal.

TABLE 14.4
DISTRIBUTION OF NET ODA ACCORDING TO LEVEL OF INCOME

	Total ODA (%)				ODA per capita US$			
	1988–2001		1996–2001		1988–2001		1996–2001	
	Spain	DAC	Spain	DAC	Spain	DAC	Spain	DAC
LLDCs	13.4	28.9	13.8	28.6	1.77	186.07	0.90	69.32
OLICs	32.7	28.8	35.7	30.9	0.87	40.04	0.46	16.15
LMICs	38.6	31.0	45.3	33.4	4.70	196.15	2.74	79.57
UMICs	15.2	4.8	5.0	3.0	2.52	41.58	0.42	9.72
HICs	-	0.06	-	0.1	0.27	109.24	0.19	67.23

Source: DAC.

Another way of analysing this issue is by considering the aid per capita given to countries at each income level. In accordance with its redistributive function, there should be an inverse relationship between the level of development and the aid per capita received. This, however, is not the case for Spain, nor for that matter for the DAC as a whole. In both cases, the LMICs stratum receives the greatest aid per capita, ranking even higher than the figure for the least developed countries. There is nevertheless a marked contrast between the two: whereas the aid per capita given by the DAC to the LDCs is almost the same as that given to the lower middle-income countries, in the case of Spain the former is barely one-third of the latter. In short, Spain's model of aid allocation is biased in favour of middle-income countries (especially lower middle-income) at the cost of the least developed.

This distribution of aid creates a conflict of interest for Spain's aid policy between the ideal of fighting poverty and the level of development in those countries that are currently the main recipients. The government has tried to construct arguments to justify the merits of converting middle-income countries into principal recipients of its aid. It has even sought to find an ally for this strategy in the international context (in particular, Britain's DFID). Two main arguments are given in this respect.

First, it is assumed that, because of the existence of a greater degree of knowledge and mutual understanding with Latin American countries, which have mid-level incomes, Spanish aid in that region can be more effective. Second, the objective of fighting poverty is regarded as compatible with directing aid towards countries of medium-level development, given the existence of pockets of poverty in these countries. The first argument seems reasonable, although the comparative advantage has to be combined with other possible criteria of distribution which should necessarily relate to a country's poverty levels. The second argument, however, is more debatable because, first, the Spanish aid is far from focusing on the poverty pockets in the middle-income countries in which it is implicated; and second, because it does not appear that international co-operation would be likely to make a similar commitment, whatever the income and poverty levels of the country. One might think medium-development countries with suitable social and redistribution policies could bring about their own eradication of extreme poverty, whereas such a possibility is beyond the hopes of the least developed countries. Clearly foreign aid has to be sensitive to the relative levels of development and poverty of the recipients. The search for greater efficacy in aid also points in the same direction. As Collier and Dollar [*1999*] show, the efficacy of aid would be improved if resources were directed towards the poorest countries, and within that group of countries, to those with the greatest institutional capacity and the best policy framework. Without a doubt, it is necessary that industrial nations support the efforts of the middle-income countries to reach a favourable integration in international markets, thus reducing their vulnerability. But probably co-operation in the technological, financial and commercial arenas would be more suitable for that purpose than ordinary ODA instruments (see Alonso et al. [2003]).

VI. SECTORAL DISTRIBUTION

The first salient feature of the composition of Spanish foreign aid is the remarkable change in its sectoral specialisation in recent years (Table 14.5). In 1991/92, more than half of bilateral aid – 58 per cent – was directed towards activities related to infrastructure and economic services, in particular, energy and transport, and up to 16 per cent was destined to productive sectors, where industry, mining and construction predominated. In contrast, aid directly related to infrastructure and social services was of relatively lower importance – 16 per cent – while other multi-sector interests, such as the environment and gender issues, accounted for only a modest 1.5 per cent of all bilateral aid. Five years later, in 1995/96, the profile of aid specialisation underwent a major change, with a reduction in the importance of economic activities to the benefit of those directly related to social services: 38 per cent of the resources were spent on infrastructure and social services, while aid destined for infrastructure and economic services and the productive sectors achieved a combined total of barely 28 per cent.

These trends have been accentuated in the past few years. Activities related to infrastructure and social services received 53 per cent of bilateral aid in 2000, clearly surpassing the DAC average. Twenty-one per cent of Spanish bilateral aid was spent on education, the sector responsible for this growth. In comparison with the DAC figures, Spanish aid related to infrastructure and economic services – 13 per cent compared with 17 per cent – and to the productive sectors – 6 per cent compared with 7 per cent – is significantly lower. The reason for these changes was the reorganisation of aid – largely through the diminished relative importance of loans – as well as the reorientation of FAD credits in accordance with the criteria agreed in the OECD. Yet it must still be said that international comparisons need to be treated with caution because not all countries have similar quotas of sector-allocable aid: aid assigned to specific sectors is more prominent in the Spanish case, at 76 per cent, compared with the DAC average of 62 per cent.

TABLE 14.5
BILATERAL AID: DISTRIBUTION ACCORDING TO SECTORS (%)

Sectors	Spain 1991/92	DAC 1990/91	Spain 1995/96	DAC 1994/95	Spain 2000	DAC 2000
Social infrastructure and services	16.7	20.6	37.7	29.0	53	33
Education	4.9	9.2	8.7	11.0	21	8
Health	4.3	2.6	15.1	3.8	12	4
Population programmes	-	2.7	0.5	1.4	1	2
Water supplies and sanitation	1.6	0.8	3.9	5.4	4	7
Government and civil society	2.8	2.7	3.7	3.0	7	5
Other infrastructure and services	3.1	2.5	5.9	4.4	9	7
Economic infrastructure and services	47.9	17.6	14.9	22.7	13	17
Transport and storage	14.4	7.5	3.4	9.6	7	9
Communications	8.1	2.1	4.1	1.6	1	1
Energy	24.9	6.9	6.4	9.0	5	3
Banking and financial services	-	0.7	0	0.6	0	1
Business and other services	0.6	0.6	1.0	1.8	0	4
Production sectors	15.9	12.1	13.7	10.6	6	7
Agriculture, forestry and fishing	0.9	7.1	6.7	7.4	4	5
Industry, mining and construction	14.8	3.3	6.7	1.6	1	2
Trade and tourism	0.2	1.6	0.2	1.3	0	0
Other	-	0.1	-	0.2	-	0
Multisector	1.5	3.0	9.3	4.5	9	9
Total sectorial	82.0	53.2	75.6	66.8	76	63
Commodity and programme aid	0.9	11.7	1.3	7.0	1	7
Actions relating to debt	-	18.1	10.2	9.3	7	8
Emergency assistance	0.5	3.3	1.8	5.0	4	8
Administrative costs of donors	2.9	3.0	4.2	4.6	6	7
Core support to NGOs	0.8	1.2	0.2	1.0	2	3
Unallocable	12.9	9.4	6.7	6.3	4	5
Total	100	100	100	100	100	100

Source: DAC.

Interpretations of such a transformation depend on the specific contents of the activities financed. Even though it may seem positive that Spanish aid is increasingly directed to the social sectors – and particularly towards education – it turns out that this aid does not pay enough attention to the needs of the poor. In fact, within the education sector Spanish aid is focused on tertiary education and scholarship programmes – no more than 12 per cent of Spanish aid in education goes to basic education. The same may be applied with regard to health disbursement, where basic health components have a lower weight in comparison with others.

These figures reveal the limited importance of basic social needs in the Spanish aid system. Statistical inconsistency, however, complicates the analysis. The Spanish authorities, using a peculiar definition of basic needs, consider that around 19 per cent of bilateral aid is spent on basic social needs. This estimate, in addition to including activities that can hardly be classified as basic, excludes FAD credits. When all the proper corrections are calculated, the quota assigned to basic social sectors diminishes considerably. The DAC estimates the percentage at 10 per cent for grants, and 7 per cent for loans – an overall average of 9 per cent [*OECD, 2002: 24*]. Even when there could be differences regarding the correct distribution of activities per sector, independent researchers estimate that the Spanish quota for the basic social sectors is between 7 and 12 per cent of bilateral ODA [*Alonso et al., 2003; Ancona et al., 2002*]. In any case, this is a percentage far below that agreed at the Copenhagen social summit (20 per cent of bilateral allocable aid).

The composition of the aid sectors described above deviates from the priority sectors defined by the *Strategic Guidelines*. These guidelines established three cross-cutting priorities: poverty reduction, the promotion of gender equality and environmental sustainability. In addition, they determined six sectoral priorities: attending to basic needs, investment in human capital, economic infrastructure promotion, environmental protection, strengthening good government practices and promoting peace. The fact that the Spanish system established some priorities was welcomed by a majority of the social actors involved in aid. However, there are no measures to guarantee that the agreed priorities will guide aid in an effective way. It could be said that the definition of priorities constitutes a system for the classification of activities rather than a framework for an effective orientation of the commitments and management processes of Spanish aid.

The first indication of the limited importance the government places on these priorities is its failure to make any progress in defining the doctrinal and strategic elements needed to guide the management of aid in most of these sectors. Furthermore, no procedures have been included for making the previous allocation of resources fit the priorities, nor are there any criteria for evaluating the effectiveness of these priorities once they are accomplished. Finally, the core institutions in charge of aid management, in particular the AECI, lack the technical capacity and expert knowledge to work in these areas, which is why proper policies and guidelines can scarcely be defined.

The three cross-cutting priorities established by the *Strategic Guidelines* show the lowest level of correspondence between actual aid policy and its formal principles. There is no indication that these priorities have sufficiently influenced decisions regarding foreign aid. This is clearly apparent as regards poverty reduction, not only because of Spanish aid's limited focus on the poorest countries, but also because there is no procedure to guarantee that aid will be directed even towards fighting 'pockets' of poverty in the countries that actually receive aid. Even in sectors which focus on social matters, such as health and education, Spanish aid specialises in areas that do not affect the poorest social sectors. As the DAC review pointed out, 'since Spain is not concentrating on the poorest countries, it particularly needs to show that the poor in other developing countries are effectively targeted' [*OECD, 2002: 27*].

If little progress has been made towards meeting these cross-cutting priorities, opinion is more divided when it comes to analysing sectoral priorities. In some of them, Spanish aid has managed to achieve sound progress, though there are other areas where it is clearly inadequate.

(a) Spanish co-operation is working acceptably in the following sectors:

- *The environment*. The design of the 'Araucaria Programme' was an interesting innovation for Spanish aid. Setting a precedent for other sectors, this was the first where work was carried out in a programmed and integral manner through the establishment of a specific strategy for the sector. The experience gained in this area allowed for the extension of the initiative to areas outside the usual Latin American sites into the 'Azahar Programme' for Mediterranean countries.
- *Strengthening institutional development*. Spanish aid has worked very hard at strengthening institutions and improving government practice. The results are less obvious, but there have been some constructive achievements in Central America involving public institutions and departments like the Ministry of Justice, the Customs Administration, the Fiscal Administration and the Ministry of the Interior. And so, Spanish aid has worked in support of a decentralised process in several Latin American countries, to strengthen municipalities and provincial departments.

(b) On the other hand, Spanish aid is clearly lacking when it comes to areas concerned with social dimensions of development.

- As noted earlier, there is no indication that Spanish aid is directing its activity and resources towards the basic needs of the beneficiary populations. In spite of the administration's inaccurate calculations, which include emergency action in the activities

programmed to address basic needs, the share of the sector – grants and loans – in total bilateral aid (9 per cent) is very far from the commitment made by Spain at the social summit in Copenhagen. Unfortunately, there is no defined policy towards aid in this field.
- Second, there has been a clear regression in the use of educational activities to invest in individuals. In the past this sector was taken up with activities related to higher education, especially through the scholarship programme. Nonetheless, part of the general scholarship programme was suspended in order to distribute its resources through the 'Carolina Foundation', a private body for the education of Latin American elites. Such a change has not only accentuated the anti-social component of the programme, depriving the most needed sectors of aid, but has also provoked a breakdown in the co-operation between Spanish universities and development assistance. In fact, the unilateral decision by the AECI to suspend the previous scholarship programme was opposed by Spain's public universities.
- Third, the confusing character of Spanish humanitarian aid is worth mentioning. The succession of natural and man-made disasters in the recent past has challenged the government's approach, of which there are three main aspects. First, the response is conditioned by foreign policy priorities and not by humanitarian criteria. Second, there is a low level of social commitment, with the intervention based almost exclusively on sending emergency aid (making the visibility of the response profitable) and, in some cases, adding FAD credits for reconstruction, which are inappropriate resources for countries in emergency situations. And lastly, there is a lack of leadership in the co-ordination of responses due to the absence of a defined plan of intervention, leading to confusion over the roles of the different actors, including the army.

In conclusion, it is doubtful that the sectoral priorities defined in the *Strategic Guidelines* have guided Spanish aid activities. In addition, the allocation of resources and the operational priorities do not seem to comply with the requirements for achieving the International Development Goals – and no mechanism exists that allows for these goals to be identified in practice and promoted.

VII. ACTORS

The central administration is the main actor in the Spanish aid system. Within the central administration, several departments had been engaged in international co-operation, some of them with small budget programmes. In order to co-ordinate these initiatives, the CICI was set up within the central administration. Despite this, the capacity for effective integration is limited by the lack of central capacity to guide the initiatives adopted independently by different ministries.

This low level of co-ordination in the central administration was accentuated by the lack of understanding between the two departments with the main responsibility for aid policy: on the one hand, the Ministry of the Economy, responsible for loans (FAD credits), debt relief operations, management of a small fund for technical assistance (FEV) and co-operation with multilateral financial institutions; on the other hand, the Ministry of Foreign Affairs, which, via the SECIPI, is responsible for bilateral grants, micro-credit programmes and contributions to multilateral non-financial institutions. The balance of Spanish aid, excluding contributions to the European Union, is mainly distributed between these two ministries. However, a visible lack of unity has traditionally prevailed between these two institutions, which have managed their own aid instruments with complete autonomy, without being subject to the criteria of an integrated policy. This is why Spanish aid could be described as a bicephalous system, consisting of two programming units that work autonomously. As a result, to the detriment of aid effectiveness, there is a very low level of integration between financial and technical aid.

The regional and municipal administrations are important public actors in the Spanish aid system. As already mentioned, this decentralised system, developed throughout the 1990s, managed to contribute close to 14 per cent of all Spanish ODA in 2002. Its evolution is a positive factor, demonstrating the vitality of Spanish society and its support for international solidarity: the DAC report of 1998 suggests that this kind of co-operation should inspire other donors (see OECD [*1998: 8*]). Nevertheless, despite this positive contribution, the lack of resources and technically skilled personnel in some of these institutions has diminished the effectiveness of their actions.

This lack of resources has meant that most of these institutions operate through support for projects presented by NGOs, contributing yet further to the dispersion of aid. Even so, interesting new forms of association between municipal governments through the so-called

Regional Funds – which pool resources and capacities with the aim of improving the quality of co-operation – have arisen.

Meanwhile, the development NGOs have achieved a leading role in the management of Spanish co-operation. Aside from participating in the co-financed programme run by the AECI – which features various levels of duration and budgetary commitment (to projects, programmes and strategies) – they also have access to the resources made available by regional and local administrations. A conservative estimate suggests that around one-third of Spanish bilateral aid is managed by NGOs. This active presence of NGOs has helped compensate for the technical weaknesses of public administrations – above all at the regional and local level – in terms of managing development aid, while at the same time giving aid activities a truly social character. Yet this involvement of the NGOs has also reinforced the dispersion of aid, highlighting concerns over 'transaction costs as well as... efficiency and co-ordination issues' [*OECD, 2002: 13*].

Despite the important part played in aid, the network of NGOs in Spain is extremely fragile. With just a few exceptions, the organisations are relatively recent, having been created in the years since the 1980s. Although they are active in the field of pressure group politics, they have a limited social base and little capacity to finance themselves, meaning that they are critically dependent on public resources. The existence of various funding sources – central, regional and municipal administrations, along with the European Union – enables them to diversify their functional dependence, yet without achieving financial autonomy. Over the years, Spanish NGOs have been able to improve their technical capacities and their levels of professionalism, but few have been able to establish clear links with the Spanish citizenry.

The role of business in the Spanish aid system is somewhat ambiguous. On the one hand, the historical importance of FAD loans, with their significant commercial orientation, has turned a portion of the business community – particularly the energy, public works and engineering sectors – into beneficiaries of aid policy. Likewise, the efforts to broaden Spain's profile in international financial institutions have as one of their key objectives an improvement in the standing of Spanish firms in markets depending on multilateral financing. Compared with this, the presence of private firms in non-reimbursable aid is relatively low, and highly focused on the activities of employers' organisations in the fields of training and institutional strengthening.

In addition, trade unions are active in aid policy, with their focus tending to be on training programmes, institutional strengthening and job creation. These unions nevertheless gain access to official funds through various NGOs that are not much different from other Spanish NGOs.

Other social actors have less presence in the aid system. Of these, the universities are possibly the most active and willing to establish a more prominent role in development aid.

VIII. THE DIRECTION OF AID POLICY

1. Prior to the Aid Law

By the end of the 1990s, Spanish aid was characterised by limited institutional consolidation and a very low technical level in its management procedures. The following were five of the basic shortcomings:

(i) The limited integration of institutions and instruments. The first limitation refers to the poor co-ordination of aid policy, especially between the Ministry of Economy and the Ministry of Foreign Affairs, the two main departments in aid management. In addition, there has been limited co-ordination with other public actors, particularly regional and municipal administrations, which run autonomous aid programmes.

(ii) Limited strategic vocation. The second defect was related to the limited strategic character of Spanish aid, demonstrated by its dispersion and frequent changes in direction. Up to 1997, the work of the OPE (Planning and Evaluation Office) had been notably distinct from the function its name implies, with its efforts directed towards two activities: the distribution and monitoring of subsidies to NGOs, and the formulation of the Annual Plan for International Co-operation (PACI). Until very recently this was the principal document used for budget forecasting, although it cannot be considered a planning document.

(iii) Weak management capacity. The limited management capacity in the Spanish system was a result of the lack of standardised procedures for managing interventions. Until very recently, the AECI's operations did not include a methodology or procedure for managing the different stages of a project – identification, formulation, implementation and evaluation. The absence of an evaluation of activities was especially significant, because it has impeded self-correction and the

learning processes derived from a critical analysis of previous experiences.

(iv) Limitations in structural framework. A fourth limitation refers to the legal nature and organisational structure of the institution responsible for managing the programmes and projects of bilateral co-operation. The AECI, as part of the public administration, has limited management flexibility and versatility. This limitation was accentuated by the deficiencies in technically specialised personnel, a feature highlighted in all DAC reports on Spanish aid.

(v) Low level of parliamentary involvement. Finally, it is important to point out the limitations of the institutional model that existed before the Co-operation Law was passed. In particular, there was little parliamentary involvement in the orientation and control of foreign aid activities.

2. With Regard to the Co-operation Law

Some of these defects were due to be taken into account in the Law of International Co-operation for Development, adopted in July 1998. This law marks an important step towards the normative institutionalisation of aid policy, because it sets out the principles, objectives, institutions and instruments of the Spanish aid system. The formulation of the law was actively followed by the social sectors involved in aid, which ultimately gave their support to the Bill. It is a law that was adopted with wide endorsement in Parliament, and which establishes a model of co-operation aligned with the international doctrines standardised by the DAC.

More precisely, the law offers solutions to some of the weaknesses and deficiencies in the Spanish aid system. In fact, it includes five interesting new features:

– First, it offers a doctrinal framework suitable for guiding aid policy. Poverty reduction is established as the main objective towards which aid efforts should be directed; the concept of human development is assumed to be the doctrinal framework for understanding development; and certain objectives are defined, including the basic features of the international agenda related to gender equality, environmental sustainability, good government and human rights protection. Given the close association between Spanish aid and commercial operations, the law's definition

marked a notable advance and laid the foundations for the possible future reorientation of aid.
- Second, the law established a framework for planning aid, generating the necessary mechanisms to give it a more strategic character. The planning cycle is defined by four basic instruments: the four-year *Strategic Guidelines* document, the *Annual Plan* that specifies the activities for each year, the *Sector Strategies* to guide aid in each specialised area, and the *Country Strategies* that define the aid activities within each priority country. Combined, these are a complete set of instruments for developing a strategic aid policy.
- Third, the law marks a step forward in its intention of integrating aid instruments. In an attempt to dissolve the bicephalous character of Spanish aid, it hands directive functions to the Ministry of Foreign Affairs, making it responsible for all aid policies, independent of the management skills of other ministries. It also declares that all aid instruments must comply with the objectives defined by the law, which clearly focuses attention on poverty reduction – an objective that would change the traditional ways of managing FAD credits, which were usually directed more to commercial goals. This is why the Ministry of the Economy and the Secretary of State for International Trade were, at that time, reluctant to approve the law.
- Fourth, the law is an attempt to co-ordinate aid actors. In addition to reinforcing the CICI and reaffirming the role of the Council for Co-operation for Development, the law introduces a new co-ordination initiative: the Inter-territorial Commission, which is aimed at promoting the co-ordination of initiatives between the central administration and the regional and municipal administrations. It also promotes the creation of a Parliamentary Commission to monitor aid policy.
- Finally, in its fourth article, the law states the necessity of promoting maximum coherence between aid policy and other government policies influencing developing countries – an aspect in which Spain appeared to be less advanced than some other donors.

The law introduced new possibilities for improving the effectiveness and quality of Spanish aid. However, as we shall see, these possibilities have not been pursued, and there have even been regressions from what was suggested by the legislation.

3. After the Co-operation Law

Between 1997, when the formulation of the law began, and the beginning of 2004, Spanish aid went through two very different stages. Despite the political continuity of the government, the general elections in 2000 marked an abrupt breakdown in the direction, purpose and *modus operandi* of the authorities in charge of bilateral aid. The first stage, lasting until 2000, is characterised by the encouragement of aid reform, as recommended by the DAC. The intention was to organise and equip aid with better technical abilities, through a more demanding definition of the project cycle and a greater focus on international criteria and standards. The establishment of the law was a result of this reform initiative, coupled with a plan to evaluate Spanish activities and instruments and an attempt to work out *Strategic Guidelines* according to DAC recommendations, focusing on poverty reduction and searching for consensus with other aid actors.

In the 2000 elections, the Popular Party gained a majority and proceeded to change aid policy drastically. The reform effort of the previous legislature was replaced by a concept of aid directly related to commercial, political and cultural foreign policy interests. Accompanying this regression was a clear critical distancing from the DAC recommendations related to poverty reduction objectives, thereby severing the dialogue with social actors.

These changes provoked the mobilisation of more than 100 aid organisations and 200 academics and experts, who drew up a document entitled *Faced with the Counter-reform of the Aid System: For an effective policy directed towards poverty reduction*. The following aspects illustrate the regression:

– First, the draft *Strategic Guidelines* agreed to by the previous legislature was replaced with a new document, very different in its principles, technical basis and strategic demands. The new document eliminates all references to principles like partnership, ownership or coherence, insisting that aid should be an instrument of Spanish foreign policy. In addition, it undermines the formulation of the poverty reduction objective that the law had defined as the main purpose of aid. The active defence of Spanish aid's orientation towards low middle-income countries, and the lack of concern shown about the weight given to basic social services in the actual spending of ODA, demonstrate this move by those responsible for aid away from the objective of poverty reduction.
– Second, the composition of the Co-operation Council was modified to increase the quota of government representatives and to

ensure that the Secretary of State should appoint experts and NGO representatives, thus eliminating the unbiased detachment previously given to the selection of representatives. By abandoning its participative role, the Co-operation Council therefore became a mainly government-nominated organisation.
- Third, the government began to reform the AECI in a manner contradictory to that demanded by the aid community. Instead of raising the institution's profile as a co-operation agency, there was a dilution of its specialisation, involving it in the promotion of cultural activities that have nothing to do with developmental aid. Furthermore, because of the need to strengthen the capacity of the AECI to design priority aid policy, it maintained the geographical structure of the organisation, but without the support for sector priorities; and instead of implementing better technology in management procedures, it allowed for increased discretion in the assignment of resources.

As a result of this regression, the dialogue between the government and a large part of the aid community has deteriorated. Tension and distrust between the central administration and other aid actors have increased, and have ultimately had a detrimental impact on the whole system.

IX. CONCLUSION

Over the course of the last two decades, Spanish international co-operation has gone through three creative stages, along with others that have marked steps backward. In the first stage (1984–86), the decision was taken to establish a specialised development aid policy, along with the basic institutions needed to manage it – a process that gave rise to the SECIPI. The second stage (1988–92) was one of institutional and budgetary consolidation: the resources for aid were increased, the AECI was created, and Spain was incorporated into the DAC. Lastly, the third stage (1997–99) was one of bringing Spain into line with international norms and improving technical capacity: the Co-operation Law was approved, the international aid doctrine was adopted (linked to the key objective of poverty reduction), and efforts were made to establish various mechanisms (Strategic Guidelines, project formulation methodology and plans for the evaluation of aid) that were needed to improve the technical components of management. Unfortunately, this third stage was interrupted.

In the 2004 legislature, the Spanish aid system underwent a retrogression in a number of areas. In terms of doctrine, the commercial and cultural interests of the donor in the conception and orientation of aid prevailed, even at the cost of shifting a critical distance from the doctrine agreed by the DAC. In the field of management, the reforms needed to improve the integration of aid policy and the technical strengthening of some of the main institutions involved in aid, such as the AECI, were postponed. In the operational field, new management tools aimed at making aid more flexible – such as aid through programmes or the sector-wide approach – were shunned in favour of an almost total domination of project-oriented aid. And lastly, in social terms, recent years have been characterised by a weakening of dialogue and of trust between the government and other actors in the aid system.

The 2004 elections, won by the socialist party (PSOE), have opened a new era for Spanish development co-operation policy. It is difficult to anticipate the line that Spanish policy with regard to development co-operation will take, but it is clear that corrections of some of the previous deficiencies have been included in the new government programme.

REFERENCES

Alonso, J.A. (dir.), D. Forte, L. Salies and M. Valdés, 2002, 'Microcréditos y pobreza: análisis del programa español' (Micro-credit and poverty: analysis of the Spanish programme), in *La realidad de la ayud, 2002–2003a*, Barcelona: Intermón-Oxfam.

Alonso, J.A. (dir.), L. González, M. Pajarín and A. Rodrígues-Carmona, 2003, 'Enfoque anti-pobreza de la cooperación española: de las declaraciones a los hechos' (Anti-poverty approach of the Spanish co-operation: from declaration to facts), in *La realidad de la ayuda, 2003–2004*, Barcelona: Intermón-Oxfam.

Ancona, C., G. Fanjul and C. Gonzálex, 2002, 'La Ayuda Oficial al Desarrollo en España en 2001 y 2002' (The Spanish Official Development Assistance in 2001 and 2002), in *La realidad de la ayuda 2002–2003*, Barcelona: Intermón-Oxfam.

Collier, P. and D. Dollar, 1999, 'Aid allocation and poverty reduction', *World Bank Policy Research Working Paper*, 2041, Washington: World Bank.

Ministerio de Asuntos Exteriores, 2001, *Plan Director de la Cooperación Española 2001–2004*, Madrid.

OECD, 1998, *Development Co-operation Review Series, Spain, 1998*, Paris: OECD DAC.

OECD, 2002, *Development Co-operation Review Series, Spain, 2002*, Paris: OECD DAC.

15

Swedish Development Co-operation in Perspective

ANDERS DANIELSON AND LENNART WOHLGEMUTH

I. INTRODUCTION

Sweden is usually characterised as a soft donor; in other words, a donor that does not attempt to use aid to blatantly pursue other objectives, that listens carefully to recipients' requests, and that tries to design aid so that it can have a large development impact. Although a convergence in aid philosophy has taken (and is taking) place among bilateral and multilateral donors, Sweden, together with a handful of other 'like-minded' countries, is still distinguished by large amounts of aid, little tying and basically no explicit aid objectives apart from the reduction of poverty.

But is Sweden's aid given simply on the recipients' terms (to borrow the title from a book in the Swedish aid debate [*Wohlgemuth (ed.), 1976*])? Certainly not. Although Sweden enjoys a good reputation in many recipient countries – for being reliable, a listener, and a generous donor – Swedish aid has its fair share of problems: there is no sharp focus on the overarching objective, the criteria for selecting development partners are not always transparent, the ability to co-ordinate aid (with other donors and in particular with the recipients' own resources) leaves much to be desired, and it is not clear how the Swedish authorities see the mechanisms by which aid is translated into poverty reduction (in spite of a number of strategies and policy papers – the latest presented in 2002 [*Sida, 2002a*]).

Note of acknowledgement. An earlier version of this chapter was presented at the EADI conference 'Perspectives on European Development Co-operation', Nijmegen, 3–7 July 2002. Participants, particularly Juhani Koponen, Gorm Rye Olsen, Helen O'Neill, Robrecht Renard and the editors of this volume provided helpful comments. Arne Bigsten, Gus Edgren, Robert Keller and Ari Kokko have patiently answered various queries. Maria Strömvik provided a crucial piece of information. We are particularly grateful to Malin Hansson, statistician at Sida, who provided most of the data. Remaining inaccuracies are our responsibility.

SWEDISH DEVELOPMENT CO-OPERATION IN PERSPECTIVE

This chapter addresses some of these issues and reviews some of the salient features of Swedish aid and how these have changed over time. The structure of the chapter is as follows. Formal objectives are identified in section II, and section III identifies solidarity as the underlying motivation for giving aid. Section IV looks at changes in Swedish aid policy over time, and section V discusses this with particular reference to Sweden's accessions to membership of the European Union (EU) in 1995. Section VI provides some data on the volume, sectoral and geographical distribution of Swedish aid, and section VII looks at administrative changes. In section VIII we attempt to see the extent to which Swedish aid policy is driven by an underlying development theory. The main priorities are identified in section IX, and the criteria for partner selection are examined in section X. Section XI offers a few concluding remarks.

Most of the discussion will be on the period after 1995, the reason being that a new aid authority – the Swedish International Development Co-operation Agency, Sida – was formed in that year. This was a merger between several old authorities – including the old SIDA (the Swedish International Development Authority) – and it changed in many ways how aid is designed and delivered. Some implications of this will be discussed in what follows.

II. OBJECTIVES

The basic objective of Swedish aid was established in 1962 in Proposition 1962:100 (the bible of Swedish development co-operation), namely to raise the living standards of the poor. This has never been seriously challenged [*OECD, 2000*], rather it has been reinforced [*Globkom, 2001; Government of Sweden, 2003*]. The main motivations for aid are moral duty and international solidarity, even though geopolitical concerns are also mentioned (particularly in Proposition 1978: 101). It is important to note that Sweden has always emphasised that foreign aid can assist in the realisation of the recipients' own development vision; aid cannot and should not be used to 'sell' the Swedish model. This, of course, has an impact on the selection of partner countries [*Rudengren, 1976*].

Formally, Swedish aid has the following six objectives (the year of adoption in parentheses) [*OECD, 1996: 13*] (The new Bill retains them but calls most of them directives rather than objectives) [*Government of Sweden, 2003*]:

- the growth of resources (1962 and 1978);

- economic and social equality (1962 and 1978);
- economic and political autonomy (1962 and 1978);
- the democratic development of society (1962 and 1978);
- the sustainable use of natural resources (1988);
- equality between men and women (1996).

These are not internally ranked and should be viewed as objectives, the fulfilment of which will contribute to the realisation of *the overall objective of poverty reduction.*

Since the early 1990s, all major donors have had poverty reduction as their primary objective. It is thus reasonable to ask whether Sweden pursues its own development agenda or runs along a joint track with other donors. It is not easy to answer this question, for a number of reasons. First, the fact that Sweden continues to give the bulk of its aid as direct bilateral aid suggests that it has objectives that would not necessarily be promoted if all aid were to be channelled through multilateral organisations. On the other hand, it is difficult to find areas where Sweden does not endorse the strategies chosen or the objectives adopted by the UN system. Second, it is quite clear that Sweden has pursued its own development agenda – at least up until the early 1980s. Since the philosophy of help for self-help permeates Swedish thinking on aid, it becomes very important to identify potential recipients with a development vision that Sweden can accept. This is one reason why considerable amounts of aid were given to countries with a socialist vision, such as Tanzania, Mozambique, Nicaragua, Vietnam, Cuba and the liberation movements in Southern Africa and Indochina. On the other hand, this tendency is not as marked in the 1990s, both because some of the countries, such as Tanzania, Mozambique and Nicaragua, have changed their development strategy and also because aid has been discontinued to others (Cuba). Even if Swedish aid was earlier characterised by a distinctly Swedish profile, the convergence in aid philosophies that has taken place among donors in the 1990s has placed Sweden more in the mainstream [*Rudengren, 1976; Annell, 1986; Hveem and McNeill, 1994; Danielson, 1999*]. To what extent this is because other donors have emulated the Swedish philosophy could be the subject of some debate.

III. SOLIDARITY AS AN UNDERLYING MOTIVE FOR AID

With no colonial heritage, Sweden, like its Nordic neighbours, had to rely on a more abstract motive for its aid to developing countries during its inception in the early 1960s. Attempts to build a new welfare state were based on solidarity with people worse off than most Swedes – a way of thinking first used and developed internally during the first half of the last century, in trying to construct a welfare state in Sweden itself. International solidarity came into focus during the explosion of independence in the late 1950s and early 1960s. The process in Sweden was an interplay between strong forces in the public led by journalists, authors and some farsighted politicians representing a number of parties, the most prominent of whom was Olof Palme (Social Democratic leader and Prime Minister for many years). This process is well described by Tor Sellström [*1999: 505–7*]:

> Swedish intellectuals and students started to raise their voices against the South African apartheid regime in the 1950s. A fundraising campaign in support of the victims of apartheid was launched. After the beginning of the 1960s, this campaign and other initiatives led, with support of the student and youth movements as well as church representatives, to the formation of a national anti-apartheid committee. Broadly based boycott campaigns against South Africa soon thereafter gave birth to active local solidarity committees and to an involvement with the entire Southern African region.
>
> ...Though it was mainly individuals and political organisations in the liberal political centre that were first active against apartheid, the humanitarian concerns soon found an echo in the ruling Social Democratic government. ...The constituent parts of the Swedish solidarity movement with Southern Africa were largely in place by the mid-1960s. A first generation of local anti-apartheid committees had been formed and the initial re-active humanitarian views had been replaced by a more pro-active and militant approach. In May 1965, the solidarity movement defined as one of its main objectives to 'convince the Swedish government, parliament and public of support to the liberation movements in Southern Africa'. The demand for an officially declared boycott against South Africa was at the same time gaining increasing support.
>
> The issue of direct official support to the liberation movements never became divisive in Sweden. The four Prime Ministers, representing different political parties, heading the government for more than twenty years (1969–1991) were all for the active Swedish involvement in Southern Africa – all became concerned with it in the 1950s or early 1960s. In 1988, the ANC leader Oliver Tambo characterised the links between Sweden and

Southern Africa as 'a natural system of relations ... from people to people ...', which is not based on the policies of any party that might be in power in Sweden at any particular time, but on ... a common outlook and impulse.

While Sellström's description is of Swedish policy towards the liberation struggle in South Africa, the same policy was applied with regard to Vietnam and its neighbours and to solidarity with the developing world in general. There is thus no question that solidarity with the oppressed and poor people in the developing world was the overriding motive for Swedish aid up until the beginning of the 1990s. During this period Sweden also led the trends and the debate to write off debts and untie all assistance. Throughout these 40 years, annual polls, organised by the Swedish Central Bureau of Statistics, showed strong public support for the level and forms of aid. The support has been expressed by the respective governments over time, in Parliament, in society at large as well as in the press. The voices from civil society including the NGOs have been loud and clear on these issues, and the private sector – also a voice to be reckoned with – has been mostly happy as long as trade and investments have not been disturbed. However, at times the private sector has been very vocal when its interests have been threatened such as in the cases of tying of aid in the early 1970s and on investment and trade restrictions to South Africa in the 1980s. In both of these cases the private sector lost out to the interest of solidarity.

With the increased aid fatigue in the early 1990s, the question of new motives for keeping up taxpayers' support for Swedish development assistance was raised. In an article entitled 'Does Aid Work? – A Review of Thirty Years of Experience', Dag Ehrenpreis [*1997*] notes that there were two main reasons for the difficulties the OECD countries had in motivating their aid at that time. First, during the Cold War, aid had been seen by some as a weapon in the struggle for influence over the resources and politics of the Third World. Now the Cold War was over. Second, the economic crisis in the donor countries made people feel there was less to distribute. Ehrenpreis goes on to identify the four motives for aid as moral/humanitarian, economic, political and commercial. In the past decade, a fifth motive has been increasingly discussed, namely the issue of common security – that support for developing countries really is in the long run support for all our common interests, an 'enlightened self-interest' as it has been referred to lately [*Globkom, 2001*].

Ehrenpreis [*1997*] refers to several studies, like those by the OECD, that rule out most of the motives mentioned above as highly inefficient, showing that, for example, tied aid increases costs so much that it becomes counter-productive. Also, the arguments that aid to the poorest countries should benefit the donor as well can easily lead to the line that it is wiser to invest in better-off countries in the vicinity of the donor rather than in really poor countries far away. Theory and practice meet in the conclusion that the only really strong and sustainable argument for continued aid to the poorest countries is the old motive of solidarity. Consequently, Globkom [*2001: 64*] stated: 'There are thus two motives for action in the context of a new Policy of Global Development (PGD) – solidarity and enlightened self-interest.'

IV. CHANGES IN SWEDISH AID POLICY OVER TIME

Sweden's aid policies have constantly been debated, reviewed and changed. In spite of all the changes, the question remains whether all these new policy directives have actually resulted in anything new in Swedish aid policy or if it is as stated by the former Director-General of SIDA, Anders Forsse [*2001*]: 'The more it changes the more it stays the same.' Another interesting question is: Who is really behind these changes – politicians or civil servants in the aid administration?

The major government Bill, which has governed Swedish aid policies since it was approved by Parliament in 1962 (Prop. 100), was based on a Government Review [*SOU, 1962: 12*] digging deep into the prerequisites for aid. Over the years, a number of major investigations have resulted in fresh government papers and in the implementation of new policy directives (cf. Carlsson [*1998: Ch. 2*] for a brief survey). The latest, Globkom [*2001*], has led to a new White Paper that has been approved by Parliament [*Government of Sweden, 2003*]. Overall, there have been surprisingly few changes in objectives and directives. Solidarity has always been the major underlying aid motive, respect for partners' sovereignty another and poverty alleviation a third. Coherence between aid and other policies with impact on the South might become another such motive [*ibid.*].

While the overriding political motives and objectives have been only marginally changed, there have been major shifts in the actual implementation of policies – mainly with regard to the donor–recipient relationship. From the strictly donor-driven project aid of the 1960s Sweden turned to a recipient orientation in the early 1970s, when country programming was introduced, and then to a period of donor domination during the 1980s, with structural adjustment and conditionality as the defining characteristics. In the late 1980s, there was a reaction to this, resulting in a renewed recipient orientation, after a major exercise to investigate the different roles of donors and recipients. Finally, in the late 1990s, partnership and ownership were introduced [Wohlgemuth (ed.), 1994]. These shifts were partly indigenous to Sweden and in part followed international trends. It is interesting to note the long-term circular reappearance of the implementation of policies.

The question of political interference in aid has been raised from time to time in Sweden, particularly when it comes to prioritising certain pet ideas and sectors. Again, as in any other sector, certain issues are important today and others tomorrow. The matter of political interference is of particular interest in Sweden, as such interference with the day-to-day running of public affairs could be questioned according to the constitution. However, political interest and engagement in aid policies are essential for the policies to be properly grounded in society and for them to be consistent with other government policies and objectives. They become cumbersome only when they occur too often and in too great detail, or if lobby groups get too strong an influence on policies.

> It is difficult to dispute the need for broad-based political support of aid policies and procedures. It is equally important to have coherence between different government policies. However, when external political pressures become too strong and incompatible objectives are being introduced, then the result is confusion [Carlsson and Wohlgemuth (eds), 2000: 11].

In the practical implementation of policies, changes have been prevalent and seemingly extensive. For instance, Sweden moved from donor-driven project aid in the 1960s, with a large technical assistance component and Swedish project co-ordinators on the project site having a final say on both large and small issues, to the present-day sectoral support and budget aid which aims to become fully integrated into the recipient economy. Aid shifted from using detailed directives at project level to the current conditionality, including issues such as macroeconomic policies as well as domestic policies regarding democratisation and human rights. Policies have also swung from domination to dialogue [Havnevik and Arkadie, 1996]. The present-day partnership relation between donor and recipient has made dialogue the hub of development co-operation. Several studies on partnership have discussed conditions and models for true dialogue [*Kifle* et al., *Kayizzi- Mugerwa* et al.,*1998*; *1997*; *MFA, 1998*]. In Sweden, partnership was established in the Bill on Swedish Africa Policy, approved by Parliament in 1998 and further discussed by Globkom [*2001*].

On the qualitative aspects of partnership this Bill lists the following prerequisites [*MFA, 1998*]:

- a basic attitude relating to sustainability and long-termism. There is need for a real change of attitude. No partnership can thrive or survive without respect between the partners;
- openness and clarity concerning the values and interests that govern co-operation. You cannot engage in a partnership without sharing values;
- an increased element of management by objectives and result orientation of aid, instead of a multitude of predetermined conditions;
- a humble, listening attitude with respect for African assumption of responsibility and awareness of the local environment;
- clarity of resource commitments, payments and reporting principles;
- desire for co-ordination among the donors.

In addition to these qualitative aspects of partnership the Bill also added the following necessary changes to be made to partnership modalities:

- African leadership and ownership, for example, holding consultative meetings to co-ordinate donors in the capitals of the recipient partners;
- improved local backing and participation. There must be respect for open political debates, the role of Parliament, and consultation with private enterprise and civil society;
- improved co-ordination. Effective African ownership requires good donor co-ordination, preferably under the recipient country's own management;
- well-developed sectoral and budget support, making the number of interactions with donors as few as possible and thereby manageable for the recipient;
- simplified procedures, minimising the numbers of reporting systems, procurement requirements, payments procedures, accounting routines, etc.;
- contractual clarity and transparency;

- increased coherence between different areas of policy. Behind this term are hidden scores of issues with tremendous long-term implications. It is not just the well-known trade and debt issues, but much else relating to everything from peace and the environment, to migration and the many issues that enable economic integration globally;
- rewards for progress;
- extraordinary debt-relief inputs for certain countries.

The aid administration has repeatedly confirmed in studies and evaluations that sustainable results of aid interventions can only be achieved if the interventions – on either the macro (political) or micro (project) level – are owned and run by the beneficiaries themselves [*DANIDA* et al., *1988; SIDA, 1988*]. Again and again, aid implementers have breached this golden rule. The reasons are manifold. Partly, it can be blamed on internal donor procedures, such as the pressure to disburse as much aid money as possible and internal methods tending to be very donor-centred, like the use of 'logical frameworks' and reporting requirements. Other reasons are more individualised. Every actor in the field wants to see results within the contract period and can therefore only with difficulty await responses from the beneficiaries, which sometimes take a long time if they are to be properly grounded among all the relevant stakeholders. These reasons were dealt with in detail in an investigation in the late 1980s and a special programme to overcome these deficiencies was developed [*SIDA, 1987*]. The issue was revisited in the second half of the 1990s in reports delivered in connection with the partnership study referred to above. In a new major study, *Ownership of Sida Projects and Programs in East Africa* [*Sida, 2002b: 49–52*], most of the conclusions are reconfirmations of the results from earlier studies:

> With regard to projects, Sida has a mixed record of involving beneficiaries in the cycle of selection, design, implementation and monitoring outcomes. With the shift towards program assistance, this experience required reinterpretation. ...
>
> Perhaps the most fundamental mistake is to presume that recipient ownership can arise from a process that is initially donor-driven ... there are circumstances in which more development assistance, especially more in the absence of implementation capacity, can make matters worse. ...

(1) If it is not recipient-owned, the funds are likely to be used ineffectively, while consuming scarce national resources for administration;
(2) fungibility can result in reducing the government's incentive to raise domestic resources for development, or provide an incentive to divert national resources for development, or provide an incentive to divert national resources to unproductive uses;
(3) by fostering aid dependency, in which macroeconomic stability becomes derivative from large aid flows.

This 2002 study ends by providing recommendations on how to deal with ownership in development assistance, starting from the proposal that 'all projects and programs should include a discussion of their ownership implications when they are proposed'. At the same time, a new action programme on 'Knowledge Development' has been developed within Sida.

Other issues discussed within the aid community, as well as in the recently developed aid research community, are aid effectiveness, aid dependency and learning in development co-operation. These discussions have, however, also in the end come to focus on the question of the relationship between donor and recipient [*Carlsson* et al., *1997*].

In summary, while the politically controlled overall policies on development co-operation have to a large extent remained unchanged, aid modalities and practices have been changed considerably over time by aid administrators. It seems, however, that the underlying problems confronting aid have remained the same during 40 years of Swedish aid, and the prescribed medicine (although the names might have changed) is made up of old ingredients. It also seems that, while policies have been driven by policy-makers, the aid administrators have been responsible for the many changes of modalities and practice.

V. SWEDEN AND DEVELOPMENT CO-OPERATION WITHIN THE EUROPEAN UNION

Sweden became a member of the European Union (EU) as late as 1995. Development co-operation is an area of high profile within the EU and the ambitions of an expanding integrated policy within this area have been strong, both within the Commission itself as well as among European integrationists. The question is, after almost a decade of membership, how much has Swedish development co-operation actually been influenced, affected and/or integrated?

The most visible and immediate consequence of Swedish EU membership is that a substantial part of the annual Swedish aid budget is handed over to the aid budget of the Commission; in 2002 this contribution amounted to 7 per cent of the total Swedish aid budget. This has, in turn, led to requests for Swedish involvement in the Commission's decision-making process, something that has been taken seriously by the concerned policy-makers in Sweden and consequently lays a heavy burden on the aid administration. Because of Sweden's long tradition as a major donor, other member states in general and aid administrators within the EU in particular felt that Sweden had a lot of experience to share. Together with the so-called 'like-minded states', Sweden became very active in pursuing a number of issues within the aid business, such as poverty reduction, gender equality, environment, democratic development and human rights.

Whether these efforts have been successful or not is difficult to say. EU aid is still assessed by many as being very old-fashioned, ineffective and driven by the donor rather than the recipient. Building its judgement on the many evaluations and studies carried out by the Commission and others, Globkom, although it saw opportunities, came to the conclusion that Sweden should consider channelling its aid funds through other more efficient channels than the EU [*Globkom, 2001*]. In its concluding remarks on EU assistance, Globkom states that:

> Should the Commission fail during the present mandate period (1999–2004) to make its administration more effective, Sweden should take the initiative and bring about a discussion of the division of responsibilities and duties, and of resources that should exist between the EU and the Member States, i.e. seriously raise the issue of a 're-nationalisation' of certain development assistance resources [*ibid.: 290*].

So far, however, the Swedish government continues to believe that it can influence the EU to improve its aid through its active participation in the decision processes.

Not much has been written about how the EU is influencing Swedish aid policies and practices. Gun-Britt Andersson, former State Secretary of Development Co-operation, stated the following in an interview:

> As donors, we continuously have to develop our capacity for analysis in order to better understand the world, relations between

people as well as the development process. In order to do this, we have to broaden our work on policies and strategies. The Swedish Foreign Ministry is presently following up on its policies towards Africa. Important tools in this process are the international work on norms and strategies brought up in connection with UN Conferences on different important issues. Preparing these conferences, all countries (developed as well as developing) work together, creating a common understanding on central issues. We see this multilateral work, which lately has been supplemented by the extensive development work within the European Union, as central for improving development in all countries. This process creates a jointly owned value system as well as ways to analyse and understand how development is going to be tackled. It's a base when we get to the dialogue on more mundane and practical issues. Also, with this kind of preparation, we create legitimacy. When we get into any development discussion – or dialogue – we can legitimate our issues by referring to joint understandings, built on agreed norms and conventions within the fields of, for example, human rights, rights of the child or environmental issues. Even when common values are not present, these norms still act as a good starting point...

...Also, the four elements within the Cotonou agreement are important for Sweden in its bilateral dialogues, particularly the paragraph according to which a dialogue on political questions should be started before the appearance of problems which could lead to clashes and abrupt discontinuation of development co-operation.

So, according to you, the Cotonou agreement is normative also for bilateral relations?

Yes, of course! We are part of that agreement, as well as part of all other EU policies towards different countries and international situations. These policies have institutionalised the dialogue and created an order for how and when assistance should be discontinued, thereby increasing the impact, even if we, in our bilateral relations, can act independently of these policies (interview by Wohlgemuth in Olsson and Wohlgemuth (eds) [*2003: 59–61*]).

The discussions leading up to the Cotonou agreement are seen as the first steps towards a joint EU policy on development co-operation. The subsequent discussions led to a joint statement signed by all EU members in November 2000 [*EU, 2000*], which contains all the important principles on how to build development co-operation coherently – a guide for the EU in its development co-operation work. While it is binding for the EU, it is hoped that it will also influence individual member states' development co-operation policies, leading to more coherent policies. The Commission was asked to assist with this harmonisation. The results have been far from satisfactory, however. The member states are still too diversified regarding both the quantity and quality of development co-operation.

Also, when it comes to the more practical implementation of development co-operation, ambitions within the EU have been to harmonise the practices of the member states. A guideline for operational co-ordination in the field was developed and published in 2001 [*EU, 2001*]. The aim was to harmonise procedures and develop co-operation between the member states in the field, particularly as regards country strategies and the overall development dialogue with the recipient. Although there has been very little co-operation within aid projects and programmes in the field, there has been a marked change in the overall development dialogue. Today, EU member states consult each other, both at headquarters and in the field, about important political questions, and they also often act in a united way. A mutual agreement can sometimes feel like a strait-jacket for an individual member country that might want to act according to its own ideas, but it gives much more weight to an intervention. The co-operation has also led to an availability of information from the other member states which was previously inconceivable. For the recipients, the most positive result has been that they can deal with a number of donors at the same time. This could, however, also be a negative aspect if the unity of the donors gives the recipient less freedom to act independently, and could cause obstacles to ownership.

In summary, there is no doubt that Sweden's membership of the EU has led to major changes within its development co-operation, both regarding policy and day-to-day activities. Whether it has led to changes in its national aid practices, which in many cases have acted as a forerunner to other EU member states' practices, is, however, doubtful.

VI. SOME NUMBERS

Swedish aid is disbursed from Sida and the Ministry for Foreign Affairs (MFA), with the former disbursing roughly two-thirds of the total aid budget. Multilateral aid – approximately 30 per cent of total aid – is almost exclusively channelled through the MFA. The majority of multilateral aid goes to the UN system, particularly to the United

Nations' Development Programme (UNDP) and the United Nations' Children's Fund (UNICEF) although the World Bank group receives one-third (the bulk of which goes to the International Development Association, IDA).

Geographically, most bilateral aid is directed to development cooperation partners in sub-Saharan Africa, particularly countries in the East and South. Of the aid going to Asia and Latin America, the lion's shares are allocated to the poorest countries in South Asia and Central America, respectively. Consequently, with respect to both the multilateral and bilateral components of the Swedish aid portfolio, there is a clear focus on the poorest countries, consistent with the explicit objectives of Swedish aid. Table 15.1 provides more details.

From the historical data on Swedish aid the following points are worth mentioning:

(i) With the increasing volumes of aid and by allowing multilateral aid to stay at around 30 per cent of the total aid budget, Swedish contributions (like those of Denmark and Norway) to certain of the UN agencies have become disproportionately large. However, this share was never met by a corresponding increase in voting power in the agencies, and with the increasing inefficiencies within some of them this became a serious problem. In the early 1990s, therefore, Sweden took the initiative of a Nordic UN project which looked into the running of these agencies and came up with serious recommendations for improvement, in particular as regards their governance and administration. Although not all the recommendations were accepted, this initiative had a major impact on the running and image of UN-administered aid.

(ii) Support for the liberation struggles in Africa as well as Indochina was an important part of the Swedish aid profile for many years and still influences the current choice of partners (see below, section IX).

(iii) The increasing number of major conflicts in the developing world during the 1990s has had a major impact on the Swedish aid profile. The support for emergencies mainly caused by conflicts has increased substantially, making Iraq, for example, a major recipient of Swedish aid at the time of the Kuwait crisis as well as Yugoslavia and Bosnia as a consequence of the Balkan wars (see below, section X).

(iv) Swedish support for NGOs has been a large and increasing part of the aid profile over the years. Most funds are channelled through some 15 large Swedish NGOs, but a significant share is disbursed directly to local NGOs.

TABLE 15.1
SWEDISH AID FLOWS, 1996–2001. US$m. (CURRENT PRICES) AND %

	1996	1997	1998	1999	2000	2001
Grand total	2,001	1,730	1,575	1,632	1,880	1,667
By disbursing organisation						
Bilateral[a]	70	70	67	70	70	72
of which:						
Sida[b]	88	86	87	88	82	86
MFA[b]	12	13	13	10	17	13
Others[b]	0	1	0	2	1	1
Multilateral[a]	30	30	33	30	30	28
of which:						
Sida[c]	0	0	0	0	0	28
MFA[c]	100	100	99	100	100	72
Others[c]	0	0	1	0	0	0
Multilateral aid by recipient (% of multilateral aid)						
UN system	44	42	40	46	46	59
Multilateral DBs	31	36	40	35	38	13
of which:						
WBG incl IMF[d]	74	68	60	62	69	0
Regional DBs[d]	20	25	28	25	22	77
Others[e]	7	4	2	1	2	3
European Union	17	18	18	19	15	24
Bilateral aid by destination (% of bilateral aid)						
Africa	35	35	36	32	30	29
of which:						
East	45	42	45	42	48	48
South	36	36	27	36	32	27
Asia	26	24	19	18	19	20
of which:						
South Asia	33	30	33	32	32	38
Latin America	10	10	9	14	12	12
of which:						
C. America	61	52	53	72	68	66
Europe	6	7	4	8	6	7

Source: Statistical Department, Sida.
Notes: MFA is Ministry of Foreign Affairs and WBG is World Bank Group. 'Others' include Nordic Africa Institute, the Swedish Institute and Swedish National Defence. The price of US$ increased from SEK6.70 in 1996 to SEK10.33 in 2001, so the 1996–98 decline in the volume of aid is increased, and the 2000 increase reduced, by expressing the amounts in US$.
[a] Per cent of grand total.
[b] Per cent of bilateral aid.
[c] Per cent of multilateral aid.
[d] Per cent of multilateral development banks (payment to WBG in 2001 delayed into next year).
[e] Includes the Nordic Development Fund and non-regional development banks.

(v) Ever since the creation of SAREC in the 1970s, development research has been a priority within the aid budget and Sweden is a leading donor in that area. Funds have also been readily available for supporting the dissemination of research and policy and methods development in Sweden as well as in other research and policy centres around the world.

VII. CHANGES IN THE AID ADMINISTRATION

Changes in aid administration have been prevalent in Sweden. When problems of effectiveness and poverty alleviation become too great, it seems as if the frustration is focused on the donor administrative machinery rather than on trying to accomplish a true dialogue with the development partner. However, in spite of all the commotion over the years, Swedish development assistance seems to have been less re-administered than most other aid establishments.

The first Swedish government aid agency was created in 1962 and was responsible for technical assistance, while financial assistance was handled by the Ministry of Finance. After a major crisis, the administration was reorganised and the Swedish International Development Authority (SIDA) was created in 1965. The Swedish tradition of strong government agencies contributed to SIDA's growth over the following ten years, making it the major Swedish institution for all matters regarding development co-operation. However, SIDA as the sole administrator of aid started to be questioned during the mid-1970s, with the research community pressing for the creation of a special agency to administer support for development research. As a result of these requests, the Swedish Agency for Research Co-operation (SAREC) was formed. Following its creation, other new agencies were established, leading to an agency for industrial co-operation (SWEDFUND) and another for technical assistance to middle-income countries, the Swedish Agency for International Technical and Economic Co-operation (BITS). A major parliamentary study on aid administration at the end of the 1970s supported this development. As late as 1991 a new agency, Swedish International Enterprise Development Corporation (SWEDCORP), was created.

The proliferation of agencies responsible for aid administration led to considerable confusion and duplication. Consequently, in 1995, after a new study and further discussions, all the agencies were again merged into one – the new Sida (this time with lower-case letters). The organisation in the field was changed as much and as many times as the national administration. After a period of considerable freedom, the Sida field offices are now clearly under the auspices of the Ministry of Foreign Affairs' organisation, that is, the embassies in the respective countries. Over the past nine years, Sida has been left alone without any major changes. What will come as a result of Globkom and the new Bill presented to Parliament in 2003 it is still too early to know.

As stated above, problems confronting Swedish aid have consistently reappeared over time. It is, however, difficult to see any correlation between problem-solving and increased effectiveness and changes in the aid administration.

VIII. 'DEVELOPMENT THEORY'

What is the development theory underlying this pattern? That is to say, how does Sweden rationalise the current aid pattern from a poverty reduction perspective? Surprisingly little has been written about these questions by Sida staff, and it appears that there is no generally accepted development theory underlying the design of Swedish aid.

Rudengren [*1976*] notes that a theory of aid, i.e. a theory of how aid can help achieve development objectives, has to be based on a theory of development, that is, a theory of resource mobilisation to achieve the objectives often manifested in development plans. During the 1960s and 1970s, the theory underlying Swedish aid changed very little. Often based on a development theory emphasising gaps or structural constraints, 'the task of [Swedish] aid is to transfer resources to developing countries in the form of capital and knowledge' [*Rudengren, 1976: 23*].

It is notable, however, that in the Swedish aid debate, development has never been couched solely in terms of income levels; the distribution of income, and its social and political aspects, have been important from Proposition 1962:100 onwards. In this sense, other donors – including the World Bank – have adjusted their thinking to the Swedish philosophy, by laying an increasing emphasis on non-economic factors and by explicitly recognising that poverty is multi-dimensional and cannot be successfully attacked through economic growth alone.

In an effort to operationalise these implicit theories for poverty reduction, Sida [*2002a*] distinguishes between four different types of effects. The first mechanism is the *direct effect*. This is usually targeted projects and sector support that is aimed directly at poor people. Examples could include water projects or infrastructure projects that are targeted at poor people. The second mechanism is projects in which *the poor are included* but are not the sole beneficiaries.

Support for primary health care or primary education would fall into this category; the poor are affected by better health care, but non-poor groups benefit as well. The third mechanism is *the indirect effect through policies and institutions*. This includes support for democracy, human rights and legal reform. Finally, Swedish aid is thought to affect poverty through support at the national level that *stimulates growth of income and employment*. Programme aid would fall into this category, as would support for trade liberalisation and other growth-enhancing measures.

In sum, then, the main motivation for Swedish aid is international solidarity. Aid is a moral obligation. The objective is to help reduce poverty in poor countries. The basic Swedish philosophy is to offer 'help for self-help', which means that the selection of countries will (at least partly) be made on the basis of whether recipient governments are pursuing development strategies that are endorsed by Sweden and also an increased emphasise on ownership.[1]

The development theory underlying aid is not explicit, but Sweden distinguishes between direct and indirect effects. Direct effects are the impact from (often targeted) projects; indirect effects operate essentially via macroeconomic stability and economic growth. Figure 15.1 is an attempt to summarise the mechanisms schematically.

It is important to understand that poverty is more than lack of income in the Swedish analysis. Non-income aspects are as important as income aspects.[2] Much of the bilateral assistance flows to projects in social sectors (see Table 15.2). This can have an immediate impact on the non-income dimensions of poverty. Economic and infrastructure projects work through higher economic growth and will thus affect income and employment opportunities; these activities will thus affect the income dimension. Programme aid, finally, works through two main channels. One is to pave the way for faster economic growth and thus reduce income poverty. The other is the direct impact on the living standards of the poor through a more stable macroeconomy; we know, for instance, that the poor are hard hit by high inflation, so stabilisation serves to improve their situation.

---

1. An argument could be made that Sweden has abandoned this principle with the support of structural adjustment programmes. This argument assumes, however, that the introduction of adjustment programmes is against the will of the recipient government. We refrain from pursuing that discussion here.
2. It should be noted that Sweden has never accepted the one-dimensional definition of poverty that is sometimes associated with multilateral institutions in the 1970s and 1980s, i.e. that poverty is simply an income problem and therefore poverty alleviation is simply a matter of economic growth. For a long time Sweden has chosen the more difficult path of insisting on poverty's multi-dimensionality. This may partly account for the difficulties encountered when it comes to operationalise policy objectives, and is possibly also an indicator of Sweden being one of the leaders in development aid thinking.

FIGURE 15.1
HOW SWEDISH AID REDUCES POVERTY

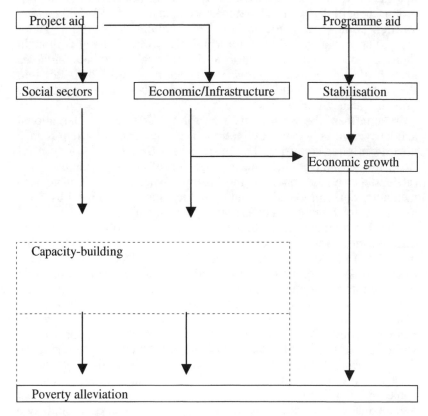

It should be noted that emergency aid – including support for refugees – grew rapidly in the early 1990s. Note also that, while Sweden in the development rhetoric often emphasises the benefits of programme aid and debt relief, this share shrank considerably during the 1990s [*Danielson and Nilsson, 1999*]. It is unclear to what extent this is consistent with Sweden's recent advances towards improved 'partnerships' with recipient countries.

TABLE 15.2
BILATERAL AID BY MAJOR PURPOSE, 1988/89–99 (%)

	1988/89	1993/94	1997/98	1999
Social	21	32	35	35
Economic	18	13	13	9
Production	15	13	9	4
Programme aid	16	5	4	5
Emergency	15	23	20	24
Other[a]	14	6	13	17

Source: OECD [*2000*], Table I.5.
Notes: Classification follows that of the OECD's Development Assistance Committee. See source for details.
[a] Includes administrative cost of donors and core support to NGOs.

IX. OPERATIONAL PRIORITIES

In the rhetoric, the main operational priority has been to maintain efficiency in the delivery of aid. This means that aid should be allocated in such a way that every dollar spent should have the maximum impact on poverty. In reality, there are several reasons why it has been difficult to stick to this priority.

Instead, much debate has been on the input side. Sweden adopted the 1 per cent objective in 1968 (i.e. the idea that 1 per cent of the gross national product (GNP) should be committed to foreign aid) and managed to achieve the target in the mid-1970s. Since the mid-1980s, Sweden has never reached that target again and, in spite of some attempts to raise the level, it has in reality been replaced by a floor of 0.7 per cent of GNP (the level agreed as a target by the UN).

TABLE 15.3
GEOGRAPHICAL DISTRIBUTION OF BILATERAL ODA (NUMBER OF COUNTRIES)

	1990/91	1994/95	2001
Africa	36	44	41
Asia[a]	21	29	27
Latin America	19	27	21
Europe[b]	1	14	14
Total	77	114	103

Source: Statistical Department, Sida.
Notes: Sweden changed in 1996 from having a fiscal year running from July to June to a calendar fiscal year. The data for 1990 and 1995 refer to ODA disbursed by all entities that comprise the new Sida, i.e. SIDA, SAREC, BITS and SWEDCORP.
[a] Including the Middle East.
[b] Including Central Asia and the Caucasus.

When SIDA was transformed into Sida in 1995, several entities dealing with foreign aid were merged. In particular, BITS – dealing with technical co-operation – and SWEDCORP – dealing with private-sector development – were combined with the old SIDA. Also aid to Eastern Europe and to the former Soviet Union was brought to the new Sida. However, the increase in the number of recipient countries reached its peak in the early 1990s, initiated by the fall of the Soviet Union and the accompanying turbulence in parts of Europe. It is clear that the early 1990s saw a significant increase in the countries receiving Swedish ODA. Table 15.3 gives the geographical distribution.

The data underlying Table 15.3 suggest two main conclusions. First, the number of countries receiving Swedish ODA increased most rapidly during the first half of the 1990s: between 1992 and 1995, it rose from 77 to 114. Second, while the most important factor explaining this increase is the inclusion of countries in Europe and Central Asia, countries in other continents as well have been added to the list: in the Caribbean, for instance, only four countries received Swedish ODA in 1990; five years later, the number was nine.[3] Similarly, the number of African countries receiving ODA increased from 32 in 1990 to 44 five years later. After 1996, however, the number of recipient countries has been relatively constant, with only a few countries entering or leaving the co-operation.

The increase in the early 1990s for Asia and Europe is a consequence of the breakdown of the Soviet Empire. To promote stability in the region, Sweden extended large amounts of aid. However, while this is understandable, it is not clear to what extent it is consistent with the principle of giving aid in order to reduce poverty in the poorest countries.

The increase in the number of countries is noteworthy, because an honoured principle in Swedish aid philosophy (on the basis of which SIDA was originally created in contrast to its predecessor NIB) has been that of concentration: provide aid to countries with whose policies with which we sympathise, and focus on areas and countries in which Sweden has a comparative advantage. The data do not suggest that this principle has been honoured.[4] We return in the next section to the principles guiding country selection.

---

3. A few countries in the Caribbean received Swedish ODA in 1995 only: the Netherlands Antilles, Saint Vincent and the Grenadines, Saint Lucia, and Saint Kitts and Nevis (received ODA also in 1994). It is possible that this is related to the 1995 UN election of new members to the Security Council for which Sweden was a candidate.

4. Rundin [*2002*] gives an interesting account of how the principles of country selection have changed since the early days (i.e. the early 1950s). We return to this issue below.

Another issue here is what is meant by 'poverty' and how the overall objective in Swedish development aid is integrated into the implementation of aid activities. The short answer is that it is not. There is a surprising absence of analyses of the impact on poverty. The OECD [*2000: 17*] notes that 'it is difficult to see the supreme objective of poverty reduction clearly and consistently reflected in Swedish aid practice', and Frühling [*2001*] surveys four of the central government directions for appropriations[5] and finds that none of them – possibly with the exception of the communication on poverty – provide guidelines on how the general objective should be interpreted in concrete aid work. His conclusion is that this shows that poverty reduction 'is not accepted or confirmed as the strategic focus or overall objective for aid activities' [*2001: 6*].

Similar criticism may be voiced regarding Sida's own work in realising the aid objective. Sida's action programmes (on the environment, poverty reduction, gender relations, and peace) give the impression, writes Frühling [*2001: 7*], that 'everything depends on everything and everything is important'. There is an absence of development theorising or ranking of goals.

The major instrument for the government is the country strategies written by the Ministry of Foreign Affairs, Sida/Stockholm and the relevant embassy, and endorsed by the government. The country strategy is probably the government's best way of concretising how aid should be used to achieve the poverty reduction objective. However, Booth *et al.* [*2001*] note that, while strategic groups are often identified in the review (the first part of the strategy document), they are often not mentioned in the second part – where the aid strategy is outlined.

Finally, neither the Budget Bill nor the annual Directions for Appropriations issued by the government provide much guidance on how to realise the poverty objective. The general impression – and here we follow Frühling [*2001*] – is that the four action plans to a large extent have replaced the six objectives of Swedish aid. The reduction of poverty in poor countries is always noted to be the overall objective, but it is easy to get the impression that this is more lip-service than serious analysis.

The problem has become even more complicated with the gradual change in views on what poverty actually is. Earlier analyses often defined poverty in terms of effects – child mortality, low income, epidemics – but this has changed to an analysis based on what are the

---

5. The statements on poverty, democracy, gender equality and environment. *Regeringsskrivelse* has been translated as directions for appropriations.

preconditions for a good life, which in turn is very much inspired by the World Bank's recent analyses of poverty. This is perhaps not surprising. The Bank has, and has had, a dominant position in development thinking, and its virtual monopoly of framing the issues and (often) delivering the answers is less challenged today than 20 years ago. The downside of this is, of course, that discussion and debates within aid organisations are more constrained – in terms of issues and concepts – than they used to be. As a consequence, capacity – often a result of learning from past experience – has not grown very fast.

Sida has struggled with this over the years and produced a number of strategies and guidelines, the latest in 2002. Like many other donor organisations, however, Sida has found it difficult to concretise the poverty objective, let alone indicators of poverty change. It is only fair to say that, while this is a problem in Swedish aid, the awareness of the problem appears to be substantial at Sida, and a great deal of effort is devoted to eliminating it. Perhaps the new demands on Sida to operationalise the new Bill of 2003 will allow Sida to develop these questions further.

Capacity development has been another central theme within Swedish aid ever since it was introduced in the early 1960s and Sida has been in the forefront of the development within this area. Although the terminology has changed over the years – sometimes to reflect real changes in emphasis – concern for the issues of institutional development and capacity-building has been directed to all sectors of activity. If no heed is taken of the need for the sustainability of activities, and the importance of capacity-building and organisational strengthening to achieving this, almost any support for development is likely to have short-lived results.

A lot of effort has been put into developing ways of working out how to learn from the past and to tackle these difficult issues. A significant reminder of the importance of this was the joint Nordic evaluation completed in 1988 as regards the sustainability of 55 completed aid projects [*DANIDA* et al.,*1988*]. It came to the conclusion that from a short-term perspective, the implementation work had been carried out extremely well. From the long-term perspective, however, the evaluators were sceptical as to the sustainability of all but a handful of projects, mainly because, when the external support had been concluded, there remained little or no internal competence in the organisations responsible for the projects to keep them running [*Johnston and Wohlgemuth, 1993*].

The Swedish effort to counter these negative results is developed in detail in Wohlgemuth *et al.*[*1998*]. Sida itself has worked out its response in a policy for capacity development [*Sida, 2001*] plus action programmes for knowledge development and a strategy for the support of the public sector [*Sida, 1991*]. However, it seems that these efforts have not been incorporated into the overall policies. Globkom thus emphasised a somewhat static view on how knowledge should be transferred rather than taking account of the dynamic relations that are required for capacities to develop, as emphasised in the action programmes referred to above.

X. CRITERIA FOR SELECTING DEVELOPMENT PARTNERS

Apart from focusing on poor countries with a development vision with which Sweden is sympathetic, the selection of partner countries is guided by three principles. The first is *concentration*. In the history of Swedish development co-operation, there have been at least three parliamentary commissions that among other things have studied the selection of partner countries; all of these have accepted the idea of focusing aid on a small number of countries. Several arguments are put forward in favour of this principle. One is that Sweden is, in volume terms, a relatively small donor. Although not stated explicitly, there seems also to be an argument that aid must reach a critical minimum before it becomes efficient. The argument makes more sense if we also take into account the fact that Sweden has traditionally chosen to work in a limited number of sectors in which it has experience and expertise. Single projects are designed as parts of larger aid efforts (such as textbook production, schools and teacher education) in which the concentration principle makes sense, particularly for a small donor. Other arguments are based on history. When Sweden started to give aid, its knowledge of poor countries was limited and it was natural to select as partners countries which it had knowledge of or ideological association with, or both. Once countries have become development partners, that co-operation is not easily broken (see below). A third possibility is that Sweden wants to be a relatively large donor in its partner countries – perhaps because large donors have more influence, both in the recipient government and among other donors. However, as Table 15.4 shows, the concentration of Swedish aid appears to have diminished over time: while the ten largest recipients received 45 per cent of all Sida aid in 1990, they received only 30 per cent in 2000.

There are at least three reasons for the increasing dispersion observed. The first is that starting in the early 1990s increasing amounts of Swedish aid have been allocated to emergencies (cf. Table

15.2). The second reason is the breakdown of states in Eastern Europe: in 2000, both Yugoslavia and Bosnia-Herzegovina qualified for the list of the ten largest recipients. The third is the merger of the aid agencies in 1995, incorporating SWEDCORP and in particular BITS in the new Sida with its many small technical assistance projects in many countries.

The second principle is that of *long-term agreements*. Sweden introduced country programming relatively early. The major argument for this was the need for the recipient to know what its future resource flows would be (which also suggests that Sweden has often tried to give aid which is as fungible as possible, preferably integrated into the recipient's budget).[6] Country programming soon evolved into country frameworks in which allocations were planned on a three-year basis. Although the concept is now abandoned, Swedish official aid still makes a distinction between the 20 or so major recipients and the rest.

The third principle is *inertia/continuity*. It may be difficult to become a Swedish development partner, but once you get there, you're likely to stay. This principle, of course, is linked to the idea of long-term relations, even though the development of Swedish aid in the 1990s (particularly the number of countries (see Table 15.2)) suggests that it has become less important.

TABLE 15.4
TEN LARGEST RECIPIENTS OF SWEDISH BILATERAL AID (% OF TOTAL ODA)

1990		1995		2000	
Tanzania	10.8	Mozambique	4.6	Tanzania	5.1
Mozambique	9.9	India	4.3	Mozambique	3.7
India	4.3	Tanzania	3.8	Honduras	3.4
Vietnam	3.9	Ethiopia	3.3	Vietnam	3.0
Ethiopia	3.5	Bosnia-Herzegovina	3.2	Yugoslavia	2.7
Angola	2.8	Vietnam	2.9	Nicaragua	2.7
Zambia	2.7	Zambia	2.7	PAA[a]	2.6
Zimbabwe	2.6	Nicaragua	2.7	South Africa	2.6
China	2.3	Zimbabwe	2.4	Bangladesh	2.6
Nicaragua	2.3	Iraq	2.4	Bosnia-Herzegovina	1.9
Total bilateral	45.1		32.3		30.3

Source: Statistical Department, Sida.
Notes: Data for 1990 and 1995 include aid from SIDA, BITS, SWEDCORP and SWEDFUND, i.e. most of the organisations that comprise the post-1996 Sida.
[a] Palestinian Administered Areas.

---

6 The application of this principle was most likely confined to countries that were ideologically close to Sweden. Tanzania in the 1970s is a case in point.

A possible fourth principle is that of *geography*. Sweden has always disbursed a large amount of its aid to sub-Saharan Africa, and particularly to Anglophone countries in southern and eastern Africa. A major reason for this was the result of Swedish foreign as well as aid policy to give support to the liberation struggle in southern Africa. The ties created during that period led to continued partnership in the form of development co-operation. This explains also the inclusion of most Portuguese-speaking countries in Africa. The same argument is also valid for former Indochina. Francophone Africa has been avoided until very recently due to language problems (the first experience in the 1960s and 1970s of aid to Tunisia laid the basis for this). Other countries have sometimes received large amounts of aid – not always on very clear principles, although, as mentioned earlier, major emergencies often lead to a large amount of aid being committed and disbursed which often leads to lasting relationships.[7]

XI. CONCLUDING REMARKS

Swedish aid is, and always has been, characterised by change on the surface. Deeper down it is not clear whether much has really changed. The volume of foreign aid has grown (but not in terms of the share of GNP), and the number of aid workers (including consultants) has certainly grown, but the ways of thinking about types of aid, the responsibilities of the recipient, the idea of 'help for self-help' and the focus on poverty remain very much the same. Much of the change is simply rhetorical [*Forsse, 2001*]. Thus, for instance, Sida recently announced that it was no longer using technical assistance. Did this mean that Sweden no longer finances courses or sends out foreign experts to work on projects? No, it means that technical assistance is no longer a separate budget line in Sida's accounts.

Of course, the sector-wide approach, budget support and the discussion on ownership did not exist in the 1970s – or at least not in those terms. But a close reading of the core literature – particularly the contributions in Frühling [*(ed.), 1986*] and Wohlgemuth [*(ed.), 1976*] – shows a clear awareness of problems and approaches which has not been superseded by the current debate. Swedish aid has changed a lot since the 1960s – in volume as well as types – but the seeds of what is now characteristic of Sweden's development

---

7. During the second half of the 1970s, Liberal Ola Ullsten – Minister of development co-operation in the conservative/liberal government – vigorously argued for continued aid to Angola, one of the few countries in the region at that time with a development vision characterised as 'scientific Marxism'.

co-operation philosophy were sown more than 30 years ago. Let us see whether operationalisation of the new Bill will change this pattern!

REFERENCES

Andersen, David et al., 2002, 'Partner Country Ownership in Development Co-operation through Sida', Synthesis report, Stockholm: Sida.
Annell, L., 1986, 'Images of Distant Countries. Reflections on the Theoretical Foundations of Sweden's Development Co-operation Policies', in Frühling (ed.).
Booth, D., T. Conway and A. Silfverstolpe, 2001, *Working with Poverty Reduction in Sida*, Stockholm: Sida/POLICY.
Carlsson, J., 1998, *Swedish Aid for Poverty Reduction: A History of Policy and Practice*, Working Paper 107, London: Overseas Development Institute.
Carlsson, Jerker, Gloria G. Somolekae and Nicolas N. van de Walle, 1997, *Foreign Aid in Africa – Learning from Country Experiences*, Uppsala: Nordic Africa Institute.
Carlsson, Jerker and Lennart L. Wohlgemuth (eds), 2000, *Learning in Development Co-operation*, EGDI 2000:2, Stockholm: Almqvist & Wiksell.
DANIDA/FINNIDA/NORAD/SIDA, 1988, *A Nordic Evaluation of the Effectiveness of Technical Assistance Personnel in Kenya, Tanzania and Zambia*, Joint Nordic Evaluation, Copenhagen: DANIDA.
Danielson, A., 1999, 'Policy Coherence Towards Developing Countries: The Case of Sweden', in Forster and Stokke (eds).
Danielson, A. and M. Nilsson, 1999, *Trends and Turns in the 1990s. Programme Aid and the Swedish Experience*, Sida Evaluation Report 99/17: 9, Stockholm: Sida.
Ehrenpreis, D., 1997, 'Fungerar Biståndet? – en genomgång av 30 års erfarenheter' (Does Aid Work? A Review of Thirty Years of Experience), in Wohlgemuth (ed.).
EU, 2000, *Joint Declaration of the Council and Commission on EU Development Policy*, Brussels: European Union.
EU, 2001, *Guideline for Operational Co-ordination*, Brussels: European Union.
Forsse, A., 2001, 'Utvecklas biståndsdebatten?' (Does the Aid Debate Develop?), background paper for the parliamentary commission on Swedish policies for global development – Globkom (mimeo).
Forster, J. and O. Stokke (eds), 1999, *Policy Coherence in Development Co-operation*, EADI Book Series 22, London: Frank Cass.
Frühling, P., 2001, 'Överordnat, jämbördigt eller mittemellan?' (Superior, equal or in-between?), background paper for the parliamentary commission on Swedish policies for global development (mimeo).
Frühling, P. (ed.), 1986, *Swedish Development Aid in Perspective. Problems, Policies and Results since 1952*, Stockholm: Almqvist & Wiksell.
Globkom, 2001, *A More Equitable World Without Poverty. Report By the Parliamentary Committee on Sweden's Policy For Global Development*, SOU: 2001: 96, Stockholm: Ministry of Foreign Affairs.
Government of Sweden, 2003, *Gemensamt ansvar: Sveriges politik for global utveckling* (Shared Responsibility: Sweden's Policy for Global Development), Proposition 2002/03: 122.
Havnevik, K. and B. van Arkadie, 1996, *Domination or Dialogue? – Experiences and Prospects for African Development Cooperation*, Uppsala: Nordic Africa Institute.
Hveem, H. and D. McNeill, 1994, *Is Swedish Aid Rational?*, Stockholm: Ministry of Foreign Affairs/SASDA.

Johnston, A. and L. Wohlgemuth, 1993, 'Capacity Building and Institutional Development – Sida's Perspectives', *IRD Currents*, No. 5.
Kayizzi-Mugerwa, S., A. Olukoshi and L. Wohlgemuth, 1998, *Towards a New Partnership with Africa – Challenges and Opportunities*, Uppsala: Nordic Africa Institute.
Kifle, H., A. Olukoshi and L. Wohlgemuth, 1997, *A New Partnership for African Development – Issues and Parameters*, Uppsala: Nordic Africa Institute.
MFA, 1998, Africa on the Move – Revitalising Swedish Policy towards Africa for the 21st Century, Government Communication Bill SKR 1997/98:122, Stockholm: Ministry of Foreign Affairs.
OECD, 1996, 'Sweden', *Development Co-operation Review Series*, No. 19, Paris: OECD/DAC.
OECD, 2000, 'Sweden', *The DAC Journal*, Vol. 4, Paris: OECD/DAC.
Olsson, J. and L. Wohlgemuth (eds), 2003, *Dialogue in Pursuit of Development*, EGDI 2003: 2, Stockholm: EGDI.
Rudengren, J., 1976, 'Svensk biståndspolitik' (Swedish Aid Policy), in Wohlgemuth (ed.).
Rundin, U., 2002, *Valet av mottagarländer för svenskt bistånd åren 1950–1990* (The Selection of Recipient Countries for Swedish Aid, 1950–1990), unpublished MA thesis, Stockholm: Department of History, Stockholm University.
Sellström, T., 1999, *Sweden and National Liberation in Southern Africa. Volume I: Formation of a Popular Opinion (1950–1970)*, Uppsala: Nordic Africa Institute.
SIDA, 1987, *Action Program for Knowledge Development*, Stockholm: SIDA.
SIDA, 1988, *Roll- och ansvarsfördelning i Sidas hantering av biståndsinsatser* (Allocation of Roles and Responsibilities in SIDA's Management of Foreign Aid), Stokholm: SIDA.
SIDA, 1991, *Making Governments Work*, Education Division, Public Administration Section, Stockholm: SIDA.
Sida, 2000, *Statistical Report*, Stockholm: Sida.
Sida, 2001, *Sida's Policy for Capacity Development*, Stockholm: Sida.
Sida, 2002a, *Perspectives on Poverty*, Stockholm: Sida.
Sida, 2002b, *Ownership of Sida Projects and Programs in East Africa*, Stockholm: Sida.
SOU, 1962: 12, *Aspekter på utvecklingsbiståndet: promemorier* (Aspects on Development Assistance: Memoranda), Stockholm: Ministry of Foreign Affairs.
Wohlgemuth, L. (ed.), 1976, *Bistånd på mottagarens villkor* (Aid on the Recipient's Terms), Uppsala: Nordic Africa Institute.
Wohlgemuth, L. (ed.), 1994, *Bistånd på utveclingens villkor* (Aid on Development's Terms), Uppsala: Nordic Africa Institute.
Wohlgemuth, L., J. Carlsson and H. Kifle, 1998, *Institution Building and Leadership in Africa*, Uppsala: Nordic Africa Institute.

16

Swiss Development Co-operation: Major Changes since the Early 1990s and Future Challenges

CATHERINE SCHÜMPERLI YOUNOSSIAN

Over the past ten years or so, Switzerland has endeavoured in both its bilateral and multilateral development co-operation activities to take global changes and new requirements into account. Through the topics it deals with, development co-operation has gradually become an integral part of the political, economic and social relations with countries receiving official development assistance (ODA). Through a more global approach, yet one that remains centred on technical co-operation and humanitarian aid, development co-operation is now better integrated with Switzerland's foreign policy.

The ever-increasing globalisation of commercial and financial relations and the emergence of global problems (environment, human rights, migration, organised crime, money laundering, drugs, etc.) have reinforced the interdependence of all countries and had significant repercussions on international co-operation. In this context, the Swiss government has presented a reorientation of its development co-operation policy. The *North–South Guidelines* [*Federal Council, 1994*] set out four strategic objectives for Switzerland with the countries of the South (see below).

This chapter aims to show how the Swiss organisations responsible for international development co-operation have adapted to the global changes, and how they have adjusted their instruments and programmes as a consequence. First, documents that guide the general development assistance policy will be reviewed. Second, the changes in the volume of official development assistance and its financing will

---

Note of acknowledgement. The author would like to thank Professor Jacques Forster and Gérard Perroulaz (scientific collaborator), both working at the Graduate Institute of Developement Studies – IUED (Geneva), for their comments and suggestions.

be examined. Then, the Swiss players involved in development co-operation will be considered, with a particular emphasis on the work and influence of non-government organisations (NGOs) and on the importance of public opinion in Switzerland. This point is important as, during the evaluation by the Development Assistance Committee of the OECD, it has been highlighted as one of Switzerland's main characteristics [*OECD, 2001*]. Because Switzerland is not a member of the European Union, it does not have to provide any share of its ODA as an obligatory contribution. This explains why this issue will not be handled in this chapter.

The chapter will move on to demonstrate the influence which the introduction of the *North–South Guidelines* and the changes in the Swiss and international contexts have had on certain development co-operation instruments. Four specific areas will be analysed: first, the way in which the Swiss Agency for Development and Co-operation (SDC) and the State Secretariat for Economic Affairs (Seco) have integrated the fight against poverty in Switzerland's bilateral development co-operation; second, Switzerland's humanitarian aid, with particular emphasis on its policy in the Balkans; third, Switzerland's multilateral co-operation and the political implications of its recent accession to the principal international organisations responsible for development; and, fourth, environmental diplomacy, a new area of Swiss foreign policy. The *North–South Guidelines* form the basis of a coherent policy towards the South. This is not without its consequences at institutional level and we shall therefore look at the way the Federal Administration has modified its practices and operations in order to take the interaction between all areas of international co-operation and the globalisation of many problems into account.[1] The evolution of the international context in the 1990s (the collapse of the communist regimes and the rise of economic globalisation), the new approaches of international development co-operation (policy coherence, political conditionality) and the new agenda for Swiss foreign policy (membership of the main international organisations) have involved a drastic change of the context in which development co-operation was operating. We shall therefore conclude with some of the new challenges which Swiss development co-operation will have to take up in the coming years.

---

1. These changes have also led in recent years to a series of adjustments to planning and management instruments. These issues will not be dealt with here.

I. DECLARED OBJECTIVES OF SWISS AID POLICY

Swiss development co-operation policy is founded on three 'pillars' which define the general guidelines for aid and lay down its operating methods: the legal bases, the Messages from the Federal Council and the reports of the Federal Council.

1. The Legal Bases

Development co-operation and humanitarian aid, together with the policy on co-operation with the East European countries and the Commonwealth of Independent States (the former Soviet bloc), depend on two legal bases:

– The *Federal Law on Development Co-operation and International Humanitarian Aid of* 1976 and its implementing regulation of 1977. This law defines the general objectives of assistance, in other words, first and foremost supporting the efforts of developing countries to improve the living conditions of their populations and enabling them to carry out their development by their own means. The law states that priority must be given to supporting the efforts of the worst disadvantaged countries, regions and population groups. Finally, it sets out the various forms of development co-operation, the main ones being technical co-operation, financial assistance, economic policy measures and humanitarian aid.

– The policy to support the transition in the East European countries and the Commonwealth of Independent States (CIS) is governed by the *Federal Decree on Co-operation with the Countries of Eastern Europe* (1995).

Solidarity with the world's poorest countries reflects Switzerland's belief that the security and prosperity of the most advanced nations can be ensured in the long term only by doing everything possible to eliminate the most glaring inequalities around the globe, in an effort to satisfy the basic requirements of all human beings, while safeguarding the rights of future generations [*Seco, website*].

2. The Federal Council's Messages on Development Co-operation

The Messages of the Federal Council (Government) set out the aims and the political and operational guidelines of the various programmes, together with a request for multi-annual funds. They are periodically submitted for the approval of the Federal Assembly (Parliament).

These Messages relate to Switzerland's five areas of development co-operation: (i) bilateral co-operation with developing countries via development co-operation (SDC) and economic and commercial measures via the State Secretariat for Economic Affairs (Seco); (ii) multilateral development co-operation; (iii) co-operation with the countries of Eastern Europe; (iv) humanitarian aid; and (v) measures to protect the global environment. Over the past decade, Parliament has approved a dozen requests for programme credits relating to international co-operation, totalling around US$14,868 million.

On the occasion of the 700th anniversary of the Confederation, these funds were supplemented by additional exceptional aid of US$488m. (CHF700 million) to finance projects with an impact on the environment and debt relief measures in the poorest countries.

TABLE 16.1
LIST OF MAIN PROGRAMME APPROPRIATIONS CURRENTLY IN FORCE (US$m.)

Type	Date of approval	$ million	Minimal duration
Financing debt reduction measures	1991	279	5 years
Ecological programmes in developing countries	1991	209	5 years
Switzerland's accession to the Bretton Woods institutions	1992	3,547	–
Swiss participation for the recapitalisation of the regional development banks	1995	677	4 years
Humanitarian aid	1996	849	4 years (1997–2001)
Economic and commercial measures	1996	777	4 years (1997–2001)
Co-operation with East European countries and the CIS	1998	621	4 years (1999–2002)
Technical co-operation with developing countries	1998	2,759	4 years (1999–2002)
Humanitarian aid	2002	1000	4 years (2002–5)
Economic and commercial measures	2003	750	6 years (2003–8)
Technical co-operation with developing countries	2003	3,400	4 years (2004–7)
Total		14,868	

Source: OECD [2001].

3. Foreign Policy Reports

The Federal Council regularly forwards reports to Parliament on the aims of its foreign policy. The most important of these in terms of Switzerland's relations with developing countries is the *Report on Switzerland's North–South Relations in the 1990s*, better known as the *North–South Guidelines* [*Federal Council, 1994*]. This report sets out guidelines for Swiss development co-operation by defining four strategic objectives covering the political, economic, social and ecological aspects of Switzerland's external relations:

– safeguarding and promoting peace and security, human rights, democracy and the rule of law
– promoting welfare
– increasing social equity
– protecting the natural environment.

The report opens up new fields to development co-operation and highlights, in particular, the importance of looking at the totality of relations with a country 'in order to guarantee an integral and coherent development policy'.[2]

Other topical reports clarify the framework for Swiss foreign policy. Indeed, the foreign policy has been extended into the fields of promoting peace, new security policy implementation and development co-operation [*Federal Council, 1999a, 1999b, 2000b*]. The issue of human rights is also at the forefront of foreign policy. The Federal Council has presented a specific report on this subject, in which it sets out the operational concept of the Swiss policy on human rights as well as the various instruments available to Switzerland to combat human rights violations. The document [*Federal Council, 2000c*] describes and interprets the diplomatic, legal, economic and development co-operation instruments in question.

Thus, having provided a general overview of Switzerland's development co-operation, let us now look at the resources it has at its disposal to implement its programmes.

---

2. Consideration of the coherence of the various policies affecting developing countries is one of the main concerns of Swiss co-operation. See Forster [*1999*].

II. VOLUME AND FINANCING OF ODA

Each year Switzerland spends approximately US$900 million on development aid, a figure which corresponds to about 0.34 per cent of Switzerland's gross national product (GNP). In 2000, the Confederation contributed US$877m., the cantons' share amounts to $8.4m. and the local authorities added another $4.8m.

1. The Confederation's Official Development Assistance

During the 1992 Earth Summit in Rio, Switzerland publicly announced its intention to increase its ODA contributions gradually, to eventually reach 0.4 per cent of GNP.[3] This goal was reaffirmed in the *North–South Guidelines*, in the Federal Council's 1999 to 2003 legislative plan [*Federal Council, 2000d*], and in the 2000 report on foreign policy [*Federal Council, 2000a*]. However, no timetable for achieving it was established. In reality, the federal finances have been facing increasing deficits since 1993 which have required overall economies to restore a degree of balance. As underlined by the federal departments concerned, the aim of gradually increasing ODA to this target has had to be abandoned, or at least postponed to better times [*SDC–Seco, 2000*].

Nonetheless, in 15 years Switzerland's financial contribution to development assistance for the countries of the South has more than doubled, from US$422m. in 1986 to $851m. in 2002, to which may be added $69m. in official assistance to the Central and Eastern European countries.[4] Table 16.2 presents an overall picture.

The substantial increase in 1992 resulted from the cost of joining the Bretton Woods institutions, with additional funds (US$200m. approximately), required to cover the costs of Switzerland's accession to the World Bank in particular. Another year of note was 1999, due to the additional funds used for the humanitarian situation in the Balkans [*SDC–Seco, 2000: 40*].

Switzerland's ODA has for several years totalled around 0.34 per cent of GNP. In 1999, Switzerland ranked seventh among the DAC members in terms of ODA/GNP ratio, which was a jump of three places in four years. The improvement in Switzerland's position is, however, due more to the decrease in the contributions of the other countries than to an increase in Switzerland's ODA [*OECD, 2001: II–32*].

---

3. Switzerland, which was not a member of the United Nations until 2002, has never signed up to the target of 0.7 per cent recommended by the UN.
4. According to the DAC directives, the amounts provided within the framework of co-operation with the countries of Eastern Europe and the CIS are recorded under 'official assistance' and not under 'official development assistance'.

TABLE 16.2
ODA FOR DEVELOPING COUNTRIES (DAC, LIST 1), 1986–2002 (US$m. AND %)

Years	ODA (US$m.)	ODA (% GNP)	Bilateral aid	Multilateral aid	Bilateral aid (%)	Multilateral aid (%)
2002	866	0.32	705	161	81	19
2001	907	0.34	644	263	71	29
2000	890	0.34	627	263	70	30
1999	984	0.35	732	252	74	26
1998	930	0.33	638	292	69	31
1997	911	0.34	575	335	63	37
1996	1,026	0.34	722	304	70	30
1995	1,084	0.34	779	304	72	28
1994	982	0.36	724	258	74	26
1993	793	0.33	635	158	80	20
1992	1,139	0.45	677	462	59	41
1991	863	0.36	727	136	84	16
1990	750	0.32	551	199	73	27
1989	558	0.30	423	134	76	24
1988	617	0.32	445	172	72	28
1987	547	0.31	389	158	71	29
1986	422	0.30	323	98	77	23

Source: SDC statistics, October 2003.

Notes: *Bilateral aid* includes contributions made by the Confederation to recipients, directly or through Swiss development organisations, universities, private enterprises (especially for highly technical projects) and aid allocated to the specific projects of international organisations. It incorporates development co-operation projects and bilateral humanitarian aid. It should also be noted that payments to the International Committee of the Red Cross are included in bilateral aid, in accordance with DAC rules.

Multilateral aid consists of general contributions to international organisations. This aid is intended for development programmes, but not to particular countries or specifically chosen and defined projects. The subscriptions are contributions to the capital of regional development banks and contributions to their special funds.

2. Decentralised Co-operation (Local Public Groups)

A number of Swiss cantons and towns also provide resources for development programmes. In 2002, the Swiss cantons provided a total of around US$8.9 million (see Table 16.3). Four cantons (Geneva, Zurich, Zoug and the two Basel half-cantons) account for approximately 75 per cent of this aid.

TABLE 16.3
ODA FROM SWISS CANTONS AND MUNICIPALITIES FOR DEVELOPING COUNTRIES, 1991–2002
(US$m.)

	1991	1992	1993	1994	1995	1996	1997	1998	1999	2000	2001	2002
Cantons	10.2	12.4	11.1	9.4	9.7	8.0	10.2	9.5	9.1	8.4	7.7	8.9
Municipalities	7.0	7.7	5.8	6.6	7.3	7.2	5.4	5.5	5.9	4.8	5.3	5.9
Total	17.2	20.1	16.9	16.0	17.0	15.2	15.6	15.0	15.0	13.2	13.0	14.8

Sources: IUED–SDC [*2002*]; SDC statistics 2003.

Very few municipalities and cantons actually administer their co-operation projects themselves. They usually disburse their funds through development NGOs. Direct decentralised co-operation aid is insignificant; for instance, in 1997 only US$0.3m., 3.7 per cent of the cantons' and municipalities' aid, was managed by the authorities themselves.[5] To facilitate the transfer of these funds from the local public donors to the NGOs, the SDC promotes the setting up of cantonal federations for development co-operation. The first federation of this type was created in Geneva Canton in 1966.

From a round table of a dozen associations involved in development co-operation at its foundation in 1966, the Geneva Federation for Co-operation and Development (GFCD) has grown to become the umbrella organisation of 51 Geneva-based local associations, grouping tens of thousands of concerned citizens functioning mainly on a volunteer basis. The GFCD offers its members a forum for the exchange of experience, a public platform for the discussion of key North–South issues, a lobby to promote international solidarity and a channel for the funding of development and information projects. It is a private non-profit organisation but its institutional and programme funding comes from public sources: the SDC, the State of Geneva and the City of Geneva [*FGC, 1996: 13*].

3. Private Development Assistance: Donations from Private Bodies

The final source of development assistance is the funds from the Swiss people. These funds, intended for development co-operation and humanitarian aid projects, are administered by Swiss development organisations. The aid collected by Swiss NGOs totalled US$162m. in 2000 ($183m. in 1999 and $169m. in 1998) [*IUED–SDC, 2002*]. The volume of aid received by development organisations is high in comparison with other countries. According to our calculations, in 2000 every Swiss citizen donated $22.5, which puts Switzerland far above the average donation of the citizens of OECD countries ($8.3).[6]

4. Economic Repercussions of ODA

Although development co-operation is intended to meet the needs of developing countries and their populations, the financial contributions referred to above have significant repercussions for the Swiss

---

5. Although these data are not available for OECD since 1997, we may assume that the situation has not changed much since then.
6. Calculations of the Graduate Institute of Development Studies (IUED) on the basis of DAC statistics.

economy. The SDC periodically commissions a study of the economic effects of development co-operation and humanitarian aid. Analysis of the data for 1998 shows that, in general, each franc (CHF) spent by the Confederation on development co-operation results in an increase in the gross domestic product (GDP) of between CHF1.40 and 1.60 [*Forster and Pult, 2000*]. Thus, the 1998 aid budget of CHF1.3 billion improved the economy to the tune of between CHF1.8 and 2 billion. There is a distinction between bilateral aid, where each franc contributes between CHF1.15 and 1.25, and multilateral aid, which accounts for between CHF1.70 and 1.90. Multilateral aid has more important repercussions on the Swiss economy because traditionally Swiss companies benefit in large measure from the acquisitions of goods and services made by the multilateral agencies (notably the World Bank and the regional development banks). Development co-operation and humanitarian aid have also had a marked effect on employment in Switzerland. The study estimates that it is responsible for between 13,000 and 18,000 jobs. It should be noted that the economic repercussions are substantial, despite the fact that Switzerland is one of the countries where aid is least tied.[7]

III. PRESENTATION OF THE MAIN GOVERNMENTAL AND NON-GOVERNMENTAL PLAYERS

1. *Federal Offices Responsible for Development Co-operation*

At federal level, two offices are responsible, under the 1976 law on development co-operation, for determining and implementing development assistance: the Swiss Agency for Development and Co-operation (SDC), which is part of the Federal Department of Foreign Affairs (DFA), and the State Secretariat for Economic Affairs (Seco), which belongs to the Federal Department of Economic Affairs (DEA). The SDC[8] is responsible for technical co-operation and financial assistance, humanitarian aid, commitments during disasters and, since 1995, technical co-operation with the countries of Eastern Europe and the CIS. Seco draws up and implements economic and commercial policy measures within the framework of development co-operation and financial co-operation with Eastern Europe and the CIS. The two offices have joint responsibility for multilateral financial aid, but the overall co-ordination of the sector is entrusted to the SDC, which

---

7. According to the DAC 2000 report, the average level of tied aid of DAC members is 11.5 per cent, while only 3.2 per cent of Switzerland's ODA is tied.
8. Until 1995, the SDC was known as the 'Directorate for Development Co-operation and Humanitarian Aid'.

manages about 80 per cent of ODA contributions and has more than 350 staff in Switzerland and working abroad. Seco is responsible for approximately 15 per cent of the aid and has about 50 staff at its disposal. In its evaluations, the Development Assistance Committee (DAC) regularly highlights the co-ordination problems that may arise as a result of the shared responsibility of the two departments. Nonetheless, in its last evaluation, the DAC underlined the efforts made to enhance co-ordination between them [*OECD, 2001: II–94*].

In addition to these two offices, one should mention the Swiss Agency for Environment, Forests and Landscape (SAEFL), which is part of the Federal Department of Environment, Transport, Energy and Communications (DETEC). Since the Rio Conference on Environment and Development of 1992, international environmental policy has gathered momentum. In addition, protection of the environment is one of the four strategic objectives of Switzerland's North–South relations. SAEFL is responsible for international environmental issues and represents Switzerland in the corresponding international bodies. It has access to financial resources for Switzerland's commitments in this field.

The adoption of the *North–South Guidelines* implies other noteworthy reorganisations. The Department of Foreign Affairs has created new divisions, whose work will complement that of the two offices responsible for development issues. Political Division III will thus deal with the activities of international organisations and the implementation of the new peace and security policy. In addition, Political Division IV is entrusted with the human rights and humanitarian policies together with international refugee policy. Political Division V is responsible for sectors which do not come under the DFA's remit but which have a bearing on foreign policy issues, including economic, financial, environmental and cultural aspects.

This new distribution of tasks is vital in achieving the new objectives, but it demands constant efforts to ensure co-ordination between the offices in order to instil the policy concerning the countries of the South with a greater sense of coherence. Conscious of these challenges, the Federal Administration is endeavouring to establish the co-ordination structures required to create closer or more restrictive links between the various policy areas. In this respect, the Interdepartmental Committee for Rio Follow-up (IDARio) should be mentioned; this unites the offices of three departments (DFA, DETEC, DEA), one of the rare examples of a supra-departmental structure.

There are many criticisms of the lack of co-ordination within the administration and calls for the creation of a neutral body to monitor and evaluate public policies concerning the South. In 2000, the *Annuaire Suisse–Tiers Monde* [*Forster (ed.), 2000: 3–117*] produced a document on the implementation of the *North–South Guidelines* within the Federal administration. Most of its contributions highlight a loss of impetus in this regard. It is important to note the lack of political will to promote better co-ordination between the various offices; yet the search for coherence remains a key political objective. Although certain offices are better co-ordinated in some sectors (as demonstrated by the IDARio), no procedures have been instituted to assess the coherence of public policies. This shortcoming is also mentioned in the assessment of the implementation of the Rio commitments in Switzerland. Among the priority points requiring action, the study's conclusions emphasise the need to 'establish neutral monitoring and control of results that enable the Government and the administration to implement a more substantial, coherent and effective sustainable development policy' [*IDARio, 2001*]. There is increasing pressure to set up a 'control body', the structure and functions of which have yet to be defined (see section IV below).

2. Non-governmental Organisations

Switzerland has a strong tradition of associations in the field of development co-operation and assistance, or Switzerland–South relations more generally. In fact, there are more than a thousand development or emergency aid NGOs, fair trade associations, solidarity groups, etc.[9] Moreover, as highlighted in the DAC report, the Swiss Confederation maintains close relations with the NGOs by providing financial support for their programmes. Some NGOs are also used by the SDC as implementing agents for its own programmes.

In 2000, the SDC allocated about US$100m. (36 per cent of Swiss bilateral aid) to NGOs [*SDC–Seco, 2001*]. It also has credit lines with some major Swiss NGOs.[10] As for the smaller NGOs, they obtain funds from the Swiss public and/or from municipalities and cantons. They may also receive funds from the Confederation, but through the cantonal federations (see section II).

Along a different line, the NGOs form an active pressure group that cannot be ignored and are considered to be true partners of the Confederation, which involves them closely in its consideration and

---

9. According to the database of the Graduate Institute for Development Studies.
10. Such as Intercooperation, Swisscontact, Helvetas, Swiss Red Cross, Program of Volunteers (UNITE), Caritas Switzerland, Bread for All, Geneva Federation for Cooperation and Development (GFCD), Swissaid, Swiss Interchurch Aid (EPER), Catholic Lenten Fund, Swiss Labor Assistance (OSEO), Pestalozzi Children's Village Trust.

development of policies [*OECD, 2001: II–27*]. In terms of raising awareness among the public, they are also extremely active.[11] The Confederation frequently holds exchange and information meetings in order to invigorate reflection on development policy issues.

One study, carried out within the framework of a national research programme, 'Non-governmental organisations and Switzerland's foreign policy', found that in relation to the drafting of policies, the NGOs often challenge and criticise the government, proposing reviews of policies or new policies, carrying out public awareness campaigns, launching petitions or using Swiss instruments for direct democracy [*Freymond and Boyer, 1998*]. In terms of the implementation of policies, however, they act first and foremost as implementing agents for mandates or tasks on the ground. It is important to be aware of the often ambiguous role of certain NGOs, which at times take the government to task and at other times are its partners. The study looks at the expectations of the Federal Administration with regard to the NGOs. The Administration recognises its need for expertise and new ideas, and seeks to involve the NGOs in the drafting and implementing of policy since, in its view, they are a good stepping stone to the public and play an essential role as intermediaries. The study concludes that relations between the Administration and the NGOs are currently progressing well and appear to be better than in other countries.

The opportunity of launching popular initiatives or rejecting, by way of referendum, certain government decisions provides the NGOs with a substantial amount of power. Associations, which are well structured in Switzerland, play a particular role in this political context. Thus, the development NGOs have on several occasions launched significant public debates on topics linked to development co-operation.

Two popular votes can be mentioned here: a referendum against Switzerland joining the Bretton Woods institutions in 1992 and the launch of a popular initiative to promote active political participation by Switzerland within the United Nations in 2002. At the beginning of the 1990s, the Federal Council proposed to the Federal Chambers that Switzerland should join the International Monetary Fund and the World Bank. Parliament supported the government's proposal, although two referendums were launched to block it. The first was the fruit of the conservative and traditionalist movements in Switzerland, which opposed any opening up of the country and raised the financial

---

11. Of the NGOs working in the area of development policy, we should mention the Swiss Coalition of Development Organisations, which encompasses the five largest NGOs (Swissaid, Bread for All, Catholic Lenten Fund, Helvetas, Caritas), and the Bern Declaration.

argument. The second was launched by certain Third World movements which had been criticising these two Bretton Woods institutions for years. Other development NGOs refused to support the referendum, arguing that lobbying would be more effective if Switzerland was a full member of the institutions. In fact, the people voted in favour of joining (by 55.8 per cent) and the doors of the Washington institutions were thus opened to Switzerland. The idea of Switzerland joining the UN was largely promoted by the political world and associations. The government had made this one of the priority objectives of its foreign policy and it was also supported by a popular initiative. In March 2002, the people had their say and came down in favour of joining, but by a small margin (54.6 per cent). In both cases, the NGOs played a decisive role by launching and fuelling the political debate.

3. Public Opinion

Every five years, the SDC and the Swiss Coalition of Development Organisations conduct an opinion poll on Switzerland's development co-operation and development policy. The 1999 results [*GfS, 1999*] show that Swiss development assistance has a solid foundation. The majority of the population (56 per cent) feel that it should continue while 20 per cent of those questioned even considered that it should be increased. Only a minority (17 per cent) were clearly in favour of a reduction in aid.

The results of the survey suggest that the majority of citizens have understood the complex relations between the global economy, poverty and environmental problems. For example, over half of all Swiss people believe that they can help personally to improve the situation in developing countries through a simpler lifestyle (56 per cent) or through financial support (52 per cent). Two-thirds of those questioned feel that Switzerland should help to adjust the structures of the global economy, for example, by fighting for better prices for raw materials or by implementing debt relief measures.

Three main factors explain the 'high' level of popular awareness of development issues and knowledge of global problems. As already noted, there is a strong tradition of forming associations in Switzerland and they are well structured to carry out awareness campaigns and launch public debates. The Confederation is not to be outdone and devotes substantial resources to information projects (around US$4m. per year). Finally, the importance of education on development in schools is widely recognised. This is demonstrated by the creation in 1997 of an independent Education and Development

Foundation, which involves the Confederation, the cantonal education departments, teacher associations and development NGOs.

4. Political Parties

Development co-operation policy is not a 'sensitive' political subject in Switzerland, as is attested by the quite minor place this issue occupies in the various political parties' plans of action. If no party declares itself against development co-operation, it is, however, the socialist and ecological parties that are the most inclined to propose an increase of ODA. In contrast, the right-wing popular party (Union Démocratique du Centre – UDC) stands clearly against an ODA increase, while not actively fighting for a drastic reduction. The UDC is very reticent about development co-operation (especially multilateral co-operation), but encourages Swiss humanitarian aid. As for the other parties, such as the Christian Democrats, the radicals or the liberals, they seldom take a position on the issue. In recent years, ODA credits were approved by Parliament without any difficulty or major political debate.

IV. STRUCTURE AND MAIN COMPONENTS OF THE AID PROGRAMME

1. Bilateral Co-operation

As mentioned above, bilateral development co-operation is in the hands of two federal offices: the Swiss Agency for Development and Co-operation (SDC) and the State Secretariat for Economic Affairs (Seco), each possessing its own instruments and geographical priorities. In this section, we shall focus on the way Swiss development co-operation has incorporated the fight against poverty into its bilateral policy, concentrating on the Seco-co-ordinated Swiss Debt Reduction Programme and the SDC projects in mountain regions.

Economic and commercial measures – Seco. The Economic Development Cooperation in the Secretariat for Economic Affairs is responsible for developing and implementing measures to promote developing countries and economies in transition. The directorate has four strategic objectives: (i) improving growth and incentive framework conditions (co-operation with the World Bank, the IMF, the EBRD and the regional development banks); (ii) mobilising private-sector financial means, know-how and technologies (funding facilities for pre-investment studies, participation in financial intermediaries, such as venture capital funds, guarantee funds, credit guarantees), plus two new instruments: the Swiss Organisation for Facilitating

Investments (SOFI) and the Swiss Development Finance Corporation (SDFC); (iii) expanding the economic and social infrastructure (mixed financing, financing assistance, credit guarantees); (iv) promoting the integration of developing countries and transition economies into world trade (promoting exports, granting preferential tariffs, encouraging the transfer of environmental technologies, promoting regional integration, improving trade efficiency). Seco administers 14 per cent of Swiss ODA.

As shown in Table 16.4, Seco's traditional instruments include: debt reduction measures (23.9 per cent), balance-of-payments assistance (15.8 per cent), mixed financing (12.9 per cent) and trade promotion (9.5 per cent). The high percentage of aid allocated to investment promotion (31.9 per cent) should be noted. In 1996, the Federal Council [*1996*] presented a reorientation of aid aimed at developing a partnership between the Confederation and the private sector with a view to encouraging investment in developing countries, together with the transfer of technology. Seco is involved in around 20 instruments promoting investment, the three main ones being the Swiss Organisation for Facilitating Investments (SOFI), the Swiss Development Finance Corporation (SDFC) and the Cleaner Production Centres. Studies are being carried out to assess the impact of these new instruments, and the conclusions will be published when the Message on economic and commercial measures is updated in 2003. Their introduction was fiercely opposed by the development organisations, which regretted that part of the development funds would be used 'to help Swiss exporters and investors rather than to support the sustainable development of the most disadvantaged countries' [*Forster (ed.), 2002*].

TABLE 16.4
ECONOMIC AND TRADE POLICY MEASURES OF SECO 1997–2001 (US$m. AND %)

Seco's Instruments	US$m.	%
Measures aimed at activating private-sector resources	135	31.9
Debt reduction measures	101	23.9
Balance-of-payments assistance	67	15.8
Mixed financing	55	12.9
Trade promotion	40	9.5
Basic products	11	2.6
Studies and others	14	3.4
Total	423	100

Sources : SDC–Seco [*1997, 1998, 1999, 2000, 2001*].

The primary objective of the Swiss Debt Reduction Programme, for which Seco has overall responsibility, is to reduce the debts of poor countries and thereby facilitate the development of disadvantaged communities. Aid is granted to countries which have demonstrated a firm commitment to a process of economic and political reform. The criteria applied to the beneficiary countries are good governance (respect for human rights, separation of judicial and political powers, freedom of the press, etc.) and a commitment to combat poverty. The Debt Reduction Programme has four types of instruments: the cancellation of bilateral public debts, the remission of non-guaranteed commercial debts, the reduction of multilateral debts and various complementary measures. The first of these is often conditional on the governments concerned setting up so-called 'counterpart' funds dedicated to funding development activities. Switzerland has been an innovator in this field, having devised a 'creative' form of debt reduction which consists of establishing a link between debt relief and social development. Over more than ten years, the Swiss programme has been instrumental in relieving the burden of the poorest and most heavily indebted countries by finding effective solutions to their insolvency problems. It has led to the writing off of debts amounting to almost US$1.2 billion [*SDC–Seco, 2001: 10*].

The SDC bilateral co-operation. SDC bilateral co-operation is managed through four channels : (i) projects run by the SDC (55 per cent, see Table 16.5); (ii) projects run by institutions and Swiss NGOs (20 per cent); (iii) contributions to Swiss organisations for specific projects (13 per cent); and (iv) contributions to international organisations for specific projects (12 per cent).

TABLE 16.5
DISTRIBUTION OF SDC BILATERAL CO-OPERATION ACTIVITIES BY SECTOR 1999–2001
(US$m. AND %)

Sectors	*US$m.*	*%*
Agriculture	144	16.8
Water, infrastructure, transport	126	14.7
Environment	98	11.4
Education, arts, culture	90	10.5
Health, population	85	9.9
Private sector, financial sector	62	7.2
Management of public sector	44	5.2
Multisectorial	208	24.3
Total	857	100.0

Sources : SDC–Seco [*1999, 2000a, 2001*].

One of the characteristics of Swiss bilateral co-operation is the commitment to mountain regions. Itself a mountainous country, Switzerland has a great deal of relevant experience to share. Some of the management methods currently practised by mountain peoples in the countries of the South are very similar to those practised in the Swiss alpine valleys in the nineteeth century. The main concern is to improve the living conditions and quality of life of the world's most deprived population groups. The problems faced by mountain communities – whether in the Andes, the Himalayas or the Alps – are much the same everywhere: they all have a fragile, difficult-to-develop environment and experience economic and political marginalisation. Having been well aware of these problems for more than 30 years, the SDC is contributing to the sustainable development of such regions in the countries of the South and East. Together with the Federal Office for Spatial Development (OSD), it took an active part in the UN International Year of Mountains in 2002 [*SDC–Seco, 2001*: 2].

In order to improve the quality of its approach to the fight against poverty, the SDC commissioned a horizontal study to analyse how the problem of poverty was tackled on the ground, the extent to which SDC staff understood the concepts of poverty and empowerment, and the measures and programmes drawn up as a result [*Egger and Egger, 1998*]. Among the main observations made by the researchers, it appears that the SDC supports numerous anti-poverty projects, particularly in rural areas. The principal concern is to generate and improve income with a view to alleviating the shortfall in basic requirements (food, drinking water, health and basic education). An indirect approach seems to be chosen most often; only a minority of projects help disadvantaged groups directly. The study also stresses that there is still a lack of knowledge about poverty and its context, and poverty assessments are rarely carried out before the beginning of a project. Promoting empowerment is one of the key instruments in the SDC's fight against poverty. This approach, which is centred on the potential of the poor people themselves, aims to provide them with the means of defending their own rights and meeting their own essential needs. However, the study notes that the term 'empowerment' is used in many different ways, depending on the programmes, and that some clarification is required. To follow up on some of the study's recommendations, the SDC produced a working paper to 'enable SDC employees to develop their basic awareness of poverty, their capacity for analysis, their knowledge of methods and their social aptitudes' [*SDC, 2000b*].

In January 1999, the SDC published its policy for social development [*SDC, 1999*], as a follow-up to the World Summit on Social Development in Copenhagen in 1996. The strategy adopted provides for intervention in 'social' sectors (primary education, development of basic care, food security). It also highlights the need to look systematically at the 'poverty' criteria, as a horizontal theme, in each technical or economic programme that does not have alleviating poverty as its main goal.

Swiss development co-operation and the fight against poverty. Combating poverty and promoting social development are the key objectives of Switzerland's technical co-operation, as set out in the 1976 law on development co-operation and humanitarian aid, which officially makes aid to the most disadvantaged countries, regions and population groups a priority. The distribution of Swiss ODA in relation to the income of the beneficiary countries shows that priority is given to the least-developed countries (approximately 42 per cent); the percentage for low-income countries is around 23 per cent, and the overall share of the two categories totals around 65 per cent. It should be noted that very little aid has been given to high-income developing countries (0.01 per cent) and that 35.8 per cent is allocated to low-income countries [*IUED-SDC, 2002: 359–63*].

It would thus appear that the plan to target the poorest countries has been fulfilled: of the 17 so-called 'priority countries' of the SDC,[12] 10 belong to the least-developed category. However, in its evaluation, the DAC made two criticisms, one concerning the geographical distribution and the other the sectoral distributions. Regarding the geographical distribution, the experts emphasise the large number of priority countries in view of the size of Switzerland's bilateral aid programme. The fact that Seco has its own priorities and is involved in no fewer than 36 countries, of which only 10 are SDC priority countries, adds to this imbalance, which unfortunately is conducive to the excessive dispersal of aid [*OECD, 2001: II.32–4*].

---

12. The SDC priority countries are the following: *Africa*: Benin (1983), Burkina Faso (1976), Mali (1977), Mozambique (1979), Niger (1977), Tanzania (mid-1970s), Chad (mid-1960s). *Latin America*: Bolivia (1969), Ecuador (1969), Peru (1970), Nicaragua/Central America (end 1970s). *Asia*: Bangladesh (1971), India (1961), Nepal (1961), Pakistan (1970), Bhutan (1983), Vietnam/Region of Mekong (1992). There are in addition four special programmes for South Africa, Madagascar, Palestine and Rwanda. The dates in parentheses refer to the beginning of the Swiss co-operation activities, whether by the Swiss co-operation agency or through NGO projects financed by the Swiss co-operation agency.

Analysis of the sectoral breakdown also gives rise to some disappointment. Switzerland is fully committed to the 20/20 initiative, which aims to ensure that 20 per cent of all aid is used for basic social services. However, Swiss ODA to this sector represents only 13 per cent of its bilateral ODA and 15 per cent of total ODA.[13] DAC's recommendations urge Switzerland to maintain its priorities for the poorest countries, but to review the list of countries concerned in order to avoid an excessive dispersal of aid and to allocate a larger portion of aid to basic social services [OECD, 2001: II–94].

Co-ordination problems between SDC and Seco. Within the framework of the reform of the whole Swiss administration launched from 1996 onwards, the organisation of the development co-operation sector was also to be reformed. Various scenarios were examined, one of them consisting of grouping both agencies (SDC and Seco) together in a single structure. Eventually the continuation of the existing organisation prevailed, and pilot committees were set up with a view to strengthening co-ordination. There is no specific document formalising a common operational strategy, as each office defines its own strategy according to its field of competence, in documents such as the *Guiding Principles* [SDC, 2000a] and the *Strategy 2010* [SDC, 2000c] or the *Guidelines of Seco* [1999] and the *Strategy 2006* [Seco, 2002]. The strategic co-ordination of Swiss development co-operation is nevertheless ensured by both offices, through a strategic committee attended by the respective directors. There are also some pilot committees which allow common programmes to be arranged aimed at the countries where both offices are involved.

In its recommendations the DAC strongly advises SDC and Seco to set up a common strategy for Swiss development co-operation [OECD, 2001: II–94].

2. Humanitarian Aid

Humanitarian aid is an important component of the Swiss aid system, because of the substantial resources involved (for years, around one-fifth of Swiss ODA has been used for humanitarian aid). Added to this is Switzerland's humanitarian tradition [Freymond, 1999] and the presence in the country of many humanitarian organisations, such as the International Committee of the Red Cross (ICRC), the High Commissioner for Refugees (HCR) and the many NGOs operating in this sector, which consolidate this aspect of Swiss foreign policy. The Humanitarian Aid of the Confederation is a department of the SDC.

---

13. On the basis of the commitments made in 1997–98 [OECD, 2001: II–94].

TABLE 16.6
GEOGRAPHICAL DISTRIBUTION OF HUMANITARIAN AID FUND, MULTIANNUAL FUND 1996–2000 (US$m. AND %)

	Africa	The Americas	Asia/Middle East	Europe + CIS	Not distributed geographically [a]	Total
US$m.	216	46	120	162	80	624
%	34.6	7.4	19.3	26.0	12.8	100

Source: Federal Council [2001].

Note: [a] The non-allocable amounts consist of contributions that are not intended for specific programmes as well as general contributions to international organisations such as the HCR, the WFP and the OCHA, and contributions for the operations of the SHA.

In its Message on humanitarian aid in November 2001, the Federal Council [2001] sets out its specifications. Humanitarian aid is intended for people who are victims of natural disasters or armed conflict, and where the own resources of a country or region hit by an intense, large-scale crisis do not enable it to confront the problem without outside help. It gives precedence to the population groups that are most vulnerable and socially most deprived, the main priority being to safeguard human life where it is at risk and to alleviate suffering. Its second priority is to facilitate reintegration and rehabilitation, and its third priority is to enhance the involvement of the people concerned and the local defence and prevention mechanisms in order to guard against the possibility of external crises. Humanitarian aid may take the form of benefits in kind, especially the provision of food, cash contributions or the provision of specialists and rescue teams [ibid.].

In 1997, the Federal Chambers approved framework credits for four years totalling around US$683 million (see Table 16.6 for the geographical distribution), to which should be added additional framework credits consisting of the federal contributions to the ICRC headquarters in Geneva (US$188m. for 1998 to 2001).

Switzerland's humanitarian aid is provided in two ways: direct interventions and support for partner organisations, both Swiss and international. The Confederation's humanitarian aid generally passes through a dense network of partners and international organisations (WFP, HCR, UNWRA, OCHA), and ICRC. The Swiss NGOs operating in this field receive one-third of its resources. According to the 2001 Message mentioned above [ibid.], the Swiss development organisations play a large part in the implementation of projects on the ground and are recognised by the Confederation as competent partners. However, the funding of development organisations' projects from humanitarian aid resources rarely exceeds 50 per cent, an aspect that helps enhance the importance of the development organisations among the Swiss population.

The principal characteristic of the Swiss system is that it has its own direct intervention system, the Swiss Humanitarian Aid Unit (SHA). This is a part-time body which brings together a pool of over 700 people who are prepared to volunteer their services, organised into various expert groups according to their knowledge and abilities.

Humanitarian aid has considerably increased in recent years as a result of support for the peace process in the Balkans. Switzerland's humanitarian aid in Albania, Macedonia and the Federal Republic of Yugoslavia (Serbia, Montenegro and Kosovo) consists of bilateral aid (humanitarian aid division and the Swiss Disaster Relief Unit), multilateral aid (World Food Programme, Focus Initiative, participation in the multinational KFOR forces) and aid provided by other departments of the Federal Administration (ODR programme to support the return of refugees staying temporarily in Switzerland and commitments by the Federal Department of Defence, Civil Protection and Sport). The Federal Council has allocated US$65m. from the humanitarian aid appropriations to the Kosovo conflict. This is accompanied by around US$59m. for operations to maintain peace and clear landmines (OSCE mission, Stability Pact for South-Eastern Europe, Swisscoy), and US$148m. for the return of refugees. These extra funds were not provided at the expense of other forms of development co-operation or other regions of the world, as additional credits were approved by Parliament [*Forster (ed.), 2000: 332*].

As pointed out in the 2001 Message of the Federal Council, humanitarian aid forms an integral part of Switzerland's foreign policy and entails close collaboration between the various players, in particular those involved with development, refugees, human rights, peace and security. It should be emphasised that humanitarian aid must not be a replacement for the other foreign policy instruments; it should only be used where other measures do not have the desired effects; in other words, where the deep-rooted causes of crises, and especially armed conflicts, require the support of other foreign policy instruments. In the DAC's evaluation of Switzerland, the experts highlighted the Swiss aid system, stating that 'it probably represents what is done best among DAC members' [*OECD, 2001: II–46*].

3. Multilateral Co-operation

Switzerland's international development co-operation has always centred around a twofold approach, bilateral and multilateral, on the basis that 'the numerous problems and challenges facing the world, and particularly developing countries, require multilateral and even global approaches and solutions' [*SDC–Seco, 2000: 85*]. Bilateral and multilateral aid account for around 70 and 30 per cent of ODA respectively (see Table 16.2).

Although not a member of the United Nations, for years Switzerland provided support for the main UN funds and programmes, as well as the regional development banks. In 2000, Switzerland supported the following institutions:

– Financial institutions, the World Bank, the IMF, the three development banks (US$171m.)
– Specialised funds, programmes and bodies of the United Nations, UNDP, UNICEF, UNFPA, WHO, etc. (US$83m.)
– Other institutions, including CGIAR, IUCN, GEF and AIF (US$8m.)

When allocating its multilateral resources, Switzerland takes account of criteria such as the role and policies of the institutions it intends to support, the complementarity of their activities with its own bilateral programmes, the opportunities to help define policies and strategies and monitor their activities, the quality of their operations on the ground and the distribution of the financial burden among donors. During the 1990s, the distribution of resources was nevertheless affected by Switzerland's growing commitments to the World Bank as a result of its accession in 1992. Contributions to the UN system are falling, but Switzerland aims to keep them at around US$59m. a year [*OECD, 2001: II–34*].

Although Switzerland has regularly provided support for international organisations, one of the key aspects of its foreign policy since 1992 has been its accession to the main international organisations. In 1992, following a popular vote (see section III), it became a member of the International Monetary Fund and the World Bank, and in 2002 the Swiss accepted the proposal to join the United Nations, although this had been rejected by more than 75 per cent of the population in 1986. These two events are extremely important for Swiss foreign policy, which is demonstrating a desire to 'open up to the outside world'.[14] Before that, one of its characteristics, distinguished by its concept of strict neutrality, was in fact its lack of participation in the main international organisations. As a full member, Switzerland can now play a much more influential political role.

---

14. In 1996, Switzerland was one of the first states to join the World Trade Organisation, as an extension of its participation in the GATT negotiations.

To conclude, we should highlight Switzerland's active participation in the round of international conferences organised under the auspices of the United Nations during the 1990s, and its strong commitment to their follow-up work, at national level, and also in relation to its development co-operation policy and practices.

V. SWITZERLAND'S ENVIRONMENTAL DIPLOMACY

Since the UN Conference on the Environment and Development at the Rio de Janeiro in 1992, environmental issues have become an important aspect of development multilateral co-operation. The Swiss government's *North–South Guidelines* of 1994 confirmed protection of the environment as one of its four main objectives at home and in the international arena. The Federal Council [*2000a*] drew up an ambitious environmental policy programme based on the development of international rules and its participation in international environmental organisations.

Switzerland is strongly committed to international environmental legislation and plays an active part in the drafting and implementation of the numerous new conventions since the Earth Summit (on climate change, biodiversity, desertification, hazardous waste, protection of the ozone layer, chemical products) where it plays an important role as a mediator. In general, it stays outside the major negotiating groups, which enables it to benefit from an image of impartiality, and by proposing compromises and new working methods, it has succeeded in relaunching negotiations that were completely blocked. In 2000, for example, its proposals to limit imports of transgenic foodstuffs contributed significantly to the conclusion of the Cartagena Protocol on Biological Safety [*SAEFL, 2002*]. In addition to its commitment to defining international environmental law, Switzerland had the ambitious plan of making Geneva the global environment capital, by accommodating, in particular, the secretariats of the climate and biodiversity conventions; but it was not successful and the two secretariats were established in other cities (Bonn and Montreal respectively).

Collaboration with international organisations is another important element of Switzerland's environmental diplomacy. It supports programmes and projects of organisations such as the UN Institute for Training and Research (UNITAR) and the Global Environment Facility (GEF) aimed at applying environmental conventions, particularly in developing countries. Switzerland is a member of the GEF, where it holds one of the 32 seats on the Executive Council, representing five countries of Central Asia and the Caucasus.[15] As a donor, it has already provided US$111m. and plans to add a further US$65m. Switzerland also supports the United Nations Environment Programme (UNEP) both financially – US$2.7m. in 2000 – and politically; in fact, it endeavours to enable UNEP to play a more substantial role in defending a coherent global environmental policy. Among the international environmental protection measures, it promotes, in particular, economic instruments and collaboration with the private sector (Cleaner Production Centres).

During the 700th anniversary of the Confederation in 1991, in response to an initiative by the NGOs, Parliament approved framework credits of US$209m. to finance ecological programmes and projects to help the global environment in developing countries. Around half of these funds were set aside for bilateral measures and the other half for multilateral aid (funds for the protection of the ozone layer, GEF, etc.). These credits ran out in 1998, but further resources were approved by Parliament with a view to continuing Switzerland's environmental activities, thus confirming its multilateral commitment in this field.

VI. CONCLUSION

During the past ten years or so, Swiss development co-operation evolved without any radical change in its orientation or practice. However, the changes which have occurred at international level (the collapse of the communist regimes and the globalisation of the economy), the new approaches defined by the international organisations (the fight against poverty, policy coherence, political conditionality), the consideration of global problems (environment, migration), as well as of new viewpoints (gender, human rights, etc.), influenced Swiss development co-operation and require Switzerland to answer new challenges.

Let us conclude by mentioning the four major challenges which await Switzerland in the field of development co-operation in the coming years. The first challenge is of a financial nature and deals with Switzerland's fulfilment of its commitment to increase its ODA. In 1992, as already mentioned, during the Earth Summit in Rio, it declared its intention to raise the level to 0.4 per cent of GNP. Ten

---

15. This group includes Azerbaijan, Kyrgyzstan, Tajikistan, Turkmenistan and Uzbekistan, also members of the same voting group as Switzerland in the World Bank.

years later this objective, though far from being achieved, is reaffirmed with the same determination. If Switzerland wants to retain its credibility in the eyes of the international community, it should meet this commitment within the time limit it set itself., i.e. it should be able to devote 0.4 per cent of its GNP to ODA by 2010.

The second challenge is of an institutional nature. Besides the existing problems of co-ordination between the traditional actors in Swiss development co-operation (SDC and Seco), frequently raised by observers and by the DAC, the implementation of the *North–South Guidelines* has widened the traditional field by bringing in new institutional actors, notably the Swiss Agency for the Environment. The Federal Administration will have to strengthen the co-ordination structures between the various offices.

The third challenge deals with better consideration of policy coherence *vis-à-vis* developing countries. This issue should be more firmly established in the organisation of the Federal Administration as well as by political support. The implementation of procedures for better evaluation of public policies achieved would be a first step in the right direction.

Finally, the fourth challenge concerns Switzerland's recent accession to the main international organisations, the International Monetary Fund and the World Bank in 1992, the United Nations Organisation in 2002. If Switzerland has succeeded in giving itself a role and a political stature by becoming a member of the executive boards of the Bretton Woods institutions upon its accession, it still has to define its political position within the United Nations.

REFERENCES

Egger, M. and J-P. Egger, 1998, *Analyse transversale. Lutte contre la pauvreté et empowerment* (Horizontal Analysis. Combating Poverty and Empowerment), vols. 1 and 2, Study carried out at the request of the SDC, Geneva and Zurich: IUED and Nadel.

Federal Council, 1994, *North–South Guidelines: Report on Switzerland's North–South Relations in the 1990s*, Bern: 7 March.

Federal Council, 1996, *Message concernant la continuation du financement et la réorientation des mesures de politique économique et commerciale au titre de la coopération au développement* (Message concerning the continuation of financing and the reorientation of economic and commercial policy measures within the framework of development co-operation), Bern: 29 May.

Federal Council, 1999a, *Rapport sur la politique de sécurité 2000* (Report on the security policy 2000), Bern.

Federal Council, 1999b, *Rapport sur les dimensions humanitaires de la politique extérieure de la Suisse* (Report on the humanitarian dimension in Swiss foreign policy), Bern.

Federal Council, 2000a, *Foreign Policy Report 2000: Presence and co-operation: safeguarding Switzerland's interests in an integrating world*, Bern, 15 November.
Federal Council, 2000b, *Rapport sur la politique de maîtrise des armements et de désarmement de la Suisse* (Report on Switzerland's arms control and disarmament policy), Bern.
Federal Council, 2000c, *Rapport sur la politique suisse des droits de l'homme* (Report on Switzerland's human rights policy), Bern, 16 February.
Federal Council, 2000d, *Rapport sur le programme de législature 1999–2003* (Report on the 1999–2003 legislative plan), Bern, 1 March.
Federal Council, 2001, *Message concernant la continuation de l'aide humanitaire internationale de la Confédération* (Message concerning the continuation of the Confederation's international humanitarian aid), Bern, 14 November.
Fédération genevoise de coopération (FGC), 1996, *Solidarité Genève-Sud – Participation genevoise à la solidarité internationale* (Geneva Participation to International Solidarity), Genève: FGC.
Forster, J., 1984, 'Swiss aid: Policy and Performance', in Stokke (ed.), *European Development Assistance*, Vol. I: *Policies and Performance*, Tilburg: EADI Book Series 4.
Forster, J., 1999, 'The Coherence of Policies Towards Developing Countries: The Case of Switzerland', in Forster and Stokke (eds), *Policy Coherence in Development Co-operation*, EADI Book Series 22, London: Frank Cass.
Forster, J. (ed.) (annually since 1981), *Annuaires Suisse–Tiers Monde* (published in French and German), Geneva: IUED.
Forster, J. and G. Pult, 2000, *Effets économiques de l'aide publique au développement en Suisse. Etude pour 1998. Rapport final* (Economic effects of official development assistance in Switzerland. 1998 study, Final report), carried out at the request of the Swiss Agency for Development Co-operation (SDC) by the Graduate Institute of Development Studies (IUED), Geneva, and the University of Neuchâtel.
Freymond, J., 1999, 'L'humanitaire dans la politique étrangère de la Suisse' (The humanitarian aspect of Swiss foreign policy), in Forster (ed.).
Freymond, J. and B. Boyer, 1998, *Les organisations non gouvernementales et la politique extérieure de la Suisse* (Non-governmental organisations and Switzerland's foreign policy), Summary report, Bern: PNR 42.
GfS-Forschungsinstitut, 1999, *Entwicklungspolitik im Spannungsfeld der Finanzierung – Bericht zur vierten Trendstudie im Rahmen der Studienreihe 'Entwicklungshilfe-Monitor'*, Bern.
IDARio, 2001, *Un avenir pour la Suisse. Le développement durable: une chance pour l'économie, l'environnement et la société* (A future for Switzerland. Sustainable development: an opportunity for the economy, the environment and society), Bern: IDARio.
IUED–SDC, 2002, *Aide suisse aux pays en développement et aux pays de l'Europe orientale, 1998–1999–2000* (Switzerland's assistance to developing countries and Eastern Europe, 1998–1999–2000), Geneva: IUED-SDC.
OECD, 2001, 'Switzerland: Development Co-operation Report', *The DAC Journal – Development Co-operation*, Vol. 1., No. 4, Paris: OECD, DAC.
SAEFL, 2002, 'La diplomatie verte de la Suisse a acquis une réputation mondiale' (Switzerland's green diplomacy has acquired a global reputation), *Environment*, No. 2, Bern.
SDC, 1999, 'Politique de la DDC pour le développement social' (SDC policy for social development), *SDC working paper* 9–F/99, Bern: DFA.
SDC, 2000a, *Guiding Principles*, Bern: DFA.
SDC, 2000b, *Poverty and Wellbeing, an orientation, learning and working tool for fighting poverty*, Bern: DFA.
SDC, 2000c, *Strategy 2010*, Bern: DFA.

SDC–Seco, 2000, *Mémorandum de la Suisse au Comité d'aide au développement (CAD) de l'OCDE sur la coopération suisse au développement* (Memorandum from Switzerland to the Development Assistance Committee (DAC) of the OECD on Swiss development co-operation, 1997–1999), Bern.

SDC–Seco (various years), *Annual Report: Development Co-operation*, Bern: DFA and DEA.

Seco, 1999, *Guidelines*, Bern: DEA.

Seco, 2000, *Development and Transition: A brief portrait*, Bern: DEA.

Seco, 2002, *Strategy 2006*, Bern: DEA.

Websites

Bern Declaration: www.evb.ch/index_f.cfm
Federal Authorities of the Swiss Confederation: www.admin.ch/ch/index.en.html
Graduate Institute of Development Studies: www.unige.ch/iued
State Secretariat for Economic Affairs (Seco): www.Seco-admin.ch/e_index.html
Swiss Agency for Environment, Forests and Landscape:
    www.umwelt-schweiz.ch/buwal/eng/
Swiss Coalition of Development Organisations: www.Swisscoalition.ch/
Swiss Development Co-operation (SDC): www.deza.ch

17

The European Union's Development Policy: Shifting Priorities in a Rapidly Changing World

GORM RYE OLSEN

I. INTRODUCTION

At the beginning of the twenty-first century, the development co-operation of the European Union covers more than 100 countries. The aims of this co-operation are laid down in the Maastricht Treaty, which establishes that the main goal of the Community's aid policy is to promote sustainable economic and social development. However, it is also laid down that the development aid of the Community has to work together with the bilateral development policies of the member states. This requires, at least in theory, some kind of co-ordination between the decision-makers in Brussels and those located in the different European capitals. It is the aim of the common aid policy to secure the gradual and harmonious integration of the developing countries into the world economy. The fight against poverty also has high priority, in combination with the aim to promote democracy, human rights and the rule of law (articles 130U–130Y).

It is important to emphasise that it is increasingly inappropriate to see the aid policy of the European Union in isolation from its other external activities. Since the end of the Cold War, it has been stressed on numerous occasions that development assistance is an integral component of the political and economic relations between individual developing countries and the Union. According to the High Representative for the EU's Common Foreign and Security Policy, Javier Solana, there is great potential in using a combination of all the Union's instruments to achieve its foreign policy goals:

> A more effective foreign and security policy begins with the political will to use all the available instruments in a co-ordinated and coherent way... Collectively these are substantial:

as the world's largest aid donor we already make an important financial contribution to aid programmes and to humanitarian and reconstruction assistance. We have a global diplomatic network. ...We can use our diplomatic, economic and financial muscle to influence the behaviour of recalcitrant parties and aggressors [*Financial Times, 29 September 2000*].

Despite the fact that development assistance is used to an increasing extent as a foreign policy instrument, this chapter focuses mainly on the development policy of the Union in the period following the end of the Cold War. Moreover, the chapter only analyses the Official Development Assistance (ODA) of the European Union and therefore so-called Official Aid (OA) is not included. It means that the chapter does not address the issue of Europan Union aid to the more affluent countries contained in the part II list of the DAC. Concretely, that leaves out the CEECs/NIS because these countries simply do not receive ODA which is reserved for countries below a certain per capita income level [*OECD, 2003: 327*].

The EU's development aid is divided into two main categories, both of which are administered by the Commission in Brussels. First, there is the aid to the countries in Africa, the Caribbean and the Pacific (the ACP) which in historical terms has been by far the most important part of the Community's development activities. During the period of the Cold War, well above 60 per cent of the EU's total development assistance went to the ACP. After the end of the Cold War, the share dropped quite dramatically to around 34 per cent in 2001, as may be seen from Table 17.1. From 1975 to 2000 the co-operation between the ACP and the European Community was governed by the four different Lomé Conventions. Since 2000 it has been laid down in the Cotonou agreement. Both the Cotonou agreement and the Lomé Conventions are financed by the European Development Fund (the EDF), which is the outcome of a separate round of negotiations.

The other category of development co-operation covers the rest of the world, meaning Latin America, the Mediterranean, Asia and 'Europe'. This part of the Community's aid is financed by the ordinary budget and is therefore subject to the annual political infighting. In 2001, around 13.5 per cent of the total aid budget was allocated to Asia, whereas Latin America received some 10 per cent as shown in Table 17.1. Among other important budget lines, the Mediterranean region (North Africa and the Middle East) accounted for 16 per cent, while 'Europe', meaning primarily the former Yugoslav republics, received as much as 27 per cent of total EU development assistance. Table 17.2 shows the growth in the volume of official development assistance in absolute figures from the European Community. In 2001, it reached US$5.961m. (in current prices) or just over 10 per cent of the global figure at $52.236m. [*OECD, 2003: 242–3*]. Adding to it the development assistance of the member states brings total 'European aid' close to 50 per cent of global development aid [*ibid.*].

Nor should it be ignored that the EU is one of the biggest donors of humanitarian assistance, with a budget for this type of intervention equivalent to 10 per cent of total Community aid or €537.79m. in 2002 [*ECHO, 2003*]. Since the beginning of the 1990s, the European Commission has had its own office for humanitarian aid, ECHO, which does not implement projects itself but channels its assistance through other organisations such as NGOs and the UN system. The ACP countries received 39 per cent of total EU humanitarian aid in 2002, with Asia coming second with 26 per cent.

TABLE 17.1
MAJOR RECIPIENT GEOGRAPHICAL AREAS FOR EUROPEAN DEVELOPMENT ASSISTANCE, 1980/81–2000/1 (% OF TOTAL ODA)

Recipient region	1980–81	1990–91	2000–1
Sub-Saharan Africa	60.4	59.4	33.6
North Africa	6.8	9.8	11.0
Europe*	3.5	5.0	26.9
South and Central Asia	17.2	5.5	7.4
Far East Asia and Oceania	5.0	5.6	5.9
Middle East	1.7	5.6	4.9
North, Central and South America	5.4	9.1	10.3

Source: OECD [*2003: 305*].
Note: * Europe = Albania, Bosnia, Croatia, Macedonia, Yugoslavia, Slovenia, Malta (as of 1 January 2003, Slovenia and Malta were removed from the DAC list of countries that can receive ODA).

TABLE 17.2
ODA FROM THE EUROPEAN COMMUNITY, SELECTED YEARS (US$m., CURRENT PRICES)

Years	1990–91	1998	1999	2000	2001
Total ODA	3.343	5.140	4.937	4.912	5.961
Grants and grant-like contributions	3.032	4.462	4.514	4.019	4.810
Of which:					
– technical co-operation	34	215	195	211	197
– development food aid[1]	507	364	382	320	350
– emergency and distress relief	525	501	677	519	526
– contributions to NGOs	100	162	184	120	-

Source: OECD [*2003*].
Notes: The figures do not add up to total ODA because minor budget lines are excluded.
[1] Up to and including 1995: includes emergency food aid.

II. THE FOCUS OF THE CHAPTER

Table 17.1 reveals two striking changes in the Community's development assistance in the post-Cold War period. On the one hand, there is a conspicuous decline in the relative allocations to sub-Saharan Africa. On the other hand, there is the striking increase in the relative allocations to 'Europe'. For obvious reasons, it is interesting to analyse and understand both these changes. However, it is *mainly* the aid relationship with sub-Saharan Africa that will be dealt with in this chapter. From the start of the European Community in 1958, sub-Saharan Africa has been its most important aid-recipient region among the poor areas of the world. As late as 1996 the European Commission described the aid relationship formalised in the successive Lomé Conventions as one of the most important facets of the European Union's external activities [*Commission, 1996*].

The focus on the European–African aid relationship is also based on the assumption that the policy reflections, the policy content and the changes in the EU's policies towards Africa to a large extent shed light on the aid policies towards other geographical regions, with the exception of 'Europe' which is considered a special case, mainly because the conflicts in the Balkans created instability close to the EU's borders. Martin Holland quite convincingly argues the same point of view emphasising that 'both historically and contemporarily the most intense dialogue has been with the ACP countries, with Asian and Latin American relations given considerably less attention … the policy towards the ACP shapes and determines the nature of EU's Third World relations globally' [*Holland, 2002: 234*].

In the wake of the end of the Cold War, new priorities began to influence EU policy towards poor countries, as manifested mainly in two respects. First, its aims widened from the original goal of promoting economic and social development towards giving increasing priority to promoting stability, security and democracy. Second, the geographical focus shifted slowly from its former emphasis on sub-Saharan Africa to giving more priority to North Africa and 'Europe', meaning the Balkans. The shift appears clearly in Table 17.1.

It is difficult to explain the dual reorientation of the EU's development policy unless it is accepted that 'European' interests in the developing world have changed and that these changes create new policy options. It is the aim of this chapter to demonstrate that the goals and the use of European development aid shifted quite considerably during the years following the end of the Cold War, and to argue that the policy shift may be explained by change in European interests from the promotion of 'development' to a more explicit interest in becoming a significant international actor. Because of the changes in the international system, it became within reach for the EU to realise this ambition that the Community had had since its initiation in the 1950s [*Cafruny and Peters, 1998a; Cameron, 1998; Bretherton and Vogler, 1999*].

The chapter is structured as follows. It starts with a brief presentation of the theoretical reflections guiding the interpretations of the empirical changes in the EU's development policy. The empirical section of the chapter begins with a brief historical overview of the Community's development aid relations with the Mediterranean 'Europe', Latin America and Asia. Then follows an introduction to the EC's relations with Africa during the Cold War, focusing on aid. The core of the analysis deals with the changes in the EU's policies towards sub-Saharan Africa in the post-Cold War era. This is divided into a number of sub-sections using the successive rounds of negotiations between the Community and Africa (i.e. the ACP countries) as the organising principle. The argument for this structure is that the interests of the Europeans crystallised during the negotiations and that this crystallisation points towards the future direction of the EU's development policy.

The analysis is not restricted to the different rounds of negotiations on the Lomé Convention and its successor the Cotonou agreement. Three issues that figured prominently on the Community's agenda during the 1990s are also scrutinised in some detail. First, the promotion of democracy and respect for human rights was a high-profile issue of the EU's external relations. Second, the question of security, and in particular conflict prevention, in sub-Saharan Africa became increasingly prominent. Third, criticism of lack of efficiency, lack of coherence and lack of co-ordination in the Community's development aid became vociferous. A separate analysis of the initiatives launched in 2000 to reform the aid delivery system is therefore carried out, together with a brief discussion of the preliminary results achieved.

III. THE ANALYTICAL FRAMEWORK

There are a number of theoretical and empirical challenges involved in analysing the development aid policy of the European Community/ European Union. In spite of these challenges, it is surprising that there are so relatively few theoretical contributions that aim at explaining the European Community's policies towards the Third World. The most elaborated contribution is John Ravenhill's theory of 'collective clientelism', which argues that a mixture of psychological, political, bureaucratic and economic factors can explain the European Community's aid relationship with the ACP countries. The relationship is basically clientelistic in nature where the European countries act as the collective patron towards the developing countries which are the 'clients of the Community' [*Ravenhill, 1985*]. Among other theoretical contributions can be mentioned the dichotomy between donor interests and recipient needs where the argument is that the core motives of the EC countries to give development to poor countries are mainly selfish donor interest such as trade, access to raw materials and security [*Bowles, 1989; Grilli, 1994*]. As possible alternatives to these approaches, Martin Holland offers a range of theories to explain the development policy of the European Union. Among others, he mentions the potential explanatory value of integration theory, neofunctionalism, liberal intergovernmentalism, etc. However, he seems to favour the integration theory as the best tool to explain the EU's international role, including its development policy [*Holland, 2002: 234–44*].

The theoretical starting point of this chapter is that two main variables in combination can explain the specific content and the changes in the development aid policy of the European Union. On the one hand, it is possible to identify so-called 'European' (national) interests that can explain the general policy features and changes. On the other hand, minor and more specific components and changes of the policy have to be explained by bureaucratic policy-making which, to a large extent, reflects the interests of the narrow elites involved in development aid [*Olsen, 1998*]. Thus, it is argued that the totality of policy content and policy changes has to be explained by the combination of general European interests and more narrow elite-based interests expressed in bureaucratic policy-making.

As a start, it must be remembered that parallel to the common aid policy directed by the European Commission in Brussels, all member states have their own bilateral aid policies. Decision-making on the common policy involves decision-making both within the member states and also within and among the EU institutions in Brussels. Andrew Moravscik has suggested a model framework for analysing this dual decision-making structure. He calls it 'liberal intergovernmentalism' [*Moravscik, 1993, 1995*]. To put it simply, liberal intergovernmentalism is a two-step sequential model of preference formation. In the first stage, national leaders aggregate the interests of their domestic constituencies and articulate national preferences regarding, in this case, European development policies. This first stage does not receive much attention here. In the second stage, national governments bring their preferences to the intergovernmental bargaining table in Brussels where agreements reflect the interests and the relative power of each member state and where supranational organisations such as the European Commission may exert limited or a great deal of influence depending on the specific policy field [*Cafruny and Peters, 1998a: 16ff.*].

Based on Moravcsik's model, it is logical and also fruitful to operate with a concept of 'European interests', meaning common interests that may very well be different from the interests of the individual member states. The defenders and also the articulators of such European interests are to a large extent the Council of Ministers together with the European Commission, the latter being responsible for the day-to-day workings of the policy towards developing countries. Because aid is the focus of this analysis, it is relevant to recall one of the crucial arguments from the aid motivation debate, namely that the amount of aid received by any low-income country is proportional to the level of interest to the donor [*McKinlay and Little, 1979: 240*], indicating that there is a straightforward relationship between the amount of aid and the strength of the donor's interest. This explanation will be given considerable importance in the following analysis.

The aid motivations literature establishes that the motivations of the European Union in giving aid have mainly been 'donor interests' and not 'recipient needs', i.e. the level of need in the poor countries [*Bowles, 1989: 7–19; Grilli and Riess, 1992: 214; Grilli, 1994: 71ff.*]. Donor interests are defined as economic interests particularly in trade and private investments, but also security in the narrow Cold War sense of the word [*McKinlay and Little, 1979: 236–50*]. John Ravenhill agrees that the European motives for giving aid have basically been selfish donor interests. However, according to Ravenhill, there is a special aspect of European donor interests at least when it comes to giving development aid to the ACP countries. It is because the European Community gains 'a certain psychological satisfaction from providing development assistance to... a large number of the world's least developed countries' [*Ravenhill, 1985: 35*].

It is the theoretical argument here that the EU's Cold War development aid policy may be explained by the existence of donor interests, i.e. selfish European concerns. When we reach the post-Cold War period, donor interests can still explain at least part of the development policy. However, because of the new international situation, the donor interests have to an increasing extent shifted away from the traditional economic and psychological interests towards what is called here 'European' interest in becoming a global actor. In addition, Georg Sørensen has argued that, in the post-Cold War era, the EU as a 'post-modern state' is mainly exposed to a special type of security threats which stem from turmoil and general instability in weak postcolonial states, for example in Africa [*Sørensen, 2002: 129ff.*]. So, according to Sørensen, in the post-Cold War period the European interests in sub-Saharan Africa and in other developing regions are to an increasing extent tied to different types of security concerns. Postcolonial areas therefore come into the focus of the Common Foreign and Security Policy (CFSP), where it seems pertinent to recall the argument of Christopher Hill that CFSP 'defence is the key to the development of the Community's place in the world' [*Hill, 1993: 318*]. To sum up, it is the first explanatory argument that, in the post-Cold War era, it increasingly became a European interest to integrate sub-Saharan Africa and other regions in the Union's overall foreign and security policy reflections because these geographical areas to a greater or lesser extent represent a security threat to the EU.

Turning to the second explanatory variable, namely the question of bureaucratic politics, it is characteristic of EU policy-making, i.e. in Moravscik's second stage, that it is highly differentiated and segmented. This means that the different policy fields are to a considerable extent isolated from each other and therefore, as Guy Peters states, within the EU there is a 'tendency of bureaucratic decision-making to occur within policy communities' [*Peters, 1992: 77, 115–21; Bulmer, 1994*]. Far-reaching decisions about the future of Europe and the overall policy of a given sector are usually made at the level of prime ministers and ministers in the relevant Councils of Ministers. Much of the day-to-day politics, however, is the result of bureaucratic interaction and bargaining with the relevant actors, such as the individual Directorates General in the Commission, national foreign policy civil servants and, to a limited extent, also pan-European NGOs.

The trend towards decentralised and bureaucratic decision-making may be further strengthened in the case of development policy because it is a low-profile area both within the European Union and in most member states. This leads to a situation where it is highly probable that a bureaucratic organisation such as the European Commission in combination with a limited number of national civil servants and occasionally pan-European NGOs exerts considerable influence on most policy initiatives related to development policy, at least as long as these initiatives are not related to the overall policy priorities of the Union [*Olsen, 1998*].

Put differently, policy-making on development assistance is to a large extent restricted to elite actors based in the national aid bureaucracies as well as in the Directorate General for Development (formerly DG VIII) in Brussels. For example, John Ravenhill claims that DG VIII has shown significant interest in maintaining and if possible expanding the aid relationship [*Ravenhill, 1985: 33, 36*]. Because of lack of strong political interests in development issues within the EU, not only decisions on day-to-day policies are in the hands of the bureaucrats. Decisions related to the long-term strategic goals and directions of the Community's development policy may also be strongly influenced by the bureaucracies. This latter point may explain the increasingly obvious similarities between the Union's aid policy and the policies of the bilateral donors as well as that of the World Bank.

The growing similarities reflect the existence of a transnational policy community on development issues which may very well have its counterparts within other policy fields [*Marcussen and Ronit, 2003*]. As in national policy communities, the participants in transnational policy communities influence each other and over the years tend to develop a strong consensus on what are the 'best' policies and also agreement on what are the solutions to identified problems. It can even be claimed that the similarities between the different donors have become stronger in the years following the fall of the Berlin Wall. If this latter is correct, it may also explain why, during the 1980s, the EU was so slow to accept the shift towards structural adjustment policies, strongly lobbied for by the World Bank and followed by an increasing number of bilateral donors. The slow adaptation of DG VIII is to a large extent explained by the powerful influence of France on it during the Cold War. France was fairly strongly opposed to the shift towards structural adjustment [*Grilli, 1994*].

Despite the fact that bureaucratic interests may explain a number of policy changes and also the continuation of policies, they cannot on their own explain the more general changes in policy, such as the remarkable shifts in the geographical focus of the Community's aid from Africa towards 'Europe', i.e. the Balkans. The reference to bureaucratic interests may also have difficulties in explaining the dramatic cuts in aid budgets. Concerning the reductions in aid to sub-Saharan Africa since the early 1990s, it is hardly in the interests of the bureaucracies to reduce the aid level as it could automatically lead to reductions in staff. On the other hand, if the reduction in aid to sub-Saharan Africa can be compensated for by an increase in transfers to the Balkans, it is still possible that the bureaucracies can see such a shift to be in their best interest.

In summary, it is the core argument of this chapter that the EU's development aid policy can be explained by a combination of 'European' and bureaucratic elite interests. During the Cold War it was a European interest to maintain a high profile on aid and this was done mainly through the Lomé Conventions. During the 1990s, it increasingly became a 'European' (national) interest to develop a CFSP both because it was a precondition for the EU playing a role in world politics and, second, because the CFSP was the answer to the new post-Cold War security situation. Thus, the first explanation suggested here is that the change in 'European interests' can explain the general policy shifts in European development aid. The second explanation put forward here is that the more specific policy initiatives and changes may be explained by bureaucratic politics, i.e. they may reflect the interests of the participants involved in the daily policy-making and implementation of development policy. It is important to underline that the two explanations may very well supplement, rather than compete with, each other.

IV. THE EUROPEAN AID POLICY FINANCED BY THE ORDINARY BUDGET

As mentioned, one part of the aid of the Europan Community is financed by the ordinary budget. It is the development assistance that is channelled to the Mediterranean, Asia, Latin America and 'Europe'. During the Cold War, this type of aid was fairly limited compared with the aid that was reserved for the ACP countries. As shown in Table 17.1, the aid to 'Europe' and the Mediterranean region did not start to increase until the end of the Cold War.

From the early 1960s, the EC aimed at developing an aid programme for the Mediterranean region. However, there was never a clear and coherent policy towards this region and many observers have described it as 'accidental at best' [*Grilli, 1994: 181; Lister, 1997: 79ff.*]. In spite of a recognition within the Community of the need to increase the dependence of the Mediterranean countries on the Community, its Mediterranean policy never became the 'twin of the Lomé'. The Southern neighbours of the EC were placed at a level clearly lower than the ACP countries in the 'pyramid of privileges' [*ibid.: 188–9*]. However, during the latter half of the 1980s and the 1990s, the Mediterranean region became the object of considerable attention inside the Community.

There were several reasons for the increased concern for the situation to the south and the east of the Mediterranean. The most important had to do with the deterioration in the socio-economic circumstances of a number of countries and with the interest of the Community to alleviate this situation. Contractions in incomes and continued population growth generated substantial reduction in per capita consumption levels. Unemployment rates, particularly among young people, soared and emigration to Europe came to represent the only vent left for surplus labour. This new focus was clearly stated in an official report published in 1989 [*Grilli, 1994: 212ff.*]. The outcome was a new policy that aimed at improving the socio-economic conditions in the southern Mediterranean countries by means of establishing a free trade area and by increasing the aid volumes [*Lister, 1997: 87ff.*].

The Barcelona Conference of November 1995 brought together 15 EU member states with 11 Mediterranean countries. The conference called for a regular political dialogue among the signatories and greater EU aid to the southern Mediterranean states plus control of immigration, crime and drugs [*Lister, 1997: 88ff.*]. On the one hand, the increase in aid meant that the Mediterranean recipient countries received the same amount of aid per capita as did the Lomé countries. But, on the other hand, compared with the aid allocated to Central and Eastern Europe, the disbursements of aid to the southern neighbours were still limited [*Lister, 1979: 89*].

The conspicuous growth in EU aid to 'Europe' mainly went to the countries in the Balkans with the Federal Republic of Yugoslavia and Bosnia-Herzegovina being the largest recipients during the latter half of the 1990s [*OECD, 2003: 305*]. The increasing aid to the poor countries in Central and Eastern Europe clearly has to be explained by the fear in the European Union that the increasing poverty would have consequences for the social stability in these countries. Also, the civil wars in the former Yugoslavia added to the fear of increasing immigration into the European Union. It was expected that allocating aid to these areas could contribute to stabilise the situation.

Also, the Europan Community has had development aid policies directed towards Latin America and Asia. Back in 1980 to 1981, India was actually the main recipient of aid from the EC whereas Bangladesh figured as number four on the top 15 list of major recipients of EC aid. No Latin American countries were found on the list. Ten years later, i.e. in 1990 to 1991, Bangladesh had dropped to number 10 and India had moved down to number 12 among the main recipients of EC development assistance. In 2000 to 2001, there were no Asian and no Latin American countries among the 15 major recipients of European aid [*OECD, 2003: 305*].

Starting with Latin America, the relations between the EC and the region 'have always been on the fringes and marginal to Europe's mainstream development', Martin Holland points out [*Holland, 2002: 52f.*]. One explanation of this is found in the historical context of the 1950s and 1960s when the Community's policies and interests were defined. Here, the absence of a direct member state colonial heritage militated against creating a preferential-type framework which was offered to Africa largely on French insistence [*ibid.: 53; Grilli, 1994: 225ff.*]. The accession to the EC of Spain and Portugal in 1985 changed the situation to a certain extent. In the years following 1985, a number of co-operation agreements between the EC and a number of Latin American countries were signed. Of greatest regional significance were the 1992 Inter-institutional Co-operation Agreement and the 1995 Inter-regional Framework Agreement signed with MERCOSUR (Argentina, Brazil, Paraguay, Uruguay with Chile as associate member), both of which were designed to strengthen regional political co-operation and lead to the progressive liberalisation of trade [*Holland, 2002: 56ff.*]. As far as the members of MERCOSUR and a number of other countries in the regions are concerned, trade is the most important issue when it comes to relations with the European Community [*Grilli, 1994: 241ff.; Holland, 2002: 58*]. Thus, compared with trade, development assistance is of much less importance to most Latin American countries in their relations with the EC.

The same applied to Asia; it was not a policy priority of the EC. The peripheral character of the relationship is all the more puzzling given that Asia in many ways shared colonial history with other parts of the developing world. It was not until 1996 that the Asia–Europe Meeting (ASEM) established a regular forum for dialogue between the two regions. It is striking that despite the strong ties between the UK and the Indian subcontinent in particular, no Asian country was permitted to join the Lomé Convention when it was negotiated in the first half of the 1970s. This missed opportunity essentially confined Asian–EU relations to the lowest of priorities for the next two decades, Martin Holland argues [*Holland, 2002: 60*]. Enzo Grilli has characterised Europe's position towards South Asia as 'mildly sympathetic ... with minimal effective involvement in terms of economic assistance' [*Grilli, 1994: 276*].

As already mentioned, India and Bangladesh were among the largest recipients of EC development assistance up until the early 1990s when the European priorities changed quite dramatically in favour of

'Europe' and the Mediterranean. Grilli noted that 'with more than two and a half times the population of sub-Saharan Africa and a substantially lower per capita income, South Asia received five times less financial aid from the Community during 1976–88' [Grilli, 1994: 280].

Given the concentration of the world's poor in South Asia, it is obvious that the declared aim of poverty eradication of the EC's aid policy is not realised to any degree worth mentioning when it comes to this particular region of the Third World. Thus, in conclusion, this brief overview of the European Community's development aid policy financed by the ordinary budget makes it appropriate to quote Martin Holland when he states that 'relations with Latin America as well as those with Asia constitute the remaining missing elements in the EU's international actor profile' [Holland, 2002: 57].

V. THE AID POLICY OF THE EUROPEAN COMMUNITY TOWARDS THE ACP COUNTRIES DURING THE COLD WAR

In comparison with the regions of the Third World mentioned above, the importance of Africa manifested itself in the Treaty of Rome, Article 131 of which established the association between the Community and the countries and territories in Africa with the aim to 'promote the economic and social development of the countries and territories and to establish close relations between them and the Community as a whole' [Grilli, 1994: 8]. The system of association in the Rome Treaty contained two main elements, namely the establishment of free trade between the two parties and the provision of economic aid to the poor territories [Grilli, 1994: 11; Lister, 1988: 20].

During the first years of the European associational co-operation with the African territories, decolonisation speeded up. For obvious reasons, many of the new states wanted to maintain preferential access to the EC markets for their exports and to continue receiving aid from the Community. There was therefore need for a new arrangement that took into account that a number of colonial territories had become independent states. Thus, in July 1963, a Convention between the six EC member states and 18 associated states was signed in the capital of Cameroon, Yaoundé, which also gave its name to the new convention. Like the Treaty of Rome, Yaoundé I and II covered both trade and aid. The principal accomplishment of Yaoundé I was, according to Marjorie Lister, that it preserved the EurAfrican entente [Lister, 1988: 55].

The next significant step in EC relations with the developing countries occurred in 1975, two years after the oil crisis and the entry of the United Kingdom, Denmark and Ireland into the Community [*Grilli, 1994: 21ff.*]. A new Convention was signed on 28 February 1975 in Lomé, the capital of Togo, extending the association to 46 countries including most of the former British colonies in Africa. What the EC members had envisaged as a simple extension of the Yaoundé terms became a much more complex and somewhat novel type of association. In particular, it is significant that Lomé I (1975–80) explicitly recognised that the associated countries in Africa, the Caribbean and the Pacific (ACP) needed stability in export revenues which gave rise to the STABEX system of stabilisation of export earnings. It was also important that the principle of reciprocity in trade between the poor countries and the nine EC members was abandoned in favour of non-reciprocity which was assumed to be an advantage for the developing countries and thus to comply with their needs [*Lister, 1988: 74–102*].

By the time it was signed, the Lomé Convention was widely considered a unique example of North-South co-operation and it was argued that it could serve as a model for a new era in North–South relations [*Lister, 1988: 79–80, 1997: 109*]. The most important feature of EC aid was that it was intended to be politically and economically neutral. Also significant, the principle of 'parity' or 'equality' contained in Lomé presupposed the active participation of the ACP states in all aspects of the aid process from project preparation via implementation to *ex-post* project evaluation. Finally, in the mid-1970s, it was considered to be positive because EC aid meant resources additional to the bilateral aid supplied by the individual EC member states [*Grilli, 1994: 91–136*]. Irrespective of these positive evaluations of the Lomé arrangement, Marjorie Lister maintains that, instead of creating a new economic order, Lomé was an elaboration of the old order [*Lister, 1988: 81*].

There is no doubt that security of the supply of raw materials was the primary interest pursued by the EC member states in relation to Africa in the early days of co-operation. However, other factors also played a role in the establishment and continuation of the relationship. Bipolarity and the competition between the superpowers during the Cold War underlined the importance for Europe of maintaining the political allegiance and friendship of the newly independent states in Africa. If 'Europe' wanted to maintain a minimum of international influence and power, it was in its political interests that the EC should keep at least one traditional sphere of influence. This line of thinking was very much inspired by the French foreign policy goals as clearly expressed by General de Gaulle [*Grilli, 1994: 335*].

In summary, during the late 1970s and the 1980s, the whole concept of associationism lost its appeal for the Europeans. This was due to at least two circumstances. First, Africa's role as a supplier of raw materials to Europe simply disappeared and with it the security argument in favour of the original idea of association lost its appeal for the European Community [*Lister, 1997: 109*]. Nevertheless, Marjorie Lister concludes: 'although Lomé was disappointing in its trade results and in many of the development projects it financed, it was not completely unsuccessful...it did accomplish its primary, if unwritten objective: the continuation of the EurAfrican association' [*Lister, 1988: 110*]. Second, the rapidly changing international situation in the second half of the 1980s with the obvious weakening of the Communist bloc also contributed to changing the attitudes of European decision-makers concerning the future of Lomé. These two circumstances formed the background for the negotiations on Lomé IV that began towards the end of the 1980s, at a time when the Cold War was drawing to an end.

VI. LOMÉ IV, 1990 TO 1995

The Community entered the negotiations on Lomé IV with the view that the new Convention should not result in major changes but only adapt to the experience of the past Conventions. The general pattern was that the Europeans put forward most of the proposals, while the ACP mainly reacted to them, which is one of the decisive reasons why the changes included in Lomé IV were mainly those the Community wanted, such as a Convention with a ten-year duration divided into two sub-sections [*Lister, 1997: 113*]. The same is true of the principle of a 'policy dialogue' between the parties that was sustained in the new Convention.

As a new feature, Lomé IV included funds earmarked for the structural adjustment programmes that were under way in many African countries. The introduction of funds for structural adjustment in the Union's development policy implied a crucial change of EU policy as it signalled complete backing for the extensive reform processes required by the two Bretton Woods institutions. During the 1980s, the European Union had officially been quite critical towards the idea fostered in Washington that the African countries should be forced to introduce structural economic reforms. In view of the strong criticism of this kind of liberal economic policy, it may be surprising that the proposal for funds for structural adjustment was not even opposed by the ACP states. It may be even more surprising because this new policy element meant a radical departure from the principle of impartiality and neutrality [*Lister, 1997: 118, 115*].

In fact, the ACP countries were lobbying actively on only one issue, namely the size of the financial contribution in the agreement. The ACP demanded 15.5 billion ECU for the five-year period but settled without further ado for the 12 billion ECU offered by the Community. This represented a minor increase of 4 per cent in current prices over Lomé III, while an increase of around 50 per cent was widely reckoned to be necessary to maintain the real value of aid in the face of inflation and the growing ACP populations. It is worth noting that the 50 per cent increase was actually favoured by the Commission, whereas a number of member states could not or would not accept an increase of that magnitude [*Lister, 1997: 112*].

In summary, the moderate financial portfolio, combined with the successive steps towards tightening the conditions for receiving assistance from the Community, may be interpreted as an expression of the decreasing European interest in Africa. It may be argued that the ACP had understood the figure of 12 billion ECU as striking a balance between what they and the Commission wanted and what the sceptical member states were prepared to accept. Aderemi Oyewumi finds that the 1980s signalled a shift in the character of the EC–African relationship from one based on a form of partnership 'to a more paternalistic regime' [*Oyewumi, 1991: 136*]. Enzo Grilli claims that, in spite of the continuous supply of aid, 'the attention of Europe is already moving away from Africa... [So if the continent remains] a priority in EC development co-operation, it is more for historical and humanitarian reasons and perhaps because of the inertia built into long-following practices than for the protection of European economic interests considered to be vital' [*Grilli, 1994: 345–6*].

VII. PROMOTION OF DEMOCRACY: HIGH PROFILE, LOW PRIORITY

Shortly after the signing of Lomé IV a number of new policy issues appeared in the international debate on aid and development. One of these was the need for democracy and respect for human rights which ended up having quite a considerable impact on the general relationship between the developing countries and the European Community and on the relationship between the EU and Africa in particular. Thus, only a few months after the fall of the Berlin Wall, European politicians began to link economic development with political freedom and democracy. They were quite outspoken about the need to change European aid policy, stressing that in future it would be hard for authoritarian regimes to obtain development assistance from Europe unless they started to democratise their governments [*ODI, 1992; Stokke (ed.), 1995; Crawford, 2001*]. As early as November 1991, the 12 EC member states adopted a resolution stating unequivocally that, in future, democracy and respect for human rights would be preconditions for receiving aid from Europe [*Council of Ministers, 1991*]. Subsequently, the Maastricht Treaty stressed that 'the development and consolidation of democracy and the rule of law and adherence to human rights and fundamental freedoms' [*Art. 130U, sect.2*] are among the important aims of the development policy of the European Union.

Despite the fact that the promotion of democracy and human rights was indisputably a high-profile issue in Europe's external relations throughout the 1990s, the budget line 'Governance and civil society' received conspicuously low priority. In 1992 only 1.5 per cent of the total budget was allocated for this purpose. In the following years it was even lower, reaching only 1.2 per cent in 1995 [*Cox and Koning, 1997: 40*]. Since then the level has risen somewhat, reaching 2.0 per cent of total Commission aid in 1999 and jumping to 6.8 per cent in 2001. A comparison between the relative share of EU development assistance supporting 'governance and civil society' and the corresponding level of US support shows the US in general allocating a substantially larger proportion of its aid to these purposes. Thus, in 1999, the US allocated 6.6 per cent of its total aid to governance and civil society, and in 2001, 13.3 per cent [*OECD, 2003: 269*]. Compared with most EU member states, the Commission's allocations for governance and civil society were also considerably lower. Nevertheless, Richard Youngs concludes an analysis of European approaches to assistance for democracy with the statement that the EU has become a significant and more unified actor in the field of democracy assistance, its political aid budgets having increased notably and its work being incorporated into more formalised policy initiatives. However, he also acknowledges that 'despite [the] progress made, the EU still needs to develop its political strategies further' [*Youngs, 2003: 137*].

VIII. THE 1995 MID-TERM REVIEW

Before Lomé IVa expired in February 1995, it was clear that the Convention had not worked well and neither the EU nor the ACP were satisfied [*Personal interviews, Brussels, January 1995; Bossuyt et al., 1993*]. An old, but crucial, complaint concerned the effectiveness of the common aid programme. For a number of years frustration had increased steadily about the lack of positive results after such a long period of development assistance to Africa. Some member states, especially Germany and the UK, were particularly critical of the joint EU aid, since several studies of integrated rural development and livestock projects had been shown to be unsuccessful, including a number of direct disasters [*Riddell, 1993: 62ff.*].

To make things worse, by the end of the five-year period only about half of the total sum of 12 billion ECU had been spent. One explanation of this was the lack of absorptive capacity due to inefficiencies in the administrative apparatus of the recipient countries. Also, the more or less conspicuous chaos in about 30 states prevented payments of aid either in part or in full. The EU's own procedures for disbursing aid were criticised for being both slow and excessively bureaucratic to the extent that the Commission's activities were mainly a set of ad hoc piecemeal responses to unfolding events [*Grilli, 1994: 350–90*]. Finally, the common aid programme has been characterised by a pronounced lack of co-ordination with the bilateral programmes, and this posed a serious challenge to the continuation of the Lomé co-operation [*Box and Goodison, 1994*].

It was in this general atmosphere of disillusion and critique that the mid-term negotiations on Lomé IVb (1995–2000) took place. The negotiations were long-drawn-out mainly because of the strong disagreements among the EU member states. Despite the fact that the European Commission proposed a financial protocol for the years 1995 to 2000, which merely covered inflation, a number of member states wanted a much smaller financial aid package and demanded a reduction in the Commission's proposal.

It took more than four months of internal EU negotiations to reach agreement among the member states on a proposal to present to the ACP countries. To reach this point, the involvement of the French President and the German Chancellor as well as the British Prime Minister was required. Finally, at the EU summit in Cannes in June 1995, the heads of state and government agreed to release 12.8 billion ECU for the years 1995 to 2000 [*Council, 1995*]. The figure was a slight improvement compared with that suggested earlier. In view of the fact that the amount needed to be at least 14 billion ECU simply to compensate for inflation, the ACP countries were understandably disappointed. Off the record, most ACP governments had begun to fear a much worse outcome [*West Africa, 20–26 November 1995: 1795*] which may explain why a few days later they issued a press release expressing satisfaction with the result of the mid-term negotiations.

There is little doubt that the negotiating positions taken by many of the EU member states during the mid-term review reflected widespread disquiet about the effectiveness of the common development assistance in general as well as strong scepticism about the effects of giving aid to Africa, in particular. Here it should not be overlooked that, parallel to the discussions in Brussels, a number of individual member states had already started to reduce their development aid budgets quite dramatically [*OECD, 1997*].

At the Cannes summit, the changes in the external political priorities of the EU were made explicit. While there was a considerable reluctance to finance Lome IVb, the European heads of state and government decided to allocate 4.7 billion ECU to the Mediterranean region for the years 1995 to 2000. The package decided on at the summit was tantamount to EU per capita aid to the Mediterranean countries becoming equal to that allocated to Africa south of the Sahara, irrespective of the fact that the Mediterranean countries are all considered middle-income countries, while most sub-Saharan African countries are still low-income countries [*Ministry of Foreign Affairs, 1996: 140*]. At the subsequent summit in Barcelona in November 1995, the EU offered to develop gradually a free-trade zone between the 15 EU member states and the 12 poor countries in the south and east of the Mediterranean before the year 2010.

However, the financial protocol in Lomé IVb is hardly the most remarkable part of the mid-term review. The non-financial elements of the agreement were probably more important, as they made it clear that the original 'partnership' between the European Union/European Commission and the ACP was changing dramatically during the 1990s. Among the significant changes were the clauses that became formal parts of the Convention and which required the ACP countries to observe the rules of democracy and respect for human rights. The introduction of such political conditionalities moved the Convention one more step away from its original concept of 'political neutrality'. The same applies to the introduction of the principle of so-called 'phased programming' of the aid provision into Lomé IVb, implying that the aid figures were only indicative, and not an entitlement as they had been in earlier Conventions. This kind of performance-based aid policy had few similarities with the original Lomé concept built on the principles of 'partnership'.

Even though the trade component of the Lomé Convention granting non-reciprocal concessions to the ACP countries was not changed in the MTR, the fall in the real value of the aid funds, combined with the number of political conditions and control initiatives, indicated 'a declining [EU] commitment to the ACP group, particularly in contrast with the growing volume to other regions', Gordon Crawford argues [*Crawford, 1996: 504*]. Table 17.1 shows clearly that the regions which started to receive relatively more aid from the EU were first of all 'Europe', i.e. the Balkans, and then North Africa and the Middle East. Moreover, Marjorie Lister claims that the MTR demonstrated that the Europeans had lost faith in the capacity of the ACP countries to take responsibility for their own development. The content of Lomé IVb also indicated that future changes were going to point towards a Convention which would be much less sensitive to ACP development concerns [*Arts and Byron, 1997; Lister, 1997: 136–42*].

In summary, the aid component in the MTR, including all the conditionalities and control initiatives, underlined the trend away from the first five years of Lomé IV, namely that the EU was losing interest in Africa and in the ACP countries in general. It is worth noting that the decrease of interest was combined with a corresponding lack of confidence in the capacity of the recipient countries to promote their own development. Nevertheless, the aid programme was continued. The European Commission, and DG VIII in particular, struggled vigorously for an agreement that could maintain the Lomé arrangement. Moreover, DG VIII also fought to obtain an agreement that was favourable to the ACP. Based on the course of the negotiations, it is possible to argue that the Commission (DG VIII) and the Council of Ministers were the bearers and defenders of the 'European' (national) interests of the Union which, in mid-1995, were to maintain some kind of organised relationship between the EU and the ACP. In addition, it is also possible to contend that the intense lobbying of DG VIII was an expression of bureaucratic interests in maintaining one's own policy field, combined with elite arguments that aid in itself was in a moral cause.

IX. NEW POLICY PRIORITIES, NEW INSTRUMENTS

The more or less radical changes in the aid regime did not occur in isolation from other developments within the EU. Parallel to the decreasing interest in the developing countries, numerous European actors lobbied vigorously in favour of strengthening the international position of the EU by developing the CFSP [*Bretherton and Vogler, 1999*]. In spite of the reluctance of many member states to give too much power to the CFSP [*Cafruny and Peters, 1998b: 299; Rhodes, 1998*], the Maastricht Treaty introduced the system of 'joint actions' as a new element in European foreign policy-making which opened up closer co-operation in foreign policy [*Piening, 1997: 40–42*]. The joint actions were clearly related to the ambition to develop the Union into a significant international actor.

It is important that one of the first five joint actions decided upon in 1993 – and the only out-of-area action – was actually directed towards Africa, namely towards South Africa. Martin Holland argues that this particular joint action is significant because it offered an opportunity for the new Union to express its global role [*Holland, 1997: 174*]. As another element in the development of the CFSP, the EU launched its first initiative on conflict prevention and resolution in Africa in 1993. It was followed by a number of declarations stressing that the Union was seriously concerned about the rising number of violent conflicts on the continent, as was stressed at the EU summit in Essen in December 1994 which called for an 'intensive political dialogue between the EU and OAU in particular regarding conflict prevention in Africa' [*Landgraf, 1998: 103*]. The same line was taken in the final document from the EU summit in Madrid in December 1995 [*Council, 1995*]. In 1996, the European Commission issued a 'communication' on conflict and conflict prevention stressing that the use of development aid and related instruments was considered to be important to the EU in relation to managing conflicts. In June 1997, a 'common position' on these questions was adopted making it clear that conflict prevention in Africa was an EU priority [*Landgraf, 1998: 110*]. Finally, in the latter half of the 1990s, the EU issued several joint actions and 'common positions' related to the violent conflicts and to conflict management in and around the Democratic Republic of Congo, Angola and Sierra Leone.

The genocide in Rwanda in 1994 was an extremely frustrating experience for the EU member states. It gave an additional impetus to the EU debates regarding how to prevent a recurrence of such tragedies [*Olsen, 1997: 312ff.*]. As one important step towards serious involvement in conflict prevention in Africa, the EU Council of Ministers agreed on a joint action on 25 March 1996 appointing a special envoy, Aldo Ajello, to the Great Lakes Region. One important aim of this appointment was clearly to give the EU visibility in the region and thereby visibility in the world [*Confidential interviews, Brussels, January 2002*] and as such was an element in the strategy to develop the EU into an important international actor.

The increasing significance of general foreign and security concerns in combination with aid to Africa was explicitly articulated in 2000 by the Portuguese Presidency in its 'Reflections Paper' concerning the EU's future relations with Africa. The paper stressed that 'development priorities should be thought of in the context of ongoing dynamics, namely those related to the reorganization of external relations [in the Commission] and the building of a European CFSP. Being realistic about development means thinking in an integrated manner about politics, security, trade as well as development aid itself' [*Cardoso* et al., *2000*].

This line of thinking led logically to the European–Africa summit in Cairo in early April 2000, which may be seen as a symbolic expression of the 'new' European thinking on Africa combining development aid, common foreign policy and, not least, common security priorities. The preamble to the summit's action plan stresses that the two parties are committed to work towards a new strategic dimension to the global partnership between Africa and Europe which precisely includes issues such as peace-building, conflict prevention, and management and resolution of conflicts in Africa [*Cairo, 2000*].

In summary, during the latter half of the 1990s it was possible to identify two trends in the EU's policies towards developing countries most clearly illustrated in its policy towards sub-Saharan Africa. One trend manifested itself in the shrinking aid to Africa already mentioned. The other trend manifested itself in the increasing preoccupation with security issues in Africa, reflecting that the national interests of the member states had changed in the years following the end of the Cold War. At the same time, there was a clear trend towards perceiving development aid as only one among a number of instruments in the EU's external policies [*Financial Times, 29 September 2000*] and in particular as an instrument for managing and preventing conflicts. This is indicated by the numerous statements, joint actions and common positions issued since the mid-1990s.

However, it is important to stress that it may be difficult to distinguish clearly between aid aimed at promoting social and economic development and aid for conflict prevention purposes. Thus, before he was forced to retire, the Commissioner for Development Assistance, Joao de Deus Pinheiro, stated that 'development co-operation is indisputably the single most important instrument for an effective policy of peace-building'. The position is also reflected in a publication supported by the EU which focuses directly on how development assistance can promote stability and how it can also contribute to conflict prevention [*Lund and Mehler, 1999*]. As a digression, the conspicuous growth in the EU's development assistance to the Balkans/Europe may well be explained by this very understanding, namely that development assistance is an adequate instrument for conflict prevention.

X. THE COTONOU AGREEMENT – 2000FF

The European commitment to carry out radical changes in the EU–ACP relationship was highlighted soon after the conclusion of the MTR when the Commission published in late 1996 a so-called 'Green Paper' discussing the future relationship between the EU and the ACP countries [*Commission, 1996*]. It soon became clear that the Commission preferred one particular choice, implying that the ACP group would be split into a number of regional and sub-regional units [*Lister, 1998: 375–90*]. The Commission's negotiating mandate for a new Convention therefore included a proposal to replace the existing general non-reciprocal trade preferences with so-called 'Regional Economic Partnership Agreements' (REPAs) that were to be signed with different ACP regions and countries [*Council, 1998*].

The main aim of the suggestion for a new trade arrangement is clearly to strengthen the regional integration among the ACP countries and also to enhance their access to the European market. In spite of the proposal of regional free trade areas implying that the old principle of non-reciprocity would come to an end, there was universal agreement that the principle in itself had not helped the ACP to increase their market share in the European Community. Rather to the contrary, the ACP share of European markets had declined quite dramatically.

The ACP countries strongly opposed the Commission's proposal to establish regional free trade groups. In particular, they were against giving up the trade preferences granted to them under the principle of non-reciprocity. Many observers of EU–ACP relations share the ACP countries' negative reaction to the EU's negotiating mandate and, in particular, are sceptical about the suggested changes in trade concessions. Patrick Watts describes the idea of regionalising the ACP as 'poorly thought-out' and 'ill-defined' [*Watts, 1998: 52–63; McQueen, 1998: 691*]. Due to the increase in this type of criticism, the Europeans felt that they had to commission a number of studies on the possible consequences of regional free trade arrangements. The result of these desk studies was in most cases the finding that the least developed countries have little to gain from regional agreements such as those proposed by the EU [*ECDPM/ 10/99, 1999; ZEF/GEMDEV/ ECDPM, 1999*]. It is beyond dispute that the EU's proposal for a revised trade component of Lomé V (Cotonou) was a radical break with the state of affairs up until then. Marjorie Lister finds that 'it is a radical departure from Lomé ... and it creates a high risk of political marginalization of each region, especially in Sub-Saharan Africa'

[*Lister, 1998: 384*]. However, it must be stressed that the future signing of regional co-operation agreements is voluntary. As a compromise reached during the negotiations on the Cotonou agreement, it was agreed that negotiations on establishing these co-operation agreements were to start in September 2002 during the Danish EU Presidency and these negotiations began with the clear aim that the regional free trade agreements were to start no later than 1 January 2008 [*IP/02/871, 2002*].

The aid component of the Cotonou agreement underlined that the European attitude towards the ACP countries at the end of the 1990s had not changed. Thus, the European Commission offered 14.33 billion ECU which was only just enough to keep up with inflation over the previous five years, indicating that the real value of the EU–ACP aid for 2000 to 2005 was under no circumstances likely to grow. Several EU member states would not even accept this moderate financial suggestion and demanded reductions. The member states finally agreed to offer the ACP 13.8 billion ECU [*Financial Times, December 1999, and personal interviews, Brussels, November and December 1999*]. The aggregate outcome of the MTR and the negotiations on Lomé V/Cotonou was thus a decline in the real value of the EU's aid package to the ACP countries and to Africa, in particular, in two consecutive periods.

Like its predecessor, the Cotonou agreement's non-financial elements represented an additional step towards tightening the conditionalities for receiving EU assistance. Anti-corruption measures were given high priority, along with the issues of good governance and 'political dialogue', with either individual developing countries or regional groupings. Compared with Lomé IV, a totally new element entered the Cotonou agreement, namely the possibility of suspending the co-operation if the level of corruption became excessive. Among the other new elements in the agreement were peace-building and conflict prevention.

It is worth noting in particular that migration was included as a special issue in the agreement's Article 13. The Cotonou agreement required the ACP countries to 'accept the return of any of their nationals' found illegally resident in the EU 'without further formalities'. While the migration provisions also apply in reverse to EU nationals in the ACP states, the clear motivation behind them came from a European Union that is highly concerned about the increased migration from the developing world [*Holland, 2002: 204*]. Finally, under the heading of the 'political dialogue' article of the Cotonou agreement, 'drugs' and the fight against drugs were also included.

To sum up, the Cotonou agreement was signed in June 2000 by the EU and 77 countries in Africa, the Caribbean and the Pacific. Its contents confirmed the development trends which had become increasingly obvious during the 1990s. The agreement was preserved but most of the original elements and characteristics of European development co-operation were gone. By 2000, it was obvious that development co-operation for the EU's decision-makers was, to a large extent but not exclusively, an instrument for the Union to influence the course of development in the recipient countries. The Cotonou agreement also indicates that the EU wants the developing countries to take on a much greater responsibility for their own development. This is clearly indicated by one of the aims of the regional partnership agreements, namely to promote regional trade among the developing countries instead of continuing the colonial trade patterns with the EU. Also, the Cotonou agreement reflects the fact that some decision-makers within the Union have increasingly lost confidence in the effects of development assistance, at least in sub-Saharan Africa. However, the most conspicuous new element is probably that the agreement has now become one among a number of instruments in the domestic political sphere, illustrated by the inclusion of Article 13 on migration.

XI. REFORM OF THE AID ADMINISTRATION

While the changes in the general framework of the EU's development co-operation were taking place, the debate became more concrete among the different decision-makers on the need to streamline EU development policy and the instruments attached to it. During the 1990s the criticism of the inefficiency of the aid system came from a number of member governments, as already mentioned. But there was also criticism from within the aid policy community itself. The collective resignation of the European Commission headed by Jacques Santer in March 1999, and, during the initial stages of the negotiations on the Cotonou agreement, the appointment of a new Development Commissioner, Poul Nielson, became the starting point for reform initiatives to address these criticisms.

Poul Nielson became responsible for DG-Development, which was primarily in charge of the aid relationship with the ACP countries, and he also became head of ECHO. Moreover, he was also responsible for development co-operation with Asia, Latin America and the Middle East. However, the Commissioner for External Relations (DG Relex), Chris Patten, was made responsible for all other relations between the latter group of countries and the EU. Trade with developing countries became the responsibility of the Commissioner for Trade, Pascal Lamy. Thus, as a consequence of the new division of labour within the Commission, no fewer than three commissioners are involved in policy-making on relations with the developing world. Concerning relations with Asia, Latin America and the Middle East there is even a dual responsibility on the European side, depending on whether the issue is development assistance in the narrow sense or other matters.

Shortly after taking up his position as Development Commissioner, Poul Nielson stated in a paper addressing the issue of the effectiveness of the EU's aid policy, that the development policy of the Union 'may indeed be affected or even contradicted by the effects of virtually all other Community policies' [*unpublished paper, 2000*]. His need to address the issues of administrative efficiency was made more urgent by the publication of the first official evaluation of the impact of the Community's development assistance. Its main conclusion was that the impact of the EU's aid aimed at poverty reduction had at best been limited. As far as the attempts to promote good governance and democracy were concerned, the impact was correspondingly small [*Commission, 1999: 16ff.*].

The highly critical paper by Commissioner Nielson provided a list of examples of the obvious contradictions in the EU's aid and trade policy, which was described as being 'far from ...exhaustive'. The new Commissioner even stated publicly that 'the Commission machine was never constructed to deliver development assistance. It was designed for producing directives, regulations, trade negotiations, to facilitate political relations between EU states. For development assistance, it doesn't work' [*Financial Times, 17 May 2000*]. Apparently, the Nielson paper was so controversial in the eyes of the new Commissioner's colleagues that it was never adopted or made officially public [*Development Today, 4, 2000: 5*]. Instead, the Commission issued a so-called 'Communication' to the Council and the European Parliament concerning the future of the Community's development policy, by which it recognised the limited impact of its aid and the numerous problems related to the aid system itself [*Commission, 2000: 15*].

The open admission of the numerous shortcomings in the EU's aid policy led to the launching of a number of comprehensive reform initiatives aimed at improving the aid administration [*Financial Times, 17 May 2000; Holland, 2002: 94ff.*]. In particular, the Commission recognised that it had to reduce the length of time it took to deliver the EU's aid commitments and also that the administrative procedures were excessively complex and therefore time-consuming. It was officially acknowledged that the aid administration laboured under chronic staffing problems which had a direct negative impact on the timeliness and quality of EC aid delivery. Where member states or the World Bank had between four and nine officials to manage €10 million of aid, the Commission had only 2.9 officials to administer the same amount of money [*Commission, 2000; Holland, 2002: 97–100*]. The admission of the urgent need for reform basically echoed the numerous calls in the past for greater co-ordination and coherence between the Community's different external policies, such as, for example, between trade and aid [*Hoebink, 1999: 323–45*].

In the Maastricht Treaty, a number of fundamental principles for the development policy of the Union were established, such as the need for co-ordination, coherence and complementarity. In the Amsterdam Treaty of 1997 a fourth principle was introduced, namely that of consistency. The Maastricht Treaty establishes that 'the Community and the member States shall co-ordinate their policies on development co-operation and shall consult each other on their aid programmes' (Article 130x). 'Collectively the introduction of these "four C's" was tantamount to recognition that previous policy frameworks had been inadequate: past experience had proved that the EU often faced profound difficulties in achieving effective policy collectively' [*Holland, 2002: 118*].

The basic rationale behind co-ordination is that it provides a practical mechanism for enhancing complementarity with the aid policy of the member states. Through the focus on co-ordinating the administrative actions of the member states, the Commission and the recipient countries seek to maximise the effectiveness of execution of EU development policy. In the years following the Maastricht Treaty, the Council issued a number of resolutions promoting co-ordination, covering topics like gender, the environment, conflict resolution, poverty, education and food aid [*ibid.: 118*]. The principle of coherence was introduced to ensure that all areas of EU policy were compatible with EU development objectives. Finally, the principle of consistency involves linking all the EU's external activities such as the CFSP, the external economic actions and development in a consistent and logical manner. However, 'agreeing on the need for consistent and coherent policies is not the same as changing national behaviour to achieve that goal' [*ibid.: 119*].

The different aims and goals of the individual EU member states' development policies have been a fundamental obstacle and have made it extremely difficult to establish inter-member-state co-ordination, coherence or complemantarity. So, the first challenge was at the level of the member states, not at the EU level. The EU member states have tended to behave in very different ways and there has been little shared knowledge on what each individual aid administration was doing. There was also a lack of institutional relations and capacities that could have been able to co-ordinate individual bilateral development programmes, despite the aspirations of Maastricht. No doubt, the problem of co-ordination has also permeated the Commission itself. As the external competences of the Union have evolved, these have resulted in a diversification, not a centralisation of institutional responsibility [*ibid.: 172*]. The historical development of EC development policy produced an excessively complex organisational structure that became one of the main obstacles to pursuing the task of co-ordination. Therefore, one possible solution was to create a separate agency within the Commission which would bring all the different professional and technical functions of the different agencies together.

The general overhaul of the aid programming process that was initiated by the new Commission aimed to ensure a better match between resources and political priorities [*Commission, 2000*]. It is also a clear objective of the reform to give more visibility to the EU's development efforts: 'The external aid programmes are a vital instrument in ensuring the EU continues to have a strong voice in the world' [*ibid.: 17*]. The specific aim was to ensure a greater focus on the poorest people in the recipient countries, based on the fact that poverty alleviation had been the overall purpose of the EU's development policy since the Maastricht Treaty was ratified. In order to secure more efficient implementation and coordination of the Community's development assistance, the Commission established a common service in 2001 called the 'EuropeAid Co-operation Office' (AIDCO). Its explicit aim is to improve the quality and flexibility of project administration, to reduce the amount of time spent on policy formulation and shift it to implementation, and to make sure that procedures are harmonised. AIDCO is responsible for the implementation of all external EU aid, both the elements financed by the European Development Fund and those financed by the Community's ordinary budget. The new implementing unit is headed by a board with Commissioner Patten as chairman and Commissioner Nielson as executive director [*Commission, 2000: 17ff.*].

Along with the establishment of AIDCO, the Commission began work on a general decentralisation of its development administration, implying that staff based in Brussels should be transferred to the delegations in the recipient countries. In the terminology of the Commission this process is called 'de-concentration' and aims at shortening the lines of communication and decision-making and not least at reducing the need for involving Brussels in the daily administration of the aid programmes [*ibid.: 20–1*].

Country strategies for each recipient country have become a crucial instrument in making the EU's development aid more efficient [*Commission, 2002*]. The preparation of these strategies is based on an analysis of the current situation in the country in question including an account of its development needs, building on close co-operation with the local authorities and a combination of the Community's own development priorities and those of the individual recipient country, in order to promote local 'ownership' of these strategies. The Commission emphasises the need to include all relevant donors, meaning, in particular, other EU bilateral donors that are present in the country in question. It also seeks to involve non-EU donors, one of the instruments in this endeavour being to use the Poverty Reduction Strategy Papers that are prepared in collaboration with the World Bank and the International Monetary Fund [*ibid.*].

No doubt, the reform of the external development assistance launched in May 2000 was a huge undertaking for the Commission. Despite the fact that it took a long time to start implementing the extensive reform initiatives [*confidential interviews, Brussels, January 2002 and Copenhagen, September 2001*], some significant results have been achieved. In a speech in Brussels in January 2003, Commissioner Nielson maintained that the following results had been accomplished after two and a half years of administrative reform. The creation of the EuropeAid Co-operation Office has been completed. A massive de-concentration of management responsibilities to nearly 50 Commission delegations all over the world has been carried out. Technical assistance offices have been dismantled, and the overhaul of budgetary rules, computer systems, contracting and tendering procedures has likewise been implemented. Nevertheless, the Development Commissioner concluded that 'the reform will continue apace in 2003 and key areas of action will be to pursue the reform ... and the completion of de-concentration' [*Nielson, 2003*]. Based solely on Nielson's statement, significant elements in the extensive reform initiatives appear to have been implemented, some of them with positive effects. It is probably safe to assume that the different initiatives have improved the effectiveness of the European aid system.

Despite the fact that it was a goal in itself to improve the aid delivery mechanisms, it should not be ignored that the reform of the aid administration was an integrated element in the overall reform of the external services of the European Commission. The Commissioner for External Relations, Chris Patten, underlined the importance of the reform of the development aid system as follows: 'The reform will radically transform the way the Commission manages external assistance programmes. It will restore its credibility as a foreign policy actor in one major field of external EU action' [*quoted in Holland, 2002: 100*]. Thus strengthening the aid system will also improve the possibilities for the European Union to operate as a significant actor in international affairs [*Commission, 2000, 2002: 22ff.*]. This official emphasis on development assistance as an important instrument in the totality of EU foreign policy instruments was made crystal clear during the European summit in Seville in June 2002.

The Seville summit decided to reform the working of the Council of Ministers, which resulted in the abolition of the Development Council that had previously been the main forum for discussions and decision-making on EU development policy. As a result of the decision, development and humanitarian issues are now to be discussed within the framework of the 'External Relations' sessions of the new General Affairs and External Relations (GAERC) Council. The GAERC covers two areas of activities and holds two separate meetings with two separate agendas. The area of relevance for this analysis deals with the EU's external actions, namely European foreign and security policy, European security and defence policy, foreign trade, development co-operation and humanitarian aid [*Carbone, 2002: 6*].

The abolition of the separate Development Council led a number of observers to fear that development issues would have to take second place to external relations priorities. The first two GAERC meetings following the Seville decision, namely in November 2002 and May 2003, did not confirm the pessimistic evaluations of these critics. At the Council meeting in November 2002 the agenda included a 'development cluster' that addressed a number of important development issues such as the relationship between trade and development, untying of aid, humanitarian crises in Africa and the follow-up to international conferences [*ibid.*]. The same was true of the GAERC meeting in May 2003 when the EU's commitment to raise its official development aid to the UN-endorsed target of 0.7 per cent of GNI was discussed. At the European Summit in Barcelona in March 2002, the EU member states had committed themselves to increasing their collective aid level from 0.33 per cent of GNI in 2001 to 0.39 per cent of GNI in 2006. The Barcelona Council also gave the Commission a mandate to monitor the progress on these commitments. Thus, for the first time the Commission was allowed to check the actions of the member states and discuss collectively how the obligations are being upheld. Finally, the ongoing reform of the Commission's development policy programme was discussed at the May 2003 summit [*Morrissey, 2003: 6*].

In summary, the appointment of a new Commission, and not least of a new Commissioner for Development Aid, became the precondition for initiating a thorough reform of the whole development aid system of the EU. The reform involved both 'strategic' interests as well as more narrow bureaucratic interests. The strategic or European interests were related to the fact that, with the ambition to establish the CFSP, an efficient aid delivery system became an important component. At the same time, there is no doubt that besides the 'European' interest in maintaining a special European aid programme that is efficient, there were strong bureaucratic interests in having a well-functioning aid delivery system since it was both frustrating and unsatisfactory for the employees in DG-Development to be exposed to constant criticism from European policy-makers and from the aid policy community. The specific content of the reform process can mainly be explained by reference to the influence of the civil servants and the few policy-makers on aid who were involved in designing the proposals for reform. So, in conclusion, the reform process and its specific content can be explained by combining the argument that European interests influenced the course of the EU's development policy with the argument that bureaucratic interests no doubt influenced this policy field.

XII. CONCLUSIONS AND DISCUSSION OF FUTURE DEVELOPMENTS

This chapter has shown that the content of the development policy of the European Union changed quite dramatically from its origins in 1958. The changes were particularly conspicuous during the 1990s and the first years of the new century. The analysis demonstrated that, in the wake of the Cold War, development aid became increasingly integrated into the general foreign and security policy of the EU, implying that it lost some of its former 'independent' status. It is the main argument of the chapter that this trend reflected a shift in the interests of the European Union following the radical changes in the international system from bipolarity to the new international system. The latter opened up the possibility of pursuing the old ambition of the European Community to become an important player on the international scene. It is the increasing strength of this so-called

'European' interest that can explain the general changes in the aims of development policy from 'pure' development objectives to a more general concern with foreign policy and security issues.

However, such European interests cannot explain the more specific changes in the EU's development aid system. They can only be explained by referring to the interests of the civil servants and decision-makers involved, including the Development Commissioner himself, who have a basic interest in maintaining the European development aid system. These actors therefore have a dual set of interests. On the one hand, in order to secure the continuation of a European aid system, they have a considerable interest in streamlining and making the aid delivery as efficient as possible. On the other hand, with the aim of maintaining a common European aid system, it is necessary to adjust and develop the instruments and purposes of development aid. The increased focus on conflict and conflict prevention is an example of an adjustment with the aim of preserving a European Community aid system. The same is true of the redirection of assistance during the 1990s from sub-Saharan Africa to 'Europe'. The two examples underline one of the theoretical arguments of this chapter, namely that the two explanations supplement each other.

Based on the analysis in this chapter, it seems safe to assume that in the future there will be an ever closer relationship between development aid and other external policy instruments, not least those belonging to the CFSP. With the pledges made in Barcelona in March 2002 to increase the relative volume of EU aid, it is reasonable to expect a slow increase in the Union's ODA/GNI ratio. For the time being, it is difficult to predict anything about the geographical allocations of this slightly increasing aid, mainly because there are no stable priorities in the EU's development policy, as has been shown here. The priorities, including the geographical ones, are subject to the political priorities which might change, depending on what the decision-makers find appropriate in relation to what they perceive as European interests. And on this general level of explanation, it is very probable that the interests of the bureaucracies involved will pull in the same direction.

The biggest unknown factor for the immediate future is the enlargement with ten new member countries, due to take place in 2004. One possible result of the enlargement is that it will put unexpected pressure on the common budgets and that such a situation will inevitably influence negatively the budget for development purposes. It will be possible to reduce the aid budgets, simply because this is not a high priority area of the EU. Another possible result of the enlargement is that there will be more financial resources available for development purposes, simply because the new member countries are committed to contributing to the EU's budget. According to calculations, 4.68 per cent of the ten accession states' contribution to the EU is to be allocated to development [*Morrissey, 2003: 7*]. Of course, their budget contributions may not be very big, but according to this calculation the accession will add new resources to the existing budget for development purposes. Having said this, it is also obvious that there is a clear understanding among all 25 EU member states that it is impossible for the ten new members to reach the EU target of 0.7 per cent of GNI in accordance with the set timetable.

REFERENCES

Arts, K. and J. Byron, 1997, 'The Mid-Term Review of the Lomé IV Convention: Heralding the Future?', *Third World Quarterly*, Vol. 18, No.1.

Bossuyt, J., G. Laporte and G. Brigaldino, 1993, *European Development Policy After the Treaty of Maastricht: The Mid-Term Review of Lomé IV and the Complementarity Debate*, Maastricht: ECDPM.

Bowles, P., 1989, 'Recipient Needs and Donor Interests in the Allocation of EEC Aid to Developing Countries', *Canadian Journal of Development Studies*, Vol. X, No. 1.

Box, L. and P. Goodison, 1994, 'Coordinate or be Coordinated: Europe–ACP Development Policies from a Maastricht Perspective', in Rhi-Sausi, J.L. and Dassù, M. (eds).

Bretherton, C. and J. Vogler, 1999, *The European Union as a Global Actor*, London and New York: Routledge.

Bulmer, S., 1994, 'The Governance of the European Union: A New Institutionalist Approach', *Journal of Public Policy*, Vol. 13, No. 4.

Cafruny, A. and P. Peters, 1998a, 'EU Foreign Policy: From Maastricht to Amsterdam', in Cafruny and Peters (eds).

Cafruny, A. and P. Peters, 1998b, 'Towards a European Foreign Policy? Evidence from the Case Studies and Implications for the Next Decade', in Cafruny and Peters (eds).

Cafruny, A. and P. Peters (eds), 1998, *The Union and the World: Economy of a Common European Foreign Policy*, The Hague: Kluwer Law International..

Cairo, 2000, *Africa–Europe Summit under the Aegis of the OAU and the EU, Cairo, 3–4 April: Cairo Plan of Action* (197/4/00 REV 4).

Cameron, F. 1998, 'The European Union as a Global Actor: Far from Pushing its Political Weight around', in Rhodes (ed.).

Carbone, M., 2002, 'A "development cluster" in the new General Affairs and External Relations Council', *The Courier ACP-EU*, No. 195, November–December.

Cardoso, F.J., W. Khüne and J.B. Honwana, 2000, *Reflection Paper. Priorities in EU Development Cooperation in Africa: Beyond 2000*, Brussels: Council of Ministers (Fax as of 26 January 2000).

Commission, 1996 (European Commission, DG VIII), *Green Paper on Relations between the European Union and the ACP Countries on the Eve of the 21st Century. Challenges and Options for a New Partnership*, Brussels, 14 November.

Commission, 1999, *Evaluation of the European Aid. ACP, MED, ALA Countries and Humanitarian Aid. Final Draft of the Synthesis Report*, 30 April.
Commission, 2000, *Communication to the Council on the Reform of the Management of External Assistance*, Brussels, 16 May.
Commission, 2002, *Commission Staff Working Paper. The European Community's Development Policy. Programme of Action 2002*, Brussels, SEC (2002) 568, 14 May.
Council, 1995, *The Conclusions of the Presidency. The European Summit*, Madrid, 15–16 December.
Council, 1998, *Directives for the Negotiations of a Development Partnership Agreement with the ACP Countries*, Brussels, 30 June.
Council of Ministers, 1991, *Communication to the Press*, Res. No. 10107, Brussels, 28 November.
Cox, A. and A. Koning, 1997, *Understanding European Community Aid. Aid Policies, Management and Distribution Explained*, London: Overseas Development Institute.
Crawford, G., 1996, 'Whither Lomé? The mid-term review and the decline of partnership', *Journal of Modern African Studies*, Vol. 34, No. 3.
Crawford, G., 2001, *Foreign Aid and Political Reform. A Comparative Analysis of Democracy Assistance and Political Conditionality*, Basingstoke: Palgrave.
ECDPM/10/99, 1999, *Lomé 2000. Debating Future Cooperation between Europe, Africa, the Caribbean and the Pacific*, Maastricht: ECDPM.
ECHO, 2003, *Responding to New Needs. ECHO 2002*, Luxemburg: ECHO.
Engel, U. and A. Mehler (eds), 1998, *Gewaltsame Konflikte und ihre Prävention in Afrika*, Hamburg: Institut für Afrika-Kunde.
Forster, J. and O. Stokke (eds), 1999, *Policy Coherence in Development Co-operation*, London: Frank Cass.
Grilli, E., 1994, *The European Community and the Developing Countries*, Cambridge: Cambridge University Press.
Grilli, E. and M. Riess, 1992, 'EC Aid to Associated Countries: Distribution and Determinants', *Weltwirtschaftliches Archiv*, Vol.128, No.2.
Hill, C., 1993, 'The Capability–Expectations Gap, or Conceptualizing Europe's International Role', *Journal of Common Market Studies*, Vol. 31, No. 3.
Hoebink, P., 1999, 'Coherence and Development Policy: The Case of the European Union', in Forster and Stokke (eds).
Holland, M., 1997, 'The Joint Action on South Africa: A Successful Experiment?', in Holland (ed.).
Holland, M., 2002, *The European Union and the Third World*, Basingstoke: Palgrave.
Holland, M. (ed.), 1997, *Common Foreign and Security Policy. Record and Reforms*, London: Pinter.
IP/02/871, 2002, *EU Adopts New Strategy for Trade Relations with African, Caribbean and Pacific Countries*, Brussels, 17 June.
Landgraf, M., 1998, 'Peace-building and Conflict Prevention in Africa: A View from the European Commission', in Engel and Mehler (eds).
Lister, M., 1988, *The European Community and the Developing World*, Aldershot: Avebury.
Lister, M., 1997, *The European Union and the South. Relations with Developing Countries*, London: Routledge.
Lister, M., 1998, 'The European Union's Green Paper on Relations with African, Caribbean and Pacific Countries', *Oxford Development Studies*, Vol. 26, No. 3.
Lister, M. (ed.), 1998, *European Union Development Policy*, Basingstoke: Macmillian Press.
Lund, M. and A. Mehler, 1999, *Peace-building & Conflict Prevention in Developing Countries: A Practical Guide*, Brussels and Ebenhausen: Stiftung Wissenshaft und Politik.

Marcussen, M. and K. Ronit, 2003, 'Internationalisering af den offentlige forvaltning i Danmark: strukturer, opgaver og relationer', in Marcussen and Ronit (eds).
Marcussen, M. and K. Ronit (eds), 2003, *Internationaliseringen af den offentlige forvaltning i Danmark. Forandring og Kontinuitet*, Århus: Århus Universitetsforlag.
McKinlay, R.D. and R. Little, 1979, 'The US Aid Relationship: A Test of the Recipient Need and the Donor Interest Models', *Political Studies*, Vol. XXVII, No. 2.
McQueen, M., 1998, 'ACP–EU Trade Cooperation after 2000: An Assessment of Reciprocal Trade Preferences', *Journal of Modern African Studies*, Vol. 36, No. 4.
Ministry of Foreign Affairs, 1996, *Strategies for Individual Organizations. Annex to the Plan of Action for Active Multilateralism*, Copenhagen: Ministry of Foreign Affairs.
Moravscik, A., 1993, 'Preference and Power in the European Community: A Liberal Intergovernmentalist Approach', *Journal of Common Market Studies*, Vol. 31, No. 2.
Moravscik, A., 1995, 'Liberal Intergovernmentalism and Integration: A Rejoinder', *Journal of Common Market Studies*, Vol. 33, No. 4.
Morrissey, D., 2003, 'Development Finds its Feet in the new Council Configuration', *The Courier ACP-EU*, No. 198, May–June.
Nielson, Poul, 2003, '2003: The Year Ahead in EU Development Policy', Speech /03/13, Brussels, 21 January.
ODI, 1992, *Aid and Political Reform*. Briefing Paper, London: Overseas Development Institute.
OECD, 1997, *Efforts and Policies of the Members of the Development Assistance Committee. Development Cooperation. 1996 Report*, Paris: OECD/DAC.
OECD, 2003, *The DAC Journal. Development Co-operation. 2002 Report*, Vol. 4, No.1, Paris: OECD/DAC.
Olsen, G. Rye, 1997, 'Western Europe's Relations with Africa since the End of the Cold War', *Journal of Modern African Studies*, Vol. 35, No. 2.
Olsen, G. Rye, 1998, 'Bureaucratic Interests and European Aid to Sub-Saharan Africa', in Lister (ed.).
Oyewumi, A., 1991, 'The Lomé Convention: From Partnership to Paternalism', *The Round Table*, No. 318.
Patten, C. and P. Nielson, 2000, 'Statement by Commissioners Patten and Nielson on the Management of EU External Assistance Programmes', IP/00/917, Brussels, 7 August.
Peters, G., 1992, 'Bureaucratic Politics and the Institutions of the European Community', in Sbrabia (ed.).
Piening, C., 1997, *Global Europe. The European Union in World Affairs*, Boulder, CO: Lynne Rienner.
Ravenhill, J., 1985, *Collective Clientelism. The Lomé Conventions and North–South Relations*, New York: Columbia University Press.
Rhi-Sausi, J.L. and M. Dassù (eds), 1994, *Coordinating the Development Aid Policies of European Countries*, Rome: Centro Studi di Political Internationalizionale, n.p.
Rhodes, C., 1998, 'Introduction: The Identity of the European Union in International Affairs', in Rhodes (ed.).
Rhodes, C. (ed.), 1998, *The European Union in the World Community*, Boulder, CO: Lynne Rienner.
Riddell, R., 1993, 'European Aid to Sub-Saharan Africa. Performance in the 1980s and Future Prospects', *The European Journal of Development Research*, Vol. 4, No.1.
Sbrabia, A.B. (ed.), 1992, *Euro-politics. Institutions and Policymaking in the 'New' European Community*, Washington, DC: The Brookings Institution.
Stokke, O. (ed.), 1995, *Aid and Political Conditionality*, London: Frank Cass.
Sørensen, G., 2002, *Changes in Statehood. The Transformation of International Relations*, Basingstoke: Palgrave.

Watts, P., 1998, 'Losing Lomé: The Potential Impact of the Commission Guidelines on the ACP Non-Least Developed Countries', *Review of African Political Economy*, No.75.
Youngs, R., 2003, 'European Approaches to Democracy Assistance: Learning the Right Lessons?', *Third World Quarterly*, Vol. 24, No. 1.
ZEF/GEMDEV/ECDPM, 1999, *Looking Beyond Lomé IV: Towards Practice-Oriented Policies*, Maastricht: ECDPM.

Journals and newspapers

Development Today, Vollen, Norway.
Financial Times, London.
West Africa, London.

Glossary

LIST OF SELECTED ACRONYMS AND ABBREVIATIONS

ABIA	Advisory Board of Ireland Aid
ABM	Anti-ballistic Missile
ABP	Area-based programme
ACDC	Advisory Council on Development Co-operation (Ireland)
ACP	African, Caribbean and Pacific countries
ACTED	Agence d'aide à la coopération technique et au développement (France)
ADA	Austrian Development Agency
AECI	Spanish Agency for International Co-operation
AFD	French Development Agency
AFDI	Agriculteurs français et développement international (France)
AFVP	Association française des volontaires du progrès (France)
AGEH	Arbeitsgemeinschaft für Entwicklungshilfe (Association for Development Co-operation – Personnel Agency of the German Catholics for International Co-operation)
AIDCO	EuropeAid Co-operation Office (EU)
AIDS	Acquired immune deficiency syndrome
AIF	Agence intergouvernementale de la francophonie (Intergovernmental agency for Francophony)
ANC	African National Congress (South Africa)
ANCI	Associazione Nazionale dei Comuni Italiani (Association of Italian Municipalities)
APPI	UNDP's Anti-poverty Partnership Initiatives
APSO	Agency for Personal Service Overseas (Ireland)
ASEM	Asia–Europe Meeting
ASTM	Action Solidarité Tiers Monde (Solidarity Action Third World) (Luxembourg)
ATP	Aid and Trade Provision (UK)
ATTAC	Association pour la taxation des transactions financières
AvH	Alexander-von-Humboldt-Stiftung (Alexander von Humboldt Foundation) (Germany)

BADC	Belgian Administration for Development Co-operation
BAe	British Aerospace
BITS	Swedish Agency for International Technical and Economic Co-operation
BMaA	Bundesministerium für auswärtige Angelegenheiten (Ministry of Foreign Affairs) (Austria)
BMZ	Bundesministerium für wirtschaftliche Zusammenarbeit und Entwicklung (Federal Ministry for Economic Co-operation and Development) (Germany)
BSF	Belgian Survival Fund
BTC	Belgian Technical Co-operation
BWI	Bretton Woods Institutions (World Bank, IMF)
CAP	Common Agricultural Policy (EU)
C2D	Debt Reduction-development Contracts (France)
CBI	Centrum voor de Bevordering van Importen uit Ontwikkelingslanden (Centre for the Promotion of Imports from Developing Countries) (Netherlands)
CCFD	Comité catholique contre la faim et pour le développement (France)
CDA	Christen Democratisch Appél (Christian-Democratic Party, Netherlands)
CDD	Debt Reduction-development Contracts (France)
CDF	Comprehensive Development Framework
CDG	Carl-Duisberg-Gesellschaft (Carl Duisberg Society) (Germany)
CDU	Christlich-Demokratische Union (Christian Democratic Union) (Germany)
CEEC	Central and East European Countries
CeSPI	Centro Studi di Politica Internazionale (Centre for Studies in International Policies) (Italy)
CFA	Communanté Financière Africaine (African Financial Community – Franc zone)
CFSI	Comité français pour la solidarité internationale (France)
CFSP	Common Foreign and Security Policy (EU)
CGIAR	Consultative Group on International Agricultural Research

GLOSSARY

CIC	Center on International Cooperation
CICI	Inter-ministerial Commission for International Co-operation (Spain)
CICID	Interministerial Committee for International Co-operation and Development (France)
CICS	Comitato Interministeriale per la Cooperazione allo Sviluppo (Inter-ministerial Committee for Development Co-operation)
CIFAD	Inter-ministerial Commission on Development Fund (Spain)
CIM	Centrum für internationale Migration und Entwicklung (Centre for International Migration and Development) (Germany)
CIPE	Comitato Interministeriale per la Programmazione Economica (Inter-ministerial Committee for Economic Planning)
CIS	Commonwealth of Independent States
CIUF	Conseil Interuniversitaire de la Communauté Française (Inter-university Council of the French Community)
CM	Micro-credit Concession Fund (Spain)
D66	Democraten '66 (Liberal-democrats Party, Netherlands)
DAAD	Deutscher Akademischer Austauschdienst (German Academic Exchange Service)
Danida	Danish International Development Agency (MFA)
DCI	Development Cooperation Ireland (MFA) (since 2003)
DDC	Department of Development Co-operation (Austria)
DD1	(UN) First Development Decade (1960s)
DDPS	Département fédéral de la défense, de la protection de la population et des sports (Federal Department of Defence, Civil Protection and Sports) (Switzerland)
DDR	Deutsche Demokratische Republik (German Democratic Republic)
DEA	Département fédéral de l'économie – DFF (Federal Department of Economic Affairs) (Switzerland)

DED	Deutscher Entwicklungsdienst (German Development Service)
DEG	Deutsche Investitions- und Entwicklungsgesellschaft (German Investment and Development Corporation)
DETEC	Département fédéral de l'environnement, des transports, de l'énergie et de la communication (Federal Department of Environment, Transport, Energy and Communications) (Switzerland)
DF	Department of Finance
DFA	Département fédéral des affaires étrangères – DFAE (Federal Department of Foreign Affairs) (Switzerland)
DFA	Department of Foreign Affairs (Ireland)
DFG	Deutsche Forschungsgemeinschaft (central public funding organisation for academic research in Germany)
DFID	Department for International Development (UK)
DGBR	Directorate General Bilateral Relations (Netherlands)
DGCID	Directorate General for International Co-operation and Development (France)
DGCS	Direzione Generale per la Cooperazione allo Sviluppo (Directorate General for Development Co-operation) (Italy)
DGDC	Directorate General for Development Co-operation (Belgium)
DGIS	Directoraat-Generaal Internationale Samenwerking (Directorate General International Co-operation – part of the Ministry of Foreign Affairs, Netherlands)
DIPCO	Dipartimento per la cooperazione (Development Co-operation Department)
DOV	Duurzame Ontwikkelingsverdragen (Sustainable Development Treaties) (Netherlands)
DPI	(UN) Department of Public Information
DRC	Democratic Republic of the Congo
DSE	Deutsche Stiftung für internationale Entwicklung (German Foundation for International Development)
DTI	Department of Trade and Industry (UK)
DÜ	Dienste in Übersee (Service Overseas) (Germany)

EADI	European Association of Development Research and Training Institutes
EBRD	European Bank for Reconstruction and Development
EC	European Commission
ECOSOC	Economic and Social Council (UN)
ECOWAS	Economic Community of West African States
EDF	European Development Fund
EEC	European Economic Community
EED	Evangelischer Entwicklungsdienst (Church Development Service, a development association of the Protestant churches in Germany)
EPAs	Economic Partnership Agreements (EU–ACP)
EPTA	Expanded Programme of Technical Assistance (UN)
ERP	European Recovery Programme
ESAF	Enhanced Structural Adjustment Facility
EU	European Union
EZE	Evangelische Zentralstelle für Entwicklungshilfe (Protestant Association for Co-operation in Development) (Germany)
FAC	Fonds d'aide et de coopération (France)
FAD	Development Aid Fund (Spain)
FAI	Fondo Aiuti Italiani (Italian Aid Fund)
FAIMO	Frentes de Alta Intensidade de Mão de Obra (High-intensive Labour Systems) (Luxembourg)
FAO	Food and Agricultural Organisation
FATF	Financial Action Task Force on Money Laundering (France)
FCD	Fonds de Coopération au Développement (Development Co-operation Fund) (Luxembourg)
FCM	Micro-credit Concession Fund (Spain)
FDI	Foreign direct investment
FES	Friedrich-Ebert-Stiftung (Friedrich Ebert Foundation) (Germany)
FEV	Fund for Technical Assistance (Spain)
FINIDA	Finnish International Development Agency (MFA)
FMO	Financierings Maatschappij voor Ontwikkelingslanden (Financing Company for Development

	Countries, Netherlands)
FNS	Friedrich-Naumann-Stiftung (Friedrich Naumann Foundation) (Germany)
FPÖ	Freiheitliche Partei Österreich (Freedom Party of Austria)
Frp	Fremskrittspartiet (Progress Party) (Norway)
FSP	Priority Solidarity Fund (France)
GAERC	General Affairs and External Relations Council (EU)
GATT	General Agreement on Tariffs and Trade
GAVI	Global Alliance for Vaccination Initiative
GDP	Gross Domestic Product
GEF	Global Environmental Facility
GFCD	Geneva Federation for Co-operation and Development (Switzerland)
GGO	Gemeinsame Geschäftsordnung der Bundesministerien (joint standing orders of the federal ministries)
GNI	Gross National Income
GNP	Gross National Product
GTZ	Gesellschaft für Technische Zusammenarbeit (Company for Technical Cooperation) (Luxembourg)
GTZ	Deutsche Gesellschaft für Technische Zusammenarbeid (German Agency for Technical Co-operation)
H&E	Humanitarian and Emergency Aid
HBS	Heinrich-Böll-Stiftung (Heinrich Böll Foundation) (Germany)
HCCI	High Council for International Co-operation (France)
HCR	High Commissioner for Refugees
HDI	Human Development Index
HIC	High-income country
HIPC	Heavily Indebted Poor Country
HIPC Initiative	Heavily Indebted Poor Countries Initiative
HIV	Human immunodeficiency virus
HMT	Her Majesty's Treasury (UK)
IARC	Ireland Aid Review Committee

GLOSSARY

IBRD	International Bank for Reconstruction and Development (window of World Bank that lends at near-commercial rates of interest)
ICRC	International Committee of the Red Cross
IDA	International Development Association (World Bank)
IDARio	Interdepartmental Committee for Rio Follow-up (Switzerland)
IDC	International Development Committee
IDGs	International development goals
IDS	Institute of Development Studies (University of Sussex, Brighton)
IFAD	International Fund for Agricultural Development
IFDA	International Foundation for Development Alternatives
IFI	International Financial Institute
ILO	International Labour Organisation
IMF	International Monetary Fund
INGO	International non-governmental organisation
IOB	Inspectie Ontwikkelingssamenwerking en Beleidsevaluatie (new name of the evaluation unit of the Ministry of Foreign Affairs, Netherlands)
IOV	Inspectie Ontwikkelingssamenwerking te Velde (old name of the evaluation unit of the Ministry of Foreign Affairs, Netherlands)
IUCN	International Union for the Conservation of Nature
IUED	Graduate Institute of Development Studies (Geneva)
KAS	Konrad-Adenauer-Stiftung (Konrad Adenauer Foundation) (Germany)
KFOR	International Security Force in Kosovo
KFS	Kofinanzierungsstelle (Austria)
KfW	Kreditanstalt für Wiederaufbau (Bank for Reconstruction and Development) (Germany)
KrF	Kristelig Folkeparti (Christian People Party/Christian Democrats) (Norway)
KZE	Katholische Zentralstelle für Entwicklungshilfe (Catholic Central Agency for Development) (Germany)

LDCs	Less developed countries
LIC	Low-income country
LLDC	Least developed countries (List of LDCs)
LMIC	Lower middle-income country
LPF	Lijst Pim Fortuyn (Conservative populist party, Netherlands)
LUF	Luxembourg's Franc
MADCT	More Advanced Developing Countries and Territories
MAI	Multilateral Agreement on Investment (UK)
MAPS	Multi-annual programme scheme
MDC	Ministry of Development Co-operation
MDGs	Millennium Development Goals (UN)
MDM	Médecins du monde (France)
MENA	Middle East and North Africa (Italy)
MERCOSUR	Southern Common Market (Lat. Am.)
MF	Ministry of Finance
MFA	Ministry for Foreign Affairs (Finland)
MFA	Ministry of Foreign Affairs
MFO	Medefinancieringsorganisatie (Co-financing organisation, Netherlands)
MFP	Medefinancieringsprogramma (Co-financing Programme) (Netherlands)
MIC	Middle-income country
MILIEV	Industry and Environment programme (Netherlands)
MKB	Midden- en Kleinbedrijf (small and medium enterprises, Netherlands)
MSF	Médecins sans frontières (France)
NAR	Nationale Adviesraad voor Ontwikkelingssamenwerking (National Advisory Council for Development Co-operation) (Netherlands)
NATO	North Atlantic Treaty Organisation
NCDE	National Committee for Development Education (Ireland)
NCDO	Nationale Commissie Duurzame Ontwikkeling (National Commission Sustainable Development, Netherlands)
NCM	Nederlandse Credietverzekerings Maatschappij

GLOSSARY

	(Dutch credit insurance company)
NEPAD	New Programme for Africa's Development (Ireland)
NGO	Non-governmental organisation
NIB	Nämnden för internationellt bistånd (The Committee on International Assistance) (Sweden)
NIC	Newly industrialised country
NIEO	New International Economic Order
NIMF	Nederlands Investerings Matching Fonds (Dutch Investment Matching Fund)
NIS	New Independent States
NOK	Norwegian kroner (currency)
NORAD	Norwegian Agency for International Development/now Norwegian Agency for Development Co-operation
NORFUND	Statens investeringsfond for næringsvirksomhet i utviklingsland (Norway's Fund for Investments in Economic Development in Developing Countries)
NOU	Norges offentlige utredninger (Norway's Official Reviews)
NUPI	Norsk Utenrikspolitisk Institutt (Norwegian Institute of International Affairs)
NVT	Nucleo di Valutazione Tecnica (Technical Evaluation Unit)
OA	Official Aid
OCHA	Office for the Co-ordination of Humanitarian Affairs (Switzerland)
ODA	Official Development Assistance
ODA	Overseas Development Administration (UK)
ODR	Office fédéral des réfugiés (Federal Office for Refugees) (Switzerland)
OECD	Organisation for Economic Co-operation and Development
OLICs	Other LICs
OND	Office National du Ducroire (National Export Credit Insurance Agency) (Belgium)
OPE	Planning and Evaluating Office (Spain)
ORET	Ontwikkelingsrelevante Export Transacties (Development relevant Export Transactions, Netherlands)

OSCE	Organisation for Security and Co-operation in Europe
OSD	Office fédéral du développement territorial – ARE (Federal Office for Spatial Development) (Switzerland)
OTC	Technical co-operating office (Spain)
ÖFSE	Österreichische Forschungsstiftung für Entwicklungshilfe (Austrian Foundation for Development Research)
ÖRF	Österreichischer Rundfunk und Fernsehen (Austrian Broadcasting Corporation)
ÖVP	Österreichische Volkspartei (Austrian People's Party)
PACI	Annual Plan for International Co-operation (Spain)
PAICV	Partido Africano da Independencia do Cabo Verde (African Party for the Independence of Cape Verde)
PAIGC	Partido Africano da Independencia da Guiné e Cabo Verde (African Party for the Independence of Guinea Bissau and Cape Verde)
PC	Programme country
PESP	Programma Economische Samenwerkingsprojecten (Programme for Economic Co-operation Projects) (Netherlands)
PIMS	Policy Information Marker System (UK)
PM	Prime Minister
PMO	Prime Minister's Office
POM	Policy Objective Marker (UK)
PRGF	Poverty Reduction and Growth Facility
PRIO	(International) Peace Research Institute, Oslo
PRISM	A Performance Reporting and Information System for Managers (UK)
PRSP	Poverty Reduction Strategy Paper
PSA	Public Services Agreement (UK)
PSD	Private-sector Development
PSF	Pharmaciens sans frontières (France)
PSO	Programma Samenwerking Oost-Europa (Programme Cooperation Eastern Europe) (Netherlands)
PSOM	Programma Samenwerking Opkomende Markten

GLOSSARY

	(Programme Co-operation Emerging Markets, Netherlands)
PUM	Programma Uitzending Managers (Netherlands Management Co-operation Programme)
PvdA	Partij van de Arbeid (Labour Party, Netherlands)
REPAs	Regional Economic Partnership Agreements (EU)
RIA	Reception and Intergration Agency (Ireland)
RLS	Rosa-Luxemburg-Stiftung (Rosa Luxemburg Foundation) (Germany)
RPF	Rwandan Patriotic Front
S 21	Shaping the 21st Century (document adopted by the DAC members in 1996)
SAEFL	Office fédéral de l'environnement, des forêts et du paysage – OFEFP (Swiss Agency for Environment, Forests and Landscape)
SAREC	Swedish Agency for Research Co-operation
SDC	Direction du développement et de la coopération – DDC (Swiss Agency for Development and Co-operation)
SDFC	Société financière suisse pour le développement (Swiss Developement Finance Corporation)
SECIPI	State secretariat for international co-operation and Latin America (Spain)
Seco	Secrétariat d'Etat à l'économie (State Secretariat for Economic Affairs) (Switzerland)
SEQUA	Stiftung für wirtschaftliche Entwicklung und berufliche Qualifizierung (Foundation for Economic Development and Vocational Training)
SER	Sociaal-economische Raad (Social Economic Council) (Netherlands)
SES	Senior Expert Service (Germany)
SHA	Corps suisse d'aide humanitaire – CSA (Swiss Humanitarian Aid Unit)
Sida	Swedish International Development Co-operation Agency (since 1995)
SIDA	The Swedish International Development Authority (before 1995)
SIK	Sparkassenstiftung für internationale Kooperation (Savingsbanks Foundation for International

	Cooperation) (Germany)
SILICs	Severely indebted low-income countries
SME	Système Monétaire Européen (EMS)
SOFI	Swiss Organisation for Facilitating Investments (always used in English)
SP	Socialistische Partij (Socialist Party, Netherlands)
SPD	Sozialdemokratische Partei Deutschlands (Social Democratic Party of Germany)
SPÖ	Sozialdemokratische Partei Österreichs (Social Democratic Party of Austria)
SSA	Sub-Saharan Africa
Stabex	Earning-guarantee scheme (of the Lomé Convention)
SUNFED	Special United Nations Fund for Economic Development
Swaps	Sector-wide approaches
SWEDCORP	Swedish International Enterprise Development Corporation
SWEDFUND	Agency for Industrial Co-operation (Sweden)
Sysmin	Production-support scheme (of the Lomé Convention)
TA	Technical Assistance
TC	Technical Co-operation Aid
TNC	Transnational Corporation
UDC	Union Démocratique du Centre (Switzerland)
UK	United Kingdom
UMC	Upper middle-income country
UMIC	Upper middle-income country
UN	United Nations
UNAIDS	United Nations Programme on HIV/AIDS
UNCDF	United Nations Capital Development Fund
UNCED	United Nations Conference on Environment and Development (Rio de Janeiro, 1992)
UNCSD	United Nations Conference on Sustainable Development
UNCTAD	United Nations Conference on Trade and Development
UNDP	United Nations Development Programme
UNEP	United Nations Environment Programme

GLOSSARY

UNESCO	United Nations Educational, Scientific and Cultural Organisation
UNFPA	United Nations Fund for Population Activities
UNGASS	UN General Assembly Special Session
UNHCR	United Nations High Commission for Refugees
UNICEF	United Nations Children's Fund
UNITAR	United Nations Institute for Training and Research
UNOPS	UN Office for Project Services
UNWRA	United Nations Relief and Works Agency for Palestine Refugees in the Near East
UPI	Unione delle Province d'Italia (Union of Italian Provinces)
US	United States (of America)
USA	United States of America
USAID	United States Agency for International Development
USSR	Union of Soviet Socialist Republics
UTC	Unità Tecnica Centrale (Central Technical Unit)
UTLs	Unità Tecniche Locali (Local Technical Units)
VLIR	Vlaamse Interuniversitaire Raad (Flemish Interuniversity Council) (Belgium)
VVD	Volkspartij voor Democratie en Vrijheid (Conservative Liberal Party, Netherlands)
WAEMU	The West African Economic and Monetary Union
WCED	World Commission on Environment and Development
WDR	World Development Report (World Bank)
WFP	World Food Programme
WHO	World Health Organisation
WTO	World Trade Organisation
ZAV	Zentralstelle für Arbeitsvermittlung der Bundesanstalt für Arbeit (Central Placement Office of the Federal Employment Office) (Germany)
ZSP	Priority Solidarity Zone

Notes on Contributors

José Antonio Alonso (b. 1953) is Professor of Applied Economics at the Complutense University of Madrid, specialising in growth theory and development economics. He has been Director of Economic Co-operation at the Instituto de Cooperación Iberoamericana (the Spanish Co-operation Agency in the 1980s) and has been an Expert at the Development Co-operation Council. He is currently Director of the Master in Development and International Aid Programme of the Complutense University. Among his more recent publications is *Lecciones de Economía Mundial. Introducción al desarrollo y a las relaciones económicas internacionales* (Madrid, 2003).

Guido Ashoff (b. 1949) holds a Ph.D. in political science and is Head of Department at the German Development Institute (GDI) in Bonn. He has written extensively on development co-operation and carried out empirical research mainly in Latin America and North Africa. Recent research subjects include policy coherence in development co-operation, country assistance strategies as a management and evaluation instrument, and the 'Peer Reviews of the OECD's Development Assistance Committee'.

Anders Danielson (b. 1957) holds a Ph.D. in economics and is Associate Professor at the Department of Economics, University of Lund. His research focus is on issues of foreign aid, external debt and the role of the state in low-income countries. His most recent books are *Is the Ugly Duckling Growing Up?* (with Arne Bigsten) and *Towards Sustainable Development in Central America and the Caribbean* (2002, editor with Geske Dijkstra).

Jean-Jacques Gabas (b. 1952) is Professor of Economics at University of Paris XI-Orsay and at Sciences-Po, Paris. He was President of GEMDEV between 1997 and 2002, a network of 52 teams of research on globalisation and development, and a member of the High Council for International Co-operation (HCCI). Previously, he was consultant for Club du Sahel/OECD (1978–98) and technical assistant in the Ivory Coast (1976–78). He is the author of several books and articles. His recent publications are *Nord-Sud: l'impossible coopération?* (Presses de Sciences-Po, Paris, 2002) and *L'Union européenne et les pays ACP. Un espace de coopération à construire* (Karthala, Paris, 1999). Since 2003, he has been a co-convenor of the EADI WG on Aid Policy and Performance.

Paul Hoebink (b. 1949) is Associate Professor at the Centre for International Development Issues Nijmegen (CIDIN), the Catholic University of Nijmegen. His research has been mainly on development co-operation policies, aid effectiveness and evaluation. He has published widely on the aid programmes of the Netherlands, the EU and other donors and is a member of several advisory boards. His recent publications include *European Development Cooperation and the Poor* (with Aidan Cox, John Healey and Timo Voipio) and *European Aid for Poverty Reduction in Tanzania* (with Timo Voipio). Since 2003, he has been a co-convenor of the EADI WG on Aid Policy and Performance.

Nathalie Holvoet (b. 1969) is an economist by training (Ph.D.), specialising in the evaluation of development activities. She is currently a junior Lecturer at the University of Antwerp where she teaches quantitative impact assessment and a course on 'gender development'. She has done consulting work for the DGIC, the EU, UNIFEM and several NGOs.

Juhani Koponen (b. 1947) is Professor and Director of the Institute of Development Studies, University of Helsinki. He has studied the history of Tanzania and more recently worked on the past and present of Finnish aid and on development assistance more generally. He is currently a member of the Executive Committee of EADI.

Oliver Morrissey (b. 1961) is Reader in Development Economics and Director of CREDIT, School of Economics, University of Nottingham, and Research Fellow at the Overseas Development Institute, London. He has published many articles in international journals, mostly on aid policy and effectiveness, trade policy reform, conditionality and adjustment. His books include *British Aid and International Trade* (with B. Smith and E. Horesh, 1992), *Evaluating Economic Liberalisation* (edited with M. McGillivray, Basingstoke: Macmillan, 1999) and *Globalisation and Trade: Implications for Exports from Marginalised Economies* (edited with I. Filatotchev, London: Frank Cass, 2001).

NOTES ON CONTRIBUTORS

Helen O'Neill is Professor Emeritus of Economics in University College Dublin where she was founder-Director of the Centre for Development Studies. She was President of EADI from 1993 to 1999. Within Ireland she has chaired the Advisory Council on Development Co-operation, the Irish Commission for Justice and Peace and the Association of Canadian Studies in Ireland. She has been a Visiting Professor in the World Bank Institute, the University of Zambia and the Budapest University of Economic Sciences. She has carried out assignments for UNIDO and the World Bank in Africa and the trans-Caucasus region and a meta-evaluation for the European Commission in Brussels where she has also been an expert to the Economic and Social Committee. She has published widely on topics in development and international relations and given guest lectures in universities in all five continents of the world.

Michael Obrovsky (b.1957) graduated from the University of Vienna in 1983 (science of communications, philosophy and history of arts). Since 1984 he has been a staff member at the Austrian Foundation for Development Research (ÖFSE). Since 1994, he has given lectures on Austrian development policy and co-operation at the Department of Political Science of the University of Vienna.

Gorm Rye Olsen (b. 1950) is Head of the Department for European Studies at the Danish Institute for International Studies. He also teaches international politics at the University of Copenhagen. His current research is on Europe's relations with and policies towards the developing world with special focus on Africa. His most recent publications have appeared in *International Politics, Journal of Modern African Studies, Civil Wars, Global Society, Disasters, Journal of Disaster Studies* and *Policy and Management.* In 2003, he was a co-convenor of the EADI WG on Aid Policy and Performance.

Robrecht Renard (b. 1947) is Professor of Development Economics at the University of Antwerp. His interests include economic evaluation and aid policies. He is Director of an international Master's programme on 'Development Management and Evaluation' at the same university. He has worked as a consultant for bilateral and multilateral organisations and for NGOs. At the end of the 1980s he served as a senior adviser to the Belgian Minister for Development Co-operation. He was a member of the EADI Executive

Committee for several years (1984–89) and Convenor of the Working Group on Aid Policy and Performance since 2003.

José Luis Rhi-Sausi (b. 1947) is currently Director of the Centro Studi di Politica Internazionale (CeSPI) based in Rome. He graduated in economics at Universidad de Nuevo León, Monterrey, México, and obtained a Master's in social sciences at FLACSO (Facultad Latinoamericana de Ciencias Sociales), Santiago, Chile. As co-ordinator of the areas on International Co-operation and Latin American Studies, he is currently directing the following programmes: CeSPI report on Italy's Development Aid, Italian Decentralised Co-operation, Small and Medium Enterprise Promotion and Local Development in Latin America, Migration and Development Relationship. He is also a Visiting Professor in several Master's courses and he has published in the area of development co-operation.

Lau Schulpen (b. 1961) is Lecturer and Researcher at the Centre for International Development Issues Nijmegen (CIDIN) of the University of Nijmegen in the Netherlands. In 1997, he defended his Ph.D. thesis entitled 'The Same Difference: A Comparative Analysis of Dutch Aid Channels to India'. Since then, he has been involved in research on private-sector development, NGOs and the development co-operation of the Netherlands. He is a member of the Netherlands' Advisory Council on International Affairs.

Lauri Siitonen (b. 1954) is a Researcher at the Institute of Development Studies, University of Helsinki. He has published on aid policies, democracy assistance and Finland's aid policy.

Olav Stokke (b. 1934) is currently a Senior Researcher at the Norwegian Institute of International Affairs, where earlier he served as Acting Director, Deputy Director and, for years, the Research Director. Since 1974, he has been the editor of *Forum for Development Studies*. He has been a member of the EADI Executive Committee for several years (1978–81, 1984–90) and Convenor of the EADI Working Group on Aid Policy and Performance (1979–2002). He has authored and edited more than 25 books and written more than 130 articles and book chapters. The most recent books are *Policy Coherence in Development Co-operation* (ed. with Jacques Forster, 1999), and *Food Aid and Human Security* (ed. with Edward Clay, 2000).

NOTES ON CONTRIBUTORS

Lennart Wohlgemuth (b. 1940) (h.c. Ph.D.) has been Director of the Nordic Africa Institute, Uppsala, since 1993. Prior to this he worked for many years for SIDA, most recently as Assistant Director-General and Head of the Sector Department. He has written and edited numerous books and articles on aid, development and development policy. His most recent book is *Dialogue in Pursuit of Development* (ed. with Jan Olsson, 2003).

Catherine Schümperli Younossian (b. 1965) is a scientific collaborator at the Graduate Institute of Development Studies in Geneva, Switzerland, since 1996. She is one of those responsible for the *Annuaire suisse de politique de développement,* a yearbook that presents the relationships between Switzerland and developing countries. She graduated (Ph.D.) from the Graduate Institute of International Studies, Geneva. She also worked for several years in a Swiss NGO, the Berne Declaration, as lobbyist concerning development policy issues.

Marco Zupi (b. 1966) is Deputy Director of the Centro Studi di Politica Internazionale (CeSPI), where he is co-ordinator of international development and political economy studies. He lectures at the Roskilde University Centre, Denmark, and the Salerno University, Italy. He graduated in development economics (MA) from Università 'La Sapienza' in Rome and has obtained post-graduate degrees in international economics from the Università Tor Vergata in Rome, in social science data analysis econometrics from Essex University, Colchester, and in international political economy (Ph.D.) from the Roskilde University Centre, Roskilde. He has published a number of articles and books in the area of development co-operation and development economics.

Index

accountability 69, 87, 103, 153, 258, 306, 307, 311, 325, 367
accounting 69, 305
Achterhuis, H. 138
ACP countries 17–19, 247, 250, 251, 260, 290, 319, 321, 574–9, 582, 585–92, 595–7
additionality 356, 359
adjustment, structural 9, 60, 62–4, 70n50, 74, 81, 103, 163, 164, 172, 220, 244, 251, 273, 322, 457–9, 465, 524, 524, 581, 587
administration 6, 15–19 *passim*, 22, 69, 70 *see also individual countries*
Afghanistan 29, 98, 188, 355
Africa 1n1,10, 18, 45, 89, 120, 165–70, 180, 194, 209, 242, 244, 248, 249, 345, 350, 463, 529, 538, 585–96 *see also individual countries*
  East 169, 336, 388
  ECA 325
  ECOWAS 209, 250
  Genoa Plan for 340
  Great Lakes region 45, 72, 593
  Horn 45, 169–70, 336, 355, 358, 362
  NEPAD 325–6, 340
  North 34, 244, 342, 345–9, 574–6 *passim*, 592
  Sahel 390; Initiative for 348, 362
  Special Partnership with 102
  SSA 167, 169, 172, 173, 176, 180, 181, 209, 244, 247, 249, 250, 268, 288, 311, 314, 341, 342, 345, 347, 355, 502, 531, 543, 575–7, 581, 591, 594, 597
  Trade and Poverty Programme 176
  WAEMU 250
Agenda 21 79, 100
agriculture 12, 14, 88, 121, 122, 175, 176, 235, 259–61, 274, 305, 306, 313, 321, 334, 442, 457, 459, 467
Ahern, Bertie 307
Ajello, Aldo 593
Albania 51n19, 347, 355, 356, 358, 367, 501, 566
Algeria 267, 250, 501
Alonso, José Antonio 15–16, 493–517
Amin, Samir 58n30, 66
Amos, Baroness 182
Amsterdam treaty 277, 599
Ancona, C. 507
Andersen, Johannes 192, 193, 203
Andersson, Gun-Britt 528–9
Andrlik, E. 132

Angola 35, 169, 170, 249, 355, 501, 503, 542, 593
Annan, Kofi 345
Annell, L. 520
van Ardenne, Minister 406, 407, 413, 415, 418, 426, 429–30, 434, 438–40, 442
Argentina 336, 355, 356, 499, 584
van Arkadie, B. 524
Arts, K. 592
Arusha agreement 137
Ashoff, Guido 10–11, 267–302
Asia 37, 180, 194, 314, 350, 574–6, 582–4 *passim*, 597, 598 *see also individual countries*
  Central 2, 51, 172, 538
  East 167, 170–2, 175, 180
  – Europe meeting 584
  South/South-East 170, 172, 173, 180, 181, 250 388, 502, 531, 584, 585
auditing 69
Australia 23, 28n10
Austria 6–7, 20, 50n18, 52n21, 113–35, 394, 395
  ADA 123, 125–6
  administration 7, 119–24, 130
  bilateral aid 115, 116, 121–2
  and co-ordination, aid 122–3, 127, 130
  DC 7, 116, 119, 122, 129–30
  distribution, aid 115, 118; prioritisation 120–1, 124, 126
  and environment 122, 126, 127, 133
  and EU 7, 20, 115, 117, 122, 124, 127–30, 133
  European Recovery Programme 119
  evaluation 130
  legislation 124–6
  multilateral aid 115, 117, 122, 130, 131
  NGOs 120, 124–30, 133
  objectives 126–7
  and poverty reduction 126–7
  public support 131–2
  and UN 117, 118, 122
  volume of aid 23, 24, 26, 28, 114–19, 125, 132, 133; targets 117–18, 125

background, aid 33–8, 53–6
BAe 174–5
balance of payments 61, 103, 411, 431, 458
Balkans 13, 113, 118, 314, 337, 345, 347, 354, 361, 369, 370, 375, 390, 536, 547, 551, 566, 576, 581–3, 592, 594

Stability Pact 293, 354, 566
Balladur, Edouard 248
Bangladesh 3, 164, 170, 210, 422, 542, 583, 604
banks 1, 61, 81, 82
  development 324, 344, 391, 500, 554
Barcelona process 2–3, 117, 250, 583, 604
basic needs strategy 59–60, 76, 304, 317, 408, 457, 506–9 *passim*
Bauer, Peter 56n27, 61n34
Belgium 7–8, 35, 136–60, 398
  administration 7–8, 137–8, 142–3, 147–9, 152–6, 158; reform 152–6, 158
  bilateral aid 7, 141, 143–6, 149–51
  BSF 148–9
  BTC 7, 138, 143, 152–4, 156, 158
  co-gestion 142–3
  composition, aid 143–9
  and debt relief 143, 146, 154–5
  DGDC 7, 138, 143, 144, 147, 148, 152, 154, 156, 158
  distribution, aid 7, 28, 149–51, 158; prioritisation 141–2, 149–51
  evaluation 147, 148, 152, 153
  and human rights 140–2 *passim*, 154, 158
  Lambermont negotiations 155–6
  motivation 7, 139, 158
  multilateral aid 144, 145, 148, 156
  NGOs 7, 138, 146–8, 156–8
  objectives 7, 140–2
  and poverty reduction 140, 151, 154–5
  public support 137, 156–7
  regional governments 8, 138–9, 155–6, 158
  volume of aid 7, 23–7 *passim*, 136, 139, 145
Benin 358
Berlage, Lodewijk 4
Berlusconi, Silvio 345, 368
Bhutan 120, 121
bipolar system 34, 36–8 *passim*, 40, 44, 54, 95, 586
Bird, Graham 62n36
Blair, Tony 8, 161, 164
Boel, Erik 192, 206
Bolivia 143, 150, 151, 501, 503
Booth, D. 539
Borre, Ole 193
von den Bosch, F. 431
Bosnia–Herzegovina 121, 229, 347, 358, 501, 503, 531, 542, 583
Bossuyt, J. 589
Botswana 176
Bottai, B. 369
Boutmans, E. 142, 154, 156
Bowles, B. 578, 579
Box, L. 418, 590
Boyer, B. 557

Brazil 353, 356, 362, 584
Brennpunkt Drett Welt 380, 399
Bretherton, C. 577, 592
Bretton Woods Institutions 17, 63, 64, 71, 81, 101, 240, 251, 322–4, 327, 479, 551, 557–8, 587 *see also individual entries*
Britain 8, 20, 34n2, 35, 50n18, 161–83, 251, 477, 584, 586, 590
  administration 162–6
  bilateral aid 8, 167–73
  Chevening Scholarships 168
  DFID 8, 161–6, 169, 173, 175–82, 504
  distribution, aid 28, 167–73, 179–80
  and education 170, 179–81; Chevening Scholarships 168
  and effectiveness 162, 163, 166
  and EU 181
  evaluation 165; PRISM 165
  multilateral aid 167
  objectives 163, 165, 180–1
  and poverty reduction 162–4, 166, 179–81
  PSA 180–1
  and trade 175–7, 182; ATP 167n2, 169–72, 174
  volume of aid 23, 24, 26–8 *passim*, 161, 167, 169, 173, 179; targets 161, 167
  White Papers 8, 161, 162, 170–1, 173–9
Brown, Gordon 161
Brundtland Report 77–8, 458
budget support 22, 103, 143, 146, 153, 155, 172, 179, 236, 287, 312, 411–13, 428, 429, 498, 524, 525
bureaucracy 18, 191, 578, 580–2, 592, 603, 604
Burkina Faso 120, 127, 150, 247, 358, 388
Burundi 7, 120, 127, 136–7, 143, 150, 157, 245, 388, 389
Byron, J. 592

Cafruny, A. 577, 579, 592
Cambodia 247
Cameron, F. 577
Cameroon 150, 245, 247, 249n2, 358
Canada 23, 28n10
capacity-building 71, 84, 141, 176, 182, 196, 207, 231, 258, 274, 313, 540–1
Cape Verde Islands 13, 120, 378, 379, 385, 388–99, 403, 501
capital 54–5, 57, 60, 175, 177
  human 100, 393, 507
  markets 499
Carbone, M. 602
Cardoso, F.J. 594
Caribbean 170, 172, 250, 347, 538
Carlsson, J. 253, 254, 257
Cartagena Protocol 568

INDEX

Central African Republic 261
Chad 249, 261
Chalker, Lynda 173
Chenery, Hollis 59
children 340, 341, 364
Chile 155, 220, 388, 389, 499, 584
China 1n1, 29, 68n47, 170, 171, 173, 230, 234, 237, 269, 288, 336, 345, 356, 499, 501, 503, 542
Chirac, Jacques 249
Christensen, J. Gronnegaard 186
Christiansen, P. Munk 186
Church, Catholic 338, 383
CIDIN 5, 30
CIS 51, 548, 554
civil society 60, 65, 93, 103, 122, 237, 258, 260, 337–8, 358, 408, 466, 589
Clay, Edward 4, 35
clientelism, collective 577–8
co-financing 138, 146–8, 315–16, 380, 386, 398, 413, 418, 434–40 *passim*, 511
Cohen, M.D. 39
coherence, policy 14, 20, 21, 84–8, 91, 93, 94, 209, 218, 277–80, 320, 455, 465–6, 479, 481–2, 514, 523, 526, 530, 547, 570 *see also* EU; Netherlands
Collier, P. 504–5
Colombia 503
Colombo, Foreign Minister 361
Colombo Plan 36n4
colonialism 9, 34–7, 242, 249, 392, 584
commodity aid 286–7, 361
common goods 40, 42, 53, 95, 97, 257, 261, 450, 453, 485
Commonwealth Development Corporation 168
Comoros 249, 388
competition 177, 229, 364
concentration, aid 151, 173, 194, 233, 244, 350, 388–9, 400, 407, 413, 421–8, 463, 469–73, 502, 538, 541
conditionality 3, 29, 63–4, 69–71, 73, 81, 84, 103, 166, 173, 180, 217, 221, 232, 322, 409. 464–5, 486, 588–9
  political 4, 66–7, 69, 70, 102, 141, 142, 173, 273–4, 465, 473, 524, 588–9, 591, 596
conflicts 44–5, 51, 73–4, 199, 271, 274, 410, 455, 593, 594, 604
  East–West 44, 48, 270
  prevention 73, 127, 142, 154, 169, 188, 209, 271, 304, 319, 338, 456, 461, 577, 593, 594, 604
  resolution 15, 19, 35, 72–4, 91, 271, 319, 326, 472, 599
Congo, 35, 45, 245, 359
  Democratic Republic 7, 136, 141, 150, 157, 245, 327, 593

consistency 599
consultative groups 94n77, 525
convergence 91, 99–101, 448, 494, 518, 520
co-ordination, aid 2, 15, 20–1, 72, 93, 94n77, 166, 198, 199, 299, 321, 397, 403, 415, 429, 484–7, 512, 514, 525 *see also* Austria: EU; Switzerland
Cornia, Giovanni Andrea 64
corruption 10, 12, 83, 220, 232, 234, 246, 260, 278, 307, 311, 336, 368, 380, 460–1, 596
Costa Rica 120
Cotonou agreement 17, 246, 247, 250, 251, 260, 261, 319, 321, 529, 574, 577, 595–7
country programming 149–51, 194–5, 291, 311–14, 388–9, 403, 422, 444, 471–2, 481, 523, 542, 601
Cox, Adrian 5n3, 589
Crawford, G. 589, 592
credits 116–18, 121, 146, 174, 221, 229, 230, 232, 237, 345, 349, 352, 354, 356, 362, 365, 569 *see also* Finland; Italy; Spain
  Helsinki agreement 229, 269, 499
  Revolving Fund 354, 360
Cuba 49, 247, 422, 520
Cumming, G. 169, 173
culture 15–16, 37, 40, 43, 45, 142, 515–17 *passim*

Dahl, Robert 65n41
Dalmeijer, R. 432
Damgaard, Erik 195
Dauge, Y. 261
Davidson, Andrew 16, 518–45
De Michelis, Foreign Minister 351
De Rita, G. 369
debt 61, 80–2, 117, 199, 220, 305, 311, 322–3, 338, 354, 441, 455, 457, 459
  Baker/Brady plans 82n61
  cancellation 12, 82, 117, 245, 323, 495, 522, 561 *see also* Italy
  refinancing 245
  relief 6–7, 12, 52n21, 132, 166, 181, 227, 244–5, 411, 421, 479, 510, 526, 536, 549, 558, 561 *see also* Belgium; Ireland; Italy
  rescheduling 52, 82, 356, 357, 359
  servicing 61, 63, 81, 118, 323, 359
  swaps 349
decentralisation 13, 18–19, 22, 257–8, 296, 339, 341, 368–70, 410, 411, 424, 444, 600–1
decolonisation 34–5, 242, 585
deficits, budget 139, 225, 282, 395
Degnbol-Martinussen, John 196, 207
Dehaene, J-L. 152
Delcourt, C. 380
democracy/democratisation 15, 36, 65, 66, 70, 87–9, 102, 103, 122, 141, 196, 203, 246,

629

260, 275, 340, 393, 410, 429, 550 *see also* EU; Finland; Ireland; Norway; Sweden
Denmark 8, 20, 28, 29, 50n18, 51n20, 184–214, 586
  administration 190–1, 196–9, 201–3
  armed forces 188
  bilateral aid 189, 194, 197–8, 210
  business interests 206–7
  community, aid 8, 185–6, 190–2, 202–7, 210, 211
  Danida 186, 187, 195, 197, 199–202, 205–7, 526, 540
  distribution, aid 189, 194–7, 199
  and environment 195, 196, 203, 204
  and EU 189, 208–10
  evaluation 197–9
  foreign policy 187–8, 191, 199, 208
  and human rights 186, 196, 203, 204
  motives 186, 187, 190
  multilateral aid/multilateralism 188, 189, 196, 198–9, 201, 210
  NGOs 204–6, 211
  objectives 8, 193–4, 196, 199, 206
  and poverty reduction 196, 209
  public support 192–3
  and UN 189, 199, 209
  volume of aid 8, 23, 26–8, 48, 184–8, 191, 195, 210–11; targets 184
dependence, aid 59, 323, 393, 453, 527
dependencia school 58, 59
Derycke, E. 141
devaluation, CFA franc 9, 244, 257
development
  decades 53, 56, 74, 92
  education 192, 310, 315, 329–32, 380, 558–9
  human 66, 75–6, 100, 102, 247, 461
  international regime 91–4
  paradigms 5, 32, 33, 38–43, 54–71, 94, 99–101, 103, 456–7
  social dimensions of 58–9, 64, 74–6, 103
  sustainable 17, 78, 87, 199, 304, 409, 418, 424, 438, 452, 455, 458, 461
  UN Declaration on Right to 67n45
Development Today 190
dialogue, policy 19, 97, 218, 231, 260, 295, 296, 316, 524–5, 529, 530, 587, 596
Diamond, Larry 65
van Dijk, M.P. 433
diplomacy 7, 73, 137,158, 201–2, 345
distribution, aid 1n1, 2, 6, 28–9, 50–3, 59, 73, 76, 78, 98 *see also individual countries*
Djibouti 249, 350
Dollar, D. 504
Dominican Republic 501, 503
Doornbos, Martin 71
Downs, Anthony 186

Due-Nielsen, Carsten 188

EADI Working Group 3, 3–5
East Timor 172, 229, 314
Ecuador 349, 388, 389, 501
Ederer, Brigitte 123, 124
education 11, 13, 60, 64, 75, 87, 90, 100, 121, 122, 232, 235, 245, 257, 274, 286, 415, 420, 599 *see also* Britain; Ireland; Italy; Luxembourg; Norway; Spain
effectiveness, aid 8, 9, 18, 21–2, 32, 86, 277, 290, 306, 321, 368, 395, 423, 504–5, 527, 537 *see also* Britain; EU; Finland; Norway
Egger, M. and J-P. 562
Eglinton, Stephanie J. 40
Egypt 121, 169, 232, 247, 269, 288, 349, 424
Ehrenpreis, Dag 522–3
El Salvador 120, 355, 388, 501, 503
Elias, Norbert 239
Elleman-Jensen, Uffe 201
emergency aid 72, 310, 352, 355, 361, 374, 410, 467, 495, 509, 531, 536, 541
employment 103, 270, 306, 431, 433, 478, 535, 554
energy 46, 61, 121–2, 220, 269, 457, 479, 505
Engberg-Petersen, Poul 61n34, 196, 201, 207
environment 17, 53, 77–9, 87, 89, 100, 133, 142, 217, 228, 232, 235, 271, 274, 334, 400, 528, 599 *see also* Austria; Denmark; Netherlands; Norway; Spain; Switzerland
Global Facility 51n20, 78, 79, 101, 148, 227, 319, 568
Johannesburg summit 78n58, 79, 92, 97, 303, 308
Rio Conference 78, 79, 100, 127, 195, 382, 551
Stockholm Conference 77
Equatorial Guinea 388, 501, 502
Eritrea 188, 210, 247, 314, 355, 356, 358, 472
Err, Lydie 384, 388, 401
Ethiopia 120, 170, 232, 247, 305, 312, 313, 322, 323, 353, 355, 358, 367, 472, 542
Europe, Central and East 20, 49–52, 54, 73, 118, 121, 122, 124, 181, 243, 250, 280, 282, 345, 347, 350, 378, 403, 467, 502, 538, 542, 548, 549, 551, 554, 574, 583
European Bank for Reconstruction and Development 324, 391
European Union 9, 11, 16–22, 28, 50n18, 88, 95, 97, 573–608 *see also individual countries*
and ACP 17–19, 250, 251, 290, 319, 321, 574–8, 582, 585–92, 595–7 *passim*

INDEX

administration 578–81, 598–9; reform 18–19, 528, 577, 597–603
and Africa 18, 585–96
AIDCO 18, 600, 601
and Asia 574–6, 582–4, 597, 598
associationism 585–7
CAP 260–1, 327, 334, 401n55, 440, 442
and coherence/co-ordination 573, 577, 599–600
common currency 208, 282
common foreign and security policy 6, 18, 19, 21, 43, 97, 208, 374, 573–4, 580, 582, 592–4, 602–4
and democracy 573, 576, 577, 588–9, 591, 598
distribution, aid 574–6, 582–5, 589, 604; priorities 576, 579
ECHO 575, 597
EDF 129, 148, 227, 260, 318–19, 497, 574, 600
and effectiveness 589–91, 598, 601
enlargement 20, 21, 604–5
and Europe 575, 576, 581–4, 594, 604
evaluation 18, 598
GAERC 19, 320, 602
and human rights 573, 577, 588–8, 591
joint actions 592–4
and Latin America 20, 574–6, 582–5, 597, 598
and Mediterranean 574, 575, 582–5, 591
and Middle East 574, 575, 592, 597, 598
motives 578, 579
neighbourhood policy 2, 13, 352, 370–2 *passim*, 376
objectives 573
and poverty reduction 573, 598, 600
summits, Barcelona Euro-Mediterranean 2–3, 583, 591, 602; Cannes 590, 591; Essen 593; Euro-Africa 594; Madrid 593; Seville 602
and trade 579, 584–7 *passim*, 595, 597
volume of aid 17, 24, 26–7, 117, 574–5, 588, 590, 596, 602, 604; targets 602, 605
experts 144, 259–60, 412
local 22, 228, 412
exports 176, 177, 229, 260, 269, 270, 278, 280, 323, 380, 383, 385, 457, 476, 478, 586
credits 115–18 *passim*, 121, 146
promotion 269, 306, 399, 431, 432, 457, 494

famine 71, 336
FAO 149, 199, 320, 419
fatigue, aid 114, 351, 497, 522
Ferrero-Waldner, Benita 117, 124

Finland 9, 20, 51n20, 215–41
administration 236–7
bilateral aid 226, 227, 231–2
and credits 221, 229, 230, 232, 237
and democracy 217, 219–21 *passim*, 228, 235
and developmentalism 215–16, 220–2, 228, 231, 233, 238–40
distribution, aid 226, 227, 229–35
and effectiveness 215, 227, 230–4, 238, 240
and EU 9, 20, 225, 227, 229, 239
evaluation 230–1, 233
and human rights 217, 219–21 *passim*, 235
and identity 225, 239–40
motives 9
multilateral aid 226, 227
NGOs 226, 237
objectives 216–22, 231
and poverty reduction 9, 216, 217, 221, 222, 231, 232, 234
public support 226
and trade 218, 219, 228, 229, 231, 235
and UN 227, 238, 240
volume of aid 9, 20, 23, 27, 215, 220, 222–7
targets 215, 225, 226, 238
fisheries 88, 321, 457
food aid 4, 142, 260, 286, 319, 341, 355, 368, 390, 392, 394–5, 599
London Convention 341
food for work 392, 397
foreign exchange 82, 428
foreign policy 15, 18, 29, 33, 36–53 *passim*, 269, 337, 339, 361, 402, 502, 509, 515, 566
see also Denmark; EU; Ireland; Netherlands; Norway
forestry 120, 127, 235, 457
Forsse, Anders 523, 543
Forster, Jacques 4, 21, 38, 85, 87, 465, 554, 556, 560, 566
France 9–10, 22, 34–5, 50n18, 98, 242–66, 584, 586
administration 236, 252–5; reform 9, 10, 246, 252–3
and Africa 10, 242, 244, 248–51, 253
bilateral aid 10, 246–52
CICID 245, 246, 252
distribution, aid 10, 246–52; *pays du champ* 246–7, 252, 259; ZSP 10, 246–8, 261
DOM–TOM co-operation 246
and EU 10, 260–1
evaluation 258–9, 261, 265
francophony 248, 250–1, 253, 261
HCCI 253–5
and Mediterranean 248, 250, 251

631

multilateral aid 243
NGOs 256
objectives 257–62
and poverty reduction 10, 257–9
volume of aid 9–10, 23–7 passim, 242–6
Freud, C. 259
Freymond, J. 557, 564
Frühling, P. 539, 543
fungibility 527

G-8 337, 361
G-24 50n18
Gabas, Jean-Jacques 9–10, 242–66
Gabon 249
Gandhi, Mahatma 34
gas 46, 61
de Gaulle, General 243, 586
gender issues 71, 79–80, 87, 90, 122, 126, 127, 133, 219, 275, 304, 306, 311, 334, 341, 400, 410, 455, 462, 505, 507, 520, 528, 599
geopolitics 9, 13, 37, 272, 273, 355, 371, 374, 473, 519
Germany 10–11, 34, 50n18, 54n22, 98, 251, 267–302, 394, 590
  administration 10, 277–80, 283, 293–7, 299; reform 10, 11, 267, 296–9
  bilateral aid 283–8, 291–4
  BMZ 11, 268, 270–2, 278, 279, 293–5, 299
  Bundesländer 283
  distribution, aid 269, 282, 284, 286–8, 291–2; prioritisation 291–2
  and EU 290, 283, 293
  GTZ 284, 285, 291, 294–6, 398
  KfW 284, 285, 291, 294–6
  motives 10, 267–71
  multilateral aid 283, 288–90, 293
  NGOs 268, 272, 277–80, 284, 285, 297
  objectives 11, 268, 273, 274
  political conditionality 273–4
  political foundations 286, 294, 296
  and poverty reduction 11, 274–7; Programme of Action 2015 11, 273, 275–7, 299
  public support 297–8
  reunification 10, 281, 298
  and UN 289, 290, 293
  volume of aid 10, 11, 23–6 passim, 28, 267, 277, 280–3, 299
Germany, East 269, 281
Ghana 165, 169, 170, 176, 197, 250, 314
Gibbon, P. 433
Gills, Barry 40, 66
Glaser, A. 249

Global Fund to Fight AIDS, TB and malaria 325, 340, 344, 418
globalisation 10, 43, 95, 176, 199, 217–19, 235, 257, 271, 299, 327, 546, 547
Globkom 519, 522, 523, 525, 528, 534, 540
Goerens, Charles 382, 384
Golooba-Muteba, F. 399
Goodison, P. 590
governance 7, 14, 15, 64–6, 68–71, 75, 83, 87–9, 102, 103, 122, 133, 141, 142, 158, 228, 236, 258, 260, 273, 340, 388, 507, 589, 596, 598 see also Ireland; Netherlands; Norway
grants 12, 231, 306, 315, 322, 342, 352, 354–6, 362, 365, 368, 420, 476, 477, 485, 494, 498
Greece 23, 28, 497, 503
Griffin, Keith 40
Grilli, E. 578, 579, 581–6 passim, 588, 590
growth, economic 54–9 passim, 76, 78, 163, 178, 220, 232, 234, 306, 454, 455, 459, 534, 535
Guatemala 120, 495, 501, 503
guidelines 3, 6, 12, 15–17 passim, 92, 220, 340–1, 362, 367 see also Spain
Guinea 245, 357
Guinea Bissau 501

Habyarimana, President 138
Hallstein Doctrine 269
Harrod–Domar model 54n24
Hartmeyer, H. 124, 128
Hatton, J.M. 256
Havnevik, K. 524
Healey, John 66
health 12, 13, 60, 64, 75, 87, 100, 144, 169, 179–81, 235, 245, 257, 260, 340, 341, 362, 415, 420, 438, 506, 535 see also Ireland; Luxembourg; Norway
Herfkens, Minister 406, 407, 412–15, 418, 419, 421, 423, 424, 428, 430, 432, 436–8, 440–2 passim
Hettne, Björn 54, 61n34
Hewitt, Adrian P. 64
HICs 51n19
Hildyard, Nicholas 78n58
Hill, Christopher 580
HIPC Initiative 6–7, 12, 82, 117, 132, 181, 244–5, 293, 305, 311, 319, 322–4 passim, 334, 338, 356
  Trust Fund 359
Hirschman, Albert O. 55n25
HIV/AIDS 12, 22n4, 90, 139, 142, 180, 199, 203, 231, 235, 261, 310, 313, 323, 325, 334, 415

INDEX

Hoebink, Paul 1–31, 378–406, 409, 410, 412, 414, 415, 427, 429, 434, 439–43 *passim*, 599
Höll, O. 123
Holland, Martin 576, 578, 584, 585, 593, 596, 598, 599, 602
Holm, Hans-Henrik 188, 189, 193, 206
Holvoet, Nathalie 7–8, 136–60
Honduras 172, 356, 358, 501, 503, 542
humanitarian aid 15, 16, 72–4, 99, 113, 118, 121, 132, 136, 142, 154, 172, 283, 315, 319, 338, 339, 355, 374, 418, 509, 546–9 *passim*, 564–6, 575, 602 *see also* Norway
Hungary 50
Hveem, H. 520
Hydén, Göran 57n29, 69n49

IDA 148, 154, 318, 319, 324, 391, 419, 531
idealism 186, 204, 206
ideology 5, 33, 34, 39, 41, 43, 46, 53, 95, 100, 101
 humanitarian paradigm 41–2, 94
 realist paradigm 38–41, 94
IDGs 282, 509
IFAD 149
IFIs 21, 56, 98, 117, 122, 154, 227, 323, 344, 500, 510
ILO 59n33, 344
IMF 17, 61–3, 155, 163, 257, 319, 322–4, 338, 442, 557, 567, 570
 ESAF 322, 327
 PRGF 322
income distribution 453, 534
India 37, 164, 170, 173, 234, 288, 305, 355, 362, 420, 422, 424, 542, 583, 584
Indian Ocean 169
Indochina 520, 531, 543
Indonesia 35, 121, 141, 171, 288, 408, 421, 422, 424, 499, 502, 503
industry/industrialisation 55, 88, 235, 269, 306, 420, 457, 459, 467, 505
 infant 58
infrastructure 12, 57, 121, 176, 177, 198, 235, 245, 361, 362, 394, 420, 433, 457, 459, 466, 467, 505, 507, 534
INRI 193
institutions 6, 15–17 *passim*, 60, 69, 83, 91, 100, 311, 467, 508, 535
interest rates 61
interests
 bureaucratic 18, 578, 581, 582, 592, 603, 604
 commercial 13, 15–16, 29, 39, 41, 173, 174, 180, 190, 206–7, 222, 269–70, 336, 342, 345, 346

 diplomatic 337, 345, 375
 European 19, 579–80, 582, 592, 603, 604
 national 34, 38, 41, 94–7 *passim*, 101, 216, 218, 268–9, 272, 374, 470, 473, 477, 515, 578–80, 592
 self- 38–41, 84, 87, 102, 270, 477, 522, 523
intervention, political 54, 56–8 *passim*, 61, 62, 70, 524, 526
investment 1, 55, 57, 175, 177–8, 272, 305, 306, 311, 325, 340, 430–4 *passim*, 457, 522, 560, 579
IOV 406, 410
Iraq 29, 97–9 *passim*, 209, 338,. 418, 495, 531, 542
Ireland 11–12, 23, 303–35, 586
 administration 308, 320, 332
 area-based programmes 312, 313
 bilateral aid 11, 310–14
 and BWIs 322–4
 DCI 308–9, 311, 313–17 *passim*, 329, 331, 332, 334
 and debt relief 322–4, 326, 334
 distribution, aid 11–12, 305–6, 312, 334;
 prioritisation 305
 and education 305, 311, 313, 314;
 development 310, 315, 329–32, Unit 331
 and EU 11, 303, 310, 318–21
 foreign policy 319–21
 and governance 306–7, 311, 313, 314, 334
 and health issues 305, 311, 313, 314
 and human rights 304, 306, 310, 311, 314, 315, 319
 MAPS 316, 317
 motives 11, 304, 334
 multilateral aid 310, 318–27
 NGOs 12, 310, 314–18, 322, 327, 334;
 Liaison Unit 316
 objectives 303–5, 310
 and poverty reduction 11, 304–6, 310, 311, 313
 public support 317, 328–32
 and trade 305, 306, 311, 320–1, 324
 and UN 310, 318, 324–7
 volume of aid 11, 23, 27, 303, 307–9, 333, 351, 360–1; targets 303, 307–9 *passim*, 325, 326, 333
 volunteers 310, 317–18; APSO 310, 315, 317–18
Israel 251, 269, 288
 Palestine conflict 250
Italy 12–13, 34, 522n21, 336–77
 administration 13, 339, 366–8, 372; reform 13, 367–8, 372, 373, 376–7

bilateral aid 342, 344, 356–60 *passim*, 365, 366, 372–6 *passim*
Clean Hands 12, 336
composition, of aid 352–6
and credits 345, 349, 352, 354–6 *passim*, 362, 365
and debt relief 337, 338, 340, 341, 343, 350, 352, 355–61 *passim*, 366
DGCS 339, 341, 366
distribution, aid 13, 342, 345–53, 355, 360–4; prioritisation 12, 347, 349, 350
and education 340, 341, 362, 369
enterprises 13, 369, 370, 376
and EU 13, 344, 374
evaluation 368
FAI 339
legislation 336, 338–40 *passim*, 345, 366, 368, 373, 375, 376
motives 336–9
multilateral aid 13, 343–5, 360, 370, 373–6 *passim*; multi-bi 344, 362, 374
NGOs 13, 369, 370, 374
objectives 13, 339–40, 371–2
and poverty reduction 12, 337, 340, 341, 350, 359, 362, 364, 367, 368, 375
public support 337–8
regions, municipalities 339, 341–2, 354, 363, 369–70
trust funds 344, 361, 362, 165
volume of aid 12, 23, 24, 26, 28, 336–8, 342–5 *passim*, 368, 372, 497; targets 338
IUED 4
SDC 563
Ivory Coast 150, 245, 247, 249, 253

Jackson, Robert H. 65n40
Jäggle, M. 124, 128
Jakobsen, Peter Viggo 188
Jamaica 422
Japan 23, 24, 28n10, 50n18, 234, 249, 250, 272
Jensen, Hanne Nexo 202
Jepma, C.J. 410
Johnston, A. 540
joint ventures 345, 350
Jordan 29, 349, 356
Jorgensen, Hans 204, 205
Jospin, Lionel 257
Jubilee 2000 initiative 12, 82, 338, 358
Juncker, Jean-Claude 378, 382, 383
justice, social 29, 42, 66, 275, 453, 455, 456, 474, 475

justification, aid 1, 36, 46–8, 139, 485 *see also* Germany; Norway, motives

Kagame, Paul 170
Kajaste, Raili 229
Kay, Cristobal 58n30
Kayizzi-Mugerwa, S. 164, 179, 525
Kenya 3, 120, 165, 169, 170, 176, 232, 233, 247, 288
Keynesianism 54, 57, 64
Kifle, H. 525
Killick, Tony 64
Kitt, Tom 314
Klingebiel, Stephan 274
knowledge 60, 102, 412, 433, 527, 541
Köhler, Volkmar 294
Koning, A. 589
de Koning, Minister 422
Koponen, Juhani 9, 215–41
Kosovo 229, 290, 503, 566
Kreisky, Bruno 123
Kruijt, D. 434
Kuwait 72, 531
Kyoto Protocol 97

Lamy, Pascal 598
Lancaster, Carol 35, 40
Landgraf, M. 593
language issues 311, 388, 543
Laos 247, 388
Larrain, Jorge 58n30
Latin America 20, 29, 37, 58n30, 66n43, 120, 170, 172, 194, 250, 345, 347, 350, 362, 369, 499–504 *passim*, 508, 509, 531, 574–6, 582–5 *passim*, 597, 598
law 99, 189
reform 535
rule of 10, 65, 69, 87, 103, 246, 258, 273, 550, 570
Lebanon 247, 349
Leftwich, Adrian 66, 68n46
Lesotho 305, 311–13 *passim*, 322
liberalisation, financial 175,177, 257
trade 14, 82, 175–7, 182, 260, 325, 442, 535, 584
liberalism, economic 88, 457–9 *passim*
political 64–71, 100, 458–61
liberation movements 35, 36, 520–2, 531, 543
Liberia 247, 327
Libya 353
LICs 1, 51n19, 151, 427, 474, 503, 563
Lipponen, Paavo 239
Lister, M. 582, 583, 585–8 *passim*, 592, 595

INDEX

Little, Ian 61n34
Little, R. 190, 579
LLDCs 1, 10, 51n19, 310, 311, 350–1, 365, 381, 389, 474, 503, 504, 563, 595
LMICs 1n1, 28, 51n19, 151, 427, 503, 504, 515
loans, commercial 1, 61, 81, 420, 498
 soft 342, 345, 349, 352, 354–6, 368, 494, 495; BWI 322
 tied 380, 390
Lomé Conventions 17, 192, 260, 319, 574, 576, 577, 582, 584, 586–92
Lund, M. 594
Luxembourg 13–14, 378–405
 administration 384–8, 398–400
 and Cape Verde 13, 378, 379, 385, 388–90, 394–9
 distribution, aid 13, 388–91
 and education 389–90, 392–7 *passim*
 and EU 401n55
 evaluation 385, 400, 401
 and health issues 389–92, 392–6 *passim*
 legislation 380–1
 Lux Development 381, 385–8, 398–401 *passim*
 motives 13, 402
 multilateral aid 379, 385, 391
 NGOs 13, 380, 382, 383, 386–7, 390, 398, 401–2
 objectives 381–2
 and poverty reduction 381, 400
 public support 389–97 *passim*, 401
 volume of aid 13, 23, 24, 26, 27, 378–82 *passim*, 388–9, 391, 401, 403; targets 378, 380, 382

Maastricht treaty 88, 139, 277, 374, 382, 573, 589, 592, 599
Macedonia 208, 421–2, 566
Madagascar 245, 247
Maghreb 259, 362, 370
Maizels, A. 190
Malawi 169, 170, 176, 210
Mali 247, 388
Malta 358
March, D. 190
Marcussen, M. 581
markets 58, 64, 113, 176, 177, 220, 499, 500, 595
Marshall Plan 35, 238, 268
Martinussen, John 200, 206
Marxism 58n30
Mattila, Jorma T. 51n20

Mattila-Wiro, Päivi 228, 229
Mauritania 245, 501, 503
Mauritius 388
Maxwell, S. 164, 178
Mayotte, 247
Maystadt, Philippe 155
Mazzali, A. 357–9 *passim*
McKinlay, R.D. 190, 579
McNamara, Robert S. 59, 76
McNeill, D. 520
McQueen, M. 595
Mediterranean Basin 13, 181, 248, 250, 251, 337, 350, 361, 369, 370, 574, 575, 582–5 *passim*, 591
Mehler, A. 594
MERCOSUR 584
Metz, M. 394, 395
Mexico 81, 499
MICs 499, 503–5 *passim*
Michel, Louis 154, 157
Middle East 2, 251, 269, 288, 345, 347–9 *passim*, 362, 369, 502, 574, 575, 597, 598
migration 131, 361, 370, 376, 415, 583, 596, 597
military aid 10, 242, 249
military power 97–9, 102
Mill, John Stuart 58
Millennium Declaration 88, 90–1, 92n73, 101
Millennium Development Goals 2, 21, 29, 59, 77, 88–91, 92n73, 14, 219, 240, 257, 275–6, 310, 323, 414, 415, 456, 478
missionaries 38, 41, 317, 333
Mitterrand, François 243, 248
Mkapa, President 175
money laundering 246, 383, 402
monitoring 92–3, 142, 147, 326
Monserrat 172
Moore, Mick 229
Moravscik, Andrew 578, 579
Moreels, R. 141–2, 152, 153, 156, 157
Morgenthau, Hans 38n5, 39, 56n26
Morocco 247, 249n2, 250, 349, 353, 389, 501, 502
Morrissey, D. 603, 605
Morrissey, Oliver 8, 83n63, 161–83
mortality, maternal/child 90, 149, 392, 395, 539
Mosley, Paul 62n36
motives 5–7 *passim*, 9, 32–43 *passim*, 53 *see also individual countries*
mountain regions 562

635

Mozambique 120, 127, 143, 169, 170, 232, 234, 245, 305, 312, 313, 322, 353, 355, 358, 501, 503, 520, 542
multinational corporations 60, 177
Mutahaba, Gelase 70–1
Myanmar 350
Myrdal, Gunnar 55n25, 62

Namibia 120, 232, 388, 501
NATO 41, 97, 250
Ndadaye, President 136
Near East 350
Nelson, Joan M. 40
neo-liberalism 61–4, 82, 95, 100
neo-patrimonialism 62–3, 69, 83
Nepal 120, 164, 170, 210, 228, 232, 233
Netherlands 14, 35, 41, 67n44, 186, 272, 325, 397, 398, 406–47, 484
  administration 411, 418
  bilateral aid 14, 413, 419–30
  and coherence 14, 407, 413–15, 440–4 *passim*
  DGIS 404, 411
  distribution, aid 14, 418, 421–8; concentration 413, 418, 421–6, 443; criteria 422–4; thematic 424–6
  and environment 410, 415, 424, 426, 433, 438
  and EU 418–19
  evaluation 411
  foreign policy 410–12, 440, 441, 443
  and governance 409, 410, 414, 423, 424, 426, 430, 433
  and human rights 410, 422, 424, 426, 438
  multilateral aid 418, 443
  NGOs 14, 407, 413, 416, 434–40, 442, 443; MFOs 434–40, 442, 443
  objectives 407, 415
  and ownership, recipient 407, 411, 413, 414, 428–9, 440, 444
  and poverty reduction 14, 406–10, 413–16, 432, 433, 435, 438, 443, 444
  private sector 14, 407, 413, 415, 430–4, 443; ORET programme 431–3
  public support 416, 443
  and trade 434, 441, 442
  and UN 406, 419, 442
  volume of aid 14, 23, 25–8 *passim*, 408, 411, 416–21; targets 408, 416
  White Papers 408–10, 415, 441
Netherlands Antilles 424
New Caledonia 244, 247
New Zealand 23, 28n10, 497

NGOs 4, 7, 12–14 *passim*, 17, 60, 72, 82 *see also individual countries*
Nicaragua 66n43, 120, 172, 232, 233, 353, 388, 495, 501, 503, 520, 542
Nielson, Poul 209–10, 597, 598, 600, 601
NIEO 1, 59, 63, 441, 458, 463
Niger 143, 150, 388
Nigeria 165, 169, 250, 251, 314
Nilsson, M. 536
NIS 50n18, 118, 121, 122, 124, 282, 574, 502
Niskanen, William A. 187
Nissanke, M.K. 190
non-aligned movement 38, 44, 270
non-intervention 485–6
Norway 15, 28, 29, 67n44, 186, 429, 448–92
  administration 481–2; reform 481–2
  bilateral aid 466–79
  and democracy 15, 453, 455, 456, 458–60 *passim*, 465, 472–4, 484–7 *passim*
  distribution, aid 466–74; prioritisation 469–74
  and education 457, 459, 461, 466
  and effectiveness 470, 471, 475–6
  Engen Committee 450
  and environment 450, 451, 454, 455, 457, 459, 485
  foreign policy 452–5, 468, 473, 479, 481–2, 485
  and governance 456, 458–60 *passim*, 465, 466, 473, 484–7 *passim*
  guidelines 467–79
  and health issues 450, 457, 459, 461, 466
  and human rights 15, 451, 454–60 *passim*, 473–4, 484–7 *passim*
  humanitarian aid 455, 456, 462, 464, 469n40, 473, 475, 482, 483
  motives 450–3, 468, 485
  multi-bi aid 469
  multilateralism 463–4, 467–9, 485
  NGOs 464, 477, 483, 484
  NORAD 460–1, 475, 481–2, 483, 484
  North–South Commission 451, 454–5, 458, 466
  objectives 453–6, 465, 473, 481, 485, 486
  and peace 450, 453–5 *passim*, 461–2, 472
  and poverty reduction 15, 452, 453, 456, 461, 463, 465, 473–4, 478–9, 485–6; Plan of Action 478–9
  private sector 457, 459, 470, 473, 477, 478
  public support 483–4
  strategies 456–67, 473, 487
  and trade 452, 455, 457, 459, 479
  and UN 462n25, 468, 469, 479, 485

volume of aid 23, 25–8 *passim*, 452, 479–81; targets 479, 480, 497
White Papers 15, 448–58, 464–8 *passim*, 470, 473–81 *passim*
NUPI 4, 5, 30, 99n81

OA 118, 123, 282, 574
OAU 593
objectives 6, 32, 36–7, 53, 67, 86, 87, 91–3, 99–100 *see also individual countries*
obligation, moral 42, 86–7 *see also individual countries*, motives
Obrovsky, Michael 6–7, 113–35
Oceania 502
ODA 6, 32n1, 50–2, 81
  volume 22–9, 47–8, 90, 92 *see also individual countries*
ODI 4, 5n3, 589
OECD 47, 83–7 *passim*, 93, 126, 130, 140–1, 177, 178, 188, 229, 402, 414, 458, 522, 539
  DAC 15, 22–6 *passim*, 29, 47–8, 50, 51n20, 65n39, 69, 70, 72, 84–8 *passim*, 92, 93, 244, 247, 280, 337, 365, 375, 378, 381, 403, 433, 442, 465, 493, 494, 497–500 *passim*, 503, 504, 551, 574; peer reviews 114–16 *passim*, 119, 122–3, 128, 132, 156, 184, 296, 309, 310, 315, 324, 332–4 *passim*, 367, 400, 414–16, 426–9 *passim*, 444, 499, 507–11 *passim*, 519, 547, 555, 563, 564, 566
oil 46, 61, 453
Olsen, Gorm Rye 8, 17–19, 184–214, 573–608
Olsen, P. Lind 202
Olsson, J. 529
O'Neill, Helen 10–11, 303–35
Osterbaan, Maaike 229
ownership, recipient 21, 83–4, 103, 126, 153, 219, 228, 292, 306, 601 *see also* Netherlands; Sweden
Oyewumi, Aderemi 588

Pakistan 29, 120, 170, 350, 422
Palestinian Administered Areas 120, 246, 247, 269, 314, 347, 349, 355, 358, 388, 472, 501, 542
Palme, Olof 521
Paraguay 501, 584
parastatals 285, 294, 296
Paris Club 356
participation, recipient 13, 21, 65, 69, 70, 76, 83, 85, 87, 103, 204, 257, 258, 273, 306, 408, 525, 526, 586
partnership 9, 13, 92n73, 126, 141, 164, 166, 170–82 *passim*, 199, 219, 229, 231, 250,

292, 306, 310, 341, 415, 426, 524–6, 536, 591
  Regional Agreements 595, 597
Patten, Chris 597–8, 600, 602
peace 42, 45, 53, 72, 89, 97, 126, 127, 139, 192, 270, 271, 274, 275, 306, 337, 338, 410, 424, 426, 507, 550, 542 *see also* Norway
  'dividend' 48
peacekeeping 15, 72, 73, 188, 209, 326
Pearson Report 56n28
Peru 232, 349, 353, 501, 503
Peters, Guy 580
Peters, P. 577, 579, 592
Petersen, N. 188
Philippines 350, 501–3 *passim*
Piening, C. 593
Pinheiro, Joao de Deus 594
Pisani, Edgard 260
Pittelkow, Ralf 192
planning 2, 83, 403, 418, 514
pluralism, political/cultural 95
  institutional 294–6 *passim*
Poland 27, 50, 118, 234, 247
'political' assistance 472–5 *passim*
political parties 65, 103, 123–4, 138–9, 156–8, 191, 203, 221, 297, 416–17, 440, 477–8, 480, 484, 559
politics, international 33–53, 64–71, 94–9
  humanitarian paradigm 41–2, 94
  realist paradigm 38–41, 94
Polynesia 244, 247
population growth 71, 271
Portugal 23, 28, 34n2, 35, 52n21, 71, 391, 394, 397, 497, 584
Poulsen, Ole Lonsman 201
poverty eradication/reduction 2, 5n3, 6, 8–18 *passim*, 29, 49, 74–7 *passim*, 80, 84, 87–9 *passim*, 92n73, 101, 103 *see also individual countries*
  APPI 344
PRSPs 2, 22, 77, 82n62, 84n64, 130, 153, 236, 245, 277, 292, 407, 413, 414, 429, 444, 498, 601
Pratt, Cranford 41–2, 63n38, 186
Prebisch, Raúl 55n25, 58n30
preferences 176, 595
private sector 60, 93, 113, 206–7, 235, 274–5, 328, 350, 522, 559–60 *see also* Netherlands; Norway; Spain
privatisation 82, 113
processing 306, 420

project aid 13, 16, 22, 142–7 *passim*, 151, 170, 172, 244, 389–90, 394–9, 401, 411, 412, 428, 517, 523, 524, 534–5
programme aid 15, 16, 22, 63, 146, 147, 172, 197–8, 207, 236, 284, 286–7, 295, 361, 411, 415–16, 420–1, 487, 517, 535
Pronk, Jan 91, 406–8 *passim*, 413, 415, 420, 422, 423, 428, 430–2 *passim*, 436, 440, 441
protectionism 54n22, 58, 176, 321
Pult, G. 554

Raffer, K. 132
Ravenhill, John 192, 577–9 *passim*, 581
raw materials supply 586, 587
recession 1, 61, 225
refugees 45, 51–2, 73, 116, 118, 122, 274, 310, 536, 566
regionalisation 43, 95, 595–7
relief and reconstruction 38, 41, 72–4, 98, 99, 118, 136, 142, 268, 315
religion 37, 38, 41
Renard, Robrecht 7–8, 136–60
Repnik, H.P. 271
reporting 114, 116, 117, 525, 526
research 20, 71, 73, 102, 163, 251, 255, 328, 533
responsibility, recipient 15, 22, 217, 462, 467, 475–6, 485–7 *passim*, 597
Rhi-Sausi, José Louis 12–13, 336–77
Rhodes, C. 592
Rhodes, R.A. 190
Riddell, Roger 56n27, 164, 178, 590
Riess, M. 579
rights, civil/political 42, 67, 70, 138
  human 7, 10, 42, 65, 67–8, 70, 80, 87–9 *passim*, 99, 102, 122, 246, 260, 271, 273, 275, 388, 393, 524, 528, 535, 550, 573, 577, 588–9, 591 *see also* Belgium; Denmark; EU; Finland; Ireland; Netherlands; Norway
roads 306, 311, 313, 392, 394, 457
Rome, treaty of 585
Ronit, K. 581
Rosberg, Carl G. 65n40
Rostow, Walt 55
Rudengren, J. 519, 520, 534
Ruggiero, Renato 340
Russia 54n22, 167
Ruttan, Vernon W. 40
Rwanda 7, 120, 136, 137, 141, 143, 146, 150, 157, 165, 169, 170, 245, 379, 388, 389, 593

Sachs, Wolfgang 79n58
San Domingo 247

Sandberg, Svante 51n20
Santer, Jacques 382, 597
São Tomé and Principe 245, 392, 501
savings 54, 55, 57, 60
Scandinavia 20, 38, 42, 221, 225, 236, 239, 240, 272, 325, 469, 484
Schoo, Minister 422
Schröder, Gerhard 299
Schulpen, Lau 14, 406–47
Schümperli Younossian, Cathrine 16–17, 546–72
security 4, 17–19 *passim*, 39, 43–8 *passim*, 72, 76–7, 89, 96, 98, 99n81, 126, 127, 217, 270–1, 374, 451, 462, 522, 550, 576, 577, 580, 582, 594
  food 149, 306, 313, 341; FAO Trust Fund for 343, 344
self-help 16, 462, 520, 535, 543
Sellström, Tor 521–2
Sen, Amartya 71, 258
Senegal 120, 247, 249n2, 358, 388, 389, 501, 503
September 11 2–3, 44, 47, 48, 52, 96, 99, 271, 293
Serbia/Montenegro 347, 355, 366
services sector 305, 390
services, social 59n33, 60, 75–6, 121, 361, 362, 389, 390, 400, 420, 457, 505–7 *passim*, 515, 563, 564 *see also individual entries*
  20/20 initiative 75, 420, 564
Severinot, Jean-Michel 249
Short, Clare 8, 161–4, 174–5, 181–3 *passim*
Sierra Leone 169, 170, 188, 245, 247, 327, 593
Siitonen, Lauri 9, 215–41
SILICs 52n21
Sinowatz, Chancellor 123
skills 55, 305
slave trade 391
Smith, M. 186, 190, 191
Smith, S. 249
Solana, Javier 573–4
solidarity 16, 41, 87, 246–8 *passim*, 251, 261, 339, 451, 453, 519–23 *passim*, 535, 548
Somalia 72, 327, 353, 355
Sorensen, Georg 62n37, 65n40, 66, 580
SOU 523
South Africa 35, 120, 155, 170, 229, 247, 251, 311, 314, 424, 501, 521, 542, 593
Soviet Union 35, 36, 43, 49, 54, 113, 225, 238, 538
Spain 15–16, 20, 493–517, 584
  administration 15–16, 493–5, 498, 510, 512–17 *passim*; reform 16, 515, 517

INDEX

AECI 494, 507, 511–13 *passim*, 516
bilateral aid 497–500, 505
business interests 15, 500, 511, 517
community, aid 16, 494–5, 510–12, 515–16
composition, of aid 497–509
distribution, aid 15, 16, 28–9, 496, 501–9, 512; prioritisation 501, 507–9
and education 505, 506, 509, 535; Carolina Foundation 509
and environment 28, 505, 507, 508
Araucaria/Azahar Programmes 508
and EU 20, 494, 497, 500
FAD credits 494, 495, 498–500, 502, 505, 509, 511, 514
legislation 15, 495, 513–16 *passim*
multilateral aid 500
NGOs 494–5, 510, 511, 515
and poverty reduction 499, 504, 507–8, 513–15 *passim*
regional/municipal governments 494–5, 510–12, 514
regression 15–16, 509, 515–17 *passim*
SECIPI 493, 494, 510, 516
Strategic Guidelines 495, 501, 502, 507, 509, 514, 516
volume of aid 15, 23, 24, 28, 493, 495–7, 500; targets 497
Stabex 260, 586
stability/stabilisation 18, 42, 53, 72, 97, 139, 163, 270, 275, 311, 410, 433, 450, 453, 458, 535, 576
Stiglitz, Joseph E. 62n36, 64, 258
Stokke, Olav 1–112, 173, 186, 448–92, 589
students 116, 122, 136, 283, 286, 310, 328, 521
subsidies 88, 177, 229, 278, 407, 435–40 *passim*, 512
Sudan 45, 246, 305, 312, 353, 355
summits 76, 113 *see also* environment; EU
  Baule 248
  Beijing 76, 113
  Cairo 326
  Copenhagen social 75, 113, 326, 507, 509, 563
  Monterrey 11, 26, 48, 83n63, 113, 117, 225, 257, 282, 299, 323, 325–6, 354, 382, 497
  UN Millennium 2, 48, 59, 88, 90–3, 99, 303, 307, 479
  Vienna 113
  World Food 341
Sundman, Folke 51n20
Surinam 424
suspension, of aid 10, 102, 246, 596

Svendsen, Knud Erik 186, 203, 204, 206
swaps 2, 14, 16, 173, 196–8, 201, 236, 286, 295, 312, 313, 407, 428–30, 443, 498
Swaziland 356
Sweden 16, 20, 50n18, 51n20, 179, 186, 234, 239, 429, 518–45
  administration 527, 530, 533–4, 538
  bilateral aid 16, 520, 532–7
  BITS 533, 538, 542
  and democracy 520, 524, 528, 535
  development theory 16, 534–7
  distribution, aid 16, 530–3, 537–43
  prioritisation 16, 537–43
  and EU 16, 519, 527–30
  legislation 523
  motivation 16, 519–23, 534–5
  multilateral aid 530–2
  NGOs 531
  objectives 519–20, 523, 530, 539
  and ownership 524–7 *passim*, 535
  and poverty reduction 518–20 *passim*, 523, 528, 534–6, 539–40
  private sector 522
  public support 522, 524
  SAREC 533
  SIDA 519, 526, 533, 538; Sida 518, 519, 526, 530, 533–4, 538–40 *passim*, 542
  SWEDCORP 533, 538, 542
  and UN 520, 530–1
  volume of aid 23, 26, 27, 530, 532, 543; targets 537
Switzerland 16–17, 50n18, 395, 398, 546–72
  administration 554–6, 559–64 *passim*, 570
  bilateral aid 554, 559–64
  cantons/municipalities 552–3
  and coordination 555–6, 564, 570
  distribution 563–5; prioritisation 565
  and environment 547, 549, 555, 568–9
  legislation 548, 563
  multilateral aid 554, 566–8
  NGOs 17, 547, 553, 556–8, 564, 565
  North–South Guidelines 16–17, 546, 547, 550, 551, 555, 556, 568, 570
  objectives 520, 548–50, 555, 559–61, 563
  and poverty reduction 562–4, 547
  public support 547, 553, 558–9
  and UN 557, 558, 567, 570
  volume of aid 17, 23, 24, 26, 28, 549, 551–4, 569–70
Sydnes, Anne Kristin 452
Syria 349, 356
Sysmin 260

639

Tambo, Oliver 521
Tanzania 3, 120, 143, 150, 165, 169, 170, 174–6, 228, 230, 232, 245, 247, 305, 310, 312, 313, 322, 359, 520, 542
targets 6, 11, 90, 91
  % of GNI 8, 11, 13, 14, 17, 20, 26–8, 48 *see also individual countries*
technical assistance 22, 56, 57, 60, 69, 101, 143, 144, 172, 176, 228, 259–60, 289, 318, 387, 400, 412, 413, 469n40, 510, 524, 533, 543, 546
  EPTA 36n4, 53, 56
technology 55, 57, 60, 126, 390, 560
Telford, John 218
terrorism 2, 47, 96, 99, 102, 271
Thailand 230
Thiel, Reinold 298
Thorn, Gaston 379
Togo 261, 379, 388
Toye, John 61
trade 14, 37, 38, 43, 49, 54n22, 55, 61, 88, 101, 199, 269–72, 340, 383, 522, 560 *see also* Britain; EU; Finland; Ireland; Netherlands; Norway
  everything but arms 321, 334, 442
trade unions 434, 478, 484, 511–12
training 70, 245, 257, 258, 274, 305, 314, 318, 369, 396, 412, 433
transition countries 49–52, 170, 314
transparency 65, 69, 87, 103, 165, 175, 178, 182, 209, 306, 307, 323, 395, 525
transport 121, 176, 177, 180, 269, 457, 505
TRIPs agreement 321
Tromp, S. 418
Truman, President 36n4, 37
Tunisia 150, 247, 250, 350, 356, 379, 388, 389, 501, 502, 543
Turkey 269, 288
tying, aid 7, 13, 29, 102, 174, 182, 207, 228, 269–70, 306, 364–6, 379, 380, 390, 395, 468, 476–8, 494, 499, 522, 523, 554;
  untying 138, 141, 142, 158, 166, 174–5, 182, 190, 270, 306, 365, 376, 403, 432, 477, 522, 602

Uganda 120, 165, 169, 170, 176, 210, 234, 245, 305, 312,, 313, 322, 355, 358
UMICs 1, 28, 51n19, 288, 427, 503
UN 2, 17, 21, 35–8, 53–6 *passim*, 59n33, 71–8 *passim*, 88, 93, 98–101 *passim*, 178, 261 *see also* Austria; Denmark; Finland; Germany; Ireland; Netherlands; Norway; Sweden; Switzerland

Commission on Social Development 326
ECOSOC 36n4, 37325, 326
EPTA 36n4, 53, 56
Millennium Assembly 2, 48, 59, 88, 90–3, 99, 303, 307, 479, 486 *see also* MDGs
Monterrey 11, 26, 48, 83n63, 113, 117, 225, 257, 282, 299, 323, 325–6, 354, 382, 497
Nordic project 531
Security Council 97, 320, 326–7, 337
Special Fund 53, 56
UNAIDS 419
UNCDF 149, 419
UNCTAD 59n32, 81, 82n62, 101
UNDP 45, 68n46, 71–2, 75, 101, 210, 257, 261, 290, 324, 362, 391, 397, 409, 419, 458, 469, 530
  Human Development Index 94n76, 305, 311, 427; *Reports* 45, 72, 75, 79
UNEP 79, 324, 569
unemployment 71, 227, 270, 583
UNESCO 290, 419
UNFPA 391, 419
UNHCR 324, 325, 391
UNICEF 59n33, 64, 149, 324, 391, 419, 531
unilateralism 96–9
UNITAR 568
universities 7, 146, 148,, 158, 305, 327–8, 512
UNOPS 362
Uruguay 499, 584
  Round 175
US 19, 29, 35–7, 40–1, 43, 46, 50n18, 52n31, 58n31, 66n43, 67n44, 68n47, 70n51, 79, 95–9, 249–51 *passim*, 338, 458, 503, 589
  volume of aid 10, 23–5 *passim*, 28n10, 47, 48, 53

values 6, 41, 42, 53, 94, 95, 97, 100, 101, 529
Vandenbroucke, F. 141
van der Velden, F. 412
Venezuela 499, 503
Vereker, John 162, 163, 166
Verhofstadt, Guy 157
Vietnam 49, 143, 150, 151, 210, 232, 234, 247, 388, 390, 503, 520, 522, 542
Vogler, J. 577, 592
volunteers 380, 381, 387, 435

Waller, Peter 68n46, 273
war *see also* conflicts
  civil 44–5
  Cold 1, 8, 17–18, 34, 36, 38, 40, 44–8 *passim*, 54, 94, 95, 184, 188, 195, 196, 270,

408, 468, 473, 481, 522, 574–82, 585–7
passim
Gulf 286
Iraq 97, 338
on terror 2, 47, 96, 99, 243, 246
World – II 34, 35
Warburg, Anne Melte Rahbaek 51n20
'Washington consensus' 82, 83n63, 100, 251
'Washington–New York consensus' 15, 100, 101, 451, 484
water supply/sanitation 122, 235, 306, 313, 314, 389–90, 394, 420, 457, 534
Watts, Patrick 595
Wendt, Alexander 216, 240
Western Sahara 188, 501
WFP 565, 566
Whaites, A. 173, 178
White, Gordon 66
WHO 59n33, 199, 391, 419
Wilhelm, Jürgen 274
Wohlfahrt, Georges 382, 390
Wohlgemuth, Lennart 16, 518–45
women 79–80, 90, 126, 196, 203, 204, 219, 306, 340, 364, 434, 438, 454, 462 *see also* gender issues
Beijing conference on 76, 79–80
UN decade for 80
World Bank 2, 17, 21, 56, 62n36, 63, 66, 68–71, 74–8 *passim*, 93–4, 98, 100, 101, 163, 173, 201, 234, 238, 240, 290, 310, 322–4, 344, 394, 397, 409, 442, 469, 531, 540, 551, 557, 570
CDF 22, 93–4, 292, 466
Development Committee 47, 48, 403
*WDR*s 74, 77, 93n75, 221
WTO 43, 101, 175–8 *passim*, 260, 320–1, 326, 334, 442

Yaoundé Conventions 585
Yemen 246, 350, 353
Young, Crawford 36n3
Youngs, Richard 589
Yugoslavia 45, 51n19, 72, 98, 113, 118, 121, 132, 354, 358, 424, 472, 501, 503, 531, 542, 575, 583

Zambia 169, 170, 228, 230, 232, 233, 305, 312, 313–14, 322, 323, 358, 542
Ziegler, Jean 380n5
Zimbabwe 120, 169, 188, 210, 324, 465, 542
Zupi, Marco 12–13, 336–77